P. Johnson

GAP JUNCTIONS

GAP JUNCTIONS

Edited by

Michael V.L. Bennett
Albert Einstein College of Medicine

David C. Spray
Albert Einstein College of Medicine

Cold Spring Harbor Laboratory
1985

GAP JUNCTIONS
©1985 by Cold Spring Harbor Laboratory
All rights reserved
Printed in the United States of America
Cover design by Emily Harste

Front cover: (*Top*) Dye and electrical coupling of cultured sympathetic ganglion neurons (see Kessler et al., this volume). (*Bottom, left*) Freeze-fracture replica of a rapidly frozen gap junction (see Hanna et al., this volume). (*Right*) Immunogold labeling of gap junctions in a membrane preparation from rat liver (see Paul, this volume).

Back cover: (*Top, left*) Three-dimensional reconstruction of a frozen gap junction obtained by Fourier methods (see Zampighi and Simon, this volume). (*Top, right*) Immunofluorescent labeling of gap junctions in a frozen section of rat liver (see Hertzberg and Spray, this volume). (*Bottom, left*) Extensive dye coupling among retinal horizontal cells in the untreated turtle retina (see Neyton et al., this volume). (*Bottom, right*) Virtual absence of dye coupling among retinal horizontal cells of turtle after treatment; only the axon and cell body of the injected cell contain dye (see Neyton et al., this volume).

Library of Congress Cataloging in Publication Data

Main entry under title:

Gap junctions.

   Includes index.
   1. Cell junctions—Congresses. I. Bennett, Michael
Michael V.L. (Michael Vander Laan), 1931-
II. Spray, David C. III. Cold Spring Harbor
Laboratory.
QH603.C4G36 1985   574.87'6   85-11329
ISBN 0-87969-187-5

Authorization to photocopy items for internal or personal use, or the internal or personal use of specific clients, is granted by Cold Spring Harbor Laboratory for libraries and other users registered with the Copyright Clearance Center (CCC) Transactional Reporting Service, provided that the base fee of $1.00 per article is paid directly to CCC, 27 Congress St., Salem MA 01970. [0-87969-187-5 $1.00 + .00] This consent does not extend to other kinds of copying, such as copying for general distribution, for advertising or promotional purposes, for creating new collective works, or for resale.

All Cold Spring Harbor Laboratory publications may be ordered directly from Cold Spring Harbor Laboratory, Box 100, Cold Spring Harbor, New York 11724. (Phone: 1-800-843-4388). In New York (516) 367-8425.

Conference Participants

William Beers, *Department of Biology, New York University, New York*
Michael V.L. Bennett, *Department of Neuroscience, Albert Einstein College of Medicine, Bronx, New York*
Peter Brink, *Department of Anatomical Science, State University of New York, Stony Brook, New York*
Stanley Caveney, *Department of Zoology, University of Western Ontario, London, Canada*
William Coles, *Department of Neurosciences, McMaster University, Hamilton, Canada*
John Dowling, *Biological Laboratory, Harvard University, Cambridge, Massachusetts*
F. Edward Dudek, *Department of Physiology, Tulane University School of Medicine, New Orleans, Louisiana*
Malcolm Finbow, *Beatson Institute for Cancer Research, Glasgow, Scotland*
Hersch M. Gerschenfeld, *Laboratoire de Neurobiologie, Ecole Normale Superieure, Paris, France*
Christian Giaume, *Institut Pasteur, Paris, France*
James E. Hall, *Department of Physiology, University of California, Irvine, California*
Robert Hanna, *State University of New York College of Environmental Science and Forestry, Syracuse, New York*
Elliot L. Hertzberg, *Department of Biochemistry, Baylor College of Medicine, Texas Medical Center, Houston, Texas*
Ross G. Johnson, *Department of Genetics and Cell Biology, University of Minnesota, St. Paul, Minnesota*
Stanley Kater, *Department of Zoology, University of Iowa, Iowa City, Iowa*
John Kessler, *Department of Neurology, Albert Einstein College of Medicine, Bronx, New York*
W.J. Larsen, *Department of Anatomy and Cell Biology, University of Cincinnati College of Medicine, Cincinnati, Ohio*
Eric Lasater, *Biological Laboratory, Harvard University, Cambridge, Massachusetts*

Rodolfo Llinás, Department of Physiology and Biophysics, New York University Medical Center, New York, New York

Cecilia W. Lo, Department of Biology, University of Pennsylvania, Philadelphia

Lee Makowski, Department of Biochemistry and Molecular Biophysics, Columbia University College of Physicians & Surgeons, New York, New York

Jacques Neyton, Laboratoire de Neurobiologie, Ecole Normale Superieure, Paris, France

Ernest Page, University of Chicago, Chicago, Illinois

David Paul, Department of Anatomy, Harvard University Medical School, Boston, Massachusetts

Camillo Peracchia, Department of Physiology, University of Rochester Medical School, Rochester, New York

Ole H. Petersen, Physiological Laboratory, University of Liverpool, Liverpool, England

John D. Pitts, Beatson Institute for Cancer Research, Glasgow, Scotland

Fidel Ramon, Centro de Investigacion y de Estudios Avanzados Del IPN, Mexico City, Mexico

Jean-Paul Revel, Division of Biology, California Institute of Technology, Pasadena, California

Hideki Sakamoto, Department of Obstetrics and Gynecology, Yale University School of Medicine, New Haven, Connecticut

Judson Sheridan, Department of Anatomy, University of Minnesota, Minneapolis, Minnesota

David C. Spray, Department of Neuroscience, Albert Einstein College of Medicine, Bronx, New York

Anne Warner, Department of Anatomy and Embryology, University College London, London, England

Klaus Willecke, Institut für Zellbiologie, Universitat Essen, Essen, Federal Republic of Germany

Jacek Wojtchak, Department of Anesthesiology, Mt. Sinai School of Medicine, New York, New York

Barbara Yancey, Division of Biology, California Institute of Technology, Pasadena, California

Guido A. Zampighi, Department of Physiology, University of California Medical School, Los Angeles, California

First row: M. Bennett, E. Hertzberg, J. Sheridan; C. Giaume, D. Spray
Second row: P. Brink; A. Warner, J. Dowling; C. Peracchia
Third row: W. Larsen; O.H. Petersen, R. Johnson; G. Zampighi
Fourth row: H. Gerschenfeld, J. Neyton; D. Spray, A. Rivera, F. Ramon

Contents

Conference Participants, v
Preface, xiii

Introduction
Michael V.L. Bennett and David C. Spray   1

Section 1   STRUCTURE

Structural Domains in Gap Junctions: Implications for the Control of Intercellular Communication
Lee Makowski   5

The Structure of Gap Junctions As Revealed by Electron Microscopy
Guido A. Zampighi and Sidney A. Simon   13

Structural Details of Rapidly Frozen Gap Junctions
Robert B. Hanna, Richard L. Ornberg, and Thomas S. Reese   23

Molecular Conformation of the Major Intrinsic Protein of Lens Fiber Membranes: Is It a Junction Protein?
Jean Paul Revel and S. Barbara Yancey   33

Biochemistry and Structure of Cardiac Gap Junctions: Recent Observations
Ernest Page and Chellakere K. Manjunath   49

Section 2 BIOCHEMISTRY

Studies of Gap Junctions: Biochemical Analysis and Use of Antibody Probes
Elliott L. Hertzberg and David C. Spray — 57

Immunochemical Investigations of Gap Junction Protein in Different Mammalian Tissues
Klaus Willecke, Otto Traub, Uwe Janssen-Timmen, Uwe Frixen, Rolf Dermietzel, Axel Leibstein, Dieter Paul, and Hartmut Rabes — 67

Comparison of the Protein Components Present in Vertebrate and Arthropod Gap Junction Preparations
Malcolm E. Finbow, T. Eldridge, J. Buultjens, Ephraim Kam, John Shuttleworth, and John D. Pitts — 77

Is Lens MP26 a Gap Junction Protein? A Phosphoprotein?
Ross Johnson, Elizabeth Frenzel, Keith Johnson, Kathleen Klukas, Paul Lampe, Daryl Sas, M. Jane Sas, Rebecca Biegon, and Charles Louis — 91

Antibody against Liver Gap Junction 27-kD Protein Is Tissue Specific and Cross-reacts with a 54-kD Protein
David L. Paul — 107

Section 3 BIOPHYSICS

The Effects of Deuterium Oxide on Junctional Membrane Permeability and Conductance
Peter R. Brink — 123

General and Comparative Physiology of Gap Junction Channels
David C. Spray, R.L. White, V. Verselis, and Michael V.L. Bennett — 139

Control of Junctional Permeability
Fidel Ramón, Guido A. Zampighi, and Amelia Rivera — 155

Electrical Uncoupling Induced by General Anesthetics: A Calcium-independent Process?
Jacek A. Wojtczak — 167

Protein from Purified Lens Junctions Induces Channels in Planar Lipid Bilayers
James E. Hall and Guido A. Zampighi — 177

An In Vitro Approach to Cell Coupling: Permeability and Gating of Gap Junction Channels Incorporated into Liposomes
Camillo Peracchia and Stephen J. Girsch — 191

Section 4 CONTROL OF FORMATION

Reduced Junctional Permeability in Cells Transformed by Different Viral Oncogenes
Michael M. Atkinson and Judson D. Sheridan — 205

Alterations in Coupling in Uterine Smooth Muscle
W.C. Cole and R.E. Garfield 215

Development and Regulation of Electrotonic Coupling between Cultured Sympathetic Neurons
John A. Kessler, David C. Spray, Juan C. Saez, and Michael V.L. Bennett 231

Determinants of Specificity of Electrical Synapses: The Making and Breaking of Connections in *Helisoma*
S.B. Kater, C.S. Cohan, and P.G. Haydon 241

Section 5   ROLE OF INTERCELLULAR COMMUNICATION AND DEVELOPMENT

Communication Compartmentation and Pattern Formation in Development
Cecilia W. Lo 251

Control of Molecular Movement within a Developmental Compartment
Stanley Caveney and Richard Safranyos 265

Antibodies to Gap Junction Protein: Probes for Studying Cell Interactions during Development
Anne E. Warner 275

Relating the Population Dynamics of Gap Junctions to Cellular Function
W.J. Larsen 289

Junctional Communication and Oocyte Maturation
William H. Beers and Paula J. Olsiewski 307

Importance of Electrical Cell–Cell Communication in Secretory Epithelia
Ole H. Petersen 315

Section 6   ELECTROTONIC SYNAPSES

Electrical Interactions and Synchronization of Cortical Neurons: Electrotonic Coupling and Field Effects
F. Edward Dudek and Robert W. Snow 325

Electrotonic Transmission in the Mammalian Central Nervous System
Rodolfo R. Llinás 337

Interaction of Electrical and Chemical Synapses
Michael V.L. Bennett, M.B. Zimering, M.E. Spira, and David C. Spray 355

Junctional Voltage-dependence at the Crayfish Rectifying Synapse
Christian Giaume and Henri Korn 367

Neurotransmitter-induced Modulation of Gap Junction Permeability in Retinal Horizontal Cells
J. Neyton, M. Piccolino, and H.M. Gerschenfeld   381

Electrical Coupling between Pairs of Isolated Fish Horizontal Cells Is Modulated by Dopamine and cAMP
Eric M. Lasater and John E. Dowling   393

Author Index, 405

Subject Index, 407

Preface

This volume represents the proceedings of a conference held at the Banbury Center of Cold Spring Harbor Laboratory in October 1984. At an earlier meeting of the American Society for Cell Biology, several of the participants had reached a consensus that it was time for a meeting on gap junctions. As substantial progress had been made in the last few years, consolidation of current status and evaluation of future prospects seemed appropriate. The privilege (and responsibility) of organizing the meeting somehow fell to the editors, and a meeting proposal was accepted by Jim Watson.

We are deeply indebted to Judy Cuddihy and to Nancy Ford and her staff in the laboratory's publications office. Without their editorial and production skills, this volume would still be slowly progressing toward a delayed and uncertain birth. We also wish to thank Dr. Steven Zornetzer of the Office of Naval Research for supplementary funding given with a remarkable lack of governmental red tape.

Michael V.L. Bennett
David C. Spray

GAP JUNCTIONS

Introduction

Michael V.L. Bennett and David C. Spray
Albert Einstein College of Medicine
Bronx, New York 10461

The term "gap junction" arose from the work of Jean Paul Revel and Morris Karnovsky who in 1967 reported the separation of the junctional membranes and the accessibility of the intercalated region to markers for extracellular space. The gap was a structural feature distinguishing these junctions in thin section from tight junctions of epithelial cells at which extracellular space was seen to be totally occluded. These latter junctions are tight in two ways since the occlusions encircle the cells and thereby prevent transepithelial leakage. The distinction between gap and tight junctions was soon made graphic by application of the freeze-fracture technique.

The term gap junction persists, although it emphasizes a morphological characteristic that seemingly contradicts the junctions' role in intercellular communication. Of course the suggestion that close appositions are sites of coupling had been made earlier. Moreover, J. David Robertson had seen the structural units of gap junctions and at least hinted at the channel location that is generally agreed upon today. The identification of gap junctions as sites of coupling came from studies of excitable tissues. Close appositions were seen where coupling (or electrical transmission) was demonstrated and were not observed where coupling was demonstrably or presumably absent. Although the concept of intramembrane proteins as channels was not yet formulated, it was intuitively obvious that close membrane appositions were suitable for coupling and that separation of membranes would short-circuit coupling. (Never mind recent demonstrations of electrical interactions across extracellular space and calculations indicating that much of the electrical coupling actually observed could occur across extracellular space, if opposed membranes were of low resistance.)

The idea that gap junctions contained simple aqueous channels depended in part on their generally being electrically linear and on their permitting intercellular movement of various dye molecules and other tracers. Although dyes such as fluorescein initially used to demonstrate coupling in fact penetrate nonjunctional membrane to some extent and might pass between cells via extracellular space, more polar tracers, for example, Lucifer Yellow, have become available subsequently. These tracers have been shown to pass between cells through a private pathway not accessible to extracellular space. Since in some cells the only direct contact between cells is at gap junctions (and dye transfer is too rapid to be mediated by exo- and endocytosis), one concludes that dye coupling is mediated by gap junctions. Because the dyes are relatively large and charged, small ions can pass from cell to cell by the same pathway: and one further concludes that gap junctions are the site of electrical coupling. Additional direct evidence for the undoubted role of gap junctions now comes from block of coupling by antibodies raised against isolated gap junctions and junctional protein.

The conference from which this volume arose was organized around several areas: structure, biochemistry, biophysical properties and gating, formation and degradation, and finally function in inexcitable and excitable tissues. At the end of the conference the subject remained fluid and fascinating. Areas of disagreement were numerous, but there appeared to be an unusual concern for finding truths rather than establishing particular points of view. Here follows a brief description of the material presented, of areas of consensus, and of remaining questions.

The structure of the channel grows clear (in liver at least). Images obtained by X-ray diffraction and low-dose electron microscopy of frozen and negatively stained junctions are converging. The channel is there. Resolvable lumps may even represent gates that control channel patency. But the best freeze-fracture yet available does not show structural change associated with channel closing. The nature of the lens junction is unclear. It differs in ultrastructure and it shows no homology with what is known of amino acid sequences of liver or cardiac gap junction proteins. It may not form an intercytoplasmic channel. Nevertheless, a plausible tertiary structure was proposed for the lens protein as a transmembrane channel, and similar arguments will no doubt be advanced as junctional proteins from other tissues are sequenced.

The biochemical characterization of gap junctions progresses, but there is remarkable disagreement among laboratories, where similar approaches give different answers. There is agreement that the channel in each tissue is an oligomer of a simple polypeptide, but the peptides' sizes range from 16 to 54 kD. Degradation, aggregation, or precursor molecules presumably account for the differences, which will take some sorting out. In the meantime, controlled proteolysis of isolated junctions is providing clues as to parts of the channel molecules accessible from the cytoplasmic surface.

The preparation in a number of laboratories of antibodies against gap junctions and their polypeptides is a major advance. Cross-reactivity among different junctions indicates at least partial homology, but differences in specificity of particular antibodies raise the possibility of tissue specificity and isoforms of the junctional protein. The availability of these antibodies opens whole new areas of the cell biology of gap junctions to investigation. Identification and expression of the gene, genes, or gene families and the origin and fate of junctional protein will surely be topics of the next gap junction meeting.

The gap junction channel, once little more than a hole connecting cells, is turning out to have much more interesting gating functions. A variety of treatments control junctional conductance and again there is tissue diversity. Single-channel recording in reconstituted systems should ultimately explain the control of the macroscopic currents. There is evidence for both all-or-none gating and for partial closure in different systems.

Formation of junctions can be promiscuous or precise. The degree of promiscuity may indicate the extent of homology among extracellular aspects of various gap junction proteins or may be dictated by the diversity of intercellular recognition molecules. Because junctions between cells form or are removed in defined media or in response to hormones or other chemical signals, analysis of mechanism will be possible. In several instances, a role of cAMP and protein phosphorylation is indicated both physiologically and biochemically. Physiological changes caused by covalent modification of the junctional channels is another likely topic for a future meeting.

The role that gap junctions play in physiological processes must be different in different tissues, as has long been obvious. Their widespread occurrence in embryos and their appearance and disappearance at particular times have strongly implied a role in transmission of signals important in development. Newly described communication compartments provide further evidence, and disruption of development by antibodies that block gap junctional communication appears to provide the clincher. We are of course not much closer to knowing what the messages are.

In excitable tissues, a role of gap junctions that has long been absolutely clear is mediation of electrical transmission, whether or not there are other actions. The presence of gap junctions between mammalian neurons is now unquestioned. Although less plastic than excitatory chemical synapses, gap junctions can exhibit a wide range of integrative properties, particularly in interactions with inhibitory synapses. Electrotonic spread can be modulated to a large extent by nonjunctional conductance changes, and junctional conductance can be modulated directly by chemical synapses ending on the coupled cells, presumably through a cytoplasmic second messenger.

Thus, this volume represents a progress report on the study of gap junctions—a more accurate title might have been of the form "Advances in...." Unresolved issues abound, but there is also a growing corpus of what can be considered fact. The study of gap junctions has become a well-established area involving a wide range of methodologies with relevance to both excitable and inexcitable tissues. Productive directions for future work are clear.

Structural Domains in Gap Junctions: Implications for the Control of Intercellular Communication

Lee Makowski
Department of Biochemistry and Molecular Biophysics
Columbia University, New York, New York 10032

The three-dimensional structure of gap junctions isolated from mouse liver has been solved to 18 Å resolution by X-ray diffraction (L. Makowski et al., in prep.). Diffraction studies of specimens in varying concentrations of sucrose (Makowski et al. 1984a) have been used to demonstrate that the channels in these specimens are closed to sucrose and are, by this criterion, in a high-resistance configuration. The three-dimensional electron density map includes a high-density feature at a position that suggests it is responsible for blocking the channel to sucrose. Opening of the channel appears to require movement of the material making up this structure, indicating that it must be active in channel gating. In this paper the structure of the gap junctions as determined by X-ray diffraction is described in detail and consequences for the control of intercellular communication are discussed. The picture of the gap junction structure obtained by X-ray diffraction is substantially different from that obtained by electron microscopy of negatively stained (Unwin and Zampighi 1980) or frozen-hydrated (Unwin and Ennis 1984) gap junctions and indicate that the mechanism of channel closing is different from that postulated by Unwin and his colleagues (see Zampighi and Simon, this volume). The differences between the image of gap junctions as seen by X-ray diffraction and electron microscopy are described and the possible reasons for these differences are discussed.

SYMMETRY AND STOICHIOMETRY

Electron microscopy and X-ray diffraction indicate that under some conditions the morphological units of the junction form a hexagonal lattice. This hexagonal lattice is essential for the success of reconstruction by X-ray diffraction or electron microscopy of the three-dimensional structure of the junction. Consequently, we have no detailed image of the junctional subunits as they exist dispersed in an unordered junctional plaque.

Each morphological unit is made up of a pair of hexamers of the $\sim$27,000 m.w. junctional protein (Caspar et al. 1977; Makowski et al. 1977). One hexamer is associated with each membrane and the hexamers are called connexons. The six proteins making up a connexon are symmetrically equivalent, related by a sixfold rotational axis. The intercellular channel is centered on the sixfold axis. The two connexons are related to one another by twofold rotational axes parallel to the plane of the junction in the center of the gap. Thus, each morphological unit has point group symmetry 622. In the hexagonal lattices, the twofold axes of a morphological unit do not appear to point directly at nearest neighbor units (e.g., Baker et al. 1983; Unwin and Ennis 1984), indicating that the twofold rotational axes are local, noncrystallographic symmetry elements. Consequently, the symmetry of the junctional lattice is p6, implying that some differences may exist in the way the units in the two membranes interact.

THREE-DIMENSIONAL ELECTRON DENSITY MAP

X-rays scatter from electrons and, consequently, X-ray diffraction produces an image of the distribution of electrons, or electron density, in a structure. The spatial arrangement of chemical components is determined by interpretation of the electron density map based on the known electron densities of the constituents. Figure 1a is a diagram of the hexagonal junction lattice. On the diagram is marked the position of a vertical section through the junctional structure. Figure 1b shows the electron density distribution in this section. This section passes through six-, three-, and twofold rotational axes and was chosen because all of the structural features in the three-dimensional electron density map can be seen in this one section. In this drawing the high electron density regions are shaded and electron density below solvent density is indicated by broken contour lines. The only two constituents of gap junctions with electron density substantially greater than solvent are protein and lipid polar head groups. These probably occupy the bulk of the regions indicated by shading in the the map. Lipid hydrocarbon groups have electron density much lower than solvent and probably occupy the regions with electron density lower than solvent indicated by broken contour lines within the bilayers. There are three regions occupied by solvent in this map: (1) the intercellular channel that extends vertically along the sixfold rotation axis and appears isolated from the rest of the solvent; (2) the solvent in the gap between bilayers; and (3) the solvent external to the junction, which, in intact tissue, would comprise the cytoplasms of the two connected cells. The channel has a diameter of 20–30 Å along most of its length but appears to narrow to a minimum diameter of about 15 Å in the extracellular half of the bilayer.

The walls of the channel are made up of an elongated region of high electron density. This region is interpreted as being the transmembrane domain of

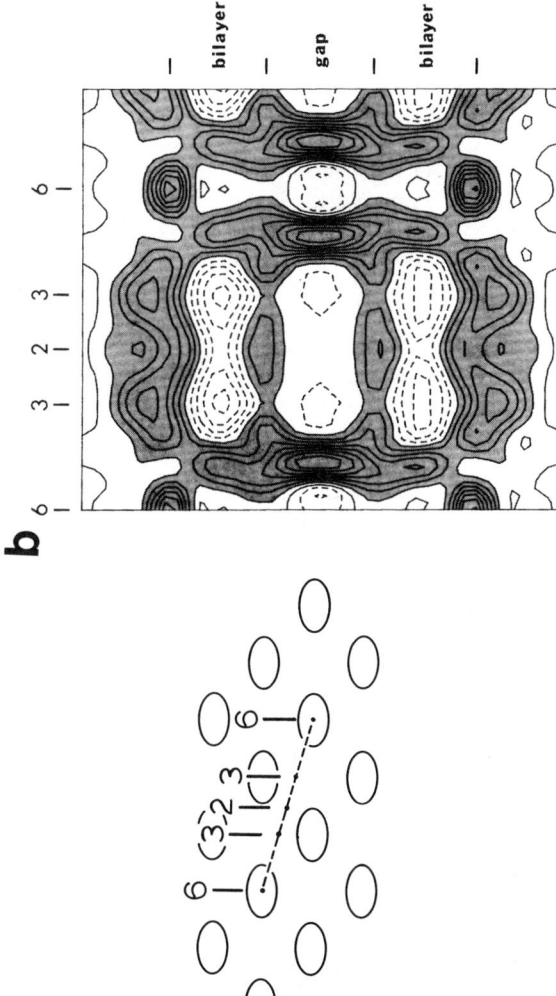

Figure 1

(a) A diagram of the hexagonal lattice of the junction indicating the position of the vertical section drawn in b. (b) A vertical section through the gap junction structure with positions of the six-, three-, and twofold axes marked. Electron density greater than that of solvent is indicated by shading and solid contours. Electron density less than the solvent level is indicated by broken contours. Protein and lipid polar head groups have high electron density and will be confined largely to the regions indicated by shading in the map. Lipid hydrocarbon chains have electron density lower than solvent and occupy regions indicated by broken contours within the bilayers. The remainder of the volume is occupied by solvent. (Revised and refined from Makowski et al. 1982, and in prep.)

the junctional protein. It has a volume corresponding to about 15,000–17,000 daltons per protein. Its electron density appears to be much greater in the gap than within the bilayer. Since polar amino acid side-chains are more electron dense than nonpolar side-chains, this suggests a segregation of polar and nonpolar amino acid residues within the junction structure. There appears to be a much higher proportion of nonpolar side-chains within the bilayer than in other portions of the junction protein. The electron-dense feature on the sixfold rotation axis near the cytoplasmic surfaces of the two membranes is also likely to be protein. Its volume corresponds to a molecular weight of roughly 5,000–10,000 daltons. Its position on the sixfold axis indicates that it is probably made up of six identical peptides (of molecular weight 1000–1500 daltons), one from each of the six proteins making up a connexon. Its position blocking the channel suggests that it may comprise a gating structure responsible for the control of channel permeability. X-ray diffraction studies of junctions in varying concentrations of sucrose (Makowski et al. 1984a) indicated that in these preparations the channel was closed to the penetration of sucrose and that a solvent region approximately 100 Å long and centered on the sixfold axis remained free from sucrose. Both the sucrose results and the three-dimensional electron density map are consistent with the idea that a structure located near the cytoplasmic surface of the membrane is blocking the channel in these preparations.

In the regions between the connexons there does not appear to be any protein within the lipid bilayer. The center of the bilayer in this region has a very low electron density as would be expected for lipid hydrocarbons. The extracellular surface of the bilayer (toward the gap) is of high electron density, as would be expected for lipid polar head groups. The cytoplasmic side of the lipid bilayer is also electron dense but its volume is much too large to be accounted for by lipid polar head groups only. Furthermore, treatment with trypsin removes a mass from the cytoplasmic surface amounting to about 4000 daltons per protein (Makowski et al. 1984a). Chemical studies indicate that trypsin removes the carboxyl terminus of the junction protein (Nicholson et al. 1983). Together with the X-ray results this suggests that the cytoplasmic domain includes the carboxyl terminus. The size and configuration of this cytoplasmic domain of the junction protein appears to be somewhat variable in preparations of gap junctions isolated in slightly different ways (Makowski et al. 1984b). It is likely that this portion of the protein is both flexible and labile.

These results provide a picture of the gap junction protein as a three-domain protein: The transmembrane domain extends across one bilayer and half of the extracellular gap, making up a portion of the channel wall. The character of the protein in the gap and in the bilayer may be somewhat different with a larger proportion of hydrophobic residues in the bilayer. The gating domain is a small (1000–1500 dalton) peptide. In these specimens, six symmetrically equivalent gating peptides appear to combine to block the transmembrane channel to penetration by sucrose. The cytoplasmic domain interacts closely with lipid polar head groups on the cytoplasmic surface of the bilayer. It is attached to the remainder of the protein by a relatively narrow connector. This portion of the protein may be flexible and appears to be sensitive to the action of proteases.

COMPARISON WITH ELECTRON MICROSCOPY

It was expected that the image of the gap junction produced by X-ray diffraction would be very similar to that obtained by electron microscopy of frozen hydrated junctions (Unwin and Ennis 1984). In fact, the two images are very different. The electron microscope image appears to be similar to the transmembrane portion of the connexon as seen by electron microscopy. Figure 2a shows a vertical section through the gap junction calculated from the data of Unwin and Ennis (1984) from their junctions in 0.05 mM Ca^{++} and which they suggest are in the closed configuration. Figure 2b shows the corresponding section calculated from their data from junctions in calcium-free solutions which they suggest may correspond to open junctions. In these maps there is no vestige of an image of a lipid bilayer, a cytoplasmic domain, nor any feature blocking the channel. The reason for suggesting that the junctions shown in Figure 2a are closed is not obvious.

The electron microscope image is subject to several distortions, the most important one being the inability to obtain data in the so-called "missing-cone" of reciprocal space. Figure 2c is an electron density map calculated from our X-ray diffraction data without using data in the region of reciprocal space corresponding to the missing cone. It is clear from a comparison of Figures 1b and 2c that the problem of the missing cone can result in substantial distortion of the structural image. The section in Figure 2c shows only the weakest image of the lipid bilayer, but the cytoplasmic domain and gating structure, although distorted, are still present. The reason for these structures being missing in the electron microscope image is not obvious. The evidence for their existence is very strong: The presence of a large cytoplasmic domain on the surface of the gap junction membranes has been confirmed by experiments in which the junctions were treated with trypsin. Trypsin removed about half of the volume of the cytoplasmic domain as imaged by X-ray diffraction (Makowski et al. 1984a); an amount consistent with the results of chemical studies (Nicholson et al. 1983). The presence of a gating structure blocking the channel was indicated by experiments in which sucrose was shown to be excluded from the channel by structures near the cytoplasmic surfaces of the membranes. The three-dimensional electron density map has confirmed the presence of these structures. These results indicate that electron microscopy of frozen-hydrated gap junctions has resulted in an incomplete image of the junction structure. This incomplete image could be the result of damage to the specimen during freezing or due to irradiation by electrons. Because of the nature of the reconstruction process that results in an average image of many unit cells and uses data from many different plaques, the destruction of a component is not required to remove it from an image; a disordering of a component is sufficient to render it invisible in the reconstructed image.

SUMMARY AND CONCLUSIONS

X-ray diffraction studies have resulted in an image of the gap junction protein as a three-domain protein made up of a transmembrane domain, a cytoplasmic domain, and a gating domain (L. Makowski et al., in prep.). X-ray dif-

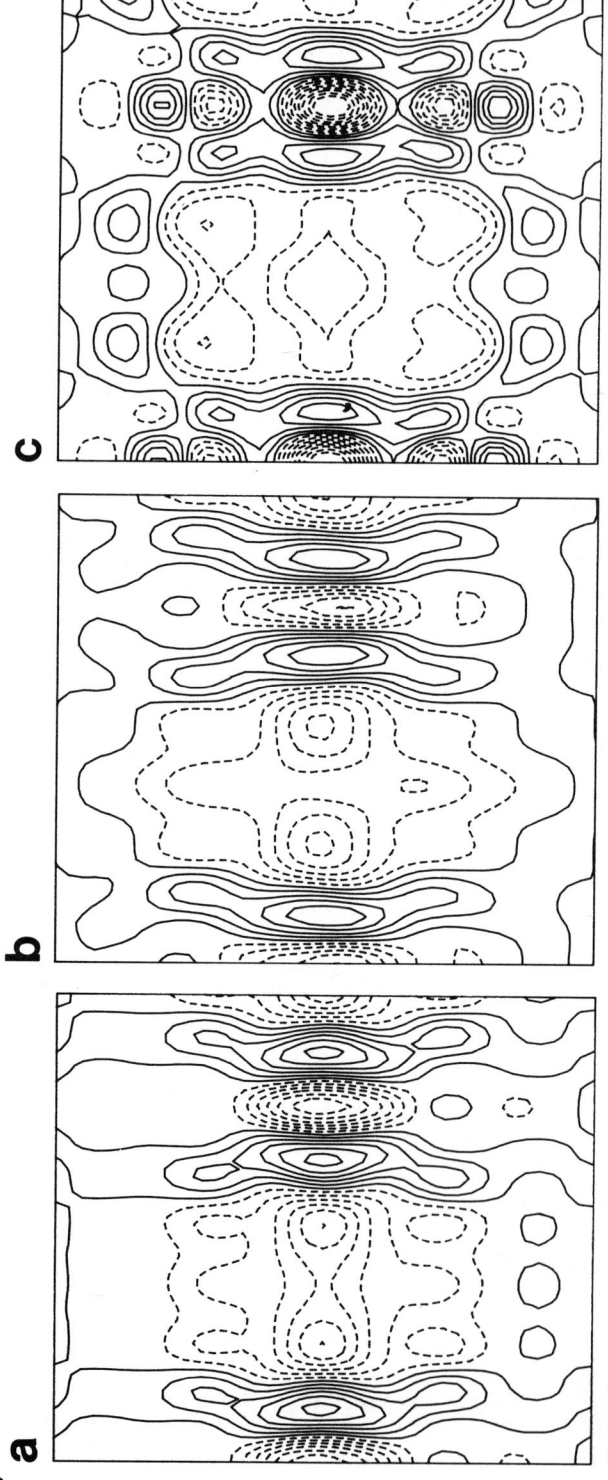

Figure 2
Vertical sections through density maps calculated from electron microscopy and X-ray diffraction. (a) From a density map for frozen hydrated gap junctions in the presence of Ca^{++} calculated from the data of Unwin and Ennis (1984), which they suggest correspond to junctions in the closed, high-resistance state. (b) A section through a density map of frozen hydrated gap junctions in the absence of Ca^{++} calculated from the data of Unwin and Ennis (1984), which they suggest correspond to a channel in the open low-resistance state. (c) A section through an electron density map calculated from X-ray data in the absence of the data in the missing cone. Removal of the missing cone of data causes substantial distortion of the density map, removing most of the features corresponding to the lipid bilayer but preserving the density corresponding to the transmembrane, cytoplasmic, and gating domains of the junction protein.

fraction from junctions in varying concentrations of sucrose has provided an assay for the permeability state of the channel and indicated that in these preparations the channel is closed (Makowski et al. 1984a). The presence of the gating structure blocking the channel near the cytoplasmic surfaces of the two membranes indicates that the mechanism for channel gating is different from that postulated by Unwin and co-workers (Unwin and Zampighi 1980; Unwin and Ennis 1984). Although no structural data yet exist for junctions that have been demonstrated to contain open channels, it is apparent from the position of the gating structure in the electron density map that it must move in order to open the channel. There is a large amount of space around the gating structure into which the six small gating peptides could move. Opening of the channel may involve the rotation of each of the gating peptides away from the sixfold axis into the unoccupied space surrounding the gate. This could occur either as a local movement or in concert with quaternary structural changes or with rotation of the cytoplasmic domain relative to other parts of the protein. Gating by a simple change in tilt of the transmembrane domains, as postulated by Unwin (Unwin and Zampighi 1980; Unwin and Ennis 1984), is not consistent with the image of the junctional protein as determined by X-ray diffraction. Motion of the domains of the junctional protein relative to one another must be a principal part of the gating mechanism.

ACKNOWLEDGMENTS

The X-ray diffraction studies discussed here have been carried out over the past decade with Dr. D.L.D. Caspar and Dr. W.C. Phillips of Brandeis University and Dr. D.A. Goodenough of Harvard University. Work in my laboratory has been supported by NIH grant GM29829 and an Irma T. Hirschl Career Scientist Award.

REFERENCES

Baker, T.S., D.L.D. Caspar, C.J. Hollingshead, and D.A. Goodenough. 1983. Gap junction structure IV. Asymmetric features revealed by low irradiation microscopy. *J. Cell Biol.* **96:** 204–216.

Caspar, D.L.D., D.A. Goodenough, L. Makowski, and W.C. Phillips. 1977. Gap junction structures I. Correlated electron microscopy and x-ray diffraction. *J. Cell Biol.* **74:** 605–628.

Makowski, L., D.L.D. Caspar, D.A. Goodenough, and W.C. Phillips. 1982. Gap junction structures III. The effect of variations in the isolation procedure. *Biophys. J.* **37:** 189–191.

Makowski, L., D.L.D. Caspar, W.C. Phillips, and D.A. Goodenough. 1977. Gap junction structures II. Analysis of the x-ray diffraction data. *J. Cell Biol.* **74:** 629–645.

―――. 1984a. Gap junction structures V. Structural chemistry inferred from x-ray diffraction measurements on sucrose accessibility and trypsin susceptibility. *J. Mol. Biol.* **174:** 449–481.

Makowski, L., D.L.D. Caspar, W.C. Phillips, T.S. Baker, and D.A. Goodenough. 1984b. Gap junction structures VI. Variation and conservation in connexon conformation and packing. *Biophys. J.* **45:** 208–218.

Nicholson, B.J., L.J. Takemoto, M.W. Hunkapiller, L.E. Hood, and J.P. Revel. 1983. Differences between the liver gap junction protein and lens MIP26 from rat: Implications for tissue specificity of gap junctions. *Cell* **32:** 967–978.

Unwin, P.N.T. and P.D. Ennis. 1984. Two configurations of a channel-forming membrane protein. *Nature* **307:** 609–613.

Unwin, P.N.T. and G. Zampighi. 1980. Structure of the junction between communicating cells. *Nature* **283:** 545–549.

The Structure of Gap Junctions As Revealed by Electron Microscopy

Guido A. Zampighi and Sidney A. Simon*

*Department of Anatomy and
Jerry Lewis Neuromuscular Research Center
UCLA School of Medicine, University of California
Los Angeles, California 90024*

**Department of Physiology, School of Medicine
Duke University, Durham, North Carolina 27710*

Over the years numerous studies have indicated that in many tissues from various animal species the passage of electrolytes and nonelectrolytes between adjacent cells is always correlated with the presence of a particular cell organelle called the "gap junction." This organelle is characterized by the intimate contact between plasma membranes of two adjacent cells and a hexagonal array of annular units located in the plane of the membranes. These features have made the gap junction an ideal candidate for the pathway by which ions and low-molecular-weight molecules diffuse between adjacent cells.

The exact location of the pathway used by these molecules, as well as the mechanisms responsible for its regulation, has been extensively studied electrically, chemically, and structurally (for reviews, see Bennett and Goodenough 1978; Loewenstein 1981). From electrical and flux measurements, it was postulated that the pathway used by these molecules was a rather large, water-filled channel located at the geometrical center of the annular units in the plane of the junction. The direct visualization of such a channel spanning the entire thickness of the junction has been the object of numerous investigations. Recent structural investigations of isolated rat liver hepatocyte gap junctions (Caspar et al. 1977; Makowski et al. 1977, 1982, 1984; Unwin and Zampighi 1980; Unwin and Ennis 1984) have provided information about the organization of the annular units forming the hexagonal array in the plane of the junction as well as its exact location across the lipid bilayers.

In this chapter, we will review the quaternary organization of the annulus as determined by three-dimensional reconstruction methods (for a review of these methods, see Amos et al. 1982). In particular, we will show that different parts of the annulus can be enhanced, depending on whether the isolated gap junctions are imaged in negatively stained or in frozen hydrated solutions. Finally, we will present information indicating that the subunits comprising each annulus can rearrange in ways that can account for changes in junction permeability.

ELECTRON MICROSCOPY AND IMAGE PROCESSING

Gap junctions were isolated from rat liver hepatocytes by methods described previously (Zampighi and Unwin 1979; Unwin and Ennis 1983). The isolated junctions are double membrane plaques of about 150 Å in overall thickness. They are composed of ordered domains of oppositely faced annular units (also called connexons [Goodenough 1976]) linked to each other in the extracellular space between both membranes, making a "gap" between the plasma membranes of about 30 Å (Revel and Karnovsky 1967).

The isolated gap junction plaques were prepared for electron microscopy by first depositing aliquots of the dispersion, as a thin film, on carbon-coated grids and then negatively staining them with uranyl acetate solutions, as described previously (Zampighi and Unwin 1979). For the study of frozen hydrated solutions of gap junctions, unstained junctions on carbon-coated grids were plunged in liquid ethane whereupon they were transferred under liquid nitrogen into an electron microscope modified to maintain the specimen at a temperature of $-120°C$ during imaging.

The image of an object obtained using an electron microscope is a two-dimensional projection of a three-dimensional object having a finite thickness (in the case of a gap junction, about 150 Å). Consequently, three-dimensional information of fine structural details is impossible to extract from a single electron micrograph. However, three-dimensional information can be obtained by tilting the junctional membranes with respect to the incident electron beam and thus creating different views. The resulting images are then processed by Fourier methods and the information from the different tilt angles is combined to represent the variation of electron density across the entire thickness of the junction (Henderson and Unwin 1975).

Two-dimensional maps of the structures calculated by the Fourier method are usually represented as a series of lines (called contours) that connect points in the plane having the same electron density. The three-dimensional information, on the other hand, is presented as a series of sections (of arbitrary separation) that are parallel to the membrane plane.

Isolated liver gap junctions were tilted at angles of between 0 and 79°C to the incident electron beam in the negatively stained junctions (Unwin and Zampighi 1980) and between 0 and 53°C in the frozen hydrated specimens (Unwin and Ennis 1984). Since it is not technically possible to tilt specimens continuously to 90°C in the microscope, there were sections (actually cones) in the three-dimensional pattern from which no information was collected. The most significant missing region was the variation in amplitude and phase along the lattice line passing through the origin (the 0,0 lattice line). Although the missing cones of information will affect the absolute levels of the contours in sections parallel to the membrane (e.g., the exact location of the boundary between lipid and protein), they do not influence the position of the protein subunits across the thickness of the bilayer.

Images of isolated gap junctions embedded in negative stain and frozen hydrated solutions are presented in Figure 1, A and B, respectively. Despite the fact that these images are presented at slightly different magnifications, it is evident that the hexagonal array of annuli is distinct from the background in negatively stained junctions and just barely visible in the frozen hydrated junctions.

Since it is established that the scattering of electrons from matter is directly proportional to the mass density of the atoms contained within, then for the particular case of the gap junctions prepared by these two methods, the contrast arising from these images can be interpreted. From these considerations, it would be expected that the images obtained from the negatively stained junctions should be more distinct than the ice-embedded ones. The very dense uranyl ions (about 11 g/cm^3) will scatter more electrons at higher angles (which will be stopped by the objective lens aperture) than the less dense matter that contains lipid, protein, and water (whose densities are all close to 1 g/cm^3 but with the protein being the largest). Since the uranyl ions are located predominantly at the membrane/water interfacial region, the contrast seen in the negatively stained images will arise mostly from the distribution of the heavy metal ions at surfaces of the junction. On the other hand, the contrast from the images of the frozen hydrated junctions will arise largely from the difference in density between protein and lipid (i.e., the middle of the membrane) and between protein and water. The junctions imaged in frozen hydrated solutions will not exhibit marked contrast between the lipid and water because of their similar mass-scattering densities. Therefore, images from frozen hydrated and negatively stained junctions could be considered complementary because they will arise from different portions of the annulus. In frozen hydrated specimens the contrast will arise from the portion of the annulus located at the center of the bilayer (i.e., the hydrophobic region), whereas in negatively stained junctions the contrast will arise from those regions of the annulus protruding from the cytoplasmic and external surfaces.

THREE-DIMENSIONAL MAPS

Three-dimensional maps calculated from one-half of the junctions embedded in negative stain and frozen hydrated are presented in Figure 2. They were constructed by tracing contour maps of sections of the connexon at different places across the bilayer on Perspex sheets which were then stacked in order to produce the three-dimensional map. In all the maps presented in Figure 2, the reader is viewing the connexon from the cytoplasmic side of the junction. It is apparent that there are significant differences between the maps calculated from junctions prepared by the two methods. The map of the negatively stained gap junction (Fig. 2A) depicts the matter (shadowed regions) clustered around the sixfold axes. The annulus is comprised of six *weakly modulated*

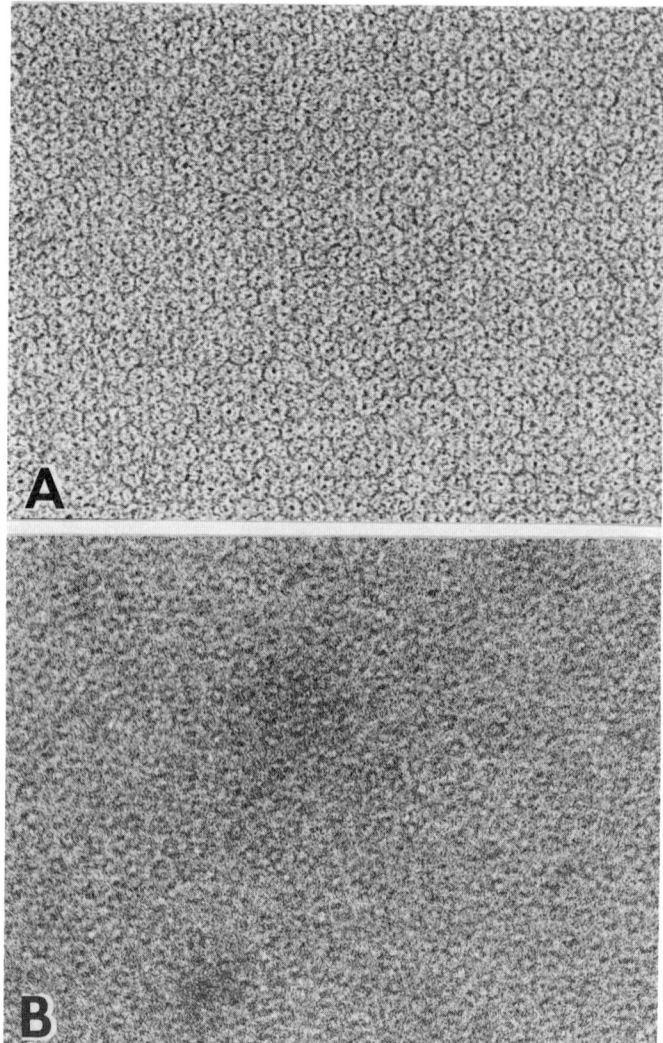

Figure 1
Electron microscopy of isolated gap junctions viewed from a direction perpendicular to the plane of the membrane. (A) The annular units negatively stained with uranyl acetate are organized on a hexagonal lattice, with a unit cell dimension of 80–85 Å. Magnification, 350,000×. (B) The image of an isolated gap in frozen hydrated solutions. The characteristic hexagonal array of hexamers is only barely visible because of the small scattering differences among ice, lipid, and protein. Magnification, 410,000×.

subunits about 70–75 Å long. The subunits surround a narrow cavity (arrow) that penetrates through the membrane and widens to a diameter of about 20 Å in the extracellular region. An important new feature of these subunits is

Figure 2

Three-dimensional maps of isolated hepatocyte gap junctions. The projections are complete maps from half of the junction (i.e., one of the membranes forming the complete junction in the tissue) and they are seen from the interior of the cell. (A) The reconstruction obtained from negatively stained gap junctions. This map emphasizes the annular appearance of the hexamer and the changes in the angular disposition of each subunit in the cytoplasmic (S) and external surfaces (S'). The small arrow labels the perimeter of the cavity. (B and C) Maps obtained from gap junctions in frozen hydrated solutions. B was obtained from junctions suspended in solutions with no calcium, whereas C is from junctions after treatment with 5×10^{-4} M $CaCl_2$. The maps emphasize the hexamer with strongly modulated subunits within the lipid bilayer. The hexamers in both B and C have the same neighbor-to-neighbor relationship in which they come closest to each other, near the middle of the lipid bilayer. This point is indicated by equivalent triangles connecting equivalent points in the maps. The maps in B and C are different, however, in the inclination of the subunit across the bilayer. The treatment by calcium results in a decrease in the angle of the subunits by about 7.5°.

that they are inclined in their passage through the lipid bilayer. This change in the angular disposition of the subunits is emphasized by the labels S and S' which correspond to the center of mass of the same subunit labeled at its cytoplasmic and external surfaces, respectively. Also, the negatively stained preparations demonstrated that the annulus (which will henceforth be called a hexamer) is asymmetrically located in the bilayer. It protrudes more from the external than from the cytoplasmic surfaces (Unwin and Zampighi 1980).

The maps calculated from the junctions prepared by frozen hydrated methods (Fig. 2B, C) show the hexamer comprised of six *strongly modulated* subunits surrounding a central, water-filled cavity that spans the entire thickness of the membrane. The question then arises as to why these two types of reconstructions show the subunits comprising the hexamers with such noticeably different degrees of modulation.

For negatively stained junctions, the annular appearance of the hexamer arises from the summation of the portions of the connexons on the cytoplasmic and extracellular surfaces, since these are the regions where the uranyl ions are deposited. Since the subunits are inclined in their passage across the bilayer, it follows that the modulations located at the cytoplasmic and extracellular faces (marked S and S') are not in register and, thus, depending on the angle of inclination of the subunits, effectively cancel each other out. On the other hand, the maps of the hexamer obtained from frozen hydrated junctions are dominated by the portion of the connexon in the hydrophobic region of the bilayer, since, as previously discussed, this is the region exhibiting the greatest electron-scattering density. The boundaries between the regions of the hexamer outside and inside the membrane (which are so well defined in negative staining) are poorly resolved here because ice and lipids have similar mean coherent electron-scattering densities. Thus, the differences in the subunit appearance between gap junctions reconstructed in negative stain and in frozen hydrated solutions could be due to the fact that the contrast in these methods arises from different portions of the hexamer (i.e., the central portion of the bilayer as opposed to its surfaces).

Another difference that was revealed upon comparing the maps from these two different sets of data is that the junctions imaged in frozen hydrated solutions were strongly chiral, i.e., the packing of the hexamers in both junctional membranes is different. A similar result was reported by Baker et al. (1983) in the study of mice liver gap junctions isolated by different methods. This was not the case with the negatively stained junctions. A probable explanation of this phenomenon proposes that the differences in the packing of the hexamers in each junctional membrane is the result of differences in the isolation procedure used. The junctions used for the three-dimensional reconstructions reported by Unwin and Zampighi (1980) were purified using minimal amounts of detergent (i.e., enough to sever the junctional plaques from the plasma membrane), and in addition, they were fractioned in a sucrose density gradient. The gap junctions used for reconstruction migrated to the 41/45% (wt/wt) sucrose interface. These junctions were visualized by thin-sectioning electron microscopy as flat sheets (Zampighi 1978). On the other hand, the junctions at the 30/37% (wt/wt) sucrose interface displayed a strong tendency to curve or vesiculate (Zampighi 1978). A result of this curvature is that the center-to-center distance between connexons in the membrane displaying the minor radius of curvature must be smaller than the distance between units in the membrane (in the same junction) with the major radius of curvature. The end result is that the hexamers cannot be packed equivalently in the two membranes and thus they will not occupy equivalent positions, making the whole structure noncentrosymmetric (e.g., it has P6 rather than P622 symmetry). This packing of the connexons in the P6 lattice for these strongly chiral junctions has been analyzed in detail by Baker et al. (1983) but interpreted in a different manner. Of course this asymmetry can be interpreted as a structural feature of the hexamers relevant to the function of the communicating channels. It can also be interpreted as a nice example of a rearrangement of the hexamers in the junctional membranes that results from the isolation procedure and, hence, it might well be totally unrelated to the organization of the gap junction in the cell. Moreover, this asymmetry will complicate the interpretation of low-angle X-ray diffraction data in those cases where P622 symmetry has been assumed (Makowski et al. 1977, 1982, 1984, and this volume).

Figure 2C is a map obtained from frozen hydrated, isolated gap junctions that were incubated in aqueous solutions containing 5×10^{-4} M $CaCl_2$ before imaging. Here, the hexamer is also constructed of six strongly modulated subunits that surround a central, water-filled cavity. However, after incubation of these junctions with 5×10^{-4} M $CaCl_2$, the inclination of the subunits is decreased such that the water-filled channel closes at the cytoplasmic surface, as we will show when discussing the quaternary states of the hexamers.

TWO QUATERNARY STATES

The information contained in the three-dimensional maps obtained from isolated gap junctions suggests a possible mechanism of junctional permeability regulation. Both negative staining and frozen hydrated methods have shown that the hexamers can exist in two conformationally distinct states. These two states are presented in Figure 3. In one state, which may correspond to the open configuration of the channel, the hexamer has the subunits tilted across the bilayer and the hydrophilic cavity at its geometrical center spans the entire thickness of the membrane (the darker regions of the subunits represent the approximate location of the bilayer). In the other state, which may correspond to the closed configuration, the subunits comprising the hexamer are less inclined across the bilayer and the water-filled channel is interrupted at its cytoplasmic portion.

The change of the structure of the hexamers upon going from the closed to the open state can be explained by a simple rearrangement of the subunits by tilting and sliding along their lines of contact. The average inclination of each subunit is reduced by the addition of 5×10^{-4} M $CaCl_2$ making it nearly perpendicular to the plane of the junction. In negative staining, the change in inclination is reduced by about 5°, whereas in the maps calculated from frozen hydrated solutions, the tilt of each subunit is reduced by about 7.5°. Because each subunit is long, the displacement at the cytoplasmic end is quite large, about 9 Å. Thus the channel-lining faces of each pair of subunits could

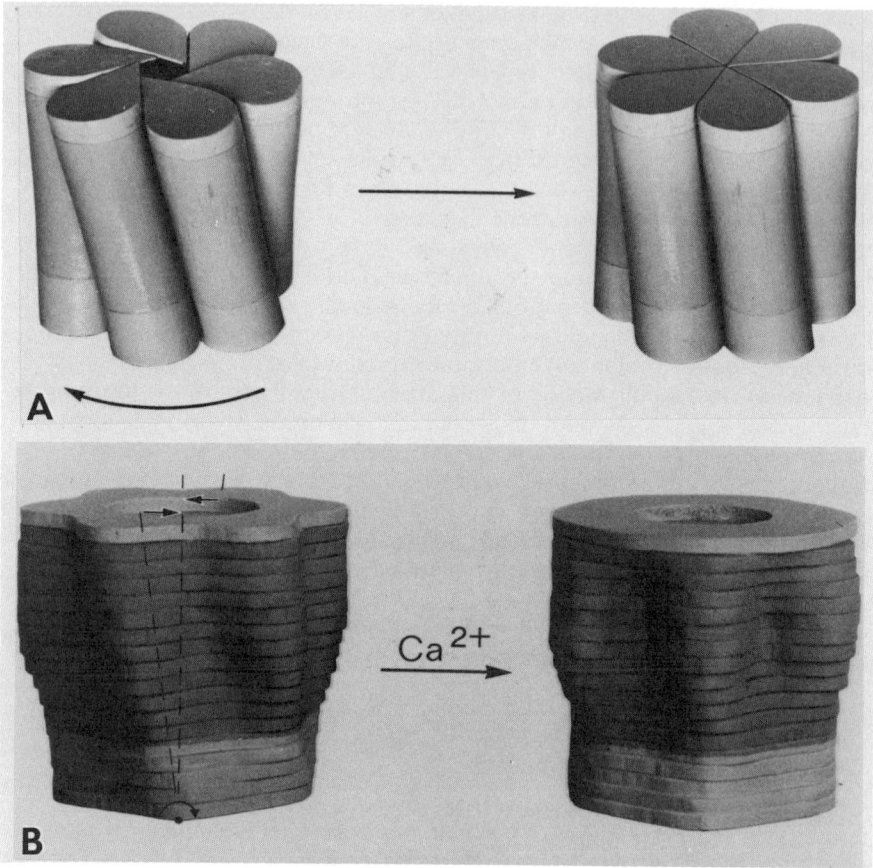

Figure 3
Wooden models of the hexamer deduced from three-dimensional reconstructions. (A) Hexamers deduced by negative staining. (B) Balsa wood models constructed from sections across the hexamer of frozen hydrated solutions. The models demonstrate that the two configurations of the hexamers seen in negative staining and frozen hydrated gap junctions are essentially equivalent. In both cases the subunits on either side of the channel tilt in opposite directions resulting in tangential displacement (pair of arrows) that is greatest at the upper, cytoplasmic surface.

move toward each other by up to 18 Å in the cytoplasmic regions of the hexamer. It was proposed that this coordinated movement among the subunits may provide the means of controlling channel permeability by the closure of the channel at its cytoplasmic end (Zampighi and Unwin 1979; Unwin and Zampighi 1980; Unwin and Ennis 1984).

COMPARISON WITH OTHER MODELS

A different model for the organization of isolated gap junctions and the change in configurations responsible for change in permeability has been presented by Makowski et al. (1977). In their original proposal the authors depicted the connexons as hexamers embedded in lipid bilayers and linked to those of the adjacent membranes. Each hexameric subunit had a dumbbell-shape about 80 Å in length located symmetrically across the bilayer, with most of its mass located outside of the membrane. The conformational change responsible for changes in the permeability of the channel was obtained by comparing low-angle X-ray patterns of gap junction pellets fixed with glutaraldehyde with patterns obtained from pellets of junctions digested with trypsin. They deduced that the dumbbell-shaped subunits can undergo a change in conformation in the portion of the hexamer at the extracellular gap. The end result of this conformational change was a collapse of the narrow extracellular gap (by about 8 Å) and the closure of the channel at that region.

The model has been greatly modified to account for improvements in the interpretation of the diffraction data (Makowski et al. 1982, 1984, and this volume). In the newest version of their model the subunits comprising the hexamers have a length of about 90–92 Å and they are asymmetrically placed in the lipid bilayers protruding more on the cytoplasmic surface than on the external surface. The most recent analysis of X-ray data taken from junctions digested with trypsin does not support the mechanism to open and close the channel proposed previously (Makowski et al. 1977). Now digestion with trypsin removes a piece of the subunits located at the cytoplasmic surface of the hexamers (Makowski et al. 1984). The channel seems to close, not at the region in the middle of the extracellular gap, but at the middle of the membrane closer to the cytoplasmic leaflets, a finding that is more consistent with the electron microscopy analysis (Unwin and Zampighi 1980; Unwin and Ennis 1984). Also, the most recent model of the junction postulates the existence of substantial portions of the hexamer occupying the cytoplasmic surface at the region, far from the channel opening, where three neighboring connexons come together (i.e., the threefold axis). However, equivalent structural features have not been detected in this region by three-dimensional electron microscopy reconstructions. Furthermore, the newest models are restricted to the description of the channels in the high-resistance, closed state and they do not show the distortion of the extracellular domain of the hexameric subunits proposed in their original work to account for changes in the functional state of the channel.

CONCLUSIONS

Knowledge about the structural organization of isolated gap junctions has been greatly advanced by three-dimensional reconstruction obtained using electron microscopy. Whereas negative stain yields information about the arrangement of the subunits at the water–membrane interface, ice embedding gives information regarding the arrangement of the subunits in the hydrophobic region of the bilayer. The preparative methods of negative staining and ice embedding complement each other to yield a consistent picture of the communicating channel in its two conformational states.

ACKNOWLEDGMENTS

These studies were supported by grants EY-04110 and NS-20669 from the National Institutes of Health and by the MDA Jerry Lewis Neuromuscular Research Center grant. We thank Dr. P.N.T. Unwin for many suggestions and Dr. E. Gogol from the Department of Structural Biology at Stanford University School of Medicine for allowing the use of his image of the frozen hydrated gap junction shown in Fig. 1B. We also thank Carl Higgins for typing the manuscript.

REFERENCES

Amos, L.A., R. Henderson, and P.N.T. Unwin. 1982. Three-dimensional structure determination by electron microscopy of two-dimensional crystals. *Prog. Biophys. Mol. Biol.* **39**: 183–231.

Baker, T.S., D.L.D. Caspar, C.J. Hollingshead, and D.A. Goodenough. 1983. Gap junction structures. IV. Asymmetric structures revealed by low-irradiation microscopy. *J. Cell Biol.* **96**: 204–216.

Bennett, M.V.L. and D.A. Goodenough. 1978. Gap junctions, electrotonic coupling, and intercellular communication. *Neurosci. Res. Program Bull.* **16**: 371–486.

Caspar, D.L.D., D.A. Goodenough, L. Makowski, and W.C. Phillips. 1977. Gap junction structures. I. Correlated electron microscopy and X-ray diffraction. *J. Cell Biol.* **74**: 605–628.

Goodenough, D.A. 1976. In vitro formation of gap junction vesicles. *J. Cell Biol.* **68**: 221–231.

Henderson, R. and P.N.T. Unwin. 1975. Three-dimensional model of purple membrane obtained by electron microscopy. *Nature* **257**: 28–32.

Loewenstein, W.R. 1981. Junctional intercellular communication: The cell-to-cell channel. *Physiol. Rev.* **61**: 829–913.

Makowski, L., D.L.D. Caspar, D.A. Goodenough, and W.C. Phillips. 1982. Gap junction structures. III. The effect of variations in the isolation procedure. *Biophys. J.* **37**: 189–191.

Makowski, L., D.L.D. Caspar, W.C. Phillips, and D.A. Goodenough. 1977. Gap junction structures. II. Analysis of the X-ray diffraction data. *J. Cell Biol.* **74**: 629–645.

———. 1984. Gap junction structures. V. Structural chemistry inferred from X-ray diffraction measurements on sucrose accessibility and trypsin susceptibility. *J. Mol. Biol.* **174**: 449–481.

Revel, J.P. and M.J. Karnovsky. 1967. Hexagonal array of subunits in intercellular junctions of mouse heart and liver. *J. Cell Biol.* **33**: C7–C12.

Unwin, P.N.T. and P.D. Ennis. 1983. Calcium-mediated changes in gap junction structure: Evidence from the low angle X-ray pattern. *J. Cell Biol.* **97**: 1459–1466.

———. 1984. Two configurations of a channel forming membrane protein. *Nature* **307**: 609–613.

Unwin, P.N.T. and G. Zampighi. 1980. Structure of the junctions between communicating cells. *Nature* **283**: 545–549.

Zampighi, G. 1978. Ph.D. thesis, Duke University, Durham, North Carolina.

Zampighi, G. and P.N.T. Unwin. 1979. Two forms of isolated gap junctions. *J. Mol. Biol.* **135**: 451–464.

Structural Details of Rapidly Frozen Gap Junctions

Robert B. Hanna
*Center for Ultrastructure Studies
SUNY College of Environmental Science and Forestry
Syracuse, New York 13210*

Richard L. Ornberg
*NINCDS, National Institutes of Health
Bethesda, Maryland 20014*

Thomas S. Reese
*Laboratory of Neurobiology, NIH, NINCDS
at Marine Biological Laboratory
Woods Hole, Massachusetts 02543*

Gap junctions have been characterized in a wide variety of tissues (for reviews, see Bennett and Goodenough 1978; Peracchia 1980; Larsen 1983); therefore, the more recent structural investigations have been concerned with the structure of the junctions in the low- and high-resistance states. Also, the structure of the individual particles that compose the gap junction has received considerable attention.

The typical gap junction, as seen in conventional freeze-fracture preparations, consists of intramembrane particles 6.5 nm in diameter that are arranged in a loose hexagonal array with a center-to-center spacing of about 10 nm (Fig. 1). Some particles appear to have a central depression (Fig. 2, arrows) which may represent the cytoplasmic "opening" of the hydrophilic channel.

To study the gap junctions in the uncoupled state, an increase in the coupling resistance of gap junctions can be brought about by various treatments: lowering of cytoplasmic pH (Hanna et al. 1978; Spray et al. 1979; Turin and Warner 1980); mechanical injury (Asada and Bennett 1971; Pappas et al. 1971; Hanna et al. 1984); DNP (Peracchia and Dulhunty 1976; Peracchia 1977; Dahl and Isenberg 1980); EDTA, followed by a return to normal saline (Pappas et al. 1971; Peracchia 1977); and immersion in low-Cl^- solutions (Asada and Bennett 1971; Pappas et al. 1971). When the gap junction is converted to the high-resistance state, many investigators have found that the interparticle spacing becomes reduced to the degree that the particles are arranged in a crystalline array (Peracchia and Mittler 1972; Peracchia and Dulhunty 1976; Makowski et al. 1977; Peracchia 1977; Baldwin 1979; Dahl and Isenberg 1980;

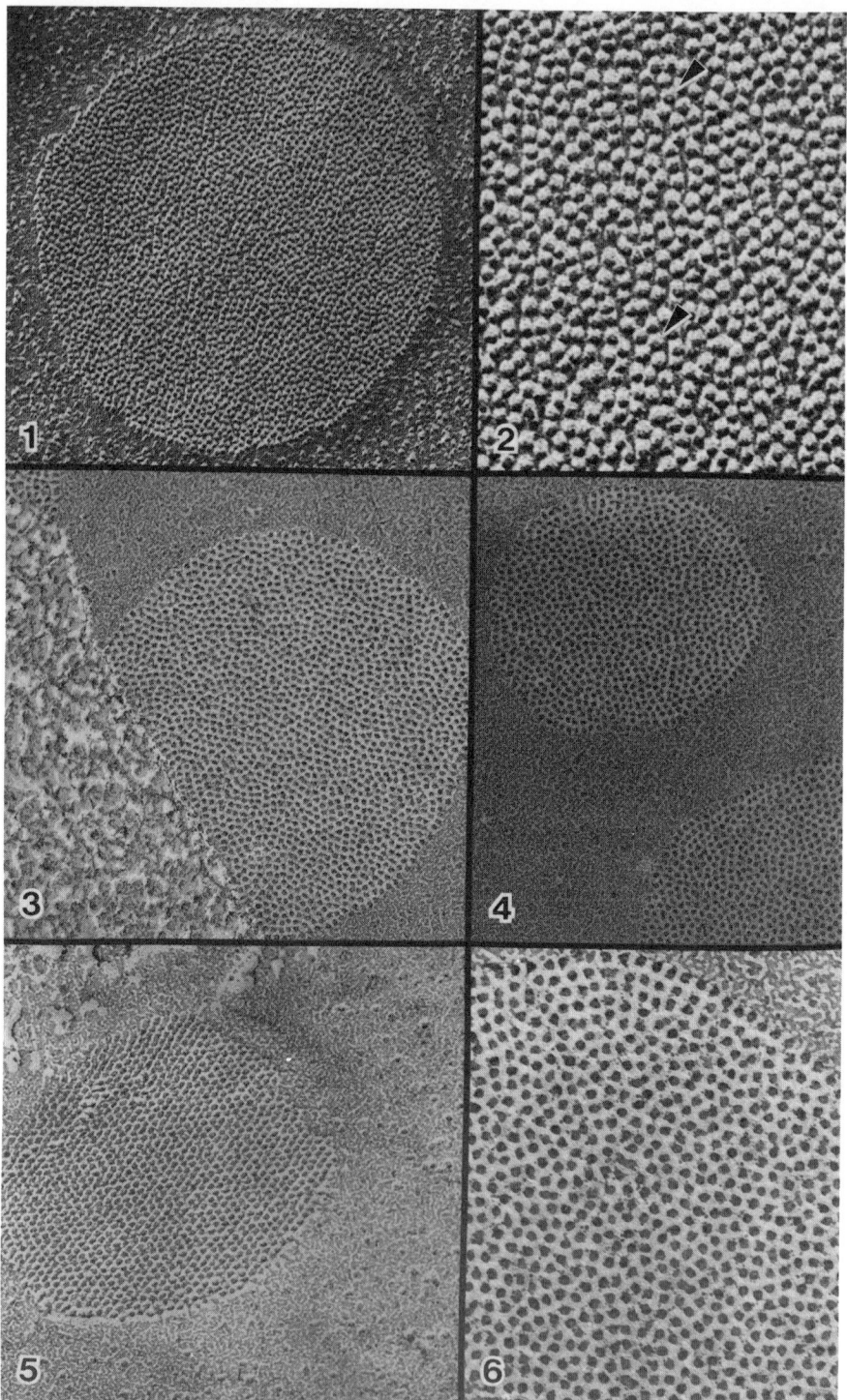

Figures 1–6 (*See facing page for legend.*)

Peracchia and Peracchia 1981). In addition, some investigators have also reported a significant reduction in particle diameters upon reaching the high-resistance state (Peracchia 1977, 1980). It has been postulated that the change in the particle organization produced the change in the conductance state (Peracchia and Mittler 1972; Peracchia and Dulhunty 1976; Peracchia 1977; Peracchia and Peracchia 1981). Recently some investigators have used the change in morphological appearance of the gap junction as an indicator of its physiological state at the time of fixation (e.g., Kistler and Bullivant 1980).

Other investigators, however, have been unable to correlate a greatly reduced center-to-center spacing of the particles (crystallization) with functional uncoupling. Sikerwar and Malhotra (1981) and Spray et al. (1979) found only a slight reduction in particle spacing upon uncoupling. Using rapid freezing, Raviola et al. (1980) found that there was a great deal of variation in particle spacing upon uncoupling. Other studies have found no significant difference in particle organization between the low- and high-resistance states (Spray et al. 1979; Hanna et al. 1981, 1984). Also in direct contrast to the crystallization of the junctional particles a dispersion of the gap junctional particles as a consequence of functional uncoupling has been reported in several systems (Lane and Swales 1978; Campbell and Albertini 1981; Lee et al. 1982; Hanna et al. 1984). Green and Severs (1984) found that the junctional particles in coupled cardiac junctions were packed in hexagonal arrays and that upon uncoupling the junction's particles displayed a variety of configurations including dispersed particles, loosely organized particles, and closely packed particles.

DISCUSSION

With a few exceptions, most of the morphological studies correlating the structure of the gap junction with its physiological state have been carried out on material that has been fixed with glutaraldehyde. This chemical fixation could possibly be a major limitation, since glutaraldehyde is known to uncouple gap junctions (Bennett 1973; Spray et al. 1981). Therefore if one is comparing the structure of gap junctions in the low- and high-resistance states, it is logical to employ rapid freezing as the method of fixation.

The tubular heart of the tunicate, *Ciona*, is a specimen ideally suited for rapid-freezing investigations of gap junctions. The tubular heart is composed

Figures 1-6
1. A conventional freeze-fracture preparation of a low-resistance gap junction from the tunicate heart. (Magnification, 75,000×.) 2. An enlargement of a portion of Fig. 1. Arrows indicate particles with a central depression. (Magnification, 230,000×.) 3. A rapidly frozen gap junction in the low-resistance state. (Magnification, 100,000×.) 4. Gap junction that had been in the high-resistance state for 5 min. (Magnification, 100,000×.) 5. Gap junction that had been in the high-resistance state for 1.5 hr. (Magnification, 140,000×.) 6. Low-resistance gap junction that has been rapidly frozen and rotary shadowed. (Magnification, 200,000×.)

of a single layer of cells that contain numerous junctions, and the heart beats synchronously in the coupled state (Kriebel 1968). In our investigations, the conversion to the high-resistance state (uncoupling) was produced by lowering the intracellular pH with CO_2-equilibrated seawater (Spray et al. 1979, 1982). This increase in junctional resistance causes the hearts to stop beating. When the hearts were returned to normal seawater, they resumed their synchronous beating and the spread of the injected dye Lucifer Yellow was again observed. The synchronous beating, or lack thereof, can be observed on the specimen holder until just seconds before freezing.

After the gap junctions had been frozen in various conductance states, the freeze-fracturing process was carried out at liquid helium temperatures ($-240°C$) with a sapphire knife in an effort to minimize the plastic deformation of the particles. To enhance further the quality of the replica, the specimens were etched briefly at $-166°C$ in a vacuum of 5×10^{-8} T and rotary-shadowed with an alloy of platinum-tantalum-iridium. The resulting replicas demonstrated extremely fine grain and produced good resolution of the intramembrane particles (cf. Figs. 2 and 6).

Gap Junction Structure

The rapidly frozen gap junctions demonstrate all of the characteristics of typical gap junctions seen with conventional freeze-fracture. The intramembrane particles are 6.5 nm in diameter with an interparticle spacing of 10 nm (Figs. 3 and 6), and the pits on the E-face always demonstrate an apparent complementarity to the particles. Those preparations in which the gap junctions had been in the high-resistance state for 5 minutes before freezing also exhibited the same characteristics. The change in coupling resistance did not produce any concomitant change in particle organization or size (Fig. 4). However, if the hearts were kept in the high-resistance state for 1.5 hours, the particles did become more tightly packed in a crystalline array (Fig. 5). Since the crystalline packing of the particles occurred long after the change to the high-resistance state, we must conclude that the particle organization does not play a role in determining the conductance state of the junction. Furthermore, the practice of using the interparticle spacing as an indicator of the state of the junction at the time of fixation must be discontinued. Junctions that exhibit the crystalline packing may indeed be in the high-resistance state; however, junctions exhibiting the "normal" or loose polygonal organization spacing may be either in the low-resistance state or in the high-resistance state.

Intramembrane Particle Structure

A factor that aids in the determination of the structure of the particles is the use of rotary shadowing. Although conventional unidirectional shadowing is used to enhance the contrast of morphological features, many details will not be observed because they are in the shadows (Fig. 2). On the other hand, when rotary shadowing is employed, a more accurate representation of the particle will be obtained because all aspects of the particle will be coated with metal (Fig. 6).

To determine the structure of the gap junctions and the intramembrane particles, one must be aware that the replicas have a very real topography and that the interpretations based on two-dimensional images can be very misleading. For example, Sikerwar et al. (1981) have shown how a symmetrical hexameric structure can appear to approach a tetrameric symmetry when the particle image is tilted 30°, an amount not uncommon in a freeze-fracture replica. The use of rotary shadowing eliminates to a large extent the perception of contour (Figs. 3–6), and therefore the replicas should be viewed as stereo pairs in order to perceive the topography and minimize the problems of misinterpretation (Figs. 7–10).

In Figures 7 and 8, the topography of both the gap junctions and the particles is evident. These junctions are not flat but rather concave, therefore, few of the particles face the electron beam directly. Some of the junctional particles are hexameres (Fig. 7, arrow 6); however, there are many particles that are not hexagonal. Some of the particles are pentagonal (Fig. 7, arrow 5) whereas others have a distinctly square shape (Fig. 7, arrow 4). Also in Figure 8, the double arrow is pointing to three polygonal particles, one of which exhibits a hexameric substructure, one a pentameric substructure, and one a tetrameric substructure. Continued observation of Figures 7 and 8 reveals many particles that consist of four and five subunits as well as six subunits.

There is, however, a consensus among those investigators employing negative staining (Unwin and Zampighi 1980; Zampighi et al. 1980; Baker et al. 1983) and X-ray diffraction (Caspar et al. 1977; Makowski et al. 1977, 1982) that the intramembrane particles are hexameres. These studies, by necessity, have been carried out on isolated gap junctions and consequently they are in the high-resistance state. The intramembrane particles of the isolated junctions are tightly packed into hexagonal arrays that would suggest a hexameric particle structure.

There are other instances in which junctional particles have been found to exhibit less than six subunits. In high-resolution replicas of gap junctions between the supporting cells in both the goldfish and the guinea pig, Hama and Saito (1977) have shown that the junctional particles consist of both pentameres and hexameres. Sikerwar et al. (1981) also demonstrated gap junction particles that exhibited less than six subunits, but they attributed the images of those particles with less than six subunits to tilt and fracture artifacts. In the calf lens, Peracchia and Peracchia (1981) and Bernardini and Peracchia (1981) have found orthogonal and rhombic packing of the junctional particles, which suggests that these particles are tetrameres.

Many of the junctional particles have what appears to be a central core or hole. Specific examples are seen in Figure 7 (arrow) and in Figure 8 (single arrows). At higher magnifications those particles that have a central core appear to be hollow columns (Fig. 9 and 10, arrows). This hollow-column appearance is similar to the structure hypothesized and depicted in several models (Makowski et al. 1977; Unwin and Zampighi 1980). It can be argued that the central core or hole was created by the use of rotary shadowing, but this would be the case only if the particles were round bumps, which they are not. Furthermore, the central core corresponds to the central depression, which can be observed in unidirectional shadowing (Fig. 2).

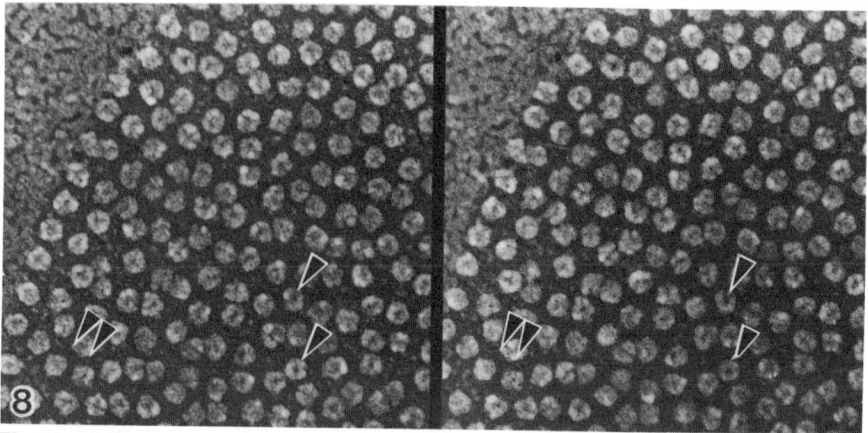

Figures 7–8
7. Stereo pair of a low-resistance gap junction. Particles consisting of several different structures can be observed: a central core or hole (arrow), hexamere (arrow 6), pentamere (arrow 5), and tetramere (arrow 4). (Magnification, 400,000×.) **8.** Stereo pair of a low-resistance gap junction. Single arrows denote particles with a central core or hole. Double arrows indicate an area where the particles have fractured very low and reveal distinct hexagonal, pentagonal, and square particles. (Magnification, 400,000×.)

It can also be seen in Figures 7 and 8 that the particles fracture at different heights. The different fracture heights are especially evident along the edge of the gap junction in Figure 8. The double arrow in Figure 8 indicates three particles whose shapes are distinctly polygonal. As is the case at the double arrow, when the particle shapes are distinctly polygonal, the particles have

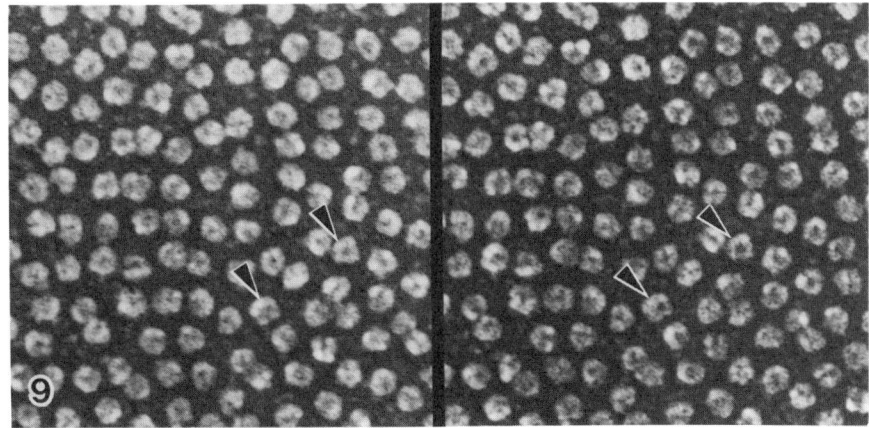

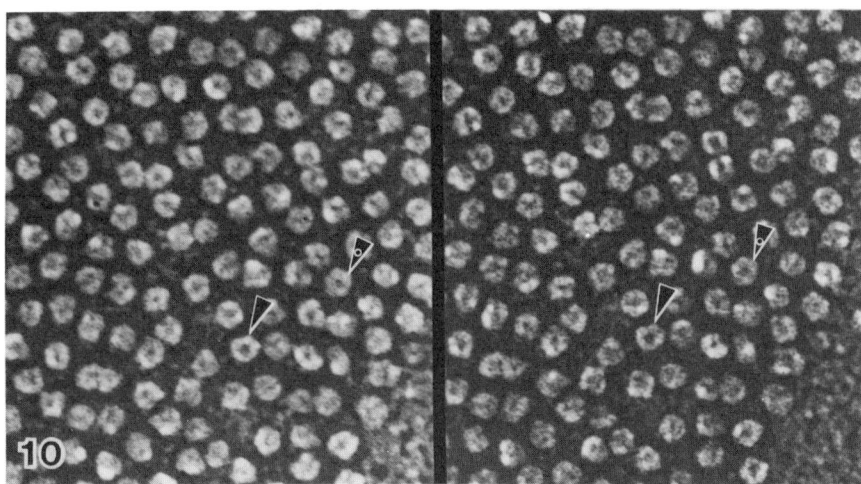

Figures 9–10
9. Stereo pair of low-resistance gap junction. Arrows indicate particles with a central core. (Magnification, 540,000 ×.) **10.** Stereo pair of a low-resistance gap junction. Arrows indicate particles with central core, one of which is pentagonal (○ arrow). (Magnification, 540,000 ×.)

usually fractured much lower than the other particles. On occasion a particle with height does present a clearly defined outline (Fig. 10, ○ arrow). However, as a direct result of the topography inherent in the replicas, most of the particles do not face the electron beam. The fact that so many of the particles are tilted with respect to the electron beam demands that the particle structures be observed with stereo pairs. This fact also reinforces the concerns of Sikerwar et al. (1981) that care must be exercised when interpreting the particle structure from conventional two-dimensional micrographs.

CONCLUSIONS

No morphological differences could be found between junctions in the low-resistance state and those junctions in the high-resistance state. Only those junctions that were allowed to remain in the high-resistance state for 1.5 hours exhibited the tightly packed or crystalline array of the intramembrane particles.

The rapid freezing and low-temperature fracture provided very fine-grained replicas which, when viewed as stereo pairs, allowed the substructure of the particles to be determined. Although some of the particles were hexameres, other particles were pentameres and tetrameres. The particles also appeared to have a central core or hole.

ACKNOWLEDGMENTS

The author would like to acknowledge Drs. David Spray and M.V.L. Bennett for their assistance in these experiments.

REFERENCES

Asada, Y. and M.V.L. Bennett. 1971. Experimental alteration of coupling resistance at an electrotonic synapse. *J. Cell Biol.* **49:** 159–172.

Baker, T.S., D.L.D. Caspar, C.J. Hollinshead, and D.A. Goodenough. 1983. Gap junction structures IV. Asymmetric features revealed by low irradiation microscopy. *J. Cell Biol.* **96:** 204–216.

Baldwin, K. 1979. Cardiac gap junction configuration after an uncoupling treatment as a function of time. *J. Cell Biol.* **82:** 66–67.

Bennett, M.V.L. 1973. Permeability and structure of electrotonic junctions and intracellular movement of tracers. In *Intracellular staining techniques in neurobiology* (ed. S.D. Kater and C. Nicholson), p. 115–133. Elsevier, Amsterdam.

Bennett, M.V.L. and D.A. Goodenough. 1978. Gap junctions, electrotonic coupling, and intracellular communication. *Neurosci. Res. Program Bull.* **16:** 373–486.

Bernardini, G. and C. Peracchia. 1981. Gap junction crystallization in lens fibers after an increase in calcium. *Invest. Opthalmol. Visual Sci.* **21:** 291–299.

Campbell, K.L. and D.F. Albertini. 1981. Freeze-fracture analysis of gap junction disruption in rat ovarian granulosa cells. *Tissue Cell* **13(4):** 651–668.

Caspar, D.L., D.A. Goodenough, L. Makowski, and W.C. Phillips. 1977. Gap junction structures. I. Correlated electron microscopy and x-ray diffraction. *J. Cell Biol.* **74:** 605–628.

Dahl, G. and G. Isenberg. 1980. Decoupling of heart muscle cells: Correlation with increased cytoplasmic calcium activity and with changes in nexus ultrastructure. *J. Membr. Biol.* **53:** 63–75.

Green, C.R. and N.J. Severs. 1984. Gap junction connexon configuration in rapidly frozen myocardium and isolated interrelated disks. *J. Cell Biol.* **99:** 453–463.

Hama, K. and K. Saito. 1977. Gap junctions between the surrounding cells in some acoustico-vestibular receptors. *J. Neurocytol.* **6:** 1–12.
Hanna, R.B., G.D. Pappas, and M.V.L. Bennett. 1984. The fine structure of identified electrotonic synapses following increased coupling resistance. *Cell Tissue Res.* **235:** 243–249.
Hanna, R.B., T.S. Reese, R.L. Ornberg, D.C. Spray, and M.V.L. Bennett. 1981. Fresh frozen gap junctions: Resolution of structural detail in the coupled and uncoupled states. *J. Cell Biol.* **91:** 125a.
Hanna, R.B., D.C. Spray, P.G. Model, A.L. Harris, and M.V.L. Bennett. 1978. Ultrastructure and physiology of gap junctions of an amphibian embryo, effects of CO_2. *Biol. Bull.* **155:** 442.
Kistler, J. and S. Bullivant. 1980. The connexon order in isolated lens gap junction. *J. Ultrastruct. Res.* **72:** 370–381.
Kriebel, M. 1968. Electrical characteristics of tunicate heart cell membranes and nexuses. *J. Gen. Physiol.* **52:** 46–59.
Lane, N.J. and L.S. Swales. 1978. Changes in the blood-brain barrier of the central nervous system in the blowfly during development, with special reference to the formation and disaggregation of gap and tight junctions. *Dev. Biol.* **62:** 415–431.
Larsen, W.J. 1983. Biological implications of gap junction structure, distribution, and composition: A review. *Tissue Cell* **15:** 645–671.
Lee, C.W., D.G. Cran, and N.J. Lane. 1982. Carbon dioxide induced disassembly of gap-junctional plaques. *J. Cell Sci.* **57:** 215–228.
Makowski, L., D.L.D. Caspar, W.C. Phillips, and D.A. Goodenough. 1977. Gap junction structures II. Analysis of the X-ray diffraction data. *J. Cell Biol.* **74:** 629–645.
———. 1982. Gap junction structures II. The effect of isolation procedures. *Biophys. J.* **37:** 189–191.
Pappas, G.D., Y. Asada, and M.V.L. Bennett. 1971. Morphological correlates of increased coupling resistance at an electrotonic synapse. *J. Cell Biol.* **49:** 173–188.
Peracchia, C. 1977. Gap junctions. Structural changes after uncoupling procedures. *J. Cell Biol.* **72:** 628–641.
———. 1980. Structural correlates of gap junction permeation. *Int. Rev. Cytol.* **66:** 81–146.
Peracchia, C. and A.F. Dulhunty. 1976. Low-resistance junctions in crayfish. Structural changes with functional uncoupling. *J. Cell Biol.* **70:** 419–439.
Peracchia, C. and B.S. Mittler. 1972. Fixation by means of glutaraldehyde-hydrogen peroxide reaction products. *J. Cell Biol.* **53:** 234–238.
Peracchia, C. and L. Peracchia. 1981. Gap junction dynamics: Reversible effects of hydrogen ions. *J. Cell Biol.* **87:** 719–723.
Raviola, E., D.A. Goodenough, and G. Raviola. 1980. Structure of rapidly frozen gap junctions. *J. Cell Biol.* **87:** 273–279.
Sikerwar, S. and S. Malhotra. 1981. Structural correlates of glutaraldehyde-induced uncoupling in mouse liver gap junctions. *Eur. J. Cell Biol.* **25:** 319–323.
Sikerwar, S., J.P. Tewari, and S.K. Malhotra. 1981. Subunit structure of the connexons in hepatocyte gap junctions. *Eur. J. Cell Biol.* **24:** 211–215.
Spray, D.C., A.L. Harris, and M.V.L. Bennett. 1979. Gap junctional conductance is a simple and sensitive function of intracellular pH. *Science* **211:** 712–715.
———. 1981. Glutaraldehyde differentially affects gap junctional conductance and its pH and voltage dependence. *J. Biophys.* **33:** 108a.
Spray, D.C., J.H. Stern, A.L. Harris, and M.V.L. Bennett. 1982. Gap junctional conductance comparison of sensitivities to H and Ca ions. *Proc. Natl. Acad. Sci.* **79:** 441–445.
Spray, D.C., A.L. Harris, M.V.L. Bennett, J.E. Brown, and R.B. Hanna. 1979. Gap junctional conductance: Dependence on internal pH. *J. Elchem. Soc.* **126:** C143.

Turin, L. and A. Warner. 1980. Intracellular pH in early *Xenopus* embryos: Its effect on current flow between blastomeres. *J. Physiol.* **300:** 489–504.

Unwin, P.N.T. and G. Zampighi. 1980. Structure and the junction between communicating cells. *Nature* **280:** 545–549.

Zampighi, G., J.M. Corless, and J.D. Robertson. 1980. On gap junction structure. *J. Cell Biol.* **86:** 190–199.

Molecular Conformation of the Major Intrinsic Protein of Lens Fiber Membranes: Is It a Junction Protein?

Jean Paul Revel and S. Barbara Yancey

Division of Biology
California Institute of Technology
Pasadena, California 91125

When the structure of electrical synpases was first described (Robertson 1963), there was little inkling that similar connections (gap junctions) also existed between nonexcitable cells. In fact it is now clear that gap junctions are so widely distributed (for review, see Larsen 1983) that cells which are not linked to each other in this way are the exception rather than the rule. In spite of some differences (for review, see Lane and Skaer 1980), junctions are generally recognizable by their characteristic morphology (Revel and Karnovsky 1967; Payton et al. 1969; Goodenough and Revel 1970), whatever their origin. The bulk of the evidence now generally accepted (see Revel et al. 1985) suggests that each is comprised of a major protein of molecular weight about 27,000 daltons (liver: Henderson et al. 1979; Hertzberg and Gilula 1979; Finbow et al. 1980; lens: Alcala et al. 1975; Nicholson et al. 1983; heart: Gros et al. 1983), although some workers now report the existence of entities of different sizes (Finbow et al. 1983; Manjunath et al. 1984; Warner et al. 1984). Gap junctions from many sources seem to share common antigenic properties (Traub and Willecke 1982; Traub et al. 1982; Dermietzel et al. 1984; Hertzberg 1984; Hertzberg and Skibbens 1984; Warner et al. 1984); however, the limited information derived from protein sequencing suggests that less than 50% of residues are conserved when junction, or putative junction, proteins in rat liver, lens, and heart are compared (Nicholson et al. 1983; Revel et al. 1984, 1985; B.J. Nicholson et al., in prep.). The number of proteins so studied is still very small, but these differences are likely to be significant since the junction proteins are strongly conserved between species in both the liver (Nicholson et al. 1981) and the lens (Zigler and Horowitz 1981). The complete primary sequence of only one protein believed to be related to gap junctions is at present known, that of the major intrinsic protein (MIP) of the eye lens fibers determined by M. Gorin and B. Yancey (Gorin et al. 1984).

DISCUSSION
Lens MIP and Gap Junctions

The status of MIP as a gap junction protein is highly controversial (Kuszak et al. 1982; Zampighi et al. 1982; Nicholson et al. 1983; Paul and Goodenough 1983a). There is good evidence that there are gap junction-like structures in the lens (Benedetti et al. 1976; Goodenough 1979; Kistler and Bullivant 1980a; Kuszak et al. 1981, 1982). Lens fibers are electrotonically coupled and exchange dye molecules (Goodenough 1979; Goodenough et al. 1980), suggesting the presence of gap-like junctions. The lens is large and avascular; efficient cell–cell junctions could ease the problems likely to arise in supporting the metabolism of the inner cells (Goodenough 1979; Schuetze and Goodenough 1982). The most abundant structures found on lens membranes are the extensive fields of somewhat loosely packed intramembrane particles seen by freeze-cleaving (Benedetti et al. 1976; Peracchia 1978; Kuszak et al. 1981). Each of these particles is of a size and appearance generally similar to connexons in gap junctions of other tissues (cf. Peracchia and Peracchia 1980; Kuszak et al. 1981; or Kistler and Bullivant 1980a to Caspar et al. 1977; and Unwin and Zampighi 1980). It would be reasonable to assume that these intramembrane particles would be composed of MIP, the major protein of lens fiber membranes (Broekhuyse et al. 1976). MIP is of a molecular weight comparable with that of junction proteins in other tissues (Broekhuyse et al. 1979). However, other closely packed structures such as the "orthogonal arrays" of Kistler and Bullivant (1980b) or Zampighi et al. (1982) also found on lens fiber membranes could themselves provide major amounts of protein. Some sets of antibodies raised against MIP react with lens fiber plasma membrane and the "thin junctions" seen in sectioned material, but do not with thicker structures that more closely resemble gap junctions (Paul and Goodenough 1983a). Other antibody preparations apparently react with both types of membranes (Bok et al. 1982; Fitzgerald et al. 1983; Sas et al. 1985). Thus, there is no agreement as to which of the several structures described by different authors on lens fiber membranes participate in metabolic exchanges, or contain MIP. It is not known whether the various structures seen on lens fiber membranes are morphologically and chemically independent, or if they represent different modes of aggregation or configuration of the same protein(s).

The known primary sequence of lens MIP correlates only marginally with the known sequence of liver or heart gap junction protein (Gorin et al. 1984; Revel et al. 1985; B.J. Nicholson et al., in prep.). Research using antibodies to compare these several proteins mostly supports the idea that they are largely different. Although some authors have argued for similarity between MIP and the liver protein on the basis of a small degree of cross-reactivity (Traub and Willecke 1982), most find no cross-reactivity using antibodies prepared against liver and lens junction fractions (Hertzberg et al. 1982; Paul and

Goodenough 1983a; Hertzberg 1984); some (but not all) of the antibodies that have been raised against liver junctions or junction protein also react with heart tissue (Traub et al. 1982; Dermietzel et al. 1984; Hertzberg 1984; Hertzberg and Skibbens 1984). The differences between MIP and liver or heart junction proteins may be related to the fact that the physiological properties of lens cell–cell channels are markedly different from those of gap junctions in other organs (see Goodenough 1979), assuming, of course, that MIP is indeed a major component of the lens junctions. The aim of this report is to discuss the models that can be imagined for the arrangement of MIP in membranes and to assess the role of MIP as suggested by the most likely configuration of the protein.

Membrane Protein Model Building

A number of approaches have been developed that allow one to predict more or less successfully the likelihood of a particular secondary structure for a given protein, on the basis of its amino acid sequence (Chou and Fasman 1978; Garnier et al. 1978; but see Nishikawa 1983 for words of caution). The data bases for these predictive methods are derived from understanding gained by X-ray diffraction studies on crystallized, water-soluble proteins; it is not evident that the same criteria can be used to predict effectively the structure of a membrane protein (Nicholson et al. 1981). In the latter case, additional constraints operate which are imposed by the necessity of accommodating part of the polypeptide chain in the hydrophobic core of membranes (Henderson 1979). Those segments of the molecule that traverse the core (transcore segment) are generally considered most likely to have an α-helical configuration (Henderson 1979), although at least one transmembrane protein, porin, is believed to be rich in β-pleated sheets (Garavito et al. 1982; Dorset et al. 1983) and gramicidin channels may be formed of β structures (see Ovchinnikov and Ivanov 1982). Helical transcore segments are at least 20 amino acids long (see Eisenberg 1984) and contain relatively few unpaired charged residues (Engelman et al. 1980). Highly charged groups are difficult to accommodate in the nonaqueous environment of the membrane core (Steitz et al. 1982).

The α-helical structure is favored for the transcore portions of the polypeptides because it maximizes opportunities for H bonding. In turns of the polypeptide chain, not all of the H-bonding capabilities can be internally satisfied, and such turns therefore occur outside the membrane core, in a polar environment. In bacteriorhodopsin, a transmembrane protein whose structure is adequately understood (Engelman et al. 1980), the Chou and Fasman (1978) or Garnier et al. (1978) approaches predict turn positions that align reasonably well with their known position, and also match the position of troughs in hydropathicity plots (Kyte and Doolittle 1982) (Fig. 1).

A Model for MIP

In trying to arrive at a model for MIP we have used a combination of approaches. Regions that are of appropriate length and amino acid composition

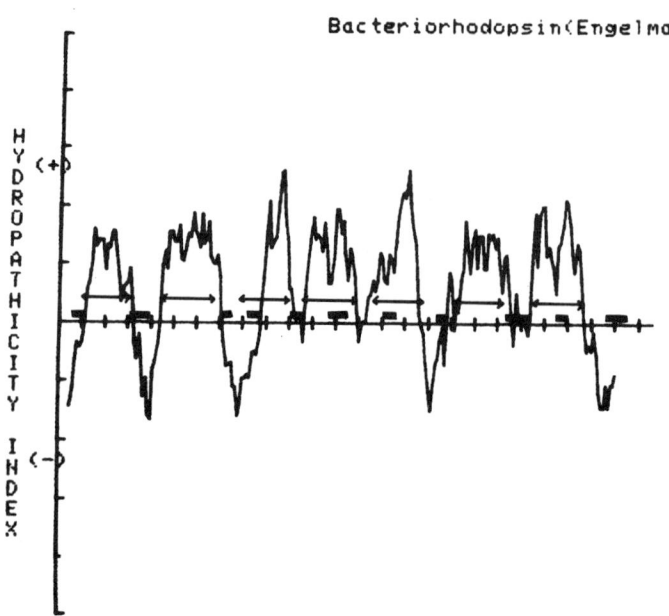

Figure 1
A plot of hydropathicity along the bacteriorhodopsin molecule, on which is superimposed the location of the helices that traverse the hydrophobic core of the membrane (double-headed arrows), and that of turns predicted by the application of the Chou and Fasman rules (heavy lines). Although there are some mismatches, predicted turns coincide quite well with regions of low hydropathicity. The only turn known to exist, but which is not predicted with a sufficiently high probability to be seen in this plot is between segments 4 and 5.

to span the membrane were identified by plots of hydropathicity generated by a program written by us, using the values suggested by Kyte and Doolittle (1982) and generally similar to their SOAP program (Fig. 2). The probable location of turns was determined as a consensus of predictions using computerized Chou and Fasman (1978) and Garnier et al. (1978) analyses, coupled with a visual search of minima in hydropathicity plots.

By using this combination of approaches one can predict that MIP makes at least five passes across the membrane (Fig. 2 and 3). There are five largely hydrophobic regions separated from each other by turns predicted by all approaches used. These regions are postulated to be α-helical for the reasons outlined by Henderson (1979), as described above. However they could be in other configurations without necessitating major changes in the model. There is very likely also a sixth region that could span the membrane, as discussed below. The carboxyterminal 40 or so amino acids of the molecule are unlikely to be transmembrane, even though they display a rather high hydropathicity, because of numerous charged groups. Only three of the postulated transcore regions of MIP need accommodate a charged group. The exact location of the

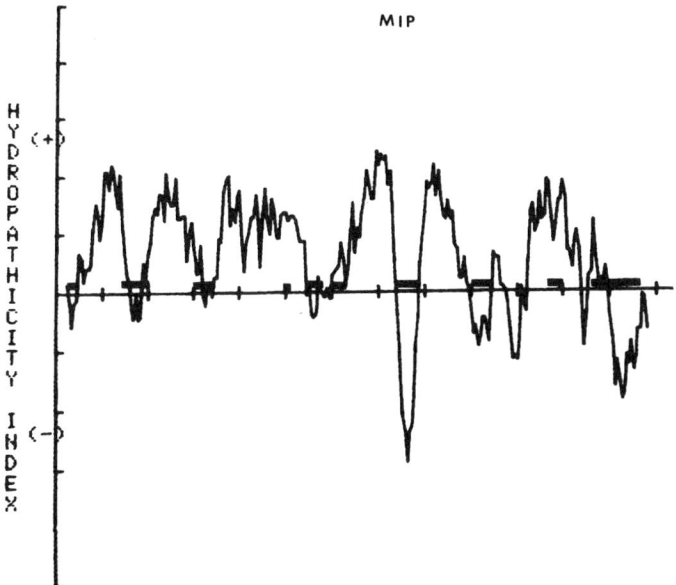

Figure 2
A similar plot to that shown in Fig. 1, but showing the MIP molecule. The position of turns is also marked by the heavy lines. The position of the transcore helices (not indicated) is derived from the consensus of considerations presented in the text.

transcore regions is, of course, not known and experimental evidence will have to be sought both to confirm their existence and to determine their precise position. For example, segment A could just as easily have been drawn from residues 12 to 32, instead of 17 to 36. This would have meant accommodating a negatively charged group (Glu-11) instead of positively charged one (Arg-33).

There is a segment of the molecule (segment F, Fig. 3) near the carboxyl terminus which is less hydrophobic than the other transmembrane segments postulated, but is long enough to be transmembrane and with only one charged group. This segment is part of a putative amphiphilic helix predicted by Fourier transform analysis (Finer-Moore and Stroud 1984) and the hydrophobic moment technique (Eisenberg et al. 1982, 1984). Analysis by the hydrophobic moment approach also suggests at least one more amphiphilic segment elsewhere in the molecule (segment C, Fig. 3), which is discussed later. The amphiphilic segment F is unlikely to penetrate the membrane core by itself because of its low overall hydrophobicity. It can be accommodated as a sixth transmembrane structure by postulating that several MIP molecules interact with each other: The hydrophilic faces of several of the amphiphilic helices could face each other, so that only their hydrophobic faces would be exposed to the core of the membrane. This is a situation similar to that found in the acetylcholine receptor (AChR). Each of the AChR molecules has one amphiphilic helix, and five of the molecules aggregate to form an aqueous pore lined by the hydrophilic faces of these helices. Several MIP molecules could interact to form a transmembrane channel and two of these (one in each cell) could interact end to end to permit cell–cell exchanges. The chosen location of the beginning of the putative amphiphilic segment (Gorin et al. 1984)

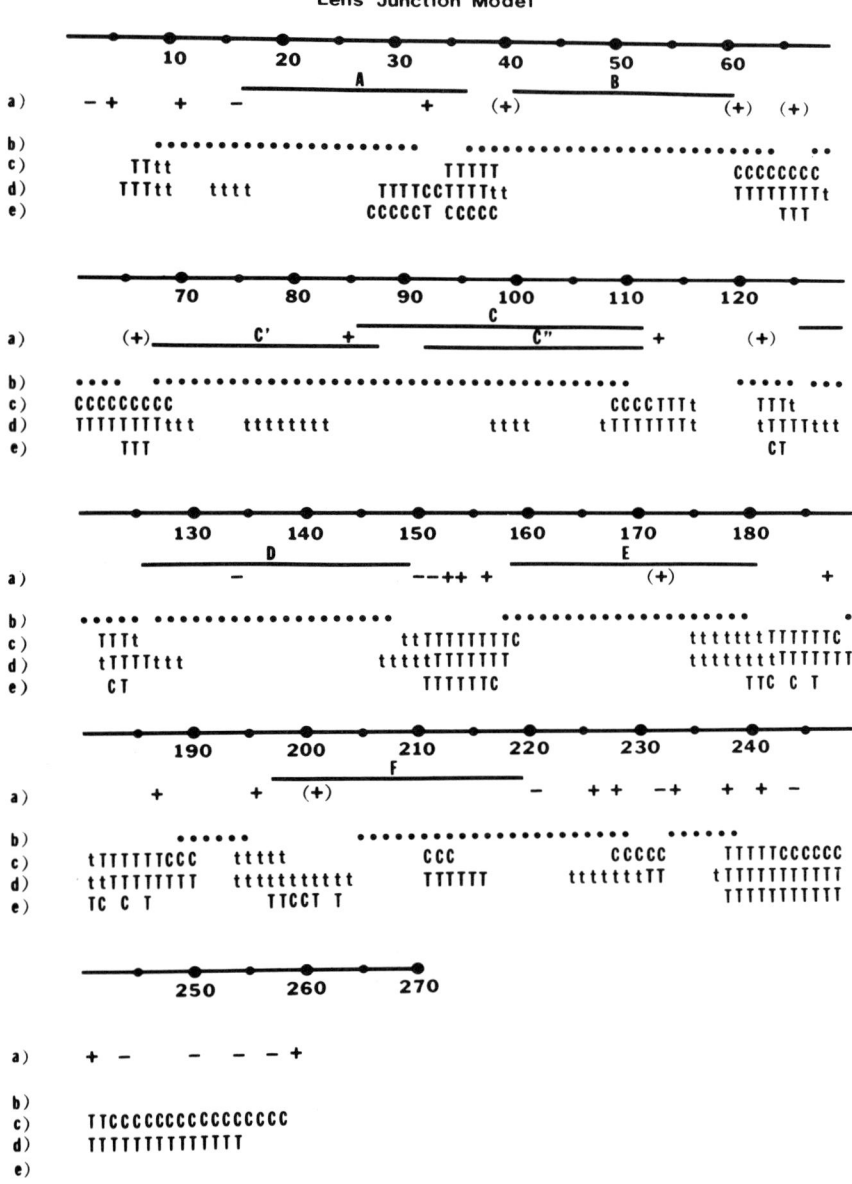

Figure 3 (See facing page for legend.)

gives a sufficient length of polypeptide to span the membrane, and involves only one charged group. Other possible starting points for segment F would have placed two or even three positively charged groups in the pore, a feature we sought to avoid because there is no evidence at present for or against charge discrimination in intercellular exchanges between lens fiber cells.

Alternatively, one could imagine amphiphilic segment F lying on the surface of the membrane with its hydrophilic groups exposed to the aqueous environment and hydrophobic residues directed toward the core of the membrane (Fig. 4), in a manner similar to that suggested by Terwilliger et al. (1982) for mellitin. A choice between these two alternate dispositions of the protein in the membrane can theoretically be made by determining which portions of the MIP molecule are exposed on what faces of the membrane. If the amphiphilic helix is part of a transmembrane channel, then both termini of the protein would be seen on the same side of the membrane, since one expects an even number of helices. In the event that the amphiphilic helix would be situated at the membrane surface, there would only be five transmembrane helices and the amino and carboxyl termini should be found on opposite sides of an individual membrane (Fig. 4b). One must note this analysis is made more complex by the possibility that segment C represents two, rather than one transcore helix.

Sidedness of the Location of Amino and Carboxyl Termini

The amino terminus is likely to be found at the cytoplasmic face of the membrane. This is suggested by in vitro translation studies in the presence of microsomal membranes (Paul and Goodenough 1983b). The aminoterminal methionine of newly synthesized and membrane-associated MIP can be removed by enzymatic treatment of the vesicles, suggesting that the amino terminus is indeed exposed at the surface which, in the intact organism, would be the cytoplasmic face of the endoplasmic reticulum membrane. A priori, one would expect that the carboxyterminal end of the molecule would also be at the cytoplasmic face of the membrane because this is where it would be left at the

Figure 3
Diagramatic representation of the correspondence between various features of the lens MIP and the proposed model. The numbers indicate the position of residues in the deduced sequence. (Note that there is overlap between the end of one line and the beginning of the next.) a) indicates predicted transcore elements of the polypeptide labeled as in Gorin et al. (1984) and also shows the position of charged amino acids. Histidine residues are shown in parentheses. Segments C' and C" represent alternate interpretations of the sequence as discussed in the text. b) indicates regions of the molecule with a hydropathicity above 0 as calculated from hydrophobicity parameters of Kyte and Doolittle (1982). c) and d) indicate predictions from Chou and Fasman analyses (1978) (using, respectively, a 15- and 29- protein data base). e) gives the location and turns or random coil made by the program PREDICT, based on the approach by Garnier et al. (1978) (R. Stroud, pers. comm.). T denotes a predicted turn, t a less likely one with other structures predicted also, and C regions likely to be in random coil configuration.

end of synthesis. It has become obvious however that this is not the case for many membrane proteins (see Blobel 1980; Sabatini et al. 1982; Strubin et al. 1984). One therefore needs to get independent data to establish the location of the carboxyl terminus. Treatment of in vitro-translated MIP with chymotrypsin (Paul and Goodenough 1983b) not only removes the aminoterminal methionine, it also reduces the molecular weight of MIP to about 20,000. Because the known sequence now suggests that the exposed amino terminus of the molecule is at most 16 amino acids (∼1700 daltons) long, the reduction in molecular weight must be due to removal of material from the carboxyl end, suggesting that both ends of MIP are on the same side of the membrane in

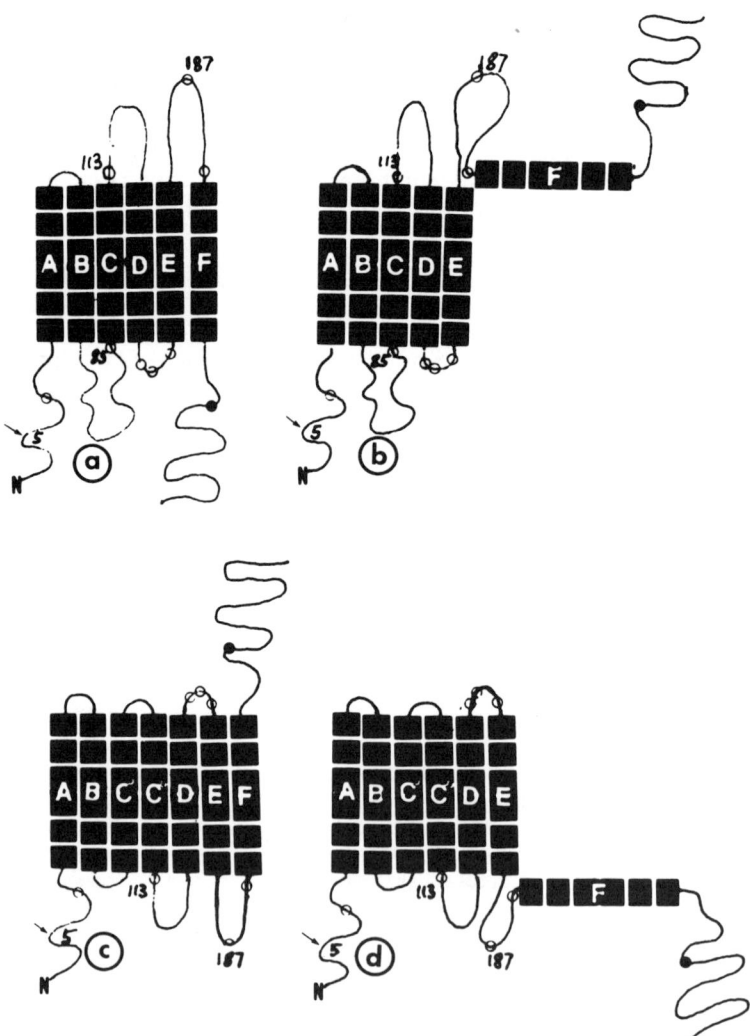

Figure 4 (*See facing page for legend.*)

this instance. Treatment of lens membrane fractions (with trypsin) also reduces the molecular weight of MIP to about 21,000 (see Kistler and Bullivant 1980b; Nicholson et al. 1983), a loss of about 5000 daltons. Sequencing of the resulting peptide suggests that only five amino acids are lost from the amino terminus of the molecule. The remaining polypeptides lost must be removed in one or more fragments from the carboxyl end of the molecule (Nicholson et al. 1983). The sequence shows that this could be achieved by cleavage at residues 233 or 238, or possibly even 226 or 228 (see Fig. 4). If the orientation of MIP in natural membranes is the same as that after in vitro translation, this would be the expected result irrespective of whether MIP is part of a junction (where only the cytoplasmic side is accessible to enzymes) or of a single membrane (where both sides of the membrane are accessible to enzymes). The data, while not definitive, are compatible with the idea that both carboxyl and amino termini are on the same side of the molecule and therefore that the amphiphilic helix must also cross the hydrophobic core of the membrane.

It could of course be argued that the in vitro system is artificial and cannot be in any case be used to decide the organization of the protein in the natural state; however appealing the model, other possibilities must be seriously considered. Assuming that the segments of the MIP molecule exposed at the ab-cytoplasmic face of the membrane were inaccessible to enzymes (whether because the membrane is part of a junction [Goodenough and Revel 1971], or due to other factors), then the data still would support the idea of a cytoplasmic carboxyl terminus to explain the results obtained after treatment with trypsin. However, should MIP not be part of a junctional structure, then both

Figure 4
Four possible models for MIP showing putative transmembrane segments indicated by capital letters and other regions of the molecule drawn as a thin line. The open circles represent the approximate position of tryptic sites predicted outside of the membrane core, except for several such sites at the carboxyl terminus of the molecule, whereas only a single group, that closest to the core, is shown by a solid dot. Arg-5, an experimentally determined tryptic cleavage point, is marked by an arrow near the amino terminus of the molecule. Cleaving between Arg-5 and the first cluster of potential sites at the carboxyl end of the molecule (solid dot) would account for peptides of the size found after enzyme attack. The other potential cleavage sites are presumed to be inaccessible to the enzyme, either because they are too close to the core of the membrane, or because they are protected between apposed membranes if MIP were part of a junction, or in some other fashion. Model a is considered most likely to represent the configuration of one MIP. It consists of six transmembrane helices. Because segment F is believed to be amphiphilic, several MIP have to cluster, forming a transmembrane aqueous pore (a connexon?), to permit this configuration. Model b places F at the surface of the membrane. In this configuration MIP need not be part of a cluster to form a pore. Models c and d show MIP with seven helical segments. The additional segment is created from the region connecting B and C as is discussed in the text and shown in Fig. 3. In model c, as in model a, a cluster of molecules should form but now amino and carboxyl termini are on different sides of the membrane. The consequences are discussed in the text. In model d, as in model b, there is no need for clustering to accommodate putative helix F. Several other variations on the models presented must be considered when testing the model.

cytoplasmic and abcytoplasmic faces of the membrane could be reached by an enzyme the size of trypsin and it would not be possible to specify the location of the termini by this type of experiment.

Clearly, further data localizing specific portions of the sequence to one side or the other of the membrane are needed to settle the issues raised. In the meantime, however, the evidence available would favor as most likely the idea that the amino and carboxyl termini are both on the cytoplasmic side of the membrane, suggesting an even number of transmembrane segments and supporting a channel (junctional?) role for MIP.

Formation of New Channels

It is interesting to note that the amphiphilic nature of helix F raises problems when one tries to imagine the configuration of newly synthesized MIP in the membrane since a single segment F would presumably be too hydrophilic to traverse the core. Several MIP, however, could have their configuration stabilized if they came to interact to form a precursor of the channel. If such protochannels form, they probably do not have the permeability characteristics of mature channels, since they represent only half of the structure normally found between cells. Half a junctional element forming a permeable pore would permit leakage, first across intracellular membranes, and, as the newly made protein progressed through the secretory apparatus of the cell, eventually would cause leaks across the plasma membrane itself. An alternative possibility is that amphiphilic helix F does not immediately take its final transmembrane position, but remains on one side of the membrane; this might be encouraged by glycosylation of a potential site (Asn Phe Thr) starting at residue 191. Since this site seems unoccupied when mature MIP is isolated, one might suppose that there is cleavage of the putative carbohydrate to permit F to become transmembrane, which would then occur if the concentration of MIP were high enough to permit the formation of oligomeric channels. These postulated steps could regulate the formation of new channels and avoid the formation of leaks harmful to cellular physiology.

Other Possible Configurations of MIP

The location of helix F is not the only problem to be resolved in arriving at a satisfactory model for MIP. As shown by the hydropathicity plot, there is a generally hydrophobic region of the molecule starting at approximately residue 67, shortly after the termination of helix B, and extending to residue 111 or so (see Fig. 2). Precise predictions for the location of transmembrane helices are difficult to make since the position of the apparent "ends" depends on several factors, including the threshold hydrophobicity chosen and the size of the "window" over which hydrophobicity values are averaged. In the model drawn (Fig. 3a) it has been assumed that Arg-85 is outside the hydrophobic membrane core because of its charge, thus leaving the polypeptide chain between residues 61 and 85 on the cytoplasmic side of the membrane. This leads to the configuration of helix C and the turns predicted for residues 112–125 as postulated by the model we favor (Fig. 3). It is possible, however, to carve out two potential transcore regions rather than one from the same segment of the protein (Fig. 4). One of these would extend from residue 69 to 88 and the other from 94 to 103. In this way one would add an extra helix to the model (helix C is replaced by helices C' and C"; Fig. 3). This seven-helix model (Fig. 4) is considered to be less likely for several reasons: (1) no turns are predicted (Fig. 3) from an analysis of the primary sequence around residue 90 where one would be expected between the two postulated transmembrane helices C' and C"; (2) Arg-85 would be forced into the hydrophobic region of the molecule, an unlikely event; and (3) Arg-187 would be exposed on the cytoplasmic side of the membrane. Unless protected by interaction with other groups, both Arg-187 and Arg-196 should be cleavable (Fig. 4). One should be able to distinguish between six- and seven-helix models by comparing the sidedness of the amino terminus and of the region of peptides around residue 187 (or others). Residue 187 should be on the same side of the membrane as the aminoterminus in the case of a seven-helix molecule (Fig. 4c,d) and on the opposite side for the favored six-helix model. The hydrophobic moment calculations of Eisenberg et al. (1982) predict an amphiphilic segment around residue 80. By a reasoning analogous to that used earlier for segment F, this might allow one to introduce Arg-85 into the aqueous pore, obviating the problems of extra charges in the hydrophobic core. This idea is not favored because the putative amphiphilic segment does not show up well in a helix net plot (while segment F does), nor is it clearly demarcated in the Fourier transform analysis. It is also unlikely that there would be room for two entire helices in the wall of the pore for each of the four to six MIP molecules believed to form the lens connexon (Peracchia and Peracchia 1980; Kuszak et al. 1982). The pore is thought to be about 10 Å in diameter, i.e., 31 Å in circumference, sufficient for six, but not for eight or twelve helices, each 4.6 Å in diameter, exclusive of side-chains. Although the putative amphiphilic regions permit one to draw interesting models, one must remember that either or both of the predicted amphiphilic segments could be at the surface of the membrane instead of forming part of the wall of a pore (Fig. 4).

MIP and the Connexon

Provisional acceptance of the six-helix model postulated for MIP implies that the protein can form a channel. It follows that one must ask whether or not the molecule as described could fit in the structure of a connexon. Typically, in liver (Unwin and Zampighi 1980), the connexon has an overall radius of about 25–30 Å, a pore radius of about 5–6 Å, and a height of about 100–120 Å. These numbers, derived from microscopy and X-ray data (Caspar et al. 1970; Unwin and Zampighi 1980), are generally compatible with permeability measurements (Meyer et al. 1981). Chothia (1984) has published figures for the volumes of amino acids in a polypeptide chain. Using his values, a volume of 36,506 Å^3 is found for each MIP molecule. By combining the approximate values for the size of the connexon with the values for the volume of MIP, one can estimate that there are between five and eight molecules of MIP per connexon. Thus, although such calculations are fraught with too many uncertainties to be taken very literally, the numbers obtained are in reasonable agreement with previous estimates of four to six protein molecules in a lens connexon (Peracchia and Peracchia 1980; Kuszak et al. 1982).

Another interesting problem is that of connexon stability. Although charged groups such as Arg-33 and Glu-134, found in potential transcore helices, might contribute to the stability of a particular MIP molecule by internal neutralization, they could also form intermolecular ion pairs within a connexon and thus contribute to connexon stability. The only other charged groups postulated by the model in the hydrophobic region of the molecule would be His-172 and His-201. Histidines may not be too difficult to accommodate, even in a hydrophobic environment, since their hydrophobicity is similar to that of the uncharged amino acid, serine. His-201 presents no problem in any case, since it is found on the hydrophilic side of the amphiphilic helix. His-172 might "peek" into the pore and therefore not present any problems in terms of stability of the hydrophobic region of the membrane. At present however there is no easy way to accommodate this residue, which would be at least partially charged in water at neutral pH (not necessarily the environment in which it is found).

If the model of an MIP is correct as drawn (Fig. 4a), there would be two loops of polypeptide extending into the abcytoplasmic space for each MIP of the connexons. These multiple loops might be involved in cell–cell interactions. One of these loops carries an arginine and a histidine residue and the other an arginine residue. Whether these charged groups are involved in the cell–cell interactions is not understood at the present time. If they are involved it is easiest to imagine that they might interact not with each other directly, but rather with negatively charged groups elsewhere in the junctional structure. One interesting possibility is that there might be an interaction with the phosphate groups of phospholipids. Phospholipids are in some way involved in junction structure (Goodenough and Revel 1970), and R. Johnson and co-workers (this volume) have actually shown that lens junctions are "split" by exposure to low pH, which might be expected to titrate such groups. The phospholipid molecules discussed may not be those involved in the bilayer itself but could be specifically associated with the extracellular bridges.

CONCLUSIONS

The role of MIP in forming junctions between lens cells is still unresolved at the present time. However, as indicated in the present discussion and the results reported by others, it is possible to imagine that it may indeed be part of gap junction-like structures between lens cells. Only further experiments aimed at establishing the orientation of MIP in the membrane and at directly testing whether MIP is involved in cell–cell communication will allow one to arrive at more complete hypotheses. At long last, however, the experiments are within reach.

ACKNOWLEDGMENTS

We would like to acknowledge our collaborators at Caltech, especially Janice Cline, Jan Hoh, and Bruce Nicholson, and at UCLA, Michael Gorin and Jo

Horwitz. The skills of Renee Thorf and Susan Mangrum, who toiled over several versions of this paper, are deeply appreciated.

REFERENCES

Alcala, J., N. Lieska, and H. Maisel. 1975. Protein composition of bovine lens cortical fiber cell membranes. *Exp. Cell Res.* **21:** 581-595.

Benedetti, E.L., I. Dunia, J. Bentzel, A.J.M. Vermorken, M. Kibbelaar, and H. Bloemendal. 1976. A portrait of plasma membrane specializations in eye lens epithelium fibers. *Biochim. Biophys. Acta* **457:** 353-384.

Blobel, G. 1980. Intracellular protein topogenesis. *Proc. Natl. Acad. Sci.* **77:** 1496-1500.

Bok, D., J. Dockstader, and J. Horwitz. 1982. Immunocytochemical localization of the lens main intrinsic polypeptide (MIP26) in communicating junctions. *J. Cell Biol.* **92:** 213-220.

Broekhuyse, R.M., E.D. Kuhlman, and A.L.H. Stols. 1976. Lens membranes. II. Isolation and characterization of the main intrinsic polypeptide (MIP) of bovine lens fiber membranes. *Exp. Eye Res.* **23:** 365-371.

Broekhuyse, R.M., E.D. Kuhlman, and H.J. Winkens. 1979. Lens membranes. VII. MIP is an immunologically specific component of lens fiber membranes and is identical with 26 K band protein. *Exp. Eye Res.* **29:** 303-313.

Caspar, D.L.D., D.A. Goodenough, L. Makowski, and W.C. Phillips. 1977. Gap junction structures. I. Correlated EM and X-ray diffraction. *J. Cell Biol.* **74:** 605-628.

Chothia, C. 1984. Principles that determine the structure of proteins. *Annu. Rev. Biochem.* **53:** 537-572.

Chou, Y. and G.D. Fasman. 1978. Empirical predictions of protein conformation. *Annu. Rev. Biochem.* **47:** 251-276.

Dermietzel, R., A. Leibstein, U. Frixen, U. Janssen-Timmen, O. Traub, and K. Willecke. 1984. Gap junctions in several tissues share antigenic determinants with liver gap junctions. *EMBO J.* **3:** 2261-2270.

Dorset, D.L., A. Engel, M. Haner, A. Massalki, and J.P. Rosenbush. 1983. Two-dimensional crystal packing of matrix porin. A channel forming protein in *E. coli* outer membranes. *J. Mol. Biol.* **165:** 701-710.

Eisenberg, D. 1984. Three dimensional structure of membrane and surface proteins. *Annu. Rev. Biochem.* **53:** 595-623.

Eisenberg, D., R.M. Weiss, and T.C. Terwilliger. 1982. The helical hydrophobic moment: A measure of the amphiphilicity of a helix. *Nature* **299:** 371-374.

Eisenberg, D., E. Schwarz, M. Komaromy, and R. Wall. 1984. Analysis of membrane and surface protein sequences with the hydrophobic moment plot. *J. Mol. Biol.* **179:** 125-142.

Engelman, D.M., R. Henderson, A.D. McLachlan, and B.A. Wallace. 1980. Path of the polypeptide in bacteriorhodopsin. *Proc. Natl. Acad. Sci.* **77:** 2023-2027.

Finbow, M., J. Shuttleworth, A.E. Hamilton, and J.D. Pitts. 1983. Analysis of vertebrate gap junction protein. *EMBO J.* **2:** 1479-1486.

Finbow, M., S.B. Yancey, R. Johnson, and J.P. Revel. 1980. Independent lines of evidence suggesting a major gap junctional protein with a MW of 26,000. *Proc. Natl. Acad. Sci.* **77:** 970-974.

Finer-Moore, J. and R.M. Stroud. 1984. Amphipathic analysis and possible formation of the ion channel in an ACh receptor. *Proc. Natl. Acad. Sci.* **81:** 155-159.

Fitzgerald, P.G., D. Bok, and J. Horwitz. 1983. Immunocytochemical localization of the main intrinsic polypetpide (MIP) in ultrathin frozen sections of rat lens. *J. Cell Biol.* **97:** 1491-1499.

Garavito, R.M., J.A. Jenkins, J.M. Neuhaus, A.P. Pugsley, and J.P. Rosenbush. 1982. Structural investigations of outer membrane proteins from *E. coli*. *Annu. Microbiol.* **133A:** 37-41.

Garnier, J., D.J. Osguthorpe, and B. Robson. 1978. Analysis of the accuracy and implications of simple methods for predicting the secondary structure of globular proteins. *J. Mol. Biol.* **120:** 97-120.

Goodenough, D.A. 1979. Lens gap junctions: A structural hypothesis for nonregulated low resistance intracellular pathways. *Invest. Ophthalmol. Visual Sci.* **18:** 1104-1122.

Goodenough, D.A. and J.P. Revel. 1970. A fine structural analysis of intercellular junctions in the mouse liver. *J. Cell Biol.* **45:** 272-290.

―――. 1971. The permeability of isolated and *in situ* mouse hepatic gap junctions studied with enzymtic tracers. *J. Cell Biol.* **50:** 81-91.

Goodenough, D.A., J.S. Dick II, and J.E. Lyons. 1980. Lens metabolic cooperation: A study of mouse lens transport and permeability visualized with freeze substitution autoradiography and electron microscopy. *J. Cell Biol.* **86:** 576-589.

Gorin, M.B., S.B. Yancey, J. Cline, J.P. Revel, and J. Horwitz. 1984. The major intrinsic protein (MIP) of the bovine lens fiber membrane: Characterization and structure based on cDNA cloning. *Cell* **39:** 49-58.

Gros, D.B., B.J. Nicholson, and J.P. Revel. 1983. Comparative analysis of the gap junction protein from rat heart and liver: Is there a tissue specificity of gap junctions? *Cell* **35:** 539-549.

Henderson, D., H. Eibl, and K. Weber. 1979. Structure and biochemistry of mouse hepatic gap junctions. *J. Mol. Biol.* **132:** 193-218.

Henderson, R. 1979. The structure of bacteriorhodopsin and its relevance to other membrane proteins. *Soc. Gen. Physiol. Ser.* **33:** 3-15.

Hertzberg, E. 1984. A detergent independent procedure for the isolation of gap junctions from rat liver. *J. Biol. Chem.* **259:** 9936-9943.

Hertzberg, E.L. and N.B. Gilula. 1979. Isolation and characterization of gap junctions from rat liver. *J. Biol. Chem.* **254:** 2138-2147.

Hertzberg, E.L. and R.V. Skibbens. 1984. A protein homologous to the 27,000 dalton liver gap junction protein is present in a wide variety of species and tissues. *Cell* **39:** 61-69.

Hertzberg, E.L., D.J. Anderson, M. Friedlander, and N.B. Gilula. 1982. Comparative analysis of the major polypeptide from liver gap junctions and lens fiber junctions. *J. Cell Biol.* **92:** 53-59.

Kistler, J. and S. Bullivant. 1980a. The connexon order in isolated lens gap junctions. *J. Ultrastruct. Res.* **72:** 27-38.

―――. 1980b. Lens gap junctions and orthogonal assays are unrelated. *FEBS Lett.* **111:** 73-78.

Kuszak, J.R., J. Alcala, and H. Maisel. 1981. Biochemical and structural features of chick lens gap junctions. *Exp. Eye Res.* **33:** 157-166.

Kuszak, J.R., J.I. Rae, B.U. Pauli, and R.S. Weinstein. 1982. Rotary replication of lens gap junctions. *J. Ultrastruct. Res.* **81:** 249-256.

Kyte, J. and R.J. Doolittle. 1982. A simple method for displaying the hydropathic character of a protein. *J. Mol. Biol.* **157:** 105-121.

Lane, N.J. and H. le B. Skaer. 1980. Intercellular junctions in insect tissues. *Adv. Insect Physiol.* **15:** 35-213.

Larsen, W.J. 1983. Biological implications of gap junction structure, distribution and composition. *Tissue Cell* **15:** 645-671.

Manjunath, C.K., G.E. Goings, and E. Page. 1984. Cytoplasmic surface and intramembrane components of rat heart gap junctional proteins. *Am. J. Physiol.* **246:** H865-H875.

Meyer, D.J., S.B. Yancey, and J.P. Revel. 1981. With a note by A. Peskoff. Intercellular communication in normal and regenerating rat liver: A quantitative analysis. *J. Cell Biol.* **91:** 505–523.

Nicholson, B.J., M. Hunkapiller, L.B. Grim, L.E. Hood, and J.P. Revel. 1981. The rat liver gap junction protein: Properties and partial sequence. *Proc. Natl. Acad. Sci.* **78:** 7594–7598.

Nicholson, B.J., L.J. Takemoto, M.W. Hunkapiller, L.E. Hood, and J.P. Revel. 1983. Differences between the liver gap junction protein and lens MIP26 from rat: Implications for tissue specificity of gap junctions. *Cell* **32:** 967–978.

Nishikawa, K. 1983. Assessment of secondary structure prediction of proteins comparison of computerized Chou-Fasman method with others. *Biochim. Biophys. Acta* **748:** 285–299.

Ovchinnikov, Y.A. and V.T. Ivanov. 1982. Helical structures of gramicidin A and their role in ion channeling. In *Conformation in biology* (ed. R. Srinivasan and R.H. Sarma), p. 155–174. Adenine Press, New York.

Paul, D.L. and D.A. Goodenough. 1983a. Preparation, characterization and localization of antisera against bovine MP26, an integral membrane protein of the lens fiber plasma membrane. *J. Cell Biol.* **96:** 625–638.

———. 1983b. In vitro synthesis and membrane insertion of bovine MP26, an integral membrane protein from the lens fiber plasma membrane. *J. Cell Biol.* **96:** 633–638.

Payton, B.W., M.V.L. Bennett, and G.D. Pappas. 1969. Permeability and structure of junctional membranes at an electronic synapse. *Science* **166:** 1642–1643.

Peracchia, C. 1978. Calcium effects on gap junction structure and cell coupling. *Nature* **271:** 669–671.

Peracchia, C. and L.L. Peracchia. 1980. Gap junction dynamics: Effect of H^+ ions. *J. Cell Biol.* **87:** 719–727.

Revel, J.P. and M.J. Karnovsky. 1967. Hexagonal array of subunits in intercellular junctions of the mouse heart and liver. *J. Cell Biol.* **33:** c7.

Revel, J.P., B.J. Nicholson, and S.B. Yancey. 1984. Molecular organization of gap junctions. *Fed. Proc.* **43:** 2672–2677.

———. 1985. Chemistry of gap junctions. *Annu. Rev. Physiol.* **47:** 263–279.

Robertson, J.D. 1963. The occurrence of a subunit pattern in the unit membranes of club endings in Mauthner cell synapse in the goldfish brain. *J. Cell Biol.* **19:** 201–221.

Sabatini, D.D., G. Kreibich, T. Morimoto, and M. Adesnik. 1982. Mechanisms for the incorporation of proteins in membranes and organelles. *J. Cell Biol.* **92:** 1–22.

Sas, D.F., M.J. Sas, K.R. Johnson, A.S. Menko, and R.G. Johnson. 1985. Junctions between lens fiber cells are labeled with a monoclonal antibody shown to be specific for MP26. *J. Cell Biol.* **100:** 216–225.

Schuetze, S.M. and D.A. Goodenough. 1982. Dye transfer between cells of embryonic chick lens becomes less sensitive to CO_2 treatment with development. *J. Cell Biol.* **92:** 694–705.

Steitz, T.A., A. Goldman, and D.M. Engelman. 1982. Quantitative application of the helical hairpin hypothesis to membrane proteins. *Biophys. J.* **37:** 124–125.

Strubin, M., B. Mach, and E.O. Long. 1984. The complete sequence of the mRNA for the HLA-DR associated invariant chain reveals a polypeptide with unusual transmembrane polarity. *EMBO J.* **3:** 869–872.

Terwilliger, T.C., L. Weissman, and D. Eisenberg. 1982. The structure of Mellitin. The form I crystals and its implications for Mellitin's lytic and surface activities. *Biophys. J.* **37:** 353–361.

Traub, O. and K. Willecke. 1982. Cross reaction of antibodies against liver gap junction protein (26K) with lens fiber junction protein (MIP) suggests structural homology between these tissue specific gene products. *Biochem. Biophys. Res. Commun.* **109:** 895–901.

Traub, O., U. Janssen-Timmen, P.M. Druge, R. Dermietzel, and K. Willecke. 1982. Immunological properties of gap junction protein from mouse liver. *J. Cell. Biochem.* **19:** 27–44.

Unwin, P.N.T. and G. Zampighi. 1980. Structure of the junctions between communicating cells. *Nature* **283:** 545–549.

Warner, A.E., S.C. Guthrie, and N.B. Gilula. 1984. Antibodies to gap junctional protein selectively disrupt junctional communication in the early amphibian embryo. *Nature* **311:** 127-131.

Zampighi, G., S. Simon, J.D. Robertson, T. McIntosh, and M. Costello. 1982. On the structural organization of isolated bovine lens fiber junctions. *J. Cell Biol.* **93:** 175–189.

Zigler, J.S. and J. Horwitz. 1981. Immunochemical studies of the major intrinsic polypeptides from the human lens membrane. *Invest. Ophthalmol. Visual Sci.* **21:** 46–51.

Biochemistry and Structure of Cardiac Gap Junctions: Recent Observations

Ernest Page and Chellakere K. Manjunath
Department of Medicine
The University of Chicago
Chicago, Illinois 60637

This article summarizes recent observations by our laboratory on the biochemistry and structure of cardiac gap junctions. Since we have recently reviewed this subject at length (Manjunath and Page 1985; Page and Manjunath 1985), we restrict ourselves here to those aspects of our work that seem of general interest to gap junctional biology. Our laboratory has studied both cardiac and liver gap junctions. Therefore we shall emphasize comparisons between cardiac and liver gap junctions to show that junctional membranes from these two sources differ in interesting ways.

CARDIAC AND LIVER GAP JUNCTIONS DIFFER IN STRUCTURE

Electron microscopy indicates that cardiac and liver gap junctions differ and that mutually consistent differences can be demonstrated by three techniques: electron microscopy of thin sections (Manjunath et al. 1984b), negative staining (Manjunath and Page 1985; Page and Manjunath 1985), and freeze-fracture combined with deep-etching (Shibata et al. 1985). When thin-sectioned mammalian heart muscle fixed with glutaraldehyde and postfixed with OsO_4 is stained with uranyl acetate, lead citrate, or both, the cytoplasmic surfaces of gap junctions between myocardial cells are seen to be covered with a layer of fuzzy material. A similar layer of fuzz was described in gap junctions from the central nervous system by Brightman and Reese (1969), but is absent from the cytoplasmic surfaces of gap junctions from at least some types of inexcitable cells. Positively stained thin sections made from pellets of gap junctions isolated from mammalian hearts also have a layer of fuzz covering the cytoplasmic surfaces (Fig. 1), provided precautions are taken to prevent proteolysis of the junctions during isolation (Manjunath et al. 1984b). The fuzzy layer cannot be removed by incubation for 1 hour in 8 M urea at 37°C or by exposure to the detergents sodium deoxycholate and N-lauryl sarcosine; by contrast, exposure to trypsin, or proteolysis by serine protease present in the granules of cardiac interstitial mast cells, strips the fuzzy layer off the cytoplasmic surfaces, leaving them smooth (Manjunath et al. 1984b, 1985) (Fig. 2). The cytoplasmic surfaces of gap junctions isolated from rat liver by a procedure similar to that used to isolate cardiac gap junctions (Manjunath et al. 1982) lack a fuzzy layer even when prepared in the presence of inhibitors of proteolysis (Fig. 3). The fuzzy layer remains on the cytoplasmic surface of unproteolyzed cardiac gap junctions after the junctions are split in the gap with 8 M urea. Therefore it can be used to define the sidedness of split cardiac gap junctions (Manjunath et al. 1984a,b). We interpret these experiments as suggesting that the fuzzy material covering the cytoplasmic surfaces of isolated cardiac gap junctions, and at least part of the corresponding layer of fuzz covering cardiac gap junctions in situ, consists of protein covalently bound to the cardiac gap junctional channel protein in the lipid bilayer of the membrane. This covalently linked material is absent in liver gap junctions from the same species (the rat).

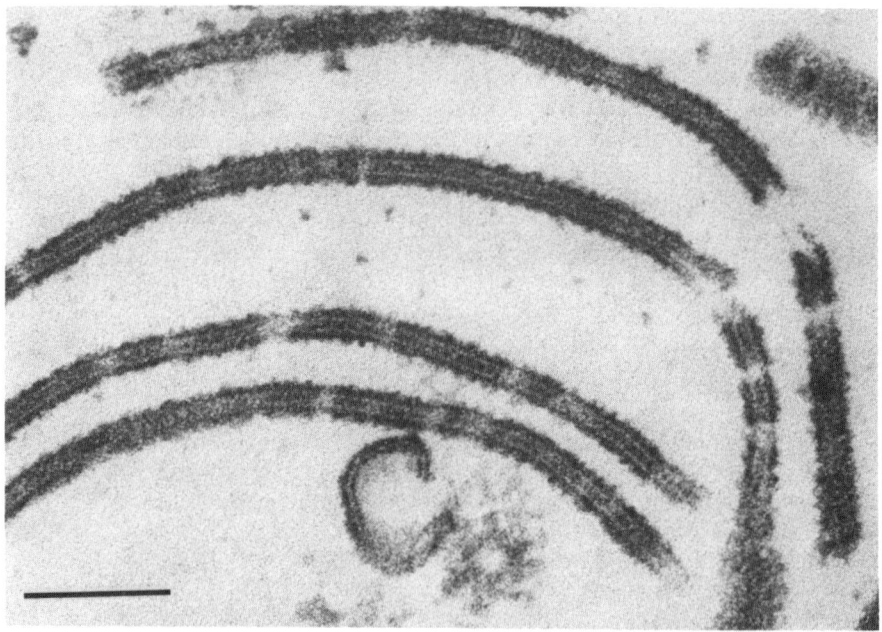

Figure 1
Unproteolyzed gap junctions isolated from rat ventricles in presence of PMSF. Note the fuzzy layer on the cytoplasmic surfaces. Isolated gap junctions were fixed with glutaraldehyde and stained with uranyl acetate and lead citrate. Scale bar, 0.1 μm.

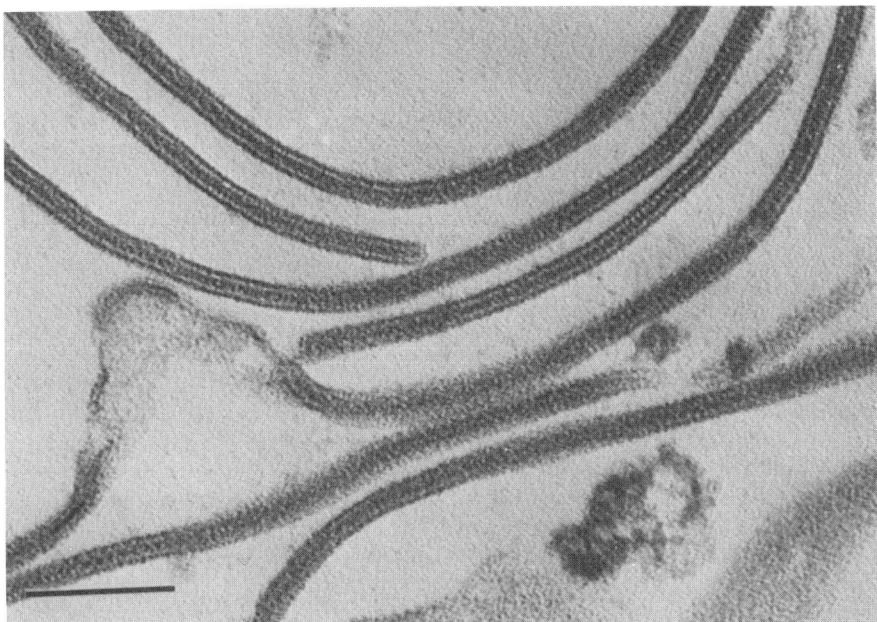

Figure 2
Proteolyzed gap junctions isolated from rat ventricles in the absence of PMSF. Note smooth cytoplasmic surfaces and narrowing of total width of the junction. Isolated gap junctions were fixed with glutaraldehyde and stained with uranyl acetate and lead citrate. Scale bar, 0.1 μm.

A second technique for directly visualizing the cytoplasmic surfaces of cardiac and liver gap junctions consists of quick-freezing the tissue or junctional pellet on a copper surface cooled with liquid helium, followed by freeze-fracture, deep-etching, and double-axis rotary replication (Shibata et al. 1984) of the gap junctions (Shibata et al. 1985). With this technique it is possible to demonstrate that unproteolyzed cardiac gap junctional pellets have particulate cytoplasmic surfaces, whereas both proteolyzed cardiac gap junctional pellets and unproteolyzed liver gap junctional pellets have smooth (nonparticulate) cytoplasmic surfaces. It has previously been shown by deep-etching and rotary-shadowing of quick-frozen material that in situ liver gap junctions have nonparticulate cytoplasmic surfaces (Hirokawa and Heuser 1982), whereas in situ cardiac gap junctions have particulate surfaces (Y. Shibata and T. Yamamoto, pers. comm.). These results support the conclusion that a cytoplasmic surface component present in unproteolyzed cardiac gap junctions is absent in liver gap junctions and proteolyzed cardiac gap junctions. These structural differences correlate with differences in protein composition that we shall discuss in the next section.

As discussed elsewhere (Manjunath and Page 1985; Page and Manjunath 1985), cardiac gap junctions both in situ and in isolated gap junctional vesicles are characterized by an arrangement of gap junctional channels made up of

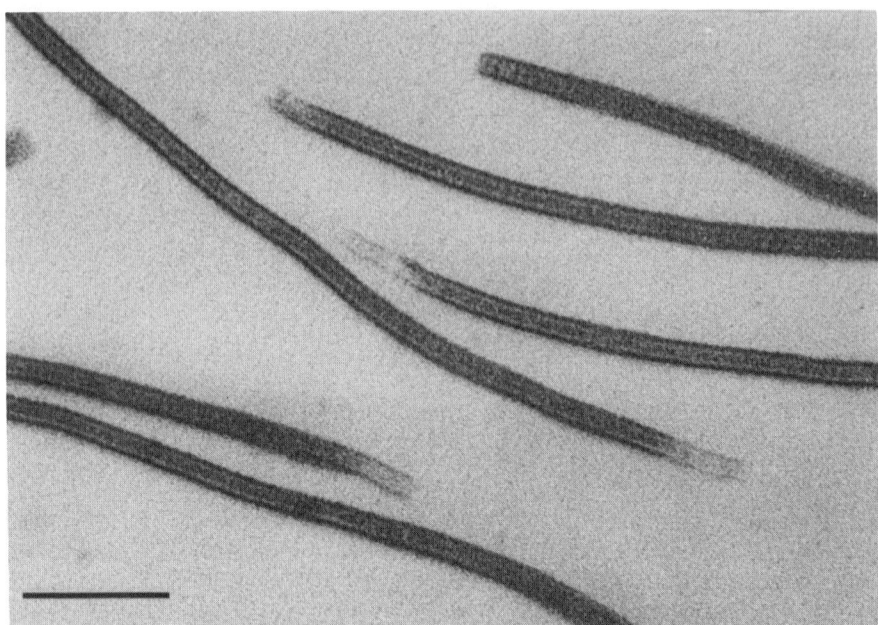

Figure 3
Presumably unproteolyzed gap junction isolated from rat liver in the presence of PMSF, leupeptin, and paramercuribenzoate. Note smooth cytoplasmic surfaces. Isolated gap junctions were fixed with glutaraldehyde and stained with uranyl acetate and lead citrate. Scale bar, 0.1 μm.

small, quasi-hexagonal arrays that are rotated with respect to one another. For in situ cardiac gap junctions, this arrangement is best seen by freeze-fracturing the junctions either after conventional glutaraldehyde fixation or by quick-freezing (Page et al. 1983); in isolated gap junctional vesicles, it is easily demonstrable by negative staining (Manjunath and Page 1985; Page and Manjunath 1985). The rotation of small junctional arrays relative to one another, as well as the undulation of the surface of cardiac gap junctions, interferes seriously with the analysis of optical diffraction patterns and with attempts in image reconstruction (Manjunath and Page 1985; Page and Manjunath 1985). By contrast, isolated liver gap junctions present large areas of membrane without such microdomains. Nevertheless, optical diffraction of negatively stained rat and rabbit cardiac liver gap junctions followed by digital image processing and reconstruction suffices to demonstrate that the channel-containing structures (connexons) are made up of six structurally similar subunits (Manjunath and Page 1985; Page and Manjunath 1985).

We have recently examined stereo-imaged complementary replicas of gap junctions in freeze-fractured sheep cardiac Purkinje fibers (Kordylewski and Page 1985). This technique yields the surprising result that P-face particles and E-face pits may not be, as usually assumed, complementary structures. The E-face pits are seen to fall between the particles, not on them. If this finding is confirmed, it would follow that the E-face pits do not lie on the same transmembrane axis as the particles. Since the particles presumably contain the cell–cell channels, the channels would not pass through the pits.

CARDIAC AND LIVER GAP JUNCTIONS DIFFER IN PROTEIN COMPOSITION

Starting with the first useful procedure for isolating cardiac gap junctions by Kensler and Goodenough (1980), we have systematically explored the relationship of the various steps in the method for isolating cardiac gap junctions to the properties of the channel proteins obtained as the end product (Manjunath et al. 1982, 1984b, 1985). Our most significant finding was that both the molecular weight of the apparent protein subunit of the gap junctional channel or connexon (as revealed by SDS-PAGE) and the ultrastructure of the gap junctional vesicles depend critically on the inhibition of alkaline serine protease activity. The protease was most effectively inhibited by phenylmethylsulfonyl fluoride (PMSF). When this inhibitor was present during the early steps of the isolation procedure, the most highly purified preparations of gap junctions from rat ventricle yielded a broad major polypeptide band of M_r 44,000–47,000 on SDS-PAGE (Manjunath et al. 1984b, 1985). In the absence of PMSF, SDS-PAGE of rat heart gap junctions gave a single major polypeptide of M_r 29,500. The cytoplasmic surfaces of junctions isolated with PMSF were covered with a fuzzy layer in thin sections (Manjunath et al. 1984b) and appeared particulate after systematic etching of freeze-fractured junctions (Shibata et al. 1985); conversely, the cytoplasmic surfaces of junctions prepared without PMSF were devoid of fuzz and appeared nonparticulate in freeze-etched membrane replicas (Shibata et al. 1985).

SDS-PAGE gives a single major subunit of M_r 44,000–47,000, even after incubation of the unproteolyzed cardiac gap junctions in 8 M urea at 37°C (which, as already noted, splits the junctions in the gap but does not remove the cytoplasmic surface fuzz or the particulate structure of the cytoplasmic surface visualized after freeze-fracture and deep-etching). We concluded that a cytoplasmic surface component of M_r 14,500–17,500 is part of the channel subunit covalently linked to the channel protein within the lipid bilayer. This protein component contributes to the fuzzy cytoplasmic surface layer and the particulate structure seen by electron microscopy, although the component defined by SDS-PAGE is not necessarily coextensive with that defined by the two ultrastructural techniques.

We studied the proteolytic cleavage of the cytoplasmic surface component in order to identify the stage of the isolation procedure critical for proteolysis, to characterize the time course of the proteolysis, and to identify the source and inhibitor specificity of the enzyme (Manjunath et al. 1985). These experiments showed that the critical step for proteolysis was the extraction of myofibrillar proteins from the cardiac homogenate by a high-ionic-strength solution (0.6 M KI). It is well known that an alkaline serine protease in the granules of mast cells within the interstitial spaces of the muscle can be released by exposure to solutions of high ionic strength (Woodbury et al. 1978). By adding the inhibitor PMSF at different intervals after exposure of the cardiac homogenate to 0.6 M KI, we were able to identify by SDS-PAGE both the intermediate products of proteolysis by the alkaline serine protease and the end-product of the reaction large enough to be retained on the gel. In this way we showed the transient appearance of intermediate bands at M_r 34,000 and 31,000, and confirmed the absence of major bands with M_r <29,500 at the completion of proteolysis after 9 hours of incubation. Chymostatin and soybean trypsin inhibitors partially inhibited cleavage of the cytoplasmic surface component by the alkaline serine protease, but were less effective than PMSF; EGTA and leupeptin were ineffective.

Unlike rat heart gap junctions, SDS-PAGE of rat liver junctions isolated with or without PMSF yields a M_r 28,000 polypeptide (Manjunath et al. 1984a, 1985); the cytoplasmic surfaces seen by electron microscopy are smooth both in thin-sectioned pellets and in deep-etched, freeze-fractured junctions (Hirokawa and Heuser 1982; Shibata et al. 1985). Although liver does contain mast cells, the isolation of liver gap junctions does not include a step in which the junctions are exposed to solutions of high ionic strength, hence the contents of mast cell granules would not be released; moreover, at the pH at which liver junctions are isolated, the proteolysis by the alkaline serine protease from mast cell granules would be very slow even if it were released. If liver gap junctions had a 14,500–17,500 cytoplasmic surface component similar to that of cardiac gap junctions, it would therefore be improbable that this component would be cleaved during isolation of the junction by an alkaline serine protease derived from mast cell granules. Low-irradiation electron microscopy (Baker et al. 1983) and X-ray diffraction (Makowski et al. 1984) of gap junctions from mouse liver suggest the presence of a cytoplasmic surface component, but the component ($M_r \approx 4000$) would seem to be substantially smaller than that identified by our biochemical and structural studies of cardiac gap junctions. Nevertheless it remains possible that liver gap junctions in their native state have a cytoplasmic surface component similar to that of cardiac gap junctions, but that this cytoplasmic surface component is somehow lost during preparation of the junctions. It is, for example, conceivable that proteolysis is so active in liver homogenates or liver cells that exogenous inhibitors of proteolysis cannot act in time to prevent cleavage, or that the inhibitors tested to date were insufficiently specific or potent. We cannot conclusively rule out these possibilities. However, on the basis of our structural and biochemical observations, we favor the interpretation that the structures and protein compositions of the cytoplasmic surfaces of cardiac and liver gap junctions differ.

This conclusion receives some support from comparison of the two-dimensional tryptic and α-chymotryptic peptide maps of rat heart and liver junctions (Gros et al. 1983), which showed little homology between the gap junctional proteins from the two tissues. This study falls short of being conclusive, because proteolyzed cardiac gap junctions (subunit of M_r 28,000) prepared without PMSF were used for peptide mapping. Thus the M_r 14,500–17,500 cytoplasmic surface component did not contribute to the pattern obtained by the technique.

Recent immunological studies with antibodies against liver gap junctions (Dermietzel et al. 1984; Hertzberg and Skibbens 1984) have led to contradictory results regarding the tissue specificity of gap junctional protein. Dermietzel et al. (1984) found no cross-reactivity of affinity-purified, fluorescent-labeled antibodies against the M_r 26,000 mouse liver gap junctional protein with frozen sections of myocardial tissue. Hertzberg and Skibbens (1984), on the other hand, reported cross-reactivity of heart sections and homogenates with antibodies raised in sheep against the M_r 27,000 rat liver gap junctional protein. It is interesting to note that the immunoreplica analysis of Hertzberg and Skibbens (1984) showed a marked reduction in the binding of antibodies against liver gap junctions to heart homogenate protein when the samples were boiled in SDS before electrophoresis, whereas no such heat-induced changes in electrophoretic behavior were observed with purified cardiac gap junctions (Manjunath et al. 1984b).

SUMMARY

Gap junctions of mammalian ventricular myocardial cells differ from gap junctions of mammalian liver in ultrastructure and protein composition. Electron micrographs of stained thin-sectioned hearts show a layer of fuzzy material covering the cytoplasmic surfaces of cardiac gap junctions in situ. This material is also present on the cytoplasmic surfaces of unproteolyzed isolated gap junctions prepared in the presence of the serine protease inhibitor PMSF, but is absent from proteolyzed cardiac gap junctions. Deep-etching of quick-frozen, freeze-fractured unproteolyzed cardiac gap junctions reveals a particulate substructure of the cytoplasmic surface that is absent in proteolyzed cardiac gap junctions prepared without PMSF. SDS-PAGE of unproteolyzed cardiac gap junctional vesicles from rat hearts yields a major protein band of M_r 44,000–47,000 which is degraded to a 29,500 peptide by the action of an alkaline serine protease present in mast cell granules. Incubation of unproteolyzed cardiac gap junctions in 8 M urea at 22–37°C splits the junctions in the gap, but does not remove the fuzzy cytoplasmic surface layer or change the M_r 44,000–47,000 major protein band obtained by SDS-PAGE. By contrast, electron microscopy of liver gap junctions shows them to have smooth cytoplasmic surfaces without subunit structure on deep-etching, whereas SDS-PAGE of liver junctions prepared with or without PMSF yields a major protein band of M_r 28,000. These observations support the conclusion that cardiac and liver gap junctions from the same species differ in the structure and protein composition of their cytoplasmic surface components.

ACKNOWLEDGMENTS

This research was supported by USPHS 10503 and 20592.

REFERENCES

Baker, T.S, D.L.D. Caspar, C.J. Hollingshead, and D.A. Goodenough. 1983. Gap junction structures. IV. Asymmetric features revealed by low-irradiation microscopy. *J. Cell Biol.* **96:** 204–206.

Brightman, M.W. and T.S. Reese. 1969. Junctions between intimately apposed cell membranes in the vertebrate brain. *J. Cell Biol.* **40:** 648–677.

Dermietzel, R., A. Leibstein, U. Frixen, U. Janssen-Timmen, O. Traub, and K. Willecke. 1984. Gap junctions in several tissues share determinants with liver gap junctions. *EMBO J.* **3:** 2261–2270.

Gros, D.B., B.J. Nicholson, and J.-P. Revel. 1983. Comparative analysis of the gap junction protein from rat heart and liver. Is there a tissue specificity? *Cell* **35:** 539–549.

Hertzberg, E.L. and R.V. Skibbens. 1984. A protein homologous to the 27,000 dalton liver gap junction protein is present in a wide variety of species and tissues. *Cell* **39:** 61–69.

Hirokawa, N. and J. Heuser. 1982. The inside and outside of gap junction membranes visualized by deep etching. *Cell* **30:** 395–406.

Kensler, R.W. and D.A. Goodenough. 1980. Isolation of mouse myocardial gap junctions. *J. Cell Biol.* **86:** 755–764.

Kordylewski, L. and E. Page. 1985. Are gap junctional pits and particles complementary structures? *Biophys. J.* **45:** 506a (Abstr.).

Makowski, L., D.L.D. Caspar, W.C. Phillips, and D.A. Goodenough. 1984. Gap junction structures. V. Structural chemistry inferred from X-ray diffraction measurements on sucrose accessibility and trypsin susceptibility. *J. Mol. Biol.* **174:** 449–481.

Manjunath, C.K. and E. Page. 1985. Cell biology and protein composition of cardiac gap junctions. *Am. J. Physiol.* (in press).

Manjunath, C.K., G.E. Goings, and E. Page. 1982. Isolation and protein composition of gap junctions from rabbit hearts. *Biochem. J.* **205:** 189–194.

———. 1984a. Detergent sensitivity and splitting of isolated liver gap junctions. *J. Membr. Biol.* **78:** 147–155.

———. 1984b. Cytoplasmic surface and intramembrane components of rat heart gap junctional protein. *Am. J. Physiol.* **246:** H865–H875.

———. 1985. Proteolysis of cardiac gap junctions during their isolation from rat hearts. *J. Membr. Biol.* (in press).

Page, E. and C.K. Manjunath. 1985. Communicating junctions between cardiac cells. In *Handbook of cardiology* (ed. H.A. Fozzard). Raven Press, New York. (In press.)

Page, E., T. Karrison, and J. Upshaw-Earley. 1983. Freeze fractured cardiac gap junctions: Structural analysis by three methods. *Am. J. Physiol.* **244:** H525–H539.

Shibata, Y., T. Arima, and T. Yamamoto. 1984. Double axis rotary replication for deep etching. *J. Microsc. (Oxf.)* **137:** 121–123.

Shibata, Y., C.K. Manjunath, and E. Page. 1985. Differences between cytoplasmic surfaces of deep etched heart and liver gap junctions. *Am. J. Physiol.* (in press).

Woodbury, R.G., M. Everitt, Y. Sanada, N. Katunuma, N. Lagunoff, and H. Neurath. 1978. A major serine protease in rat skeletal muscle: Evidence for its mast cell origin. *Proc. Natl. Acad. Sci.* **75:** 5311–5313.

Studies of Gap Junctions: Biochemical Analysis and Use of Antibody Probes

Elliot L. Hertzberg
Verna and Marrs McLean Department of Biochemistry
Baylor College of Medicine, Houston, Texas 77030

David C. Spray
Department of Neuroscience
Albert Einstein College of Medicine, Bronx, New York 10461

Until recently, the evidence that gap junctions are responsible for cell–cell communication has been primarily correlative: The transfer of small molecules between cells correlated with the presence of gap junctions and vice versa. Study of gap junctions has classically depended upon either morphological analysis or physiological measurement of gap junctional communication. Hence, isolation of gap junctions was undertaken to analyze the molecular properties of this structure and generate antibody probes specific to its protein component. Such efforts were the focus of a number of laboratories for several years, during which a 27-kD polypeptide slowly gained acceptance by most investigators as the major constituent of isolated liver gap junctions. However, only in the last year or so have a few laboratories succeeded in generating antibodies to this polypeptide (Traub et al. 1982; Hertzberg 1984).

A major impediment to progress in the field has been the low yield of gap junction protein by traditional isolation procedures, the basis for which is the enrichment of gap junctions in residues of detergent-extracted plasma membranes (Evans and Gurd 1972; Goodenough and Stoeckenius 1972; Henderson et al. 1979; Hertzberg and Gilula 1979). Recently, we have described an alternative approach to gap junction isolation in which alkali replaces detergent for extraction of plasma membranes (Hertzberg 1984). The gap junctions isolated in this manner are comprised of the same 27-kD polypeptide recovered by detergent resistance (Hertzberg and Gilula 1979). The primary advantage of the new approach is its increased yield: as much as 2–3 mg of purified gap junctions from 40 rat livers, as compared with 300–500 μg by the earlier procedures.

With the greater quantities of gap junctions made available by the new isolation procedure, antibodies to the rat liver gap junction polypeptide were generated, affinity-purified, and used to prove the involvement of gap junctions in cell–cell communication as well as to probe further, in vitro and in vivo, the communication process and its regulation.

CHARACTERIZATION OF ANTI-GAP JUNCTION ANTIBODIES

Injection of isolated gap junctions into a sheep elicited an immune response. Analysis of the immune sera by double diffusion and Western blots indicated that a complex response to multiple polypeptides had occurred, most likely due to the presence of small quantities of highly immunogenic proteins in the isolated gap junctions. Antibodies specific for the 27-kD gap junction polypeptide were therefore prepared by affinity purification. The formation of precipitin arcs upon double-diffusion analysis of these antibodies against solubilized isolated gap junctions suggested a polyclonal response to at least two determinants on the gap junction polypeptide (Hertzberg 1984).

Western blot analysis (Fig. 1) demonstrated specificity of the affinity-purified antibodies for the 27-kD gap junction polypeptide and its aggregation products in rat liver homogenates, plasma membranes, and isolated gap junctions. In this experiment, SDS was added to the liver homogenate and electrophoresis begun within 3 minutes of the death of the rat in order to minimize proteolytic degradation. The absence of any higher-molecular-weight polypeptides in the homogenate that bound antibody suggests that the 27-kD gap junction polypeptide is not a proteolytic by-product of the preparative procedure.

Indirect immunofluorescence analysis on frozen sections of liver (Fig. 2) demonstrated antibody binding to punctate regions of the plasma membrane with a distribution consistent with that expected based upon thin-section and freeze-fracture analysis of this tissue. In collaboration with Dr. Ross Johnson, we recently obtained a more direct demonstration of antibody binding to isolated gap junctions by immunoelectron microscopy.

TISSUE SPECIFICITY OF THE GAP JUNCTION POLYPEPTIDE

The degree of molecular homology between gap junctions from different tissues has become a highly contentious question. Early studies of gap junctional communication in heterologous coculture (Michalke and Loewenstein 1971; Epstein and Gilula 1977; Pitts 1980) had led to the notion that gap junctions in different tissues might be comprised of highly homologous polypeptides. However, a number of recent comparative analyses of gap junctions isolated from lens fiber cells, heart, and liver appeared to indicate distinct differences in polypeptide composition (Hertzberg et al. 1982; Gros et al. 1983; Nicholson et al. 1983).

To resolve this issue, the ability of antibodies raised against the rat liver gap junction polypeptide to bind to polypeptides in a host of different rat tissues was studied (Hertzberg and Skibbens 1984). Indirect immunofluorescence lo-

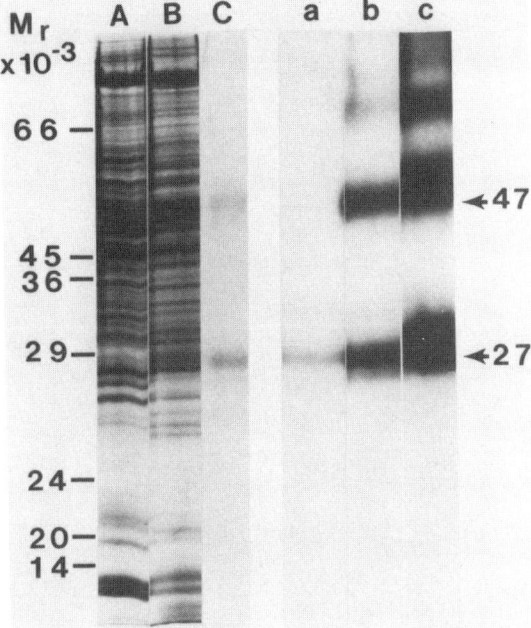

Figure 1
SDS-polyacrylamide gel and Western blot analysis of rat liver homogenate, plasma membranes, and purified gap junctions. Samples, resolved by electrophoresis on 12.5% SDS-polyacrylamide gels, were: rat liver homogenate (A,a), rat liver plasma membranes (B,b), and isolated gap junctions (C,c). Subsequent to electrophoresis, the gel was stained with Coomassie Brilliant Blue (A–C) and an unstained duplicate electrophoretically transferred to nitrocellulose (a–c). While identical sample loads were used for homogenate (a) and plasma membranes (b), only 5% of the gap junction load in C was used for transfer to nitrocellulose (c). Binding of antibody to the nitrocellulose transfer was detected by autoradiography subsequent to sequential incubation with rabbit anti-sheep IgG and ^{125}I-labeled Protein A. The mobilities of molecular weight markers, the 27-kD gap junction polypeptide, and its 47-kD dimer are indicated.

calization of antibody binding in frozen sections of rat ventricle and exocrine pancreas revealed a fluorescence pattern consistent with the distribution of gap junctions in these tissues. Moreover, a 27-kD polypeptide that binds these antibodies was identified by Western blots in extracts of liver, pancreas, heart, brain, stomach, kidney, and adrenal glands. These observations have been extended to include ovary and uterine tissue (in collaboration with Drs. Joanne Richards and David Bullock, respectively). Significant binding of antibodies to the horizontal cell layer of rabbit retina has also been detected (in collaboration with Dr. Dominic M.-K. Lam), consistent with the high content of electrotonic synapses in this region of the retina.

Among all the tissues examined by this approach, only lens fiber cells failed to cross-react. A consensus has been developing that the junctions from this tissue are indeed unique. Thus, with the exception of lens fiber cells, it ap-

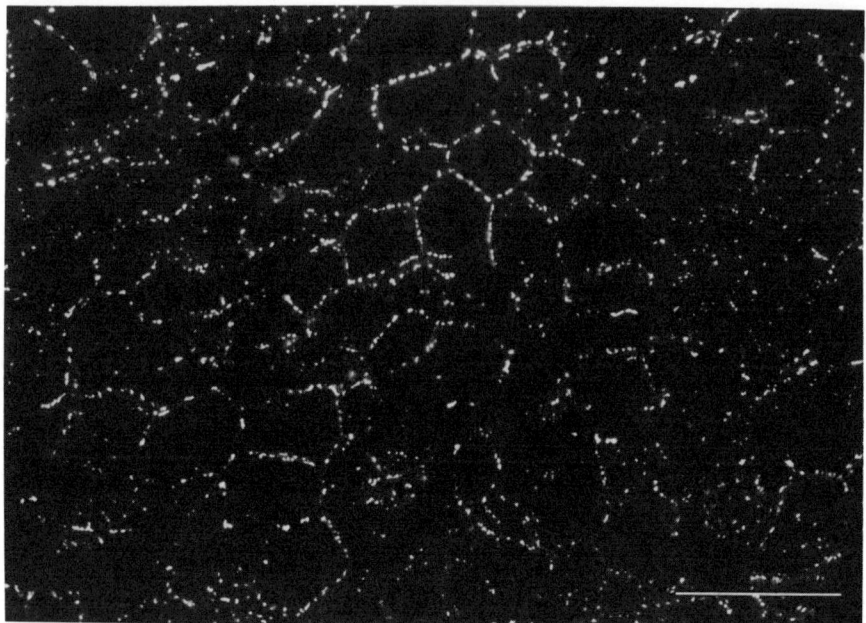

Figure 2
Indirect immunofluorescence localization of antibody binding to frozen sections of liver. Cryostat sections of liver were prepared. Subsequent to incubation with antibody, localization of antibody binding was achieved by incubation with fluorescein-conjugated rabbit anti-sheep IgG. Bar, 50 µm.

pears likely that the gap junctions in all rat tissues are comprised of highly homologous, or identical, 27-kD polypeptides.

SPECIES SPECIFICITY OF THE GAP JUNCTION POLYPEPTIDE

The degree to which antibodies to the rat liver gap junction polypeptide cross-react with gap junction polypeptides in other species was assessed by indirect immunofluorescence on frozen sections of liver and by Western blots. The Western blots indicated binding to 27-kD polypeptides in livers of representatives of all vertebrate classes. Indirect immunofluorescence demonstrated binding to punctate regions of the plasma membranes in these livers, although substantial variation in the pattern of fluorescence (intensity and size of gap junctions) was observed. In general, binding of antibody was strongest to samples derived from mammalian species, consistent with the use of rat liver gap junctions as the antigen and the source of purified polypeptide for affinity purification of antibodies. No cross-reaction of these antibodies has been detected in invertebrate tissues, including crayfish hepatopancreas, *Aplysia* buccal ganglia, and homogenates of the nematode, *Caenorhabditis elegans*.

ANTIBODY CROSS-REACTION WITH CELL LINES

Western blots were used to analyze extracts of a number of cultured cell lines. Binding was observed to a 27-kD polypeptide in BRL cells (rat liver-derived), HIT cells (pancreatic B cell-derived), and Cl-1D (mouse fibroblast L-cell-derived). The presence of the gap junction polypeptide in Cl-1D is of particular interest because it is a communication-deficient cell line in which gap junctional communication can be induced upon incubation with cAMP (Azarnia et al. 1981). The amount of the 27-kD gap junction polypeptide appeared to be similar in Cl-1D grown in the presence or absence of cAMP, suggesting that the defect leading to a loss of gap junctions is not simply the absence of the gap junction polypeptide. Although the basis of the defect is not yet known, it is possible that analysis of Cl-1D and other communication-deficient cell lines may provide information regarding the molecular events leading to gap junction formation.

INHIBITION OF GAP JUNCTIONAL CONDUCTANCE BY MICROINJECTION OF ANTIBODY

We have studied the effect on gap junctional conductance produced by microinjection of antibodies to the rat liver gap junction polypeptide into a number of different cell types, including primary cultures of hepatocytes, myocardial cells, and superior cervical ganglion neurons, under conditions in which electrotonic coupling via gap junctions was induced (Hertzberg et al. 1985). In all cases, microinjection of antibodies rapidly inhibited gap junctional conductance as assayed by either electrical coupling (Fig. 3c,d) or fluorescent dye transfer (dye coupling) (Fig. 3a,b). These experiments indicate that antibody binds to determinants on the cytoplasmic surface of the gap junction, inhibiting passage of dye and ions through the gap junction channel. Although the mechanism by which this inhibition occurs is not yet clear, it indicates that a highly homologous, if not identical, region of the gap junction polypeptide is present in cells derived from all three germ layers: liver (endoderm); heart (mesoderm); and neurons (ectoderm). These experiments provide further confirmation of the homology of gap junction polypeptides in different tissues and of the role of gap junctions in cell–cell communication, as had been indicated by earlier correlative ultrastructural and physiological studies.

RECONSTITUTION OF GAP JUNCTION CHANNELS

With the increased quantity of isolated gap junction protein available, reconstitution of gap junctional conductance in vitro from purified proteins was explored as a means to analyze the gap junction channel and its gating under entirely defined conditions. The initial approach used was the patch-clamp technique in which a "patch" of membrane is sucked onto a microelectrode, forming a high-resistance seal around the edge of the electrode, permitting

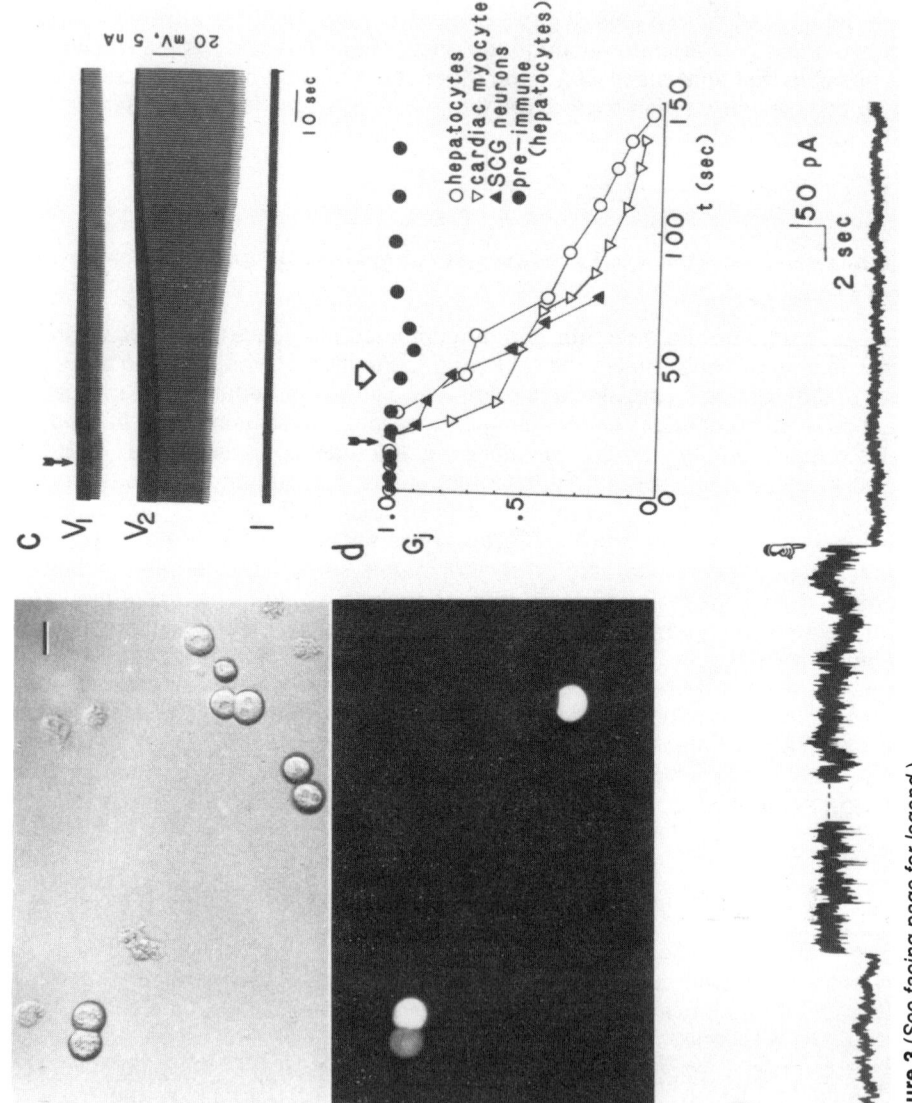

Figure 3 (See facing page for legend.)

measurement of ion flux through any channels present in that patch of membrane.

Isolated liver gap junctions were sonicated with brain phospholipids and probed with patch clamp electrodes. Single channels were found that were sensitive to pH, consistent with the inhibition of gap junctional conductance in liver at reduced pH (pK of 6.4) (Spray and Hertzberg 1985). Inhibition of channel activity was reversible upon raising the pH. Channel activity was irreversibly inhibited, however, upon incubation with affinity-purified antibodies (Fig. 3e).

The availability of inhibiting antibodies provides a crucial control, not previously available, to demonstrate that channel activities being assayed in vitro are indeed gap junctional. This reconstitution of gap junctional conductance from purified components, with sensitivity to pH and antibody, should prove to be a potent system for analysis of the gating of the gap junction channel.

PHOSPHORYLATION OF THE GAP JUNCTION POLYPEPTIDE

In collaboration with Mr. Juan Saez and Dr. Paul Greengard, isolated gap junctions have been used as substrate for in vitro phosphorylation using purified protein kinases. Phosphorylation of the 27-kD gap junction polypeptide was observed using the catalytic subunit of the cAMP-dependent protein kinase, the Ca^{++}/calmodulin-dependent protein kinase, and the phospholipid-dependent C kinase. While it is not yet clear as to how phosphorylation of the gap junction protein influences its activity, it seems likely that use of the patch-clamp reconstitution technique with junctions phosphorylated in vitro will provide a means to understand the many in vivo studies in which cyclic nucleotide-dependent alterations of gap junctional conductance have been observed.

Figure 3
Inhibition of gap junctional conductance in primary cell cultures and isolated gap junctions by antibodies to the 27,000-dalton liver gap junction polypeptide. (a and b) Injection of one of a pair of hepatocytes with antibody (cell on lower right, a and b) but not preimmune IgG (cell on top left) blocks intercellular spread of coinjected Lucifer Yellow. Light micrograph in a, fluorescence micrograph in b. (c) Electrical coupling between a pair of hepatocytes is reduced by antibody injection. Voltages (V_1, V_2) are recorded from two cells with current (I) injected alternately into each cell through an active bridge circuit. At the arrow, antibody was injected into one cell leading to a reduction of gap junctional conductance (heavy portion of traces). (d) Injection of antibody, but not preimmune IgG, into primary cell cultures reduces gap junctional conductance (G_j, normalized to initial maximum values). The brief reduction in G_j after injection of preimmune IgG is attributable to injury. (e) Inhibition of gap junctional conductance of isolated gap junctions by addition of antibody. Current through gap junction channels was recorded using the patch clamp technique. At the beginning of the record there is a jump in current of about 1 nA, which is maintained, but fluctuates, for 3 min (dashed line) before antibody is added (downward pointing hand). Thereafter, current is much reduced. Driving force is − 75 mV. (a–d reprinted, with permission, from Hertzberg et al. 1985).

CONCLUSIONS

The availability of relatively large quantities of isolated gap junction has permitted the generation of antibodies to the 27-kD gap junction polypetpide and the initiation of reconstitution studies and analysis of the covalent modification of the gap junction polypeptide. These antibodies have provided further evidence that gap junctions are a site of cell–cell communication. Immunochemical and immunocytological analyses with the antibodies have demonstrated that a highly homologous or identical polypeptide constitutes gap junctions in all rat tissues except for lens fiber cells and that this polypeptide is widely distributed among vertebrate organisms.

It is clear that these antibodies will prove useful in studying the biogenesis of gap junctions, the mechanism of regulation of gap junctional communication, and the significance of this form of cell–cell communication in the regulation of cell growth, development, and differentiation. Furthermore, they should permit access to cDNA clones, either by screening of expression libraries or by preparation of cDNA from an enriched mRNA fraction. Such clones could then be used to analyze gap junction genes and their expression in developmental systems. It seems likely that, using these combined physiological, immunological, biochemical, and molecular biological approaches, rapid progress will be made in understanding the molecular mechanism and significance of gap junctional communication.

ACKNOWLEDGMENTS

We would like to thank Robert V. Skibbens and John M. Coale for their assistance in carrying out these experiments. These studies were supported by grants from the National Institutes of Health (GM 30667 to E.L.H. and NS 16524 and NS 19830 to D.C.S.), the Robert A. Welch Foundation (Q-961 to E.L.H.), and a Grant-in-Aid from the American Heart Association and a McKnight Development Award (to D.C.S.).

REFERENCES

Azarnia, R., G. Dahl, and W.R. Loewenstein. 1981. Cell junction and cyclic AMP: III. Promotion of junctional permeability and junctional membrane particles in a junction-deficient cell type. *J. Membr. Biol.* **63:** 133–146.

Epstein, M.L. and N.B. Gilula. 1977. A study of communication specificity between cells in culture. *J. Cell Biol.* **75:** 769–787.

Evans, W.H. and J.W. Gurd. 1972. Preparation and properties of nexuses and lipid-enriched vesicles from mouse liver plasma membranes. *Biochem. J.* **128:** 691–700.

Goodenough, D.A. and W. Stoeckenius. 1972. The isolation of mouse hepatocyte gap junctions. Preliminary chemical characterization and X-ray diffraction. *J. Cell Biol.* **54:** 646–656.

Gros, D.B., B.J. Nicholson, and J.-P. Revel. 1983. Comparative analysis of the gap junction protein from rat heart and liver: Is there a tissue specificity of gap junctions? *Cell* **35:** 539–549.

Henderson, D., H. Eibl, and K. Weber. 1979. Structure and biochemistry of mouse hepatic gap junctions. *J. Mol. Biol.* **132:** 193–218.

Hertzberg, E.L. 1984. A detergent-independent procedure for the isolation of gap junctions from rat liver. *J. Biol. Chem.* **259:** 9936–9943.

Hertzberg, E.L. and N.B. Gilula. 1979. Isolation and characterization of gap junctions from rat liver. *J. Biol. Chem.* **254:** 2138–2147.

Hertzberg, E.L. and R.V. Skibbens. 1984. A protein homologous to the 27,000 dalton liver gap junction protein is present in a wide variety of species and tissues. *Cell* **39:** 61–69.

Hertzberg, E.L., D.C. Spray, and M.V.L. Bennett. 1985. Reduction of gap junctional conductance by microinjection of antibodies against the 27,000-dalton liver gap junction polypeptide. *Proc. Natl. Acad. Sci.* **82:** 2412–2416.

Hertzberg, E.L., D.J. Anderson, M. Friedlander, and N.B. Gilula. 1982. Comparative analysis of the major polypeptides from liver gap junctions and lens fiber junctions. *J. Cell Biol.* **92:** 53–59.

Michalke, W. and W.R. Loewenstein. 1971. Communication between cells of different types. *Nature* **232:** 121–122.

Nicholson, B.J., L.J. Takemoto, M.W. Hunkapiller, L.E. Hood, and J.-P. Revel. 1983. Differences between liver gap junction protein and lens MIP 26 from rat: Implications for tissue specificity of gap junctions. *Cell* **32:** 967–978.

Pitts, J.D. 1980. The role of junctional communication in animal tissues. *In Vitro* **16:** 1049–1056.

Spray, D.C. and E.L. Hertzberg. 1985. Biophysical properties of rat liver gap junction channels. *Biophys. J.* **47:** 505a.

Traub, O., U. Janssen-Timmen, P.M. Druge, R. Dermietzel, and K. Willecke. 1982. Immunological properties of gap junction protein from mouse liver. *J. Cell. Biochem.* **19:** 27–44.

Immunochemical Investigations of Gap Junction Protein in Different Mammalian Tissues

Klaus Willecke, Otto Traub, Uwe Janssen-Timmen,
Uwe Frixen, Rolf Dermietzel,* Axel Leibstein,*
Dieter Paul,† and Hartmut Rabes‡

*Institut für Zellbiologie, and *Institut für Anatomie*
Universität Essen, 4300 Essen 1, Federal Republic of Germany

†*Pharmakologisches Institut, Abteilung für Toxikologie, Universität Hamburg*
and Institut für Toxikologie und Aerosolforschung
der Fraunhofer Gesellschaft, Hannover, Federal Republic of Germany

‡*Pathologisches Institut, Universität München*
Munich, Federal Republic of Germany

Gap junction structures originally were discovered by electron microscopy as specialized domains in apposed plasma membranes of contiguous cells. It was realized that the earlier-discovered cell–cell channels for intercellular electrical conductance and permeability were likely to be mediated by gap junction structures (for review, see Bennett and Goodenough 1978; Loewenstein 1981). Later it was shown that gap junction plaques and intercellular passage of metabolites, as well as of microinjected fluorescent dye, were present or absent coordinately in cultured mammalian cells (Gilula et al. 1972). These findings supported the notion that gap junctions constitute protein channels for direct exchange of ions and metabolites between cells in contact.

All attempts to isolate the protein(s) that form the gap junction channel used criteria of electron microscopy for monitoring the biochemical purification of gap junction plaques. After SDS-electrophoretic analysis of purified gap junction plaques, a protein of 26–27K apparent molecular weight was suggested to be the main and perhaps the only component of liver gap junctions (Henderson et al. 1979; Finbow et al. 1980; Hertzberg 1980). The slightly different molecular weights determined for this protein in different laboratories are probably due to different standards of molecular weight or species differences, since this protein had been isolated from rat and mouse liver. A recent report (Finbow et al. 1983; see also this volume), however, described a 16K protein as a constituent of mouse and rat liver gap junctions. The biochemical relationship (degradation product?) of the 16K to the 26–27K protein remains to be clarified.

The production of antibodies to the liver 26–27K protein or to purified gap junction plaques provided a new approach to the characterization of gap junctions (Traub et al. 1982; Hertzberg 1984). Not only do these polyclonal antibodies react with the 26–27K protein by immunoblot after SDS electrophoresis but they also react specifically with gap junction plaques, as shown by immunoelectron microscopy and fluorescence microscopy of isolated liver plasma membranes or of tissue sections (Janssen-Timmen et al. 1983). Thus these polyclonal antibodies could be used to follow changes in the content of the 27K protein in rat livers after partial hepatectomy or cholestasis due to temporary bile duct ligation (Traub et al. 1983). It is likely that more refined immunochemical tools will lead to deeper insights into the structure-function relationship of gap junctions as well as into the biosynthetic assembly and genetic control of these intercellular channels. While pursuing these research goals we report here on recent experimental results obtained in our laboratory. Part of our ongoing work studies the following questions:

1. To what extent are gap junction proteins in different tissues homologous molecules?
2. Are there posttranslational modifications of liver gap junction protein?
3. Does the amount and localization of gap junction protein depend on the proliferative activity of the liver tissue?

RESULTS AND DISCUSSION

Gap Junctions in Several Tissues Share Antigenic Determinants with Liver Gap Junctions

The rabbit antiserum directed against mouse liver 26K protein was affinity-purified on mouse liver gap junction plaques bound to CNBr-activated Sepharose 4B. These affinity-purified antibodies were used for indirect immunofluorescence labeling of several tissue sections from mouse or rat (Dermietzel et al. 1984). Discrete fluorescent spots on apposed membranes of contiguous cells were seen in liver, pancreas (exocrine part), kidney, small intestine (epithelium and circular smooth muscle), Fallopian tube, endometrium, and myometrium of delivering rats. No specific reaction was seen on myocardium (especially, intercalated discs), ovaries (granulosa cells and stroma), and lens (epithelium and fiber cells). It is noteworthy that weak immunofluorescence was found on the endocrine pancreas (i.e., Langerhans islets) after glibenclamide treatment of mice and rats. This treatment had previously been reported to cause an increase in insulin secretion and of the size as well as the number of gap junctions in the islets of Langerhans (Meda et al. 1979). Furthermore, the specific fluorescence in the myometrium of delivering rats was much stronger than in the myometrium of nonpregnant animals. Garfield et al. (1977) and Cole and Garfield (this volume) had found by morphometric analysis a dramatic increase of gap junction abundance in the myometrium of rats during delivery. After treatment with gold Protein A, we demonstrated specific labeling of gap junction plaques by immunoelectron microscopy on ultrathin frozen sections through liver and the exocrine part of the pancreas. Moreover the affinity-purified liver 26K antibodies were shown by immunoblot to react with proteins of similar molecular weight in pancreas and kidney membranes (Dermietzel et al. 1984). Therefore gap junctions from several morphogenetically different tissues must have similar antigenic sites in common, but the extent of homology cannot yet be ascertained. Different amounts of the same gap junction protein in several tissues may cause the different extent of antibody binding. Alternatively nonidentical gap junction proteins may share regions of structural homology. This alternative is consistent with reports that junctional proteins isolated from liver, lens, and heart are different molecules that may share antigenic determinants (Hertzberg et al. 1982; Traub and Willecke 1982; Gros et al. 1983; Hertzberg and Skibbens 1983). We interpret our failure to demonstrate cross-reactivity of the liver 26K antibodies with intercalated discs of rat myocardium (where gap junctions are abundant) to mean that those antigenic determinants that are recognized by our antibodies in liver and other tissues are absent or inaccessible in myocardial tissue. This interpretation is consistent with the finding that liver and myocardial gap junction proteins may share antigenic sites that can be recognized by other antibodies (Hertzberg and Skibbens 1983).

The Gap Junction Protein Is Phosphorylated in Cultured Hepatocytes

Figure 1 illustrates the results of immunoprecipitation using affinity-purified rabbit 26K antibodies. Primary liver hepatocytes were prepared by collagenase treatment of livers from 19-day-old mouse embryos, incubated for attachment on plastic tissue culture vessels in Dulbecco's modified medium plus 10% fetal calf serum, and subsequently cultured in chemically defined, MX-83 medium containing transferrin, insulin, and epidermal growth factor (D. Paul, in prep.). Under these conditions the hepatocyte cultures proliferated and reached confluence after 3 days. At this time the cultures (2×10^6 cells) were labeled with [^{35}S]methionine or [^{32}P]phosphate (300 µCi/3 ml of culture medium) for 4 hours, harvested by scraping, collected by centrifugation, and lysed according to Furth et al. (1982). The supernatants were subjected to a standard protocol of immunoprecipitation using Protein A agarose (Kessler 1976). Figure 1 shows autoradiographs after SDS electrophoresis of the immunoprecipitated material. All immunoprecipitations were carried out in parallel with affinity-purified 26K antibodies and rabbit IgG from nonimmunized animals. The results clearly indicate specific enrichment after precipitation with 26K antibodies of a 26K band that was labeled either with [^{35}S]methionine or [^{32}P]phosphate. We conclude that the 26K gap junction protein exists in a phosphorylated form in these cultured hepatocytes. It has to be tested in future experiments whether phosphorylation of the gap junction protein affects the function of the cell–cell channel for intercellular communication. In this context it is interesting that there is preliminary evidence for in vitro phosphorylation of MP 26, the lens fiber junction protein with a cAMP-dependent protein kinase from bovine lens (Johnson and Johnson 1982; Johnson et al. this volume).

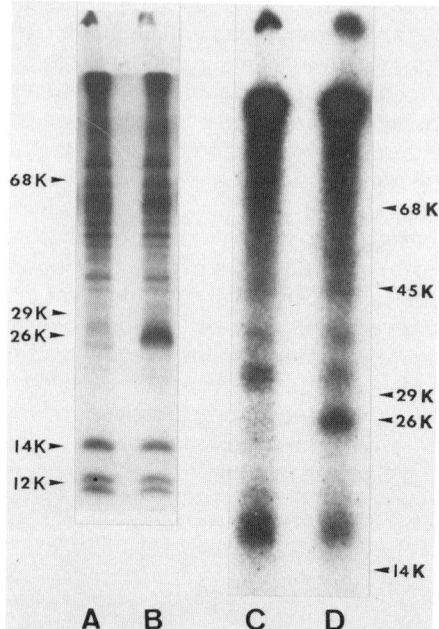

Figure 1
Immunoprecipitation of gap junction protein from cultured mouse hepatocytes. The experimental conditions for labeling, obtaining cell extracts, and immunoprecipitation are summarized in the text. The autoradiographs after SDS electrophoresis are shown. Lanes A and B are extracts from [^{35}S]methionine-labeled cells immunoprecipitated with rabbit IgG from nonimmunized animals (A) or with 26K antibodies (B). Lanes C and D are extracts from ^{32}P-labeled cells immunoprecipitated with control rabbit IgG (C) or with 26K antibodies (D). The molecular weight markers were: bovine serum albumin (68K), ovalbumin (45K), carboanhydrase (29K), chymotrysinogen (26K), and ribonuclease A (14K).

A Monoclonal Antibody Recognizes the Liver 26K Gap Junction Protein after Immunoblotting and Reacts with the Cytoplasmic Domain of Gap Junctions

In order to probe the structure of gap junctions with immunochemical reagents of highest specificity, we isolated several rat monoclonal antibodies after immunization of rats with purified mouse liver gap junction plaques and fusion of rat myeloma cells (210 RCY 3Ag 1.2.3. obtained from C. Milstein) with spleen cells from immunized animals. Eleven cell culture supernatants reacted with purified mouse liver gap junction plaques in an ELISA assay (Traub et al. 1982). Only one monoclonal antibody (12-1C5), however, recognized mouse liver 26K protein after SDS electrophoresis and immunoblot of purified gap junction plaques or of mouse liver membranes (Fig. 2). In both cases the strong 26K band and trace amounts of the "dimer" band around 47K (Henderson et al. 1979; Traub et al. 1982) were detected.

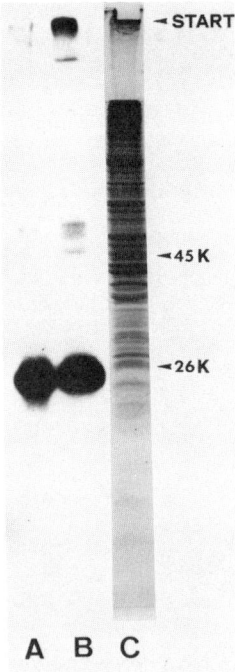

Figure 2
Immunoblot (autoradiograph) after SDS electrophoresis of purified gap junction plaques (A) and total membrane proteins (B) from mouse liver. For comparison lane C shows a Coomassie Blue-stained gel after SDS-electrophoresis of B. The monoclonal antibody 12-1C5, rabbit anti-rat IgG, and ^{125}I-labeled Protein A were used for these studies (cf. Traub et al. 1982).

The monoclonal antibody was isolated from cell culture supernatants by affinity chromatography and used for immunofluorescence microscopy. We found staining of discrete spots on apposed plasma membranes of cross-sectioned mouse liver (Fig. 3a), a pattern very similar to the results obtained with affinity purified polyclonal liver 26K antibodies (Dermietzel et al. 1984). Furthermore we found that the antibody reacted specifically with discrete plasma membrane areas of the exocrine part of pancreas and seminiferous epithelium of the testis (Fig. 3b) but not with intercalated discs of myocardium (results not shown). Finally immunoreactivity could be documented with the monoclonal antibody in cultured mouse hepatocytes (Fig. 4a). Most of the fluorescent spots are located on apposing plasma membranes of hepatocytes grown in monolayer. However, we also noticed fluorescent spots over the cytoplasmic compartment of the cells. These spots may represent intracellular membrane vesicles containing gap junction protein. No immunofluorescence was seen when cultured hepatocytes had not been permeabilized by fixation with ethanol prior to immunostaining. This finding supports the notion that the monoclonal antibody reacts with gap junctions at the cytoplasmic domain of these structures. Interestingly, the specific immunoreactivity appeared first in the center of hepatocyte colonies after 2 days in culture (cf. Fig. 4a). This observation suggests that prolonged cell–cell contact may regulate the assembly of gap junction plaques. We also found specific immunofluorescence on apposing plasma membranes of cultured MDCK cells, an established dog renal epithelial cell line. Using the same concentrations of monoclonal anti-

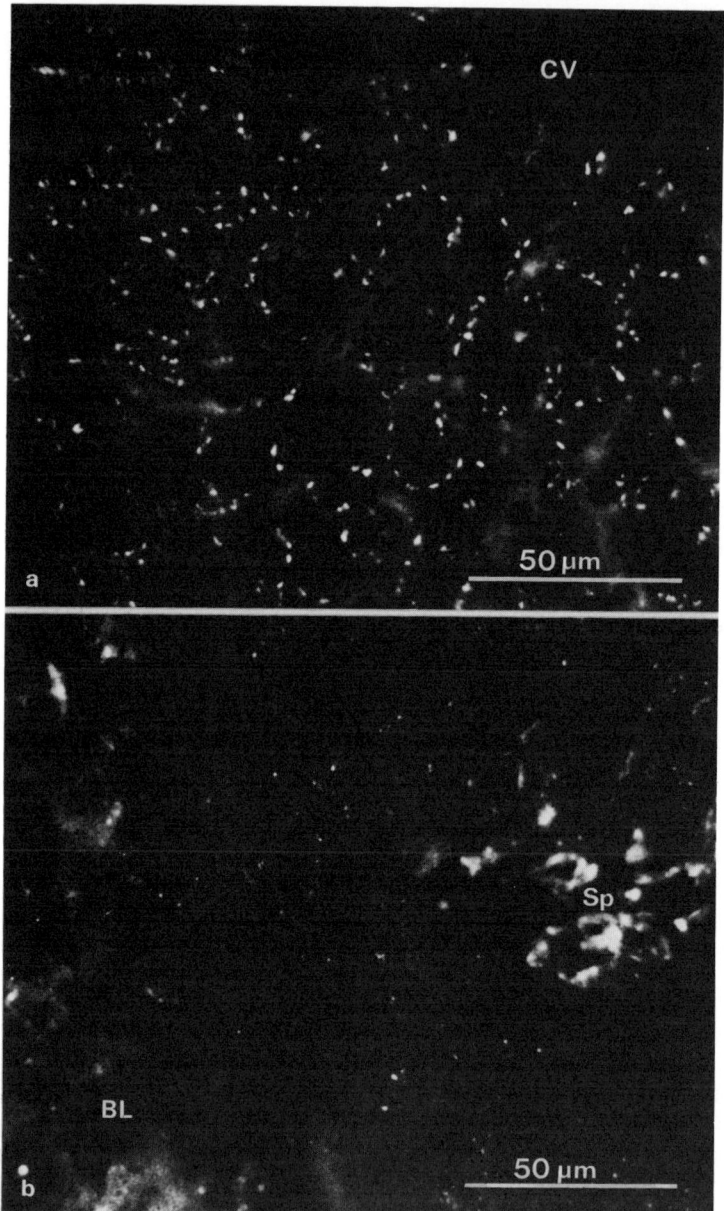

Figure 3
Immunoreactivity of tissue sections after staining with rat monoclonal antibody directed to liver 26K gap junction protein and goat anti-rat IgG-fluorescein isothiocyanate. (a) Mouse liver. CV, Central vein. (b) Seminiferous epithelium of mouse testis. The specific immunoreactivity of small fluorescent spots probably represents gap junctions between Sertoli cells. The strong unspecific fluorescence of spermatozoa heads is also seen after control incubations with rat IgG. BL, Basal lamina; Sp, spermatozoa.

body, the reaction with MDCK cells was at least 10-fold weaker than with cultured hepatocytes. Immunoelectron microscopy with gold Protein A as a marker (cf. Dermietzel et al. 1984) demonstrated that the monoclonal antibody recognized the cytoplasmic side of cross-sectioned mouse liver gap junctions.

Differences of the 26K Gap Junction Protein in Hepatoma and Normal Liver Tissue

Although several hepatoma and other cancer cells were shown to be defective in cell–cell coupling, some tumorigenic cell lines must have functional gap junctions since they show electrical and metabolic coupling (for review, see Loewenstein 1979). Specific antibodies against liver gap junctions can be exploited to compare the characteristic staining pattern obtained for normal liver with that of hepatoma tissue. Figure 4b illustrates a representative example of these preliminary experiments. We used the monoclonal antibody 12-1C5 for immunofluorescence analyses of cryosections through rat livers bearing hepatomas that had been induced by treatment with diethylnitrosamine. So far we have analyzed several cases where the hepatoma and surrounding liver tissue could easily be distinguished since no or very few fluorescent spots were detected on apposed plasma membranes of the tumor tissue. Further studies will have to prove whether the apparent lack or reduction of immunoreactivity applies to all cells of a given hepatoma in situ. Furthermore, we would like to know whether the decrease of immunoreactivity in hepatoma tissue corresponds to a deficiency of intercellular communication or to a change in cell-surface properties that interferes with the formation of gap junction plaques.

SUMMARY AND CONCLUSIONS

The use of antibodies directed against liver gap junction 26K protein or liver gap junction plaques has opened several new approaches to characterize gap junctions. Since gap junction proteins in different tissues share antigenic determinants, they must have structurally homologous regions. Several results suggest that there are structural differences as well between gap junction proteins in different tissues.

The experimental proof for the existence of a multigene family for gap junction protein in the mammalian genome, however, requires sequence comparison of the corresponding proteins or genes. Investigations of cultured, proliferating hepatocytes promise to answer several questions with regard to posttranslational modifications of the 26K gap junction protein. The biochemical details and any functional role of the newly detected phosphorylation of the 26K protein in cultured hepatocytes have to be analyzed. Furthermore, the possible regulation of gap junction biosynthesis and plaque assembly by cell–cell contact and/or cell proliferation can be studied in this system. The use of highly specific monoclonal antibodies to the 26K gap junction protein has revealed (quantitative?) differences in gap junction plaques between hepatoma and normal liver tissue.

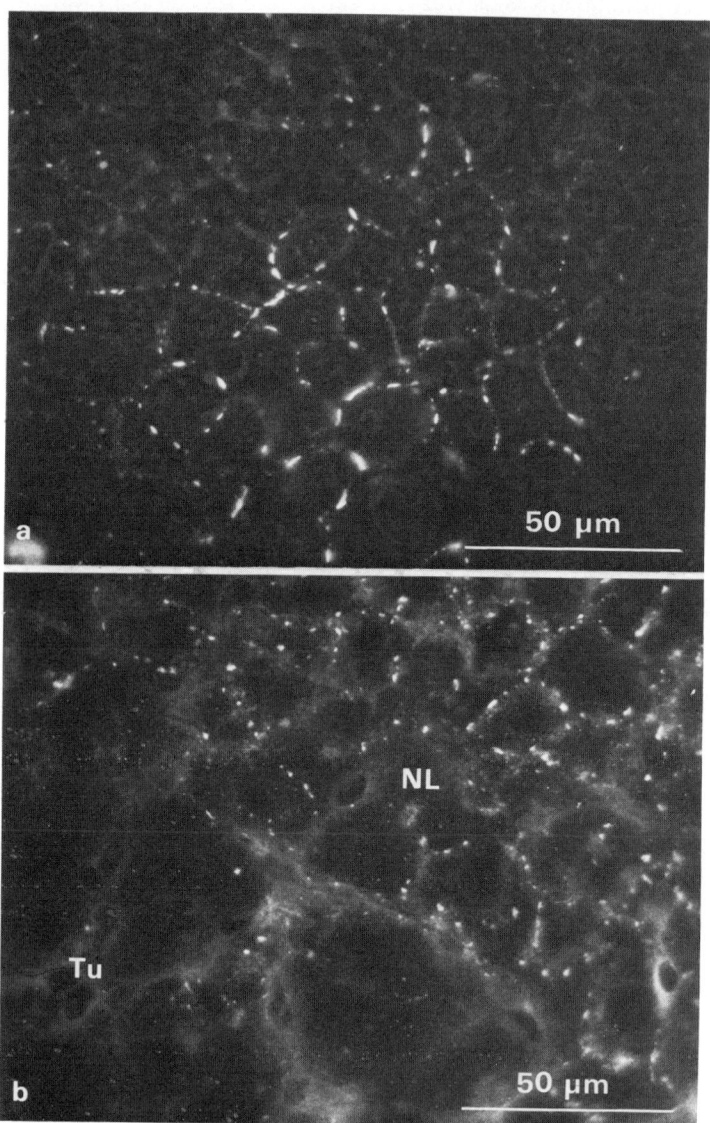

Figure 4
(a) Immunofluorescence in a colony of mouse hepatocytes after 2 days in culture. At this time gap junction immunoreactivity is confined to hepatocytes in the center of the cell clone that covers the whole microscopic view field (phase-contrast micrograph not shown). (b) Immunofluorescence on thin section of rat liver bearing a hepatoma induced by diethylnitrosamine. The micrograph shows the border area between hepatoma (Tu) and normal liver tissue (NL) (cf. Fig. 3a). The monoclonal antibody 12-1C5 and goat anti-rat IgG-fluorescein isothiocyanate were used for these studies.

ACKNOWLEDGMENTS

This work was supported by the Ministerium für Wissenschaft und Forschung, Düsseldorf, and Stiftung Volkswagenwerk, and Fonds der Chemischen Industrie, to K.W., and research grants of the Deutsche Forschungsgemeinschaft (SFB 114) to R.D., and (P266) to D.P. and of the Bundesumweltamt, Berlin, and the Ministerium für Forschung und Technologie, Bonn, to D.P.

REFERENCES

Bennett, M.V.L. and D.A. Goodenough. 1978. Gap junctions, electrotonic coupling and intercellular communication. *Neurosciences Res. Program Bull.* **16:** 372-486.

Dermietzel, R., A. Leibstein, U. Frixen, U. Janssen-Timmen, O. Traub, and K. Willecke. 1984. Gap junctions in several tissues share antigenic determinants with liver gap junctions. *EMBO J.* **3:** 2261-2270.

Finbow, M.E., J. Shuttleworth, A.E. Hamilton, and J.D. Pitts. 1983. Analysis of vertebrate gap junction protein. *EMBO J.* **2:** 1479-1486.

Finbow, M., S.B. Yancey, R. Johnson, and J.-P. Revel. 1980. Independent lines of evidence suggesting a major gap junctional protein with a molecular weight of 26,000. *Proc. Natl. Acad. Sci.* **77:** 970-974.

Furth, M.E., L.J. Davis, B. Fleurdelys, and E.M. Scolnick. 1982. Monoclonal antibodies to the p21 products of the transforming gene of Harvey murine sarcoma virus and of the cellular *ras* gene family. *J. Virol.* **43:** 294-304.

Garfield, R.E., S. Sims, and E.E. Daniel. 1977. Gap junctions: Their presence and necessity in myometrium during parturition. *Science* **198:** 958-960.

Gilula, N.B., O.R. Reeves, and A. Steinbach. 1972. Metabolic coupling, ionic coupling and cell contacts. *Nature* **235:** 262-265.

Gros, D.B., B.J. Nicholson, and J.-P. Revel. 1983. Comparative analysis of the gap junction protein from rat heart and liver; is there a tissue specificity of gap junctions? *Cell* **35:** 539-549.

Henderson, D., H. Eibl, and K. Weber. 1979. Structure and biochemistry of mouse hepatic gap junctions. *J. Mol. Biol.* **132:** 193-218.

Hertzberg, E.L. 1980. Biochemical and immunological approaches to the study of gap junctional communication. *In Vitro* **16:** 1057-1067.

———. 1984. A detergent-independent procedure for the isolation of gap junctons from rat liver. *J. Biol. Chem.* **259:** 9936-9943.

Hertzberg, E.L. and R.V. Skibbens. 1983. Antibodies to the 27,000-dalton rat liver gap junction polypeptide cross-react with a similar protein in heart. *J. Cell Biol.* **97:** 1259 (Abstr.).

Hertzberg, E.L., D.J. Anderson, M. Friedlander, and N.B. Gilula. 1982. Comparative analysis of the major polypeptides from liver gap junctions and lens fiber junctions. *J. Cell Biol.* **92:** 53-59.

Janssen-Timmen, U., R. Dermietzel, U. Frixen, A. Leibstein, O. Traub, and K. Willecke. 1983. Immunocytochemical localization of the gap junction 26K protein in mouse liver plasma membranes. *EMBO J.* **2:** 295-302.

Johnson, W.R. and G.R. Johnson. 1982. Bovine lens MP26 is phosphorylated in vitro by an endogenous cAMP-dependent protein kinase. *Fed. Proc.* **41:** 755 (Abstr.).

Kessler, S.W. 1976. Cell membrane antigen isolation with the Staphylococcal protein A–antibody adsorbent. *J. Immunol.* **117:** 1482-1490.

Loewenstein, W.R. 1979. Junctional intercellular communication and the control of cell growth. *Biochim. Biophys. Acta* **560:** 1-65.

———. 1981. Junctional intercellular communication: The cell to cell membrane channel. *Physiol. Rev.* **61:** 829–913.

Meda, P., A. Perrelet, and L. Orci. 1979. Increase of gap junctions between pancreatic B cells during stimulation of insulin secretion. *J. Cell Biol.* **82:** 441–448.

Traub, O. and K. Willecke. 1982. Cross reaction of antibodies against liver gap junction protein (26K) with lens fiber junction protein (MIP) suggests structural homology between these tissue specific gene products. *Biochem. Biophys. Res. Commun.* **109:** 895–901.

Traub, O., P.M. Drüge, and K. Willecke. 1983. Degradation and resynthesis of gap junction protein in plasma membranes of regenerating liver after partial hepatectomy or cholestasis. *Proc. Natl. Acad. Sci.* **80:** 255–259.

Traub, O., U. Janssen-Timmen, P. Drüge, R. Dermietzel, and K. Willecke. 1982. Immunological properties of gap junction protein from mouse liver. *J. Cell. Biochem.* **19:** 27–44.

Comparison of the Protein Components Present in Vertebrate and Arthropod Gap Junction Preparations

Malcolm E. Finbow, T. Eldridge J. Buultjens,
Ephraim Kam, John Shuttleworth, and John D. Pitts

The Beatson Institute for Cancer Research
Wolfson Laboratory for Molecular Pathology
Glasgow G61 1BD, Scotland

An understanding of the roles of cell–cell communication mediated by gap junctions is dependent in part on the identification and subsequent analysis of the junctional proteins. A key to the isolation of gap junctions from other plasma membrane components came almost 20 years ago when Benedetti and Emmelot (1967) discovered that gap junctions were not disrupted by detergents. Many studies have since been carried out to identify the protein component of gap junctions from rodent liver or heart.

Over the years many different-sized and apparently unrelated proteins have been proposed as the major structural component of liver or heart gap junctions. The reason for the confusion in this area of research is that the only course open in the initial identification of the structural protein is to assume that it will be the major staining band after SDS-PAGE of preparations which are judged by electron microscopy to be highly enriched in gap junctions. The shortcoming lies in interpreting the purity of the final preparation. Gap junctions retain their basic structural identity throughout the isolation procedures, but this is unlikely to be true for other less ordered structures. As a result it is difficult to identify other contaminants and, consequently, to assess their abundance. The problem is compounded by the very limited size of the sample that can be analyzed by electron microscopy.

Therefore we chose to approach the isolation of gap junctions in a different manner. Rather than restrict isolation to one or two tissues from the same species where contaminating proteins may be the same, we prepared junctions from a wide variety of tissues and cultured cells from different species and different phyla. This permitted the identification of common or similar-sized proteins with related properties as candidate gap junctional proteins. The assumption that similar proteins would be present in gap junctions from different sources is not unjustified. Gap junction structure does not vary greatly from one cell type to another (except in lens fibers), and functional gap junctions can be formed between very different cell types and between cells of different species, for instance, fish cells and mammalian cells in culture. Correlative tests, not dependent upon the subjective nature of electron microscopy, were also devised to establish further the junctional origin of candidate proteins.

THE TRITON EXTRACTION METHOD

All previously published procedures have required the preparation of plasma membranes as the starting point for detergent solubilization. This approach has two drawbacks. First, it results in low yields of junctions due to losses during plasma membrane preparation—perhaps as much as 90% (Hertzberg and Gilula 1979). And second, it limits the isolation of gap junctions to those tissues where plasma membranes can be prepared in bulk (e.g., liver and heart). We therefore sought a method of junction preparation that would be versatile, allowing isolation of junctions from varied sources, and would give high yields permitting detailed study, the preparation of antibodies, and the sequencing of the component proteins. We decided to use nonionic detergents to facilitate organelle fractionation.

Tissue homogenates are dissolved in 1% Triton X-100, 0.15 M NaCl which releases the gap junctions from the plasma membranes but which does not disrupt nuclei, cytoskeletal elements, and extracellular matrix components. These large components are easily removed by a low-speed centrifugation to leave the gap junctions in the supernatant. The gap junctions are pelleted (along with other small detergent-resistant structures) by high-speed centrifugation. Our own experience (Finbow et al. 1980), and that of others (Goodenough 1976), has shown that the use of proteolytic enzymes increases the morphological purity of the final preparations. Therefore we included in many experiments a trypsin treatment of the high-speed, detergent-resistant material. The use of trypsin during isolation has no detectable effect on the protein we have identified as being associated with gap junctions. This protein is only extensively degraded by trypsin after SDS extraction when gap junctions have been solubilized (see below). A 6 M urea/0.5% N-lauroyl sarcosine extraction step is also carried out before ultracentrifugation on conventional sucrose step gradients.

The morphological purity of the final preparations can be further increased with only a small reduction in yield by using a postmitochondrial pellet fraction (Finbow et al. 1980) as the starting point for Triton X-100 extraction. The SDS-PAGE profiles of the final preparations are the same regardless of which point Triton X-100 extraction is carried out (i.e., on crude homogenate, postmitochondrial pellet, or isolated plasma membranes).

THE 16K AND 18K PROTEINS OF ISOLATED VERTEBRATE AND ARTHROPOD GAP JUNCTIONS – DISTRIBUTION AND YIELD

The use of the Triton method has allowed the preparation of gap junctions from such diverse sources as lobster, crayfish, and *Limulus* hepatopancreas; from mouse liver, kidney, brain, heart, lung, and uterus; from liver of other mammalian, avian, and amphibian species; and from a wide variety of tissue culture cells. We have found that a single protein of M_r 16,000 (16K) is associated with gap junction fractions prepared from vertebrate sources (Finbow et al. 1983). An 18K protein is the major component of gap junctions prepared from the lobster hepatopancreas (Finbow et al. 1984).

The yield and the purity of the final preparations varies with the source. To date, the purest preparations and the highest yields come from mouse liver and kidney and lobster hepatopancreas. Analysis by electron microscopy using thin-section, freeze-fracture, and negative staining shows that gap junctions are the major component (Figs. 1 and 2). The repeated, routine analysis of many preparations by both electron microscopy and SDS-PAGE on the same junction preparations from different tissue sources has shown that the amount of the 16K protein goes hand-in-hand with the yield of gap junctions in the final preparation. Those tissues that give high yields of gap junctions, such as mouse liver and kidney, likewise are rich sources of the 16K protein. Conversely, those tissues that yield low numbers of gap junctions, such as mouse brain and heart, are poor sources of the 16K protein. (Although gap junctions are abundant in heart, the yield is poor, probably due to the difficulty of getting complete homogenization.)

From recent morphometric and structural studies, it is now possible to calculate a theoretical yield of gap junctions per gram wet weight of rat liver. The maximum weight of each connexon is $\sim$170,000 daltons (Unwin and Ennis 1984) and there are $\sim$180,000 connexons per rat liver hepatocyte (Revel et al. 1980; Meyer et al. 1981). As there are 2.4×10^8 cells/g wet weight of rat liver (Leslie 1955), of which 80% of the cells are hepatocytes, this gives a theoretical yield of $\sim$10 µg junctional protein per gram wet weight of rat liver. The yield of the 16K protein falls well within this theoretical yield, being consistently 2–2.5 µg/g wet weight of rat liver as measured by Coomassie Blue R-250 staining on SDS-PAGE compared to standard proteins. The yield from mouse liver is twice this value. The yield of the 18K protein is 4–5 µg/g wet weight lobster hepatopancreas.

CORRELATION OF THE 16K AND 18K PROTEIN WITH GAP JUNCTIONS

Any attempt to relate protein bands seen by SDS-PAGE to a particular structure is subjective. The assessment of purity by electron microscopy is not quantitative. Therefore, before deciding that a particular protein is of gap junctional origin, it is necessary to develop experiments that do not rely as heavily on the subjective nature of electron microscopy. We have attempted to use

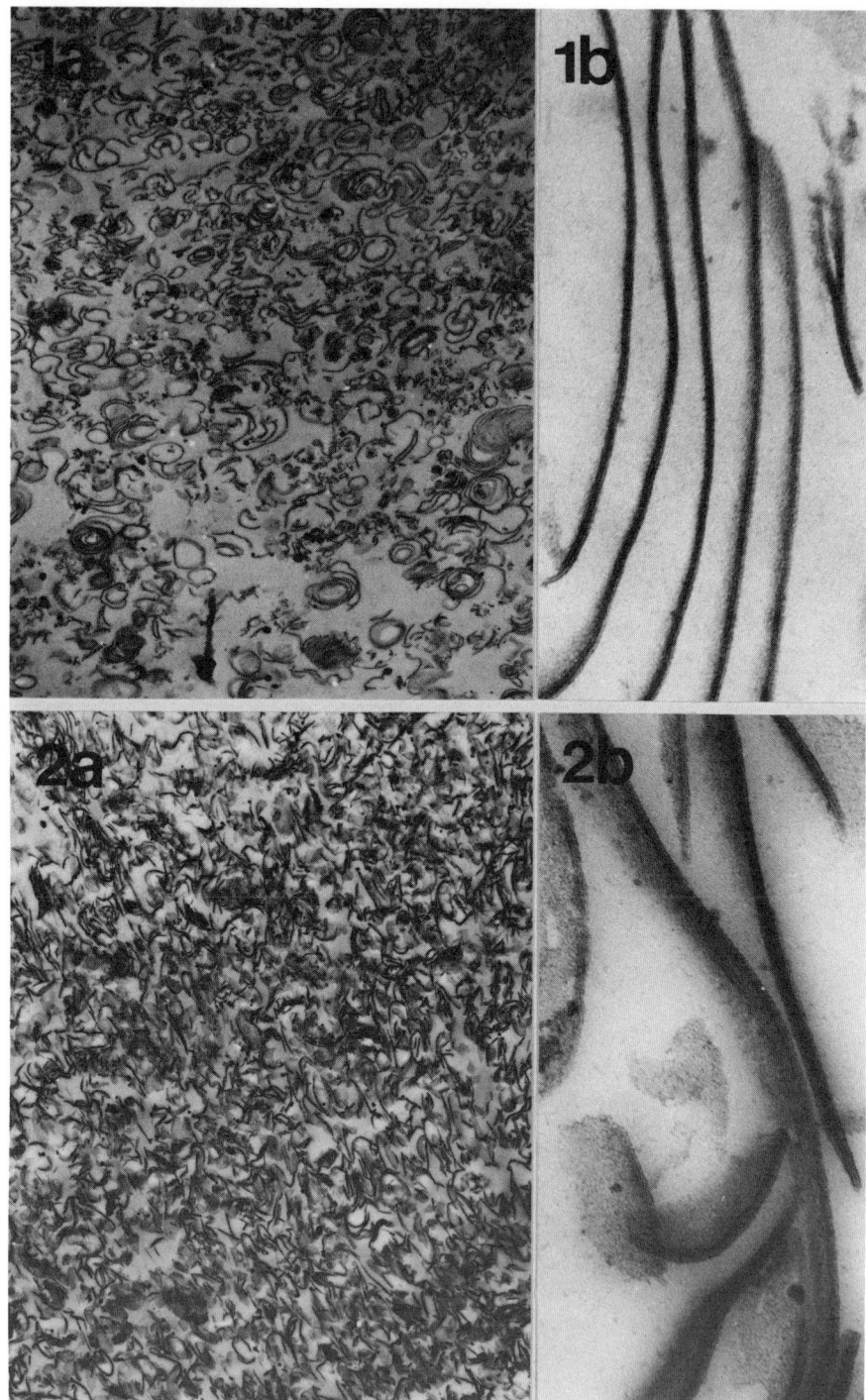

Figures 1 and 2 (*See facing page for legend.*)

physical, chemical, and biological properties of gap junctions to relate the 16K and 18K proteins, respectively, to vertebrate and arthropod gap junctions.

Being quasi-crystalline structures, isolated gap junctions should have a narrow range of bouyant densities. Earlier studies using continuous sucrose gradients gave a mean bouyant density of 1.170 g/cm^3 for rat and mouse liver gap junctions (Goodenough 1976; Hertzberg and Gilula 1979). We chose to use potassium iodide gradients because of their lower viscosity. The fractions containing gap junctions were located by electron microscopy. Mouse liver gap junctions and lobster hepatopancreas gap junctions both form very sharp peaks in these gradients with bouyant densities of 1.195 g/cm^3 and 1.260 g/cm^3, respectively (Finbow et al. 1983, 1984). These densities are consistent with gap junctions being mixtures of protein and phospholipid. The slightly higher bouyant density of the arthropod gap junctions presumably reflects the larger connexon size and therefore a greater contribution of protein. Analysis by SDS-PAGE shows that these peak fractions contain greater than 95% of the 16K protein or 18K protein, consistent with these proteins being of gap junctional origin.

SDS is the only commonly used detergent believed to solubilize gap junctions completely (Goodenough and Stoeckenius 1972). As the junctions are solubilized, the constituent protein should be released in a soluble form. We have followed the process of SDS solubilization of both mouse liver and lobster hepatopancreas gap junctions by electron microscopy and found that they differ. Whereas mouse liver gap junctions are gradually solubilized over a concentration range between 0.1% and 0.5% SDS, lobster hepatopancreas gap junctions are solubilized in a cooperative manner from 0.1% to 0.2% SDS. Coincident with these morphological changes, the 16K protein gradually disappears from the pelletable fraction and appears in the SDS-soluble fraction between 0.1% and 0.5% SDS, whereas the 18K protein is wholly in a soluble form by 0.2% SDS. The differences in the SDS solubility of mouse liver and lobster hepatopancreas gap junctions may be due to differences in, for example, lipid composition.

Two instances are known where the morphometrically measured area of gap junctions can be greatly reduced under controlled experimental conditions. One is during the regeneration of weanling rat liver after partial hepatectomy and the other is after treatment of V79 tissue culture cells (Chinese hamster lung fibroblasts) with the tumor promoter 4β-phorbol 12-myristate 13-acetate (TPA) (Yancey et al. 1979, 1982). The yield of the 16K protein is reduced by 90% at the time during liver regeneration when the junctional area is reduced by greater than 95% (Finbow et al. 1983). With the cultured V79 cells (a system that is more amenable to detailed analysis and in which functional studies to test junctional communication have been carried out using the same culture medium), the yield of the 16K protein falls by 90%, again in parallel with the loss of gap junctions. This recovers to 70% of the original value when TPA is

Figures 1 and 2
Low-power (1a and 2a) and high-power (1b and 2b) thin-section micrographs of mouse liver and lobster hepatopancreas gap junction preparations. Magnification 1a and 2a, 6800×; 1b and 2b, 105,000×.

removed and functional coupling has been restored. When cell lines are used whose communication properties are not impaired by TPA, there is no significant reduction in the yield of the 16K protein.

Although none of these experiments, taken in isolation, provide unequivocal evidence for the junctional origin of the 16K and 18K proteins, together they provide a strong foundation for the proposal that the 16K and 18K proteins are the major structural proteins of the vertebrate and arthropod gap junctions, respectively.

THE RELATIONSHIP BETWEEN THE 16K AND 18K JUNCTIONAL PROTEINS

Apart from their similar size, the vertebrate 16K and arthropod 18K junctional proteins share other properties. They are both relatively insensitive to trypsin during isolation, a finding that is consistent with the proposed arrangement of the junctional protein (Makowski et al. 1977; Unwin and Zampighi 1980). Denaturation of the protein by boiling in SDS buffers prior to SDS-PAGE results in both the 16K and 18K proteins forming multimeric aggregates, principally the 26K and 28K dimers, respectively. The 18K protein dimerizes more readily and is generally present in the 28K form after solubilization at ambient temperatures (Fig. 3). Neither protein is affected on SDS-PAGE by the presence of reducing agents.

Differences between the two proteins can be seen by two-dimensional peptide mapping. There are no major common ^{125}I-labeled tryptic or chymotryptic

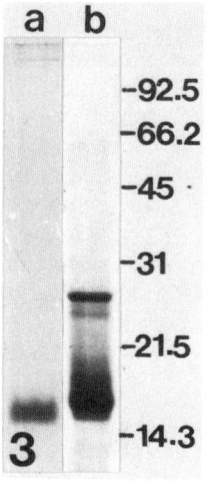

Figure 3
SDS-PAGE of samples from mouse liver (lane a) and lobster hepatopancreas (lane b) gap junction preparations. These samples are taken from the same junction preparations used for the thin-section analyses shown in Figs. 1 and 2. The 18K protein runs diffusely (except when loaded at low concentrations), probably caused by a tendency of the protein to aggregate during SDS-PAGE. A proportion of the 18K protein is also migrating as the 28K dimeric species. The Biorad low-molecular-weight standards were used to measure M_r.

peptides in maps of the lobser hepatopancreas 18K protein and the mouse liver 16K protein (Finbow et al. 1984; Fig. 4). This is not surprising as two-dimensional peptide mapping tends to emphasize differences rather than similarities between homologous proteins. A further difference between the arrangement of the two proteins is also revealed by ^{125}I-labeling and subsequent peptide mapping. Labeling with ^{125}I using the chloramine-T method can be carried before or after solubilization in SDS solutions. Little difference can be detected in the two-dimensional map of the mouse liver 16K protein prepared by either method of labeling. However, for the lobster hepatopancreas 18K protein, some peptides are labeled much more heavily when iodination is carried out before solubilization in SDS solutions. This suggests that some tyrosine residues exposed to an aqueous environment when the 18K protein is in the plaque form are not as accessible in SDS solutions.

Although homology cannot be seen at the level of two-dimensional peptide maps, the 16K vertebrate protein and 18K arthropod proteins are related immunologically. A rabbit antiserum has been raised against the chicken liver gap junction preparation. To facilitate the production of sufficient antigen for raising a response, a simplified method of junction preparation was used. As a result, the antigen contained not only the 16K protein but also contained a similar amount of a pair of proteins of M_r 32,000. On immunoblots this antiserum reacts with the chicken liver 16K and 32K proteins and also with the mouse liver 16K and lobster hepatopancreas 18K proteins. The antiserum is specific, reacting only with the 16K protein in immunoblots of mouse liver plasma membranes. To ensure that the antibodies to the chicken 32K proteins were not the ones reacting with the mouse liver 16K and lobster hepatopan-

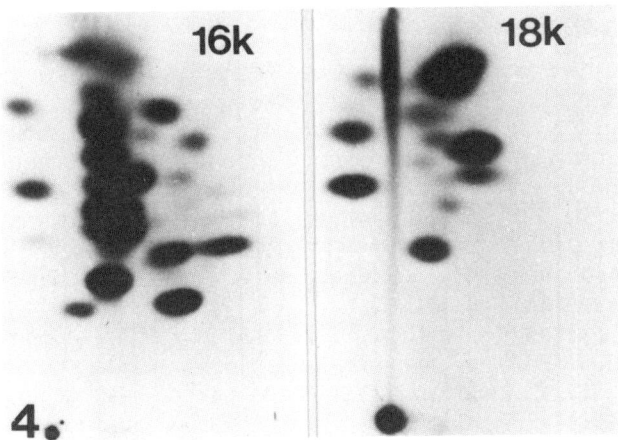

Figure 4
Two-dimensional maps of the ^{125}I-labeled chymotryptic peptides from mouse liver 16K protein and lobster hepatopancreas 18K protein. The junctions were labeled with ^{125}I after being solubilized in 1% SDS and mapped as described previously (Finbow et al. 1984). The electrophoretic separation (pH 2.1) is shown from left to right and the chromatographic separation is shown from bottom to top.

creas 18K protein, the antiserum was affinity purified. Rather than use the purified 16K protein from the chicken liver gap junction preparation, we chose to use SDS-PAGE-purified mouse liver 16K and lobster hepatopancreas 18K proteins for affinity purification using a modification of the method of Olmsted (1981). This overcomes the potential problem of degradation products of other proteins present in the chicken junction preparation that might contaminate the SDS-PAGE-purified chicken liver 16K protein. The two affinity-purified preparations were then used on a panel of protein blots of the original chicken junctions, mouse liver junctions, and lobster hepatopancreas junctions. Both affinity-purified antibody preparations react with the chicken liver and mouse liver 16K proteins and with the lobster hepatopancreas 18K protein, but neither react with the 32K chicken protein (Finbow et al. 1984). The presence of common immunological determinants in the vertebrate 16K protein and arthropod 18K protein suggest that they are evolutionarily related. If this is the case, then this finding predicts that gap junctional proteins from other phylogenetic sources would also have the same antigenic determinants. Preliminary evidence confirms this prediction.

Enriched gap junction fractions recently have been prepared from the octopus (mollusc) hepatopancreas. These fractions contain a major protein component M_r 19,000 which forms a 29K dimer on boiling in SDS sample buffer prior to SDS-PAGE. The rabbit anti-chicken gap junction serum reacts on immunoblots with the octopus 19K protein and its multimeric forms.

IMMUNOCYTOCHEMISTRY WITH ANTISERA AGAINST GAP JUNCTIONS

Antisera raised against the structural protein of gap junctions can be used to (1) correlate by electron microscope immunocytochemistry and junctional protein with the gap junction, (2) localize particular antigenic determinants of the connexon that may have structural or functional importance, and (3) specifically inhibit junctional communication either by preventing formation or blocking existing junctional channels. Furthermore, as different phylogenetic forms of the gap junction protein have common antigenic determinants, it will be important from a functional standpoint to determine where in the gap junctional structure they might be.

Two rabbit antisera have been raised against isolated lobster hepatopancreas gap junctions that have allowed immunocytochemical localization of the 18K protein to these junctions. These two antisera contain antibodies that bind strongly to the readily accessible cytoplasmic faces of isolated lobster hepatopancreas gap junctions. This can be seen by extensive decoration with Protein A-labeled gold (one Protein A-gold particle per six connexons, which is close to the theoretical decoration limit) of the isolated junctions after preincubation with the antiserum. Also, addition of the antisera to suspensions of isolated lobster hepatopancreas gap junctions results in agglutination, which can be followed by phase-contrast light microscropy. Dilutions of the sera, or affinity-purified preparations, can be quickly tested for agglutination activity to give a sharp end point, which is a valuable measure of antibody titer. The level of fluorescence after adding fluorescein-conjugated sheep anti-rabbit IgG, is high over clumped junctions, but the fluorescence level over occasional isolated junctions is modest even though the level of primary antibody is saturating.

On immunoblots the two rabbit anti-lobster hepatopancreas gap junction sera recognize the 18K protein (and its multimeric forms) and a 52K glycoprotein, even though this glycoprotein is present at very low levels in the lobster hepatopancreas gap junction preparations (Finbow et al. 1984). The anti-52K glycoprotein activity is not completely removed by two rounds of affinity purification on SDS-PAGE-purified 18K protein, but can be absorbed out by first using SDS-PAGE-purified 52K glycoprotein. The resulting affinity-purified antiserum still retains binding activity to isolated lobster hepatopancreas gap junctions (by Protein A-gold labeling) and reacts only with the 18K protein (and its multimeric forms) on immunoblots of lobster junction preparations.

The rabbit anti-lobster gap junction sera do not bind detectably to or agglutinate mouse liver gap junctions. This suggests that there are differences in the cytoplasmic surfaces of these phylogenetic forms of gap junctions. The rabbit anti-chicken liver gap junction serum also does not bind to isolated mouse liver or lobster hepatopancreas gap junctions at a detectable level when analyzed by Protein A-labeled gold electron microscope cytochemistry. This could be due to two reasons. First, the antigenic determinants may lie buried in the plasma membrane or the extracellular face, and are not therefore accessible. Or second, the antiserum may recognize the denatured junctional protein (as on immunoblots) but not the protein in its native configuration.

Interestingly though, the rabbit antisera raised against the lobster hepatopancreas gap junctions and chicken liver gap junctions both rapidly block dye coupling (within 1–2 min) between cultured BRL cells (rat liver epithelial cells) when microinjected into cells by pressure or iontophoresis, although none of the sera detectably affect junction formation or function when added to the medium. These studies show that there are several antibody activities present in each polyclonal antiserum: those that recognize the denatured protein on immunoblots, those that bind to the cytoplasmic faces of isolated gap junctions as seen by Protein A-gold labeling and agglutination, and those that can perturb functional cell–cell communication. As yet, it is difficult to relate these activities or to identify which are due to common sets of antibodies.

RELATIONSHIP OF THE 16K PROTEIN TO OTHER VERTEBRATE PROTEINS THOUGHT TO BE OF GAP JUNCTIONAL ORIGIN

Other proteins have been proposed as the major structural component of rodent liver gap junctions. The most commonly accepted protein has an M_r in the range 26,000–28,000, depending on the molecular weight markers used in different laboratories (Henderson et al. 1979; Hertzberg and Gilula 1979; Finbow et al. 1980; Nicholson et al. 1981; Traub et al. 1982) (for convenience it will be referred to as the 27K protein). The 27K protein aggregates to a presumptive dimeric form (47K) on boiling in SDS sample buffer prior to SDS-PAGE (Hertzberg 1980; Nicholson et al. 1981) and is absent from detergent-insoluble fractions at the time gap junctional area is greatly reduced in the regenerating weanling rat liver (Finbow et al. 1980).

The 27K protein is readily degraded by trypsin before SDS solubilization of preparations containining junctions (Henderson et al. 1979; Finbow et al. 1980) and so it has been suggested that the 16K protein is a degradation product of the 27K protein (Hertzberg 1984). However, the 27K protein is generally thought to be degraded to a 10K limit polypeptide(s) (Nicholson et al. 1981), whereas the 16K protein appears to be unaffected by even extensive trypsin treatment of the junctions at any time during isolation. This 10K limit polypeptide of the 27K protein is believed to be the 10K "connexin" originally described by Goodenough (1974, 1976) in mouse liver gap junction fractions prepared from collagenase and trypsin-treated plasma membranes (Nicholson et al. 1981, 1983). However, although the migration of the 27K protein on SDS-PAGE is unaffected by the presence of reducing agents, the 10K connexin migrates as multimeric forms in the absence of reducing agents and as the 10K monomeric form when they are present (Goodenough 1974, 1976).

The yield of 27K protein can be as high as 10 μg/g wet weight rat liver (Hertzberg 1984) despite the known low recovery of plasma membranes in the preliminary step of the isolation (as low as 7.8%; Hertzberg and Gilula 1979). It appears therefore that the estimated recovery of 27K protein (100 μg/g wet weight or more), unlike that of the 16K protein (2–2.5 μg/g wet weight), is considerably in excess of the theoretical abundance (10 μg/g wet weight, see above). Further work clearly is required to reconcile this anomaly if the 27K protein is to remain a candidate for the structural protein of the gap junction.

While all three proteins (27K, 16K, and 10K) are present in detergent-resistant, gap junction-containing fractions of plasma membranes, their relationship with each other was not clear. Therefore, gap junctions have been prepared by the different published methods to compare the abundance of the three proteins and to see which is common to all preparations. SDS-PAGE of the junction fractions from rat liver prepared by the "27K method" of Hertzberg and Gilula (1979), which does not use proteases, shows major bands at 27K (monomer) and 47K (dimer) and minor bands at 24K and 16K. The "connexin method" of Goodenough (1976; as modified by Finbow et al. 1980), in which mouse liver plasma membranes are first treated with collegenase and trypsin before detergent extraction, shows the 10K connexin in the presence of β-mercaptoethanol and in its absence multimeric 24K and 36K forms. A 16K band of similar intensity is also present but is unaffected by reducing agents. The "Triton method" produces the 16K protein with no detectable 27K or 10K bands in preparations from either rat or mouse liver (much of the 27K protein is lost with the Triton-insoluble material at the first low-speed centrifugation step).

Two-dimensional peptide mapping of the rat liver 27K protein, rat and mouse liver 26K proteins, and the mouse liver 10K connexin shows that (1) the rat and mouse 16K proteins are very closely related, (2) the 27K proteins contain none of the major tryptic or chymotryptic ^{125}I-labeled peptides present in the

16K protein, and (3) the 10K mouse connexin has no common tryptic ^{125}I-labeled peptides with the rat 27K protein or the 16K protein, but has many major peptides common with a minor 24K component present in rat liver junction fractions prepared without proteases by the method of Hertzberg and Gilula (1979). From these data we must conclude that the 16K protein cannot be a proteolytic fragment of the 27K protein. The 10K connexin also is not a degradation product of the 27K protein but appears to be derived from a 24K protein. The absence of a major polypeptide species derived from the 27K protein in "connexin" preparations of junctions from mouse or rat liver suggests that protease treatment of plasma membranes results in the loss of the 27K protein and its proteolytic fragments from the detergent-insoluble (junctional) fraction.

Immunoblotting data provides further evidence that the 16K protein is not related to the 27K protein and 10K connexin. The rabbit anti-chicken gap junction serum fails to react with the rat liver 27K protein (prepared by the "27K method") or the mouse liver 10K connexin (prepared by the "connexin method") on immunoblots. However, it does react with the 16K protein bands present in both these preparations. This shows that the 16K protein is not only produced by the Triton method but is present in fractions containing gap junctions, regardless of the method of preparation. The authenticity of the 16K protein present in these two preparations has been confirmed by peptide mapping.

In conclusion, although proteolysis during isolation can never be excluded (until cDNA sequence information is available), all the present data show that any hypothetical larger molecule from which the 16K protein is derived is not the 27K protein.

CONCLUSION

There is as yet no general agreement on the identification of the major gap junction protein. Two explanations are possible from the presently available and apparently conflicting data: (1) there are a number of classes of gap junctions present, even in the same cell type, and each class is composed of distinct species of proteins, or (2) only one of the proteins is a structural element of gap junctions and the other proteins are either peripherally associated or contaminants.

If the first explanation is accepted, then one would need to speculate that there are three (or more) classes of gap junctions in rodent liver; one class composed from the 27K protein, a second class composed from the 16K protein, and a third class composed from the 10K connexin (or connexin-like) protein. The gap junctions isolated by the three methods are morphologically indistinguishable, which argues against each preparative procedure giving rise to separate classes of junctions with different-sized proteins. Furthermore, if each class of gap junction is made from a different-sized protein, then it should be possible to separate each class of junction because they will have characteristic bouyant densities. (Lobster hepatopancreas gap junctions, bouyant density of 1.260 g/cm^3, can be separated from mouse liver gap junctions, bouyant density of 1.195 g/cm^3.) However, measurements of bouyant density by ultracentrifugation on continuous sucrose or potassium iodide gradients provide no evidence for such heterogeneous populations of gap junctions (Goodenough 1976; Hertzberg and Gilula 1979; Finbow et al. 1983).

Clearly, the second suggestion is the most likely. It is possible that one of the presently available antisera may help to resolve the issue. Another approach is to devise preparative methods that remove one or two (or all) of the three candidate proteins and yet yield preparations that are still highly enriched for gap junctions. The 27K protein can be removed by a protease treatment and the 10K connexin is lost by changing the detergent solubilization conditions; neither modification results in any detectable effect on the quantity or quality of the final junction preparation. On the other hand, the 16K protein is present in junction preparations regardless of the method of purification. We propose that the 16K protein is the major structural protein of the vertebrate gap junction, a contention that is supported by the finding that an immunologically related 18K protein appears to be the major structural component of arthropod gap junctions.

ACKNOWLEDGMENTS

This work was supported by the Cancer Research Campaign.

REFERENCES

Benedetti, E.L. and P. Emmelot. 1967. Hexagonal arrays of subunits in tight junctions separated from isolated rat liver plasma membranes. *J. Cell Biol.* **38:** 15–24.

Finbow, M.E., J. Shuttleworth, A.E. Hamilton, and J.D. Pitts. 1983. Analysis of vertebrate gap junction protein. *EMBO J.* **2:** 1479–1486.

Finbow, M.E., S.B. Yancey, R. Johnson, and J.-P. Revel. 1980. Independent lines of evidence suggesting a major gap junctional protein with a molecular weight of 26,000. *Proc. Natl. Acad. Sci.* **77:** 970–974.

Finbow, M.E., T.E.J. Buultjens, N.J. Lane, J. Shuttleworth, and J.D. Pitts. 1984. Isolation and characterisation of arthropod gap junctions. *EMBO J.* **3:** 2271–2278.

Goodenough, D.A. 1974. Bulk isolation of mouse hepatocyte gap junctions. *J. Cell Biol.* **61:** 557–563.

———. 1976. In vitro formation of gap junction vesicles. *J. Cell Biol.* **68:** 220–231.

Goodenough, D.A. and W. Stoeckenius. 1972. The isolation of mouse hepatocyte gap junctions. *J. Cell Biol.* **54:** 646–656.

Henderson, D., H. Eibl, and K. Weber. 1979. Structure and biochemistry of mouse hepatic gap junctions. *J. Mol. Biol.* **132:** 193–218.

Hertzberg, E.L. 1980. Biochemical and immunological approaches to the study of gap junctional communication. *In Vitro* **16:** 1057–1067.

———. 1984. A detergent-independent procedure for the isolation of gap junctions from rat liver. *J. Biol. Chem.* **259:** 9936–9943.

Hertzberg, E.L. and N.B. Gilula. 1979. Isolation and characterisation of gap junctions from rat liver. *J. Biol. Chem.* **254:** 2138–2147.

Leslie, I. 1955. The nucleic acid content of tissues and cells. In *The nucleic acids* (ed. E. Chargaff and J.N. Davidson), p. 1–50, vol. II. Academic Press, New York.

Makowski, L., D.L.D. Caspar, W.C Phillips, and D.A. Goodenough. 1977. Gap junction structures. *J. Cell Biol.* **74:** 629–645.

Meyer, D.J., S.B. Yancey, J.-P. Revel, and A. Peskoff. 1981. Intercellular communication in normal and regenerating rat liver: A quantitative analysis. *J. Cell Biol.* **91**: 505–523.

Nicholson, B.J., M.W. Hunkapillar, L.B. Grim, L.E. Hood, and J.-P. Revel. 1981. Rat liver gap junction protein: Properties and partial sequence. *Proc. Natl. Acad. Sci.* **78**: 7594–7598.

Nicholson, B.J., L.J. Takemoto, M.W. Hunkapillar, L.E. Hood, and J.-P. Revel. 1983. Differences between liver gap junction protein and lens MIP 26 from rat. *Cell* **32**: 967–978.

Olmsted, J.B. 1981. Affinity purification of antibodies from diazatized paper blots of heterogeneous protein samples. *J. Biol. Chem.* **256**: 11955–11957.

Revel, J.-P., S.B. Yancey, D.J. Meyer, and B. Nicholson. 1980. Cell junctions and intercellular communication. *In Vitro* **16**: 1010–1017.

Traub, O., U. Janssen-Timmen, P.M. Druge, R. Dermietzel, and K. Willecke. 1982. Immunological properties of gap junction protein from mouse liver. *J. Cell Biochem.* **18**: 27–44.

Unwin, P.N.T. and P.D. Ennis. 1984. Two configurations of a channel forming membrane protein. *Nature* **307**: 609–613.

Unwin, P.N.T. and G. Zampighi. 1980. Structure of the junction between communicating cells. *Nature* **283**: 545–549.

Yancey, S.B., D. Easter, and J.-P. Revel. 1979. Cytological changes in gap junction during liver regeneration. *J. Ultrastruct. Res.* **67**: 229–242.

Yancey, S.B., J.E. Edens, J.E. Trosko, C.C. Chang, and J.-P. Revel. 1982. Decreased incidence of gap junctions between chinese hamster V79 cells upon exposure to the tumor promoter 12-O-tetradecanoyl-phorbol-13 acetate. *Exp. Cell Res.* **139**: 329–340.

Is Lens MP26 a Gap Junction Protein? A Phosphoprotein?

Ross Johnson, Elizabeth Frenzel,
Keith Johnson, Kathleen Klukas, Paul Lampe,
Daryl Sas, M. Jane Sas, Rebecca Biegon,*
and Charles Louis†
Department of Genetics and Cell Biology
*Department of Anatomy and †Department of Veterinary Biology
University of Minnesota, St. Paul, Minnesota 55108

Gap junctions have now been demonstrated in a wide variety of specialized animal tissues where they are thought to provide pathways for the intercellular movement of small molecules. Thus, these junctions likely play a critical role in diverse cellular functions, and, consequently, they have been retained during the evolution of multicellular systems. This volume documents work on a number of these tissues.

As one evaluates different systems in some detail, variations in junctional structure are noted (Peracchia 1980), as are varied permeability properties of the cell–cell channels. For example, although these channels apparently close in response to reduced cytoplasmic pH in a number of cases, the effective pH varies with cell type and not all cells are found to be pH sensitive (Spray and Bennett 1985).

A frequent question asked as a result of this work on different systems is, "To what degree are the proteins of gap junctions conserved in an evolutionary sense?" Although a definitive answer is not yet available, it appears now that gap junction proteins from different tissues are not identical, in view of peptide mapping (Gros et al. 1983) and protein sequence data (B. Nicholson et al., in prep.) from the liver and heart. In addition, whereas antibodies raised against junctional proteins from the liver have been reported to cross-react with proteins from a number of tissues (Dermietzel et al. 1984; Hertzberg and Skibbens 1984), exceptions to the cross-reactivity have also been observed in these studies. Furthermore, different laboratories (utilizing different antibodies) have not always agreed on whether a particular tissue will cross-react with an antibody to liver junctional protein. Finally, there is a report that antisera raised against chicken liver junctions recognize components of arthropod gap junctions (Finbow et al. 1984). The situation clearly requires detailed analyses of varied tissues.

One of the tissues receiving considerable attention, because of the extensive intercellular junctions observed (Benedetti et al. 1976), is the vertebrate lens (Bloemendal 1981). Although these junctions share some fundamental structural similarities with gap junctions in other tissues, various observations raise questions regarding the nature and function of lens junctions. For example, cell–cell channels in the lens are not readily closed (Rae et al. 1982; Schuetze and Goodenough 1982). Furthermore, the major membrane protein MP26, presumed to be a major component of the lens junction, has not been found to cross-react substantially with antisera to liver gap junction protein (Hertzberg and Skibbens 1984; Dermietzel et al. 1984; however, also see Traub and Willecke 1982). In addition, antisera to MP26 have not been found to react with liver gap junction protein (Hertzberg et al. 1982). Thus, skepticism has existed regarding the idea that MP26 is a gap junction protein in the lens.

It is important to resolve this issue for two reasons. First, we may learn whether different cell–cell channels have evolved independently, possibly relating to different cellular functions or different regulatory mechanisms. Second, we may be better able to interpret existing data on the sequence of MP26 (Gorin et al. 1984) and on the permeability of artificial membranes containing lens membrane components (Peracchia and Girsch; Hall and Zampighi; both this volume).

DISCUSSION

To evaluate thoroughly the question, "Is MP26 in the lens a gap junction protein?", a number of features should be considered.

Is MP26 a Protein of Lens Intercellular Junctions?

Since MP26 is an abundant protein in the membrane of the lens fiber cell, it was suggested early as a component of the extensive lens junctions (Benedetti et al. 1976). More recently, this relationship has been questioned because investigators have not consistently localized MP26 to lens junctions with immunolabeling methods (Bok et al. 1982; Paul and Goodenough 1983). In addition, it has been reported that two junctional types, differing in thickness, exist in the bovine lens (Zampighi et al. 1982). This raised the possibility that MP26 resided in only one junctional type and that only one type, with or without MP26, included cell–cell channels (Paul and Goodenough 1983).

To help resolve this question, we developed a monoclonal antibody, which we showed to be specific both for MP26 in the calf and for an analogous protein in the chicken (Sas et al. 1985). This antibody, which we termed B2, clearly labeled the cytoplasmic surfaces of the thicker lens junctions, not labeled in a previous study with antibodies to MP26 (Paul and Goodenough 1983). In addition, this monoclonal antibody also appeared to label the thinner junctions, as reported in the other study. Our studies utilized indirect labeling at the level of electron microscopy with both horseradish peroxidase and colloidal gold markers (Fig. 1; Sas et al. 1985), with the degree of labeling in both control and experimental preparations being analyzed quantitatively. Preliminary studies from this laboratory (J. Sas and R. Hybertson, unpubl.) with a rabbit antiserum directed against electrophoretically purified MP26 (Keeling et al. 1983) have supported the findings made with the monoclonal antibody. Thus, we consider MP26 to be a component of both junctional types seen in the lens.

Is MP26 a Transmembrane Protein?

The major proteins of intercellular junctions would be expected to span the lipid bilayer, with domains on the external side of the junctional membrane for recognition and junction formation, and those on the cytoplasmic side for possible regulatory roles. Since MP26 clearly has domains on the cytoplasmic side, as noted above, detection of external regions on MP26 would document its transmembrane nature. No experimental demonstration of the transmembrane feature has been reported. However, a recent model for MP26, based on the sequence of MP26 as suggested by cDNA analysis, proposes that MP26 spans the lens membrane six times (Gorin et al. 1984; see also Revel and Yancey, this volume).

We have several lines of evidence that indicate MP26 is a transmembrane protein. The studies concentrate on, but do not exclusively rely on, lens membranes treated with a range of pH values from 2.0 to 7.0. Lens junctional membranes, analyzed with thin-section electron microscopy methods, are found to separate at pH 2.5–3.0 (Sas and Johnson 1983), with no junctional profiles being seen below this level. We have also investigated the effect of acid treatment on the proteolytic sensitivity of MP26. At neutral pH, a variety of proteases cleave MP26 to only 21,000–22,000 M_r fragments. However, after treatment at $\leq$pH 2.5, chymotrypsin, for example, degrades MP26 to a number of smaller fragments (Fig. 2).

We have also raised an MP26-specific monoclonal antibody against acid-treated membranes (Sas and Johnson 1983). This monoclonal, termed A4, binds to control membranes to only a limited extent in enzyme-linked, immunosorbent assays (ELISAs; Engvall and Perlmann 1972). When binding is evaluated after treatment at a variety of pH values, we again observe a critical difference at a reduced pH, with antibody binding increasing dramatically below pH 3.0. Studies at the level of electron microscopy are in progress to examine the possibility of A4 binding on the external surface of junctional membranes.

All of these results could be explained either by the physical separation of junctional membranes, granting proteases and antibodies access to new sites, or by substantial, acid-induced conformational changes. We prefer the former interpretation, since (1) some A4 binding is detected without acid treatment, (2) the binding of the cytoplasmic side monoclonal antibody is not altered by pH, and (3) the buoyant density of the treated membranes is not altered.

Another line of evidence supporting the transmembrane nature of MP26 is derived from experiments where exogenous proteases were applied to lens

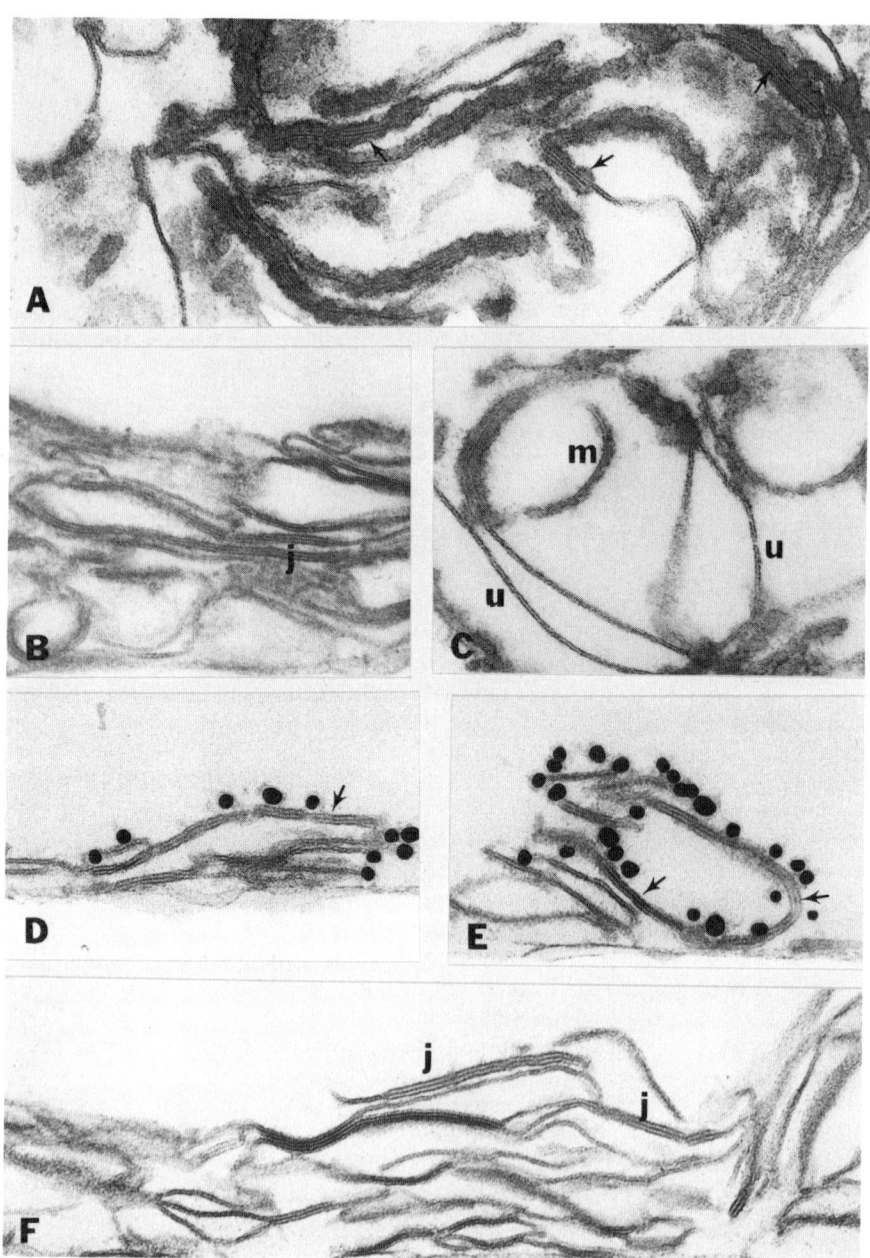

Figure 1 *(See facing page for legend.)*

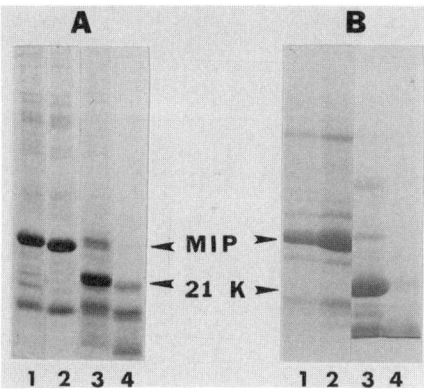

Figure 2
SDS-polyacrylamide gel electrophoresis of lens membranes from protease experiments. (A) Bovine membranes; (B) chicken membranes, with the MIP (MP26 in the calf or MP28 in the chicken) being obvious in each case. (Lane 1) control membranes; (lane 2) membranes treated at pH 2 without proteases; (lane 3) chymotrypsin without acid treatment; and (lane 4) chymotrypsin after acid treatment. Note the loss of the 21,000 M_r fragment (21K) in lane 4.

cultures containing MP26. When this treatment was followed by immunotransfer analysis of all the culture proteins, we found that monoclonal B2 recognized both intact MP26 and a minor fragment produced by cleavage in the center of the molecule. One interpretation is that the exogenous proteases did not enter the cells but cleaved a limited, nonjunctional form (a precursor?) of MP26 on the extracellular side of the membrane.

If the antibodies B2 and A4 bind on opposite sides of the membrane, they should bind to distinctly different sites on the MP26 molecule. To map these sites within the protein, we tested various protease-derived fragments and chemical cleavage products of MP26 with B2 and A4 (Sas et al. 1985 and in prep.). Monoclonal B2 was shown to bind approximately 10 residues from the carboxyl terminus (Fig. 3) by protease treatments of lens membranes and by use of a cyanogen bromide fragment of MP26 (Takemoto et al. 1983). The binding site for monoclonal A4 was mapped to the center of the MP26 mole-

Figure 1
Electron micrographs of isolated bovine lens junctions labeled with monoclonal antibody B2, shown to be specific for MP26. In A and C, the antibody is detected with a secondary antibody conjugated to horseradish peroxidase; in D and E, a colloidal gold marker is used. In B and F, respective controls for the two labels are illustrated, both involving unconditioned hybridoma medium. Other controls giving similar results included myeloma medium and an unrelated monoclonal antibody to ovalbumin. (→) Labeled junctions; (j) control junctions that are not labeled; (u) unlabeled membranes; and (m) a nonjunctional membrane with label on only one side. Magnifications are all 92,000×. (Reprinted, with permission, from Sas et al. 1985.)

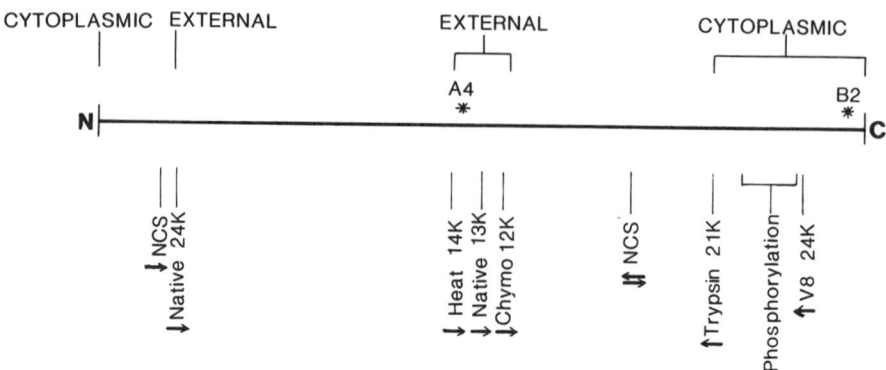

Figure 3
Diagram of MP26. The mapping of two monoclonal antibody binding sites (A4 and B2) and numerous cleavage sites are included. The arrows below each cleavage site designate the side(s) from which the observed fragments were obtained. Above the MP26 line the proposed membrane sidedness of the molecule is indicated, with several lines of evidence suggesting that the region including the monoclonal A4 site projects externally. NCS refers to cleavage, presumably at tryptophans, with N-chlorosuccinimide (D.F. Sas et al., in prep.).

cule (Fig. 3), since a fragment of approximately 14,000 M_r bound both B2 and A4, whereas 12,000 and 13,000 M_r fragments bound only B2. The A4 site is close to the location of protease cleavage seen in the lens cultures and close to a "native" cleavage site found in untreated lens membranes (D.F. Sas et al., in prep.), lending support to the transmembrane nature of MP26. The native cleavage may occur in vivo as an age-related process.

Is MP26 a Channel Protein?

Three types of studies provide evidence that MP26 is able to form channels in the lens membrane. The first involves parallel ultrastructural and electrophysiological studies on frog and rat lenses (Kuszak et al. 1984). It is reported in these two species, where junctions cover different percentages of fiber cell surfaces, that the degree of coupling in the lenses varies correspondingly. In addition, this analysis is designed to determine whether the majority of the numerous channels observed are in an open state.

The second line of support is derived from the suggested protein sequence of bovine MP26, based on nucleotide sequencing of a cDNA clone (Gorin et al. 1984). In a proposed model for the arrangement of MP26 within the lens membrane, both termini are placed on the cytoplasmic side and six transmembrane regions are included. A key feature of the model is the recognition of the amphipathic character of one transmembrane domain, indicating that it may line a membrane channel (also see Yancey and Revel, this volume). This bears a striking resemblance to the subunits of the acetylcholine receptor (Finer-Moore and Stroud 1984).

A third set of studies has evaluated the permeability of artificial lens membrane systems with liposomes containing lens membrane proteins or electrophoretically purified MP26. Utilizing methods that measure single-channel conductances in black lipid membranes (Hall and Zampighi, this volume) and liposome swelling in the presence of sugars (Nikaido and Rosenberg 1984; Peracchia and Girsch, this volume), these investigators have accumulated data on different channel properties. The former investigations have revealed high conductance channels, which are not ion-selective, consistent with the idea of size-selective gap junction-type channels. The latter studies have used channel permeability to sugars to size channels and to study channel dynamics. While entirely consistent with the concept of MP26 channels, the data are not yet conclusive on this point. Future experiments comparing the properties of the artificial channels with those demonstrated in the lens fiber cell will be most instructive.

The implication in these studies is that, if MP26 is a channel protein, MP26 can then specifically function as a protein of cell–cell channels, because of evidence that MP26 is localized within the lens intercellular junctions (see above).

Does MP26 Function in Combination with Gap Junction Proteins?

As expected of a typical gap junction system, it is known that ions and fluorescent tracers can pass between adjacent lens fiber cells (Rae 1979; Schuetze and Goodenough 1982). If MP26 forms cell–cell channels between fiber cells, one can then logically ask two related questions. Do channels formed with MP26 provide for the observed intercellular permeability properties in the lens? In addition, does MP26 constitute the lens version of proteins found in the gap junctions of varied tissues?

As an initial approach to these questions, we sought to determine whether cells rich in MP26 were able to form permeable junctions with other types of cells, which were known to contain gap junction protein, but presumably no MP26. As noted above, gap junction proteins are apparently not identical in different tissues, but they are reported to function together, as assayed by heterocellular gap junction formation (Michalke and Loewenstein 1971; Epstein and Gilula 1977). There is also in vivo evidence that different cell types form gap junctions (Johnson et al. 1973). Therefore, we have examined the formation of permeable junctions between cultured chick embryo lens cells and either chick embryo heart cells or Novikoff rat hepatoma cells.

These studies rely on the extent to which the lens cultures provide an effective model for lens fiber cells. We have made a number of observations that indicate that the cultures, indeed, provide a useful model. When trypsin-dissociated lens cells are plated on collagen (Menko et al. 1984), the differentiated "lentoids," which appear within 5–7 days, resemble fiber cells in a variety of ways: (1) only scattered cytoplasmic organelles are seen, (2) cell volume is increased, (3) DNA synthesis is limited, (4) both crystallins and MP26 are synthesized and reach levels comparable to the embryonic lens, (5) large lens junctions are revealed with freeze-fracture and the distribution of junctional sizes in culture is remarkably similar to that in the embryonic lens, and (6)

Lucifer Yellow is found to move between lentoid cells in essentially all cases (Menko et al. 1982). The extent of lentoid formation is reflected by the striking similarity between the protein profiles of 10-day-old chick embryo lenses and cultures rich in lentoids, in particular the presence of δ-crystallin (Fig. 4).

To assay for formation between lentoids and heart cells, we relied on the fact that the acquisition of synchrony in the beating of myocardial cells is known to reflect gap junction formation (Griepp and Bernfield 1978). When heart cell aggregates were plated on differentiated lens cultures, two striking results were obtained. First, heart cell aggregates became synchronized within 10 hours when they were separated by differentiated lentoid cells, at distances of up to approximately 200 μm. However, undifferentiated lens cells, which are only poorly dye-coupled to each other, do not appear to be effective in synchronizing aggregates. Second, essentially all of the aggregates in contact with lentoids stopped beating within 24 hours, whereas those on undifferentiated lens cells continued to beat for several days and heart aggregates attached to only the plastic substrate in lens cultures were found to beat for more than a week. It would appear that lentoid cells are able to form permeable junctions with cells from the heart aggregates and, as a result, indirectly

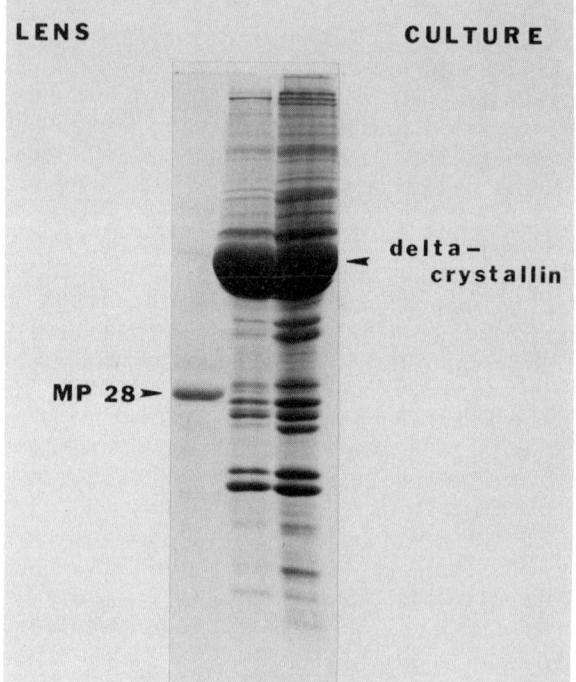

Figure 4
SDS-polyacrylamide gel electrophoresis of chick embryo lens culture proteins. Proteins from a well-differentiated culture are compared with total proteins from a 10-day embryonic lens and MP28 from adult chicken lens membranes. Note the large amount of δ-crystallin in both the culture and the lens, with other striking similarities as well.

synchronize beating (Goshima and Tonomura 1969). The cessation of beating on lentoids may reflect electrical shunting to the lens cells, due to effective coupling between these cells and the myocardial cells.

One problem with the lens–heart experiments is that the distances between synchronized aggregates may be substantial, allowing for a large number of different pathways for coupling between the two heart aggregates. To avoid this possible problem, Novikoff hepatoma cells were plated on lens cultures and electrical coupling and dye transfer were used to assay for permeable junctions in Judson Sheridan's lab (University of Minnesota). In this case, with a small group of 15-μm-diameter Novikoff cells, the testing of permeability at a lentoid cell surface can be direct. In several instances, the formation of permeable junctions between differentiated lentoid cells and Novikoff cells was detected with electrical coupling and dye transfer (Fig. 5).

Both the heart and Novikoff studies provide strong support for the formation of permeable junctions between lentoid cells and other, unrelated cell types. In the heart aggregate experiments, ultrastructural studies are in progress to visualize the heterocellular junctions and to analyze the aggregate-lentoid interface. One interpretation is that MP26 functions in combination with the gap junction proteins of these other cells to form heterocellular, permeable junctions. However, it is also possible that the coupling to lentoid cells is based on the presence of limited lens epithelial-type gap junctions, which disappear during lens fiber development (Schuetze and Goodenough 1982). Freeze-fracture studies on our cultures show that some rather well-developed lentoids retain limited epithelial junctions (with hexagonal particle packing), while others display no obvious junctions of this type. Thus, MP26 must be seriously considered as a protein capable of forming permeable junctions in combination with Novikoff gap junction protein, for example. Additional studies including different immunological approaches are needed to resolve this matter.

Is MP26 a Phosphoprotein?

With the possibility that MP26 forms cell–cell channels in the lens, there is interest in the means by which these channels might be regulated. Since membrane channels in different cells can be regulated by phosphorylation-dephosphorylation cycles (Castellucci et al. 1980), we examined the possibility that MP26 was phosphorylated within the lens fiber cell. We have obtained convincing evidence that, indeed, MP26 is a phosphoprotein. Moreover, we have analyzed various aspects of the lens cAMP-dependent protein kinase system and the phosphorylation of isolated membranes by this enzyme.

When fragments of fetal bovine lenses were incubated in [^{32}P]orthophosphate, we found several lens membrane proteins to be phosphorylated, including MP26 (Johnson and Johnson 1984). Protease studies localized the labeling on MP26 to a region approximately 20 amino acids long, which was separated from the carboxyl terminus by another 20 residues. This labeled region would likely be on the cytoplasmic side, based on the monoclonal B2 localizations described above. We need to determine whether the in vivo labeling of MP26 is stoichiometric. Sequencing studies indicate that six serine residues are found in the labeled region and, in several cases, the surround-

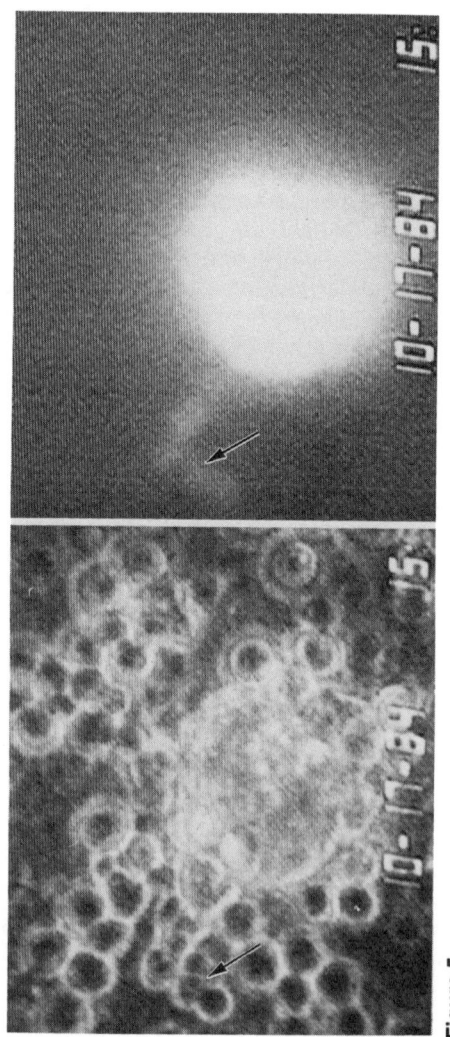

Figure 5
Lucifer Yellow transfer from lentoid to Novikoff hepatoma cells. Photograph of fluorescence image on TV monitor shows clear transfer several minutes following injection into the 100-μm-diameter lentoid (B). In A, a phase-contrast image is included. Note the unusual bend in the chain of Novikoff cells attached to the lentoid cells (arrows). Neither cell type takes up dye from the medium in control experiments. Electrical coupling was also observed between lentoid cells and Novikoff cells.

ing amino acid residues create short sequences (Table 1) which are similar to those "recognized" by cAMP-dependent kinases in other systems (Van Roy et al. 1983). It is noteworthy that two of the serines of interest in the labeled region are in a proposed protein turn (Ser-240 and Ser-243, J.P. Revel, pers. comm.). The predicted turn is based on sequence analysis (Chou and Fasman 1974) and suggests a possible mechanism for modifying channel permeability, whereby the protein turn is altered, as is the association of the carboxyterminal domain with the channel. In terms of other phosphorylatable residues, two threonines are found in the last 12 positions at the carboxylterminus and one tyrosine (position 219) is found right at the membrane surface. It is unlikely that the threonines, based on their locations, can account for the labeling detected in the lens fragments. However, we cannot exclude the phosphorylation of residues other than serine.

In view of these results, it is important to determine convincingly which kinase system(s) is(are) involved in the labeling, in order to evaluate the functional aspects and implications. With the suggestion that serine residues are involved, our initial efforts have been directed to cAMP-dependent protein kinases. We have purified a type-I cAMP-dependent protein kinase from bovine lenses and have used it to phosphorylate isolated lens membranes (Johnson et al. 1985; Louis et al. 1985). MP26 was phosphorylated in the same region and to a similar degree, as in the study with lens fragments above (Fig. 6). The labeling pattern for other lens membrane proteins was similar as well. However, it is not yet known whether cyclic nucleotides influence physiological events involving MP26. We are also examining the possible role of protein kinase C in the phosphorylation of MP26, having found that this enzyme generates labeling comparable to that obtained with the cAMP-dependent kinase (Lampe et al. 1985). Studies on the phosphorylation of MP26 become increasingly important with the suggestion that elevated levels of cytoplasmic cAMP may lead to a rapid permeability decrease in junctions linking horizontal cells in the retina (Gerschenfeld and Neyton; Lasater and Dowling; both this volume).

Table 1
Possible Phosphorylated Serines in MP26

Amino acid position of serine[a]	Sequence
229	Arg-Leu-Lys-_Ser_-Val-Ser
231	Lys-Ser-Val-_Ser_-Glu-Arg
235	Glu-Arg-Leu-_Ser_-Ile-Leu
240	Leu-Lys-Gly-_Ser_-Arg-Pro
243	Ser-Arg-Pro-_Ser_-Glu-Ser
245	Pro-Ser-Glu-_Ser_-Asn-Gly

[a]It has been proposed that in MP26 the last 44 of the 263 amino acid residues reside on the cytoplasmic side of the membrane (Gorin et al. 1984). Of six serine residues found along this length, the first four involve surrounding sequences similar to those recognized by cAMP-dependent protein kinases (Van Roy et al. 1983).

102 R. Johnson et al.

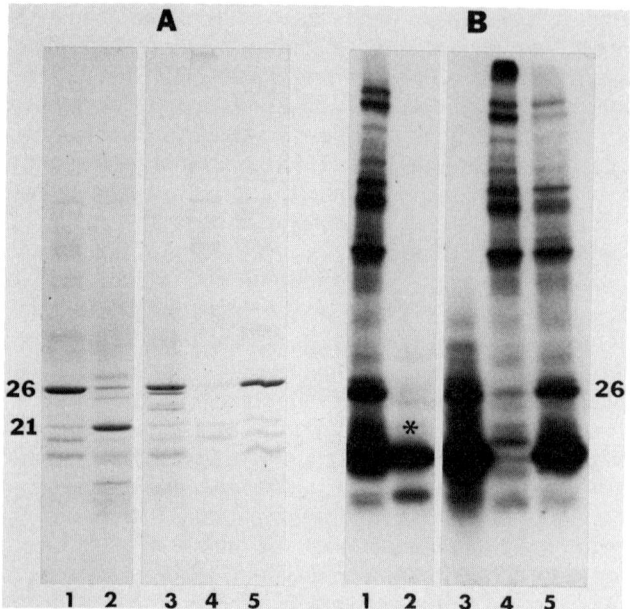

Figure 6
Phosphorylation of bovine lens membranes with a lens cAMP-dependent protein kinase. A protein kinase holoenzyme preparation (in contrast to a preparation of catalytic subunits) was used to label membranes, seen with Coomassie Blue staining in A and as an autoradiograph in B. Lane 1 contains a complete reaction mixture with cAMP, [γ-^{32}P]ATP and control membranes; lane 2 is the same, but the membranes have been treated with chymotrypsin; lane 3, urea treated after labeling; lane 4, sample boiled in SDS after labeling to aggregate MP26; and lane 5, sample rinsed with 0.5% Triton X-100 after labeling. Note that MP26 is phosphorylated and the labeling is not altered by urea or Triton. However, the label at 26,000 M_r is lost with chymotrypsin and reduced with boiling, which is known to aggregate MP26 and keep it from entering the resolving gel. The 21,000 M_r digestion product (*) is not labeled.

CONCLUSION

Lens intercellular junctions have been shown to contain MP26, a major lens membrane protein, by means of indirect labeling at the level of electron microscopy with a monoclonal antibody shown to be specific for MP26. Studies with other monoclonal antibodies, proteases, and low pH-induced junction "splitting" have provided support for the transmembrane nature of MP26. Since several lines of evidence suggest that MP26 is a channel protein, we studied gap junction formation between cultured lens cells and either heart or hepatoma cells. Differentiated lens fiber-like cells were able to form junctions in these instances, as evidenced by the indirect acquisition of synchrony in heart aggregate beating on lens cell surfaces or the transfer of dye from lens to hepatoma cells. Additional studies are required, but these efforts indicate that MP26 may function in combination with gap junction proteins in the hepatoma cells, for example, to produce permeable junctions. That is, MP26 must be seriously considered as a lens gap junction protein. To evaluate potential regulatory schemes for MP26, we demonstrated the phosphorylation of MP26 within lens fragments and also showed that lens cAMP-dependent protein kinases were able to phosphorylate MP26. Thus, a variety of data indicates that MP26 may be a component of cell–cell channels in the lens, perhaps lens gap junction channels, and that this protein is regulated in some way by means of phosphorylation.

ACKNOWLEDGMENTS

We acknowledge Kae Ebling for processing the manuscript, Luke Nguyen for the photography, and Judson Sheridan and Ron Hybertson for comments on the manuscript. This work was supported by NIH grants CA28548 (to R.J.), CA16335 (to J.S. and R.J.), and EYO5011 (to C.L.).

REFERENCES

Benedetti, E.L., E. Dunia, C.J. Bentzel, A.J.M. Vermorken, M. Kibbelaar, and H. Bloemendal. 1976. A portrait of plasma membrane specializations in eye lens epithelium and fibers. *Biochem. Biophys. Acta* **457**: 353–384.

Bloemendal, H., ed. 1981. *Molecular and cellular biology of the eye lens.* Wiley, New York.

Bok, D., I. Dockstader, and J. Horwitz. 1982. Immunocytochemical localization of the lens main intrinsic polypeptide (MP26) in communicating junctions. *J. Cell Biol.* **92**: 213–220.

Castellucci, V., E. Kandel, J. Schwartz, F. Wilson, A. Nairn, and P. Greengard. 1980. Intracellular injection of the catalytic subunit of cyclic AMP-dependent protein kinase simulates facilitation of transmitter release underlying behavioral sensitization in *Aplysia. Proc. Natl. Acad. Sci.* **77**: 7492–7496.

Chou, P.Y. and G.D. Fasman. 1974. Prediction of protein conformation. *Biochemistry* **13**: 222–245.

Dermietzel, R., A. Leibstein, U. Frixen, U. Janssen-Timmen, O. Traub, and K. Willecke. 1984. Gap junctions in several tissues share antigenic determinants with liver gap junctions. *EMBO J.* **3**: 2261–2270.

Engvall, E. and P. Perlmann. 1972. Enzyme-linked immunosorbent assay, ELISA. III. Quantitation of specific antibodies by enzyme-labeled anti-immunoglobulin in antigen-coated tubes. *J. Immunol.* **109**: 129–135.

Epstein, M.L. and N.B. Gilula. 1977. A study of communication specificity between cells in culture. *J. Cell Biol.* **75**: 769–787.

Finbow, M.E., T.E.J. Buultjens, N. Lane, J. Shuttleworth, and J. Pitts. 1984. Isolation and characterization of arthropod gap junctions. *EMBO J.* **3**: 2271–2278.

Finer-Moore J. and R.M. Stroud. 1984. Amphipathic analysis and possible formation of the ion channel in an acetylcholine receptor. *Proc. Natl. Acad. Sci.* **81**: 155–159.

Gorin, M.B., S.B. Yancey, J. Cline, J.P. Revel, and J. Horwitz. 1984. The major intrinsic protein (MIP) of the bovine lens fiber membrane: Characterization and structure based on cDNA cloning. *Cell* **39**: 49–59.

Goshima, K. and Y. Tonomura. 1969. Synchronized beating of embryonic mouse myocardial cells mediated by FL cells in monolayer culture. *Exp. Cell Res.* **56**: 387–392.

Griepp, E.B. and M. Bernfield. 1978. Acquisition of synchronous beating between embryonic heart cell aggregates and layers. *Exp. Cell Res.* **113**: 263–272.

Gros, D.B., B.J. Nicholson, and J.-P. Revel. 1983. Comparative analysis of the gap junction protein from rat heart and liver: Is there a tissue specificity of gap junctions? *Cell* **35:** 539-549.

Hertzberg, E.L. and R.V. Skibbens. 1984. A protein homologous to the 27,000 dalton liver gap junction protein is present in a wide variety of species and tissues. *Cell* **39:** 61-69.

Hertzberg, E.L., D.J. Anderson, M. Friedlander, and N.B. Gilula. 1982. Comparative analysis of the major polypeptides from liver gap junctions and lens fiber junctions. *J. Cell Biol.* **92:** 53-59.

Johnson, K.R. and R.G. Johnson. 1984. MP26 is phosphorylated in bovine lens cells. In *International cell biology 1984* (ed. S. Seno and Y. Okada), p. 314. Academic Press, Japan.

Johnson, K.R., S.S. Panter, and R.G. Johnson. 1985. Phosphorylation of lens membranes with a cyclic AMP-dependent protein kinase purified from the bovine lens. *Biochim. Biophys. Acta* (in press).

Johnson, R., W. Herman, and D. Preus. 1973. Homocellular and heterocellular gap junctions in *Limulus:* A thin-section and freeze-fracture study. *J. Ultrastruct. Res.* **43:** 298-312.

Keeling, P., K.R. Johnson, D.F. Sas, K. Klukas, P. Donahue, and R.G. Johnson. 1983. Arrangement of MP26 in lens junctional membranes: Analysis with proteases and antibodies. *J. Membr. Biol.* **74:** 217-228.

Kuszak, J., Y.K. Shek, K.C. Carney, and J.L. Rae. 1984. Comparative analysis of crystalline lens gap junctions. In *Proceedings of the 42nd Annual Meeting of the EM Society of America* (ed. G.W. Bailey), p. 114-117. San Francisco Press, California.

Lampe, P., M. Bazzi, G. Nelsestuen, and R. Johnson. 1985. Phosphorylation of bovine lens membrane proteins by protein kinase C. *Fed. Proc.* **44:** 1225.

Louis, C.F., R. Johnson, K. Johnson, and J. Turnquist. 1985. Characterization of the bovine lens plasma membrane substrates for cAMP-dependent protein kinase. *Eur. J. Biochem.* (in press).

Menko, A.S., K.A. Klukas, and R.G. Johnson. 1984. Chicken embryo lens cultures mimic differentiation in the lens. *Dev. Biol.* **103:** 129-141.

Menko, S., K. Klukas, B. Quade, T.-F. Liu, and R. Johnson. 1982. Lens gap junctions in differentiating cultures of chick embryo lens cells. *J. Cell Biol.* **95:** 106a.

Michalke, W. and W.R. Loewenstein. 1971. Communication between cells of different types. *Nature* **232:** 121-122.

Nikaido, H. and E.Y. Rosenberg. 1984. Functional reconstitution of lens junction protein into proteoliposomes. In *Proceedings of the 42nd annual meeting of the EM society of America* (ed. G.W. Bailey), p. 130-131. San Francisco Press, San Francisco.

Paul, D.L. and D.A. Goodenough. 1983. Preparation, characterization and localization of antisera against bovine MP26, an integral protein from lens fiber plasma membrane. *J. Cell Biol.* **96:** 625-632.

Peracchia, C. 1980. Structural correlates of gap junction permeation. *Int. Rev. Cytol.* **66:** 81-146.

Rae, J.L. 1979. The electrophysiology of the crystalline lens. *Curr. Top. Eye Res.* **1:** 37-90.

Rae, J.L., R.D. Thomson, and R.S. Eisenberg. 1982. The effect of 2,4-dinitrophenol on cell to cell communication in the frog lens. *Exp. Eye Res.* **35:** 597-609.

Sas, D.F. and R.G. Johnson. 1983. Lens intercellular junctions are split by low pH treatment. *J. Cell Biol.* **97:** 84a.

Sas, D.F., J. Sas, K.R. Johnson, A.S. Menko, and R.G. Johnson. 1985. Junctions between lens fiber cells are labeled with a monoclonal antibody shown to be specific for MP26. *J. Cell Biol.* **100:** 216-225.

Schuetze, S.M. and D.A. Goodenough. 1982. Dye transfer between cells of the embryonic chick lens becomes less sensitive to CO_2 treatment with development. *J. Cell Biol.* **92:** 694–705.

Spray, D.C. and M.V.L. Bennett. 1985. Physiology and pharmacology of gap junctions. *Annu. Rev. Physiol.* **47:** 281–303.

Takemoto, L.J., J.S. Hansen, B.J. Nicholson, M.W. Hunkapiller, J.-P. Revel, and J. Horwitz. 1983. Major intrinsic polypeptide (MIP) of lens membrane: Biochemical and immunological characterization of the major cyanogen bromide fragment. *Biochem. Biophys. Acta* **731:** 267–274.

Traub, O. and K. Willecke. 1982. Cross reaction of antibodies against liver gap junction protein (26 K) with lens fiber junction protein (MIP) suggests structural homology between tissue specific gene products. *Biochem. Biophys. Res. Commun.* **109:** 895–901.

Van Roy, F., L. Fransen, and W. Fiers. 1983. Improved localization of phosphorylation sites in simian virus 40 large T antigen. *J. Virol.* **45:** 315–331.

Zampighi, G., S. Simon, J. Robertson, T. McIntosh, and M. Costello. 1982. On the structural organization of isolated bovine lens fiber junctions. *J. Cell Biol.* **93:** 175–189.

Antibody against Liver Gap Junction 27-kD Protein Is Tissue Specific and Cross-reacts with a 54-kD Protein

David L. Paul
Department of Anatomy
Harvard Medical School
Boston, Massachusetts 02115

Gap junctions are present between cells in many tissues and in a wide range of species (Bennett and Goodenough 1978). The gap junctions in these tissues all share a basic structural similarity but exhibit a variation in their detailed morphologies, as revealed by electron microscopy. Thus the question arises: Are the junctions in these different tissues composed of the same, similar, or completely different protein components?

One approach to this question has been to develop procedures by which gap junctions can be isolated from different tissues and compared directly. It has been possible to isolate gap junctions directly from a few tissues and species: mouse and rat liver (Goodenough and Stoeckenius 1972; Henderson et al. 1979; Finbow et al. 1980); mouse, rat, and rabbit heart (Kensler and Goodenough 1980; Manjunath et al. 1982; Gros et al. 1983); and bovine lens fibers (Alcala et al. 1975; Goodenough 1979).

Detailed biochemical comparisons have been performed on peptides enriched by these procedures. Hepatocyte gap junction preparations contain a low number of protein components which are dominated by a single protein with reported molecular weights of 26–29 kD. Lens fiber membranes display a 26-kD protein, called MP26, as the major component. The identity of the MP26 as a structural component of the fiber junction is a matter of controversy (see Bok et al. 1982; Paul and Goodenough 1983a). Peptide mapping of liver 27 kD and lens MP26 reveals no obvious homology (Hertzberg et al. 1982; Gros et al. 1983) and this finding is supported by partial amino acid sequencing (Nicholson et al. 1983). In addition, antisera and monoclonal antibodies generated against MP26 show no affinity for liver 27-kD protein (Hertzberg et al. 1982 but see also Traub and Willecke 1982; Fitzgerald et al. 1983; Paul and Goodenough 1983a). Myocardial gap junctions have been isolated by several investigators and a number of different major proteins have been reported. These include 30–34 kD (Kensler and Goodenough 1980), 27 kD (Hertzberg and Skibbens 1984), 45 kD and 29 kD (see Page, this volume), and 29 kD (Gros et al. 1983). The heart 29 kD has been peptide mapped and is reported to show no homology with the hepatic 27-kD protein (Gros et al. 1983).

In this study, we have used antibodies to identify homologies between protein components of gap junctions in tissues where direct isolation is not practical. Antibodies were generated against the 27-kD protein associated with isolated gap junctions from calf liver. Immunohistochemistry revealed that some but not all other tissues exhibited proteins related serologically to the liver 27-kD protein. In tissues where labeling was observed, the location of the label was consistent with the distribution of gap junctions. Immunoblotting and in vitro synthesis showed that the liver 27-kD protein was probably synthesized as a 54-kD precursor.

RESULTS

Preparation and Characterization of Antibodies

Gap junctions from calf liver were prepared by a modification of the alkali extraction technique described by Hertzberg (1984). Tissue was subjected to several freeze-thaw cycles before dispersal in a Tissue-Tek homogenizer. The membrane pellet was treated with 0.5 mM diisopropyl fluorophosphate (DFP) just prior to NaOH washing. Purified 27-kD protein was obtained from the isolated junctions by preparative SDS-PAGE. The protein profiles of isolated gap junctions and purified 27K protein are shown in Figure 1 (lanes 2 and 4). The

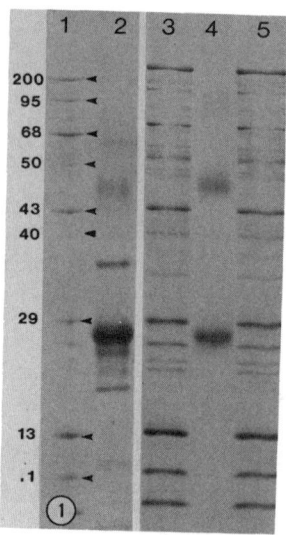

Figure 1
Coomassie Blue-stained SDS-PAGE of samples used to produce and purify antisera. (Lane 1) Molecular-weight standards: myosin (200K), phosphorylase A (95 kD), bovine serum albumin (68 kD), G globulin (50 kD), actin (43 kD), aldolase (40 kD), carbonic anhydrase (29 kD), RNase (13 kD), cytochrome c (11 kD). (Lane 2) Isolated calf hepatic gap junctions. (Lane 3) Standards. (Lane 4) Gel-purified 27-kD protein used for affinity purification. Higher-molecular-weight bands are aggregates caused by dissolving in SDS. (Lane 5) Standards.

higher-molecular-weight band at about 48 kD (lane 4) is an aggregate of the 27 kD that occurs upon heating or concentrating the protein in SDS solution (Henderson et al. 1979).

Antiserum was raised conventionally in rabbits by injecting whole isolated junctions. Anti-27-kD antibodies were produced by affinity purification of the serum against isolated 27-kD protein coupled to Sepharose (see Paul and Goodenough 1983a for details). The serum and the affinity-purified antibodies were characterized by immunoblot in Figure 2. Lanes 1–5 show the protein profile of the samples used to produce the blot. These are whole-tissue homogenates from kidney, liver, heart, and pancreas (lanes 1–4) and isolated rat liver junctions (lane 5). The crude antiserum labeled both 27-kD and 48-kD polypeptides in the isolated liver junctions (lane 10) but also labeled a large number of proteins in each of the tissue homogenates (lanes 6–9). The affinity-purified anti-27-kD antibody labeled the 27 kD, the 48-kD dimer, and a couple of lower-molecular-weight bands which are presumed breakdown products of the 27 kD (lane 15). All labeling in the tissue homogenates was eliminated

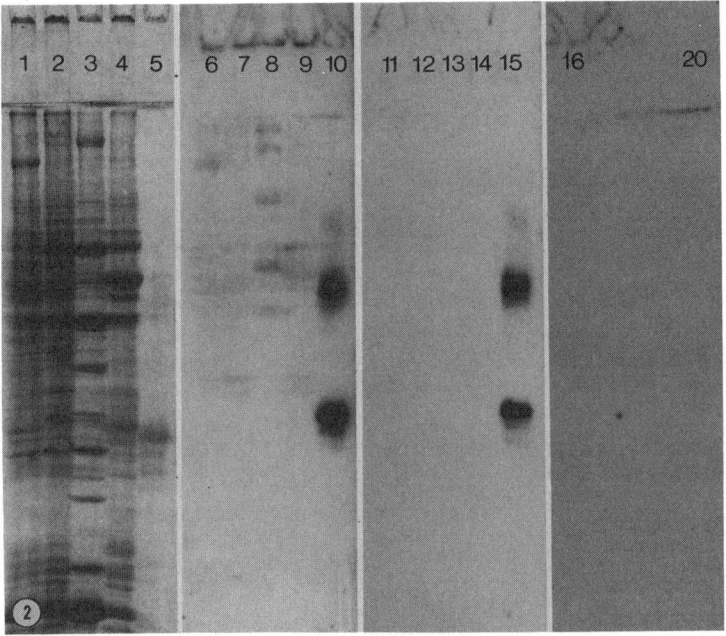

Figure 2
Characterization of specificity of anti-27-kD antibody by immunoblot. (Lanes 1–5) Coomassie-stained SDS-PAGE containing whole-tissue homogenates from liver (lane 1), kidney (lane 2), heart (lane 3), pancreas (lane 4), and isolated rat liver gap junctions (lane 5). (Lanes 6–10) Immunoblot of lanes 1–5 labeled with antiserum produced by injecting isolated hepatic gap junctions. (Lanes 11–15) Immunoblot of lanes 1–5 labeled with same antiserum after affinity purification using SDS-PAGE-purified 27-kD protein. (Lanes 16–20) Immunoblot of lanes 1–5 using preimmune serum. Binding of antibody to blot is detected by ^{125}I-labeled Protein A.

by affinity purification (lanes 10–14). Preimmune Ig showed no labeling of any proteins (lanes 16–20).

Immunolocalization of 27K Antigen in Liver

As a further characterization, anti-27K antibodies were used for immunohistochemistry on frozen sections of paraformaldehyde-fixed rat liver (Fig. 3). The primary antibody was visualized with a rhodamine-labeled, goat anti-rabbit secondary antiserum. The staining consisted of numerous punctate or thread-like regions of intense fluorescence outlining the lateral aspects of the cells. This staining was consistent with the size, shape, and distribution of gap junctions in this tissue as revealed by numerous studies by electron microscopy (EM). Localization of antibody binding was used to demonstrate that gap junctions were responsible for the staining pattern observed at the level of light microscopy. The EM study was performed on crude plasma membranes isolated from rat liver. Anti-27-kD antibody binding was detected using colloidal gold coupled to a goat anti-rabbit (Fig. 4A). Gold particles were observed only on the cytoplasmic aspects of the gap junctional regions (arrows). No specific staining was detectable on either cytoplasmic or extracellular surfaces of nonjunctional membranes (arrowheads). Preimmune Ig showed a low, nonspecific background of gold (Fig. 4B).

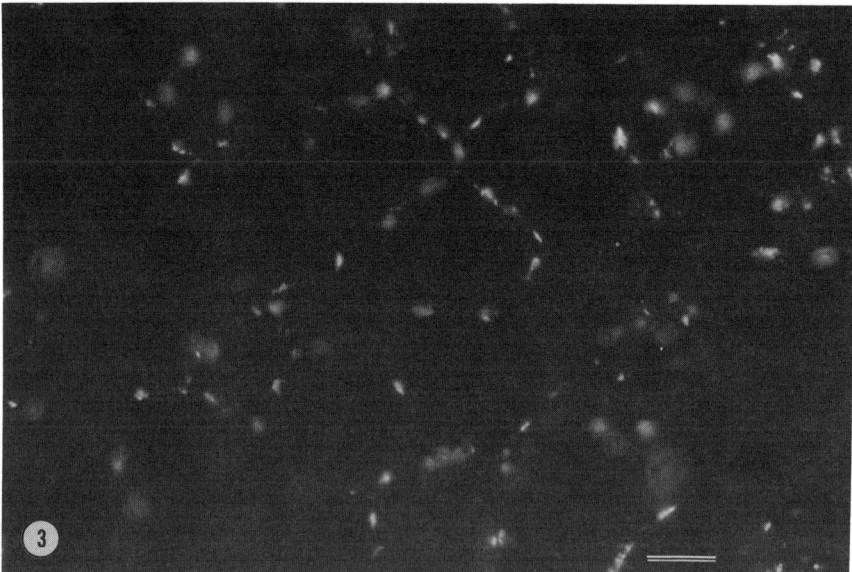

Figure 3
Immunofluorescence localization of 27-kD related antigen in formaldehyde-fixed frozen sections of rat liver. Specific labeling consisted of punctate or thread-like areas along the lateral aspects of the cell borders (arrowheads). Bar, 10 μm.

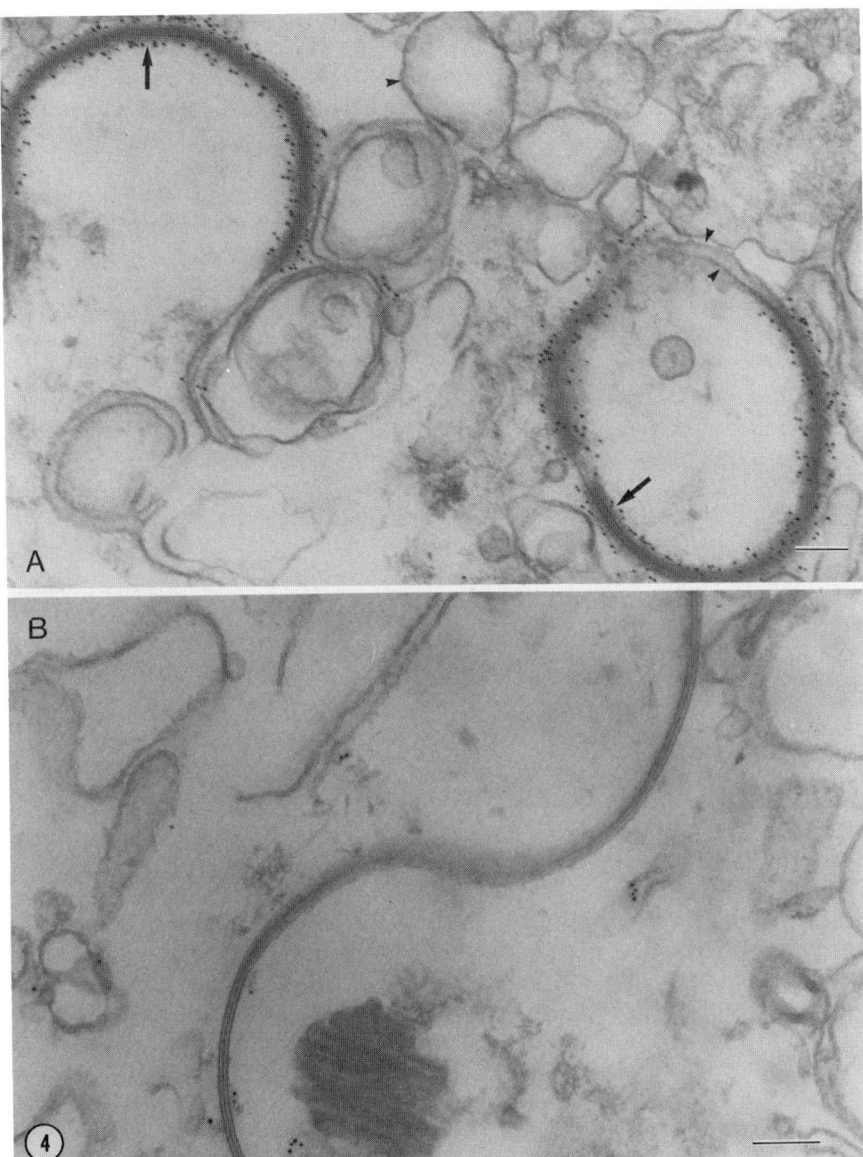

Figure 4
Localization by electron microscopy of 27-kD related antigen in rat liver plasma membranes. (A) Membranes stained with anti-27 kD followed by colloidal gold-conjugated goat anti-rabbit. Staining is observed only on the cytoplasmic surfaces of junctional membrane (arrows). No specific stain is detected on nonjunctional membranes (arrowheads). (B) Preimmune control. Bar, 100 nm.

Anti-27-kD Antibody Is Tissue Specific

Anti-27-kD antibody was used to survey a panel of tissues for the presence of cross-reacting material. In each case, formaldehyde-fixed frozen sections of tissue were stained for immunofluorescence as before. Rat pancreas (Figs. 5A,B) displayed both endocrine and exocrine tissue; the exocrine acinar cells showed numerous punctate regions of fluorescence reminiscent of liver; and the endocrine cells showed no specific labeling.

In rat myocardial tissue, gap junctions are most abundant at the intercalated disks (McNutt and Weinstein 1970). Although the intercalated disks were very poorly resolved in our phase-contrast micrograph (Fig. 6A), rat myocardium that was longitudinally sectioned in this manner displayed no detectable labeling on structures resembling intercalated disks or along cell borders (Fig. 6B). Some faint staining was observed to correspond to Z lines, but the nature of this staining was not known.

In rat stomach (Figs. 7A,B), the surface mucous cells lining the gastric pits were labeled strongly by the anti-27-kD antibody. When the plane of section was favorable (long arrows), the labeling was seen to consist of punctate or thread-like regions of fluorescence that outlined cell borders. The epithelia of the neck regions of the gastric glands that underlie the pits showed no detectable labeling (short arrows).

Frozen sections of rat lens, cerebellum, medullary kidney, and intestinal smooth muscle were also examined by immunofluorescence. No specific labeling was detected in any of these tissues (data not shown).

Monoclonal Antibody Reacts Only with Liver Gap Junctions

A monoclonal antibody (Mc83A) that bound to liver gap junctions was produced. Mc83A was obtained by immunizing Balb/c mice with intact isolated rat liver junctions. The schedule of immunization and the cell fusion technique were according to Hudson and Hay (1980). Antibodies against junctions were selected by their pattern of immunofluorescent labeling on frozen sections of unfixed rat liver. Mc83A exhibited a pattern of immunofluorescent labeling indistinguishable from that of affinity-purified anti-27 kD (data not shown).

The localization of Mc83A binding sites in isolated rat liver membranes was visualized using colloidal gold immunocytochemistry (Fig. 8A,B). Similarly to affinity-purified anti-27-kD antibody (Fig. 4), specific binding was observed only on the cytoplasmic surfaces of the gap junctions. Mc83A did not label any nonjunctional membrane above background. Since Mc83A did not react with SDS-denatured proteins, immunoblotting from SDS gels could not be used to identify the antigen. However, the distribution of this antigen in rat liver was identical to that observed for 27-kD protein using affinity-purified antibody. Since few if any proteins not related to the 27 kD are present in preparations of isolated junctions, it is reasonable to suggest that the antigen recognized by Mc83A may be the 27-kD protein. Regardless, the distribution of this antigen makes it very likely that it is a component of the hepatic gap junction.

Mc83A reacted with liver from several species (mouse, rat, human). How-

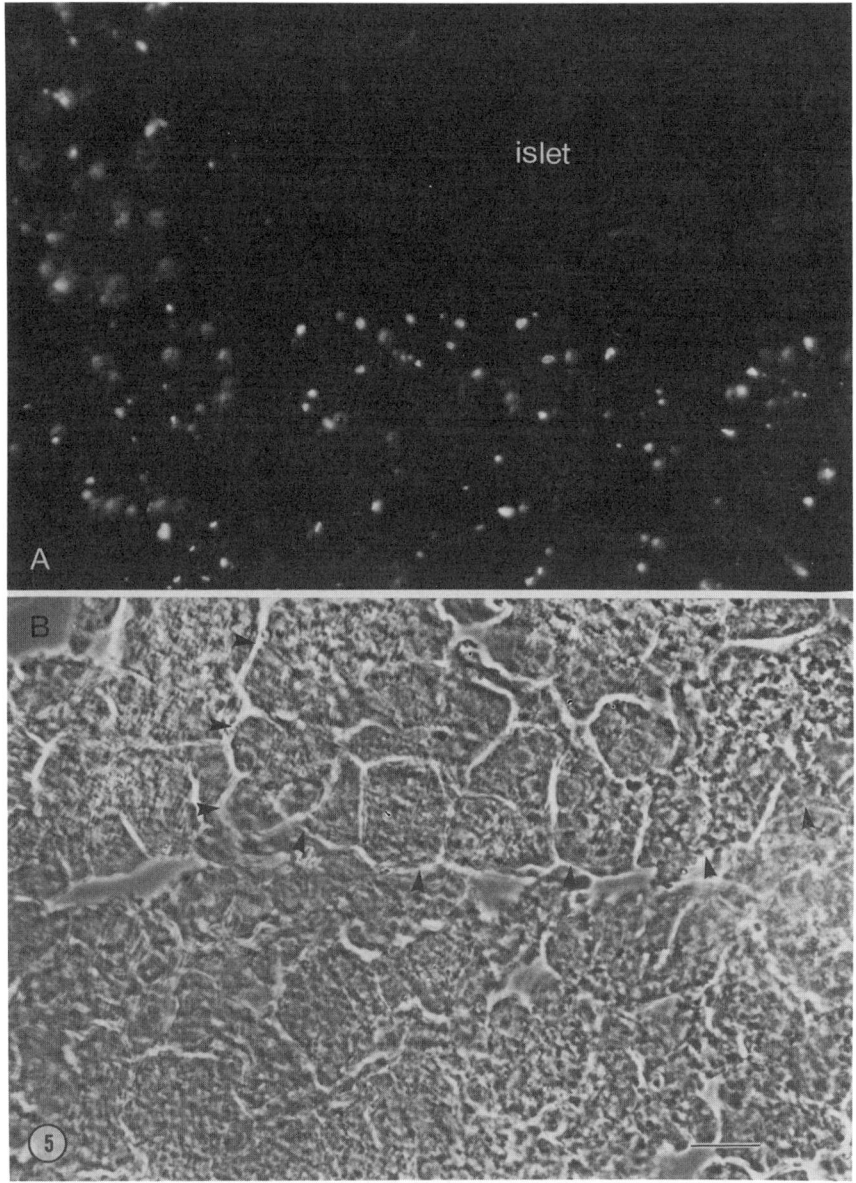

Figure 5
Immunofluorescence localization of antigens related to liver 27 kD in rat pancreas. (A) Section was stained with affinity-purified anti-27-kD antibody followed by rhodamine-conjugated goat anti-rabbit. Specific immunofluorescence consisted of punctate areas that outlined acinar cells in exocrine tissue. No specific labeling was observed in endocrine tissue (islet). (B) Phase-contrast micrograph of same field of view. Edge of islet is indicated with arrowheads. Bar, 10 μm.

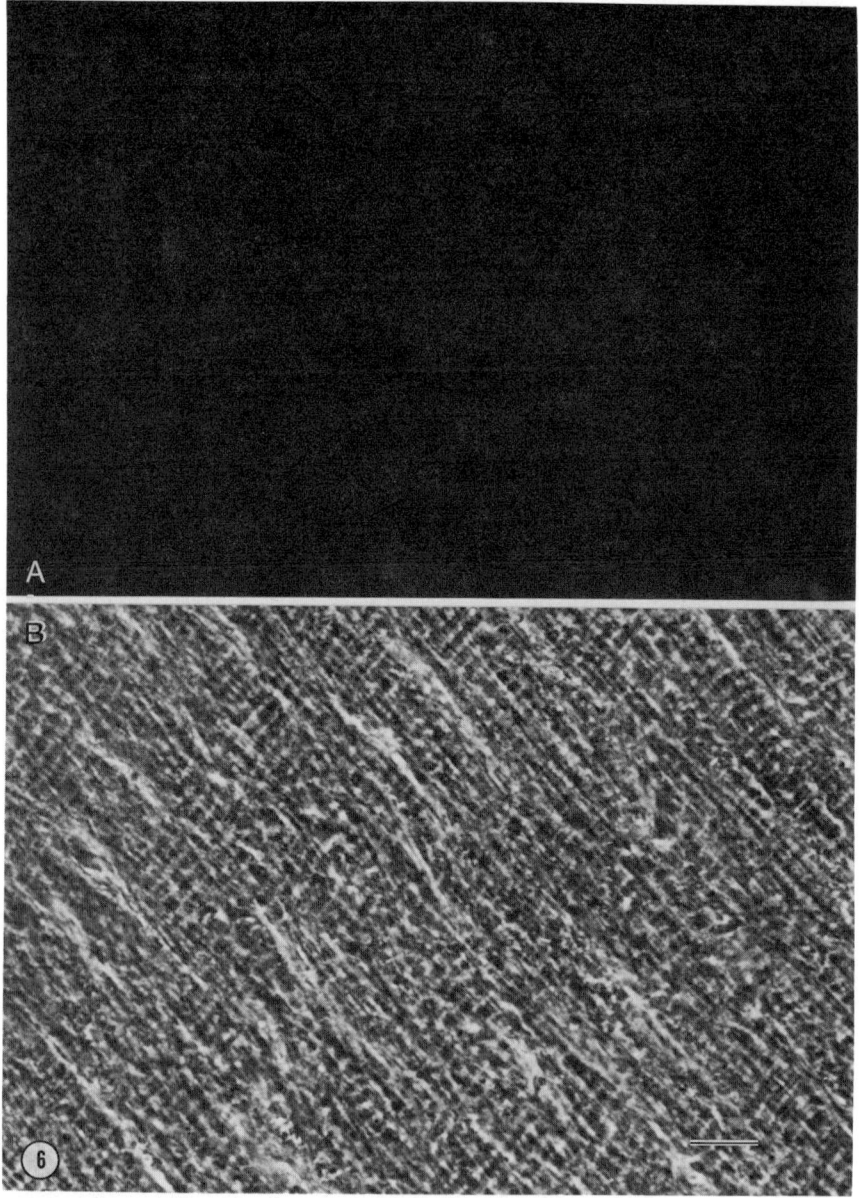

Figure 6
Immunofluorescence localization of antigens related to 27 kD in rat myocardium. (A) No specific fluorescence labeling was observed in this tissue. (B) Phase-contrast micrograph of same section. Bar, 10 μm.

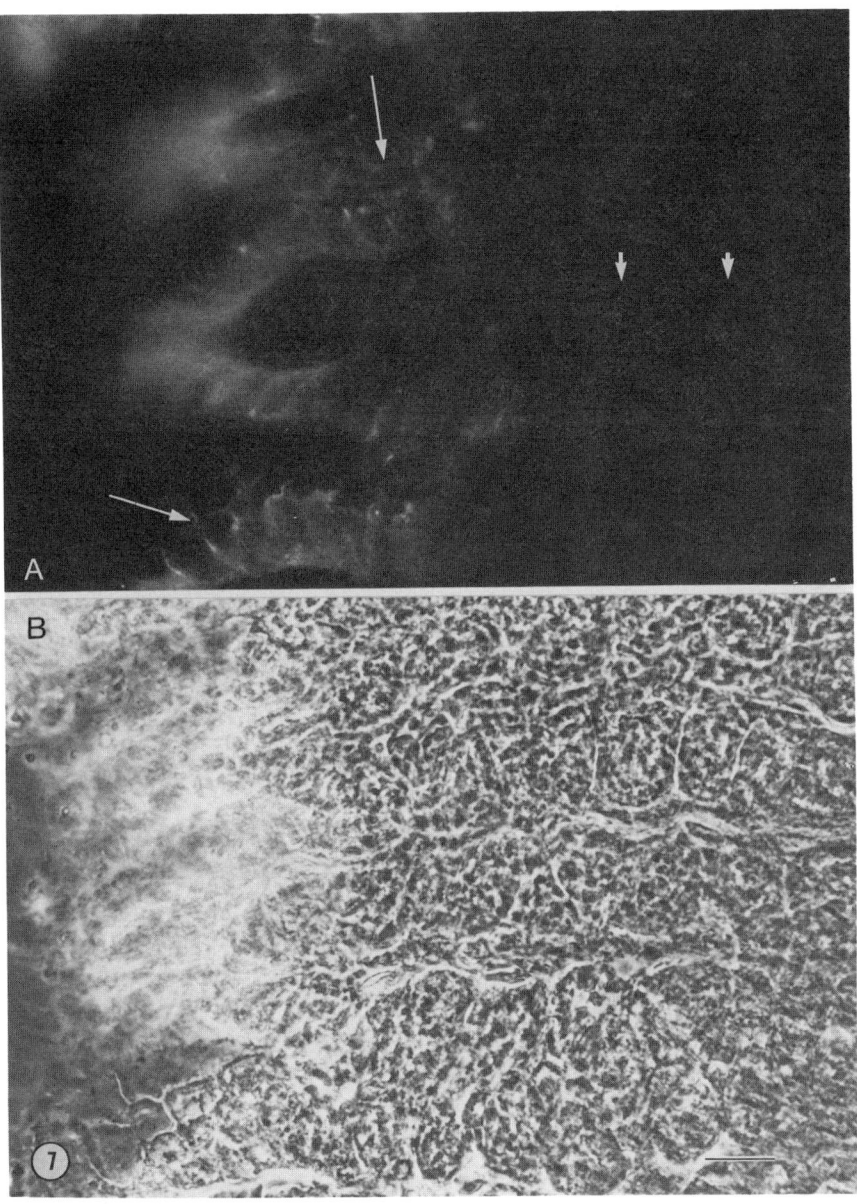

Figure 7
Immunofluorescence localization of antigens related to 27 kD in rat stomach. (A) Labeling by anti-27 kD antibody is confined to the surface mucous cells, which outlined cell borders when the plane of section was favorable (long arrows). The tubular glands exhibited no detectable labeling (short arrows). (B) Phase-contrast micrograph of same section. Bar, 10 μm.

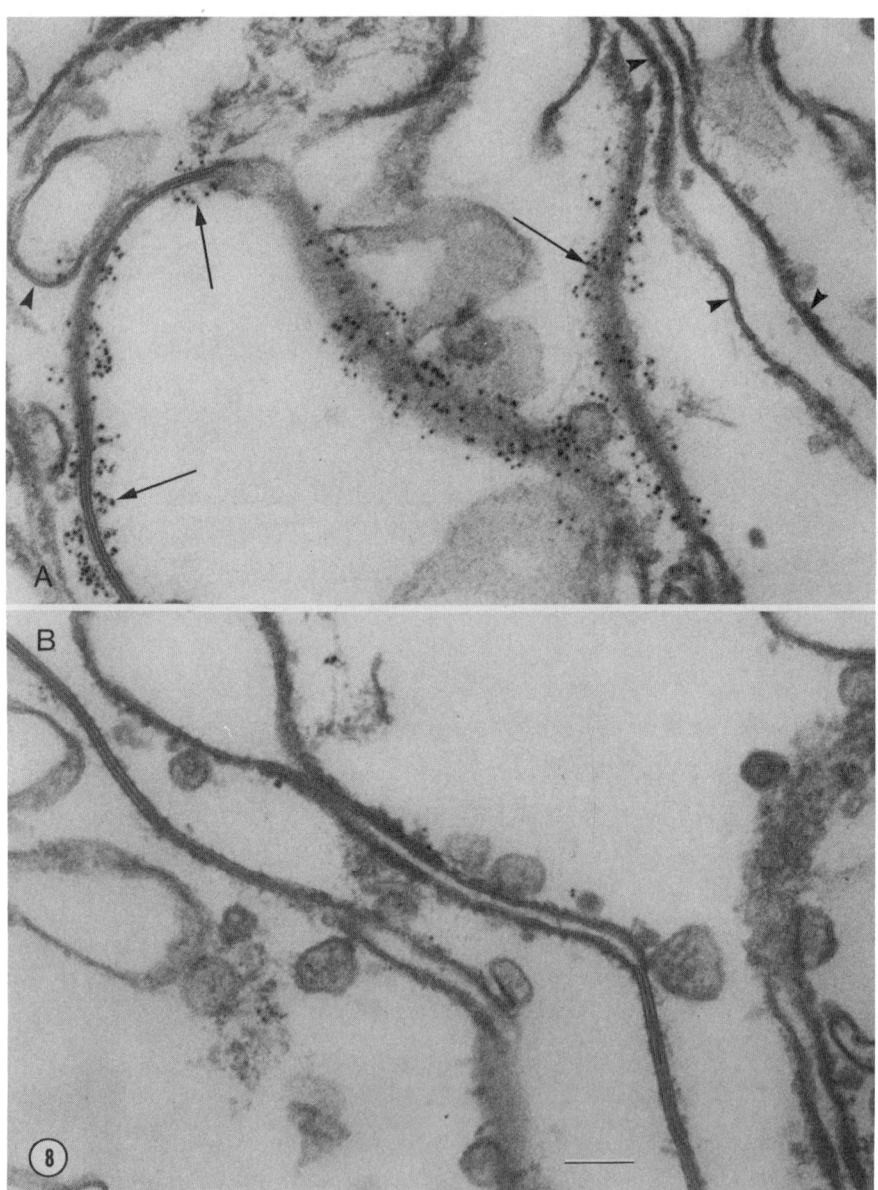

Figure 8
Localization by electron microscopy of monoclonal antibody (Mc83A) binding sites in rat liver plasma membranes. (A) Membranes stained with Mc83A followed by colloidal gold-conjugated goat anti-mouse. Staining was observed only on the cytoplasmic surfaces of junctional membrane (arrows). No specific stain was detected on nonjunctional membranes (arrowheads). (B) Nonimmune monoclonal IgM control. Bar, 100 nm.

ever, it did not react with any of the other tissues used to screen the affinity-purified antibody (lens, brain, kidney, pancreas, heart, stomach). This result suggested that the hepatic gap junction may possess at least one antigenic determinant that is completely unique to that tissue.

Anti-27-kD Antibody Cross-reacts with a 54-kD Protein in Whole Tissues

The immunofluorescence results suggested that rat pancreas and stomach contained proteins serologically related to the liver 27 kD, whereas rat heart did not. It should have been possible, therefore, to identify the gap junction proteins produced by pancreas and stomach by immunoblotting techniques. In pancreas, however, no cross-reacting proteins were detected in previous immunoblots (Fig. 2, lane 14). In our hands, the liver 27-kD protein was sensitive to endogenous proteases released during the isolation procedures (Fallon and Goodenough 1981). Therefore, measures to control proteolysis were employed before continuation of immunoblot analysis. Whole tissue was frozen immediately after dissection in liquid Freon cooled with liquid nitrogen. The tissue was pulverized under liquid nitrogen with mortar and pestle, then lyophilized. The dehydrated, powdered tissue was prepared for SDS-PAGE by resuspending in boiling SDS sample buffer freshly treated with DFP. Insoluble material was removed by centrifugation (10,000g × 10 min) and the supernatant was applied immediately to the gel.

Immunoblot analysis of total tissue homogenates of pancreas and heart prepared in this manner is shown in Figure 9. A Coomassie Blue-stained gel of

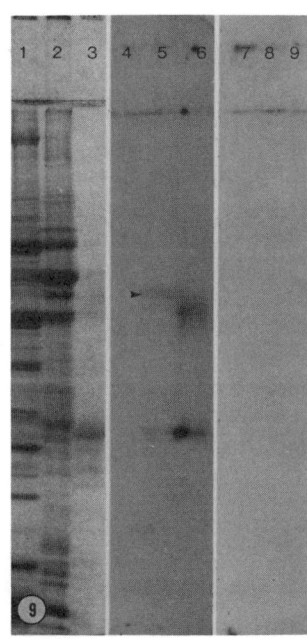

Figure 9
Immunoblot of whole-tissue homogenates from heart and pancreas prepared with proteolysis inhibition. (Lanes 1–3) Representative (see text) Coomassie-stained SDS-PAGE of samples used to prepare the immunoblot which were heart (lane 1), pancreas (lane 2), and isolated rat liver gap junctions (lane 3). (Lanes 4–6) Immunoblot labeled with anti-27-kD antibody. No bands were detected in heart. Bands were detected at 27 kD and 48 kD in both pancreas and isolated gap junction. A major band at 54 kD was detected in pancreas only (arrowhead, lane 5). (Lanes 7–9) Preimmune IgG labeled no bands on the blot.

whole heart (lane 1), whole pancreas (lane 2), and isolated rat liver gap junctions (lane 3) is provided for reference only. The gel that was used for transfer to nitrocellulose actually was loaded with 3× more heart and pancreas sample and 225× less gap junction sample than displayed in the Coomassie-stained gel. The labeling produced by the anti-27-kD antibody is shown in lanes 4–6. No specific labeling of any band was detected in the heart homogenate (lane 4). In pancreas homogenate, specifically labeled bands were detected at 27 kD, 48 kD, and 54 kD (lane 5). The 27-kD and 48-kD bands comigrated with 27 kD and 48 kD in liver gap junctions (lane 6). The 54-kD band was the most abundantly labeled band in the pancreas homogenate. No corresponding band has been reported in isolated junctions from any tissue. Preimmune IgG produced no specific labeling of any bands (7–9). A 54-kD protein was also labeled in homogenates of similarly prepared rat liver (data not shown).

In Vitro Synthesis of Liver Gap Junction Protein

The nature of the 54-kD cross-reacting protein detected by immunoblotting was investigated using in vitro synthesis. Poly(A) RNA from rat liver was translated in a reticulocyte lysate cell-free system in the presence or absence of dog pancreatic microsomes. Translation mixtures were immunoprecipitated with anti-27-kD antibody or preimmune IgG, and the precipitate was displayed on SDS gels and visualized by fluorography (see Paul and Goodenough 1983b for details).

The total products of rat liver mRNA translation are shown in Figure 10 (lane 1). This mixture was precipitated with preimmune IgG (lane 2), anti-27-kD antibody (lane 3), and anti-27-kD antibody plus 5 μg of cold, gel-purified 27-kD protein (lane 4). Anti-27-kD antibody specifically immunoprecipitated a major product of about 54-kD and a minor product migrating at about 48 kD (lane 3). The preimmune IgG precipitated no detectable products (lane 2). Gel-purified 27-kD protein (see Fig. 1, lane 4) completely competed the specific immunoprecipitation of both the 54-kD and 48-kD bands (lane 4). This suggested that both the 48-kD and 54-kD proteins were related serologically to the 27-kD gap junction protein.

To investigate the mechanism of membrane integration of the 54-kD protein, in vitro synthesis was performed in the presence of dog pancreatic microsomes (lanes 5–7). These microsomes have been shown to support the in vitro membrane insertion of a variety of membrane proteins (Sabatini et al. 1982). The addition of microsomal membranes to the translation mix resulted in a decrease in the amount of 54 kD and a comcomitant increase in the amount of the 48 kD (lane 5). Sucrose density gradient centrifugation was used to test the membrane association of the newly synthesized proteins. Most of the 48-kD and 54-kD proteins cosedimented with the microsomal membranes (lane 6), while a small amount of both proteins remained with the soluble components of the lysate at the top of the gradient (lane 7). If microsomes added after synthesis had been completed (1 hr) were subjected to gradient isolation, the 48-kD and 54-kD proteins remained at the top of the gradient and did not cosediment with the membranes (data not shown). These data suggested that

Figure 10
In vitro synthesis of rat liver gap junction protein. Immunoprecipitated protein was analyzed by SDS-PAGE and fluorography. (Lane 1) Total products of translation of rat liver poly(A) RNA. (Lane 2) Immunoprecipitation of total products by preimmune IgG. (Lane 3) Immunoprecipitation by anti-27-kD antibody. A major product was observed at 54 kD and a minor product at 48 kD. (Lane 4) Immunoprecipitation by anti-27-kD antibody in the presence of cold, gel-purified 27-kD protein; 27-kD protein effectively competed the specific immunoprecipitation of both the 54-kD and 48-kD species. (Lane 5) In vitro synthesis and immunoprecipitation in the presence of dog pancreatic microsomes. The amount of 54-kD protein was reduced concomitant with an increase in the amount of the 48-kD species. (Lanes 6-7) Liver mRNA was translated as in lane 5 in the presence of microsomes that were separated from soluble components of the lysate on a sucrose gradient after the completion of synthesis. Microsomes were separately immunoprecipitated (lane 6) from the remainder of the lysate (lane 7).

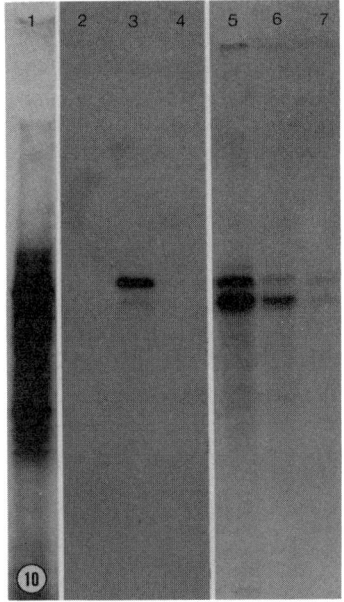

the 48-kD and 54-kD proteins became membrane associated when the microsomes were present at the onset of translation.

DISCUSSION

We have shown that antibodies produced against the 27-kD protein associated with hepatic gap junctions bound only to gap junctions in that tissue and not to nonjunctional membranes of any sort. Therefore the antibodies were used to detect the presence of serologically related proteins in other tissues. When this was investigated using immunohistochemistry, it was found that cross-reacting material was detected in some tissues but not others. In the pancreas, we observed labeling of structures in acinar cells with a distribution consistent with that of gap junctions. However, we observed no detectable labeling in islets. Gap junctions have been detected in both exocrine and endocrine pancreas by electron microscopy and both ionic coupling and dye transfer have been observed (Friend and Gilula 1972; Meda et al. 1979). However, endocrine pancreatic junctions are smaller and less numerous than junctions between acinar cells. Thus, there were several possible explanations for our failure to detect immunofluorescent staining of pancreatic islet cell junctions: (1) The junction may have been too small to detect by immunofluorescence; (2) the islet junctions could have been destroyed differentially by the fixation, sectioning, or staining procedures; and (3) the islet junctions could have been composed of protein(s) serologically distinct from the liver 27 kD.

The same set of possibilities may have accounted for our observations in antibody-labeled sections of stomach. Although the junctions joining surface mucous cells are large and numerous (Claude and Goodenough 1973; S. Ito, pers. comm.), the junctions of the mucous neck and parietal cells are small and less common. In fact, an analysis of the electrophysiological properties of the gastric glands suggests a very low level of electrical coupling (Diamond and Machen 1983).

Although gap junctions are both large and abundant in myocardium (McNutt and Weinstein 1970), our antibodies failed to label them. Recently, Hertzberg and Skibbens (1984) reported the immunofluorescent labeling of myocardial gap junctions with an antiserum raised in sheep against rat liver gap junctions. They reported positive junctional labeling in brain and kidney as well. Our antibodies did not label these tissues.

A simple model consistent with all current observations would hold that gap junctions in different tissues were comprised of proteins that were in some measure conserved, and therefore exhibited common antigenic determinants. In addition, these junctional proteins were also in some measure divergent, since they displayed antigenic determinants that were restricted to only a few tissues or were even specific to one tissue.

The results of immunoblotting and in vitro synthesis suggested that the 27-kD liver gap junction was initially synthesized in a higher-molecular-weight (~54-kD) form. In this regard, Warner et al. (1984) have reported the presence of 54-kD proteins in whole-tissue homogenates of rat liver and *Xenopus* oocytes that cross-react with antibody affinity-purified against rat liver 27-kD protein. On the other hand, Hertzberg and Skibbens (1984) did not detect a 54-kD protein in immunoblots of alkali-washed membranes from a variety of tissues. We found that inhibition of proteolysis was critical for detection of the 54-kD protein (see Results). Therefore it is possible that the failure to observe 54-kD protein in alkali pellets was due to the release of endogenous protease by the alkali wash. Another possibility is that the 54-kD protein was not sedimented after alkali washing of membranes.

A simple explanation for the relation of 54 kD to 27 kD is that of precursor to product. However, it remains to be demonstrated that the serological relation between the 54-kD and 27-kD proteins indicates a precursor-to-product relationship. It is clear that the integration of 54-kD protein into microsomal membranes in vitro does not result in processing of the 54-kD protein to 27-kD. A cleavage of the 54-kD protein to 48 kD may accompany membrane integration, and may be a first step in the processing of the 54-kD to the 27-kD form. If the 54-kD form is processed to 48 kD at the level of the rough ER, then it is not clear why the 54-kD protein is relatively more abundant in immunoblots of whole tissue. Peptide mapping of the 54-kD, 48-kD, and 27-kD species may clarify this issue and is in progress.

ACKNOWLEDGMENTS

I would like to thank Drs. J.H. Hartwig and M. Neutra for help with the colloidal gold immunohistochemistry, Dr. D. Perlman of Brandeis for providing microsomes, Dr. P. Breitfield for sharing a vastly improved immunoprecipitation procedure with us, and Dr. D.A. Goodenough for insight and enthusiasm as always. This research was supported by GM18974 and EYO2430 from the National Institutes of Health.

REFERENCES

Alcala, J., N. Lieska, and H. Maisel. 1975. Protein composition of bovine lens cortical fiber cell membranes. *Exp. Eye Res.* **21**: 581–595.

Bennett, M.V.L. and D.A. Goodenough. 1978. Gap junctions, electrotonic coupling and intercellular communication. *Neurosci. Res. Program. Bull.* **16**: 373–486.

Bok, D., J. Dockstater, and J. Horwitz. 1982. Immunocytochemical localization of the lens main intrinsic polypeptide (MIP26) in communicating junctions. *J. Cell Biol.* **92**: 213–220.

Claude, P., and D.A. Goodenough. 1973. Fracture faces of zonulae occludentes from tight and leaky epithelium. *J. Cell Biol.* **58**: 390–400.

Diamond, J.M. and T.E. Machen. 1983. Impedance analysis in epithelia and the problem of gastric acid secretion. *J. Membr. Biol.* **72**: 17–41.

Fallon, R.F. and D.A. Goodenough. 1981. Five-hour half-life of mouse liver gap junction protein. *J. Cell Biol.* **90**: 521–526.

Finbow, M., S.B. Yancey, R. Johnson, and J.P. Revel. 1980. Independent lines of evidence suggesting a major gap junctional protein with a molecular weight of 26,000. *Proc. Natl. Acad. Sci.* **77**: 970–974.

Fitzgerald, P.G., D. Bok, and J. Horwitz. 1983. Immunocytochemical localization of the main intrinsic polypeptide (MIP26) in ultrathin frozen sections of rat lens. *J. Cell Biol.* **97**: 1491–1499.

Friend, D.S. and N.B. Gilula. 1972. Variations in tight and gap junctions in mammalian tissues. *J. Cell Biol.* **53**: 758–776.

Goodenough, D.A. 1979. Lens gap junctions: A structural hypothesis for non-regulated low resistance intercellular pathways. *Invest. Opthalmol.* **18**: 1104–1122.

Goodenough, D.A. and W. Stoeckenius. 1972. The isolation of mouse hepatocyte gap junctions. Preliminary chemical characterization and X-ray diffraction. *J. Cell Biol.* **54**: 646–656.

Gros, D.B., B.J. Nicholson, and J.P. Revel. 1983. Comparative analysis of gap junction protein from rat heart and liver: Is there a tissue specificity of gap junctions? *Cell* **35**: 539–549.

Henderson, D., H. Eibl, and K. Weber. 1979. Structure and biochemistry of mouse hepatic gap junctions. *J. Mol. Biol.* **132**: 193–218.

Hertzberg, E.L. 1984. A detergent independent procedure for the isolation of gap junctions from rat liver. *J. Biol. Chem.* **259**: 9936–9943.

Hertzberg, E.L. and R.V. Skibbens. 1984. A protein homologous to the 27,000 dalton liver gap junction protein is present in a wide variety of species and tissues. *Cell* **39**: 61–69.

Hertzberg, E.L., D.J. Anderson, M. Friedlander, and N.B. Gilula. 1982. Comparative analysis of the major polypeptides from liver gap junctions and lens fiber junctions. *J. Cell Biol.* **92**: 53–59.

Hudson, L. and F.C. Hay, eds. 1980. *Practical immunology.* Blackwell, Oxford, England.

Kensler, R.W. and D.A. Goodenough. 1980. Isolation of mouse myocardial gap junctions. *J. Cell Biol.* **86**: 755–764.

Manjunath, C.K., G.E. Goings, and E. Page. 1982. Isolation and protein composition of gap junctions from rabbit hearts. *Biochem. J.* **205**: 189–194.

McNutt, R.S. and R.S. Weinstein. 1970. The ultrastructure of the nexus. A correlated thin section and freeze cleave study. *J. Cell Biol.* **47**: 666–688.

Meda, P., A. Perrelet, and L. Orci. 1979. Increase in gap junctions between pancreatic B-cells during stimulation of insulin secretion. *J. Cell Biol.* **82:** 441–448.

Nicholson, B.J., L.J. Takemoto, M.W. Hunkapillar, L.E. Hood, and J.P. Revel. 1983. Differences between the liver gap junction protein and lens MIP26 from rat: Implications for tissue specificity of gap junctions. *Cell* **32:** 967–978.

Paul, D.L. and D.A. Goodenough. 1983a. Preparation, characterization and localization of antisera against bovine MP26, an integral protein from lens fiber plasma membrane. *J. Cell Biol.* **96:** 625–632.

———. 1983b. In vitro synthesis and membrane insertion of bovine MP26, an integral membrane protein from lens fiber plasma membrane. *J. Cell Biol.* **96:** 633–638.

Sabatini, D.D., G. Kreibich, T. Morimoto, and M. Adesnik. 1982. Mechanisms for the incorporation of proteins into membranes and organelles. *J. Cell Biol.* **92:** 1–22.

Traub, O. and K. Willecke. 1982. Cross-reaction of antibodies against liver gap junction protein (26K) with lens fiber junction protein (MIP) suggests structural homology between these tissue specific gene products. *Biochem. Biophys. Res. Commun.* **109:** 895–901.

Warner, A.E., S.C. Guthrie, and N.B. Gilula. 1984. Antibodies to gap-junctional protein selectively disrupts junctional communication in the early amphibian embryo. *Nature* **311:** 127–131.

The Effects of Deuterium Oxide on Junctional Membrane Permeability and Conductance

Peter R. Brink
*Department of Anatomical Sciences
School of Medicine, SUNY at Stony Brook
Stony Brook, New York 11794*

One of the principal tenets of gap junction-mediated intercellular communication has been an aqueous channel connecting intracellular spaces of adjacent cells (Loewenstein 1976; Bennett 1977; Bennett and Goodenough 1978). The high conductivity of gap junctions and the rapid transfer of hydrophilic fluorescent probes through the junctions lends strong support to the notion that an aqueous channel connects adjacent cell interiors (Weidmann 1966, 1970; Asada and Bennett 1971; Sheridan 1974; Brink and Barr 1977; Spray et al. 1981; see also Sheridan in this volume and Weingart 1974). Voltage and pH gating mechanisms have been demonstrated in gap junctions of amphibian and teleost blastomeres (Turin and Warner 1977; Spray et al. 1981, 1982, 1984). Makowski et al. (1984a) analyzed the extent of penetration of sucrose into intercellular channels of isolated mouse liver junctions and found that the sucrose did not invade the channel completely but was absent from its most interior portions, prompting the authors to argue that the isolation procedure caused channel closure via regulatory gates. The ultrastructure of the inner portions of the intercellular channel has not as yet been completely elucidated to reveal if indeed there are gating sites placed more interiorly than those already found near the cytoplasmic orifice (Makowski et al. 1984b) or if the diameter of the channel remains constant. The electrophysiological, diffusional, and morphological data yield a composite image of the integral component of the intercellular junction as a group of proteins (connexon) spanning two adjacent membranes that contains a channel that is discriminatory with respect to solute size and charge with heterologous gating mechanisms.

The aqueous nature of the channel can be further characterized by monitoring the effect of heavy water exchange with normal water on both junctional membrane conductance and permeability of junctions to fluorescent probes. Some of the physical properties of D_2O are inconsequential with respect to H_2O, such as bonding length, dipole moment, and molecular dimension (Arnett and McKelvey 1969). Others, such as viscosity, are significantly different (Nemethy and Scheraga 1964; Arnett and McKelvey 1969). There is a 20% increase in viscosity in D_2O over H_2O at 20°C (1.1 versus 0.9 centapoise) and at 5°C the increase is 30%. The viscosity effect is best explained by considering the present models for water structure. Water molecules form clusters of varying size and lifetimes (approximately 0.1 nsec) and these clusters increase in size and duration with cooling. In D_2O these phenomena are exaggerated, presumably due to the increased strength of deuterium bonding relative to hydrogen (250 cal/mole; Nemethy and Scheraga 1964; Hahin and Strichartz 1981). In a related phenomenon, the hydration shell around a solute also increases in D_2O. These clustering and hydration effects are termed solvent isotope effects (Schauf and Bullock 1979). Primary and secondary isotope effects involve the exchange of deuterium for hydrogen on structural components in the system being studied. Hydrogen-deuterium exchange has the potential to change tertiary and quaternary structure of proteins and therefore alter the function of those components (Schauf and Bullock 1979; Brink 1983). Both solvent isotope effects and primary isotope effects have been observed for the diffusion of the fluorescent probe dichlorofluorescein (Cl_2F) through gap junctions (Brink 1983). Another interesting effect is the hindrance or block of dye diffusion observed at low temperatures in both D_2O and H_2O whereas junctional conductance and the propagation of action potentials across gap junctions are not blocked with cooling (Brink 1983; Brink et al. 1984b). Examples of solvent isotope effects, primary isotope effects, and diffusion blocks will be given below.

The study of solvent isotope effects can yield, if not quantitative, qualitative information on the role of solute and channel hydration in determining the permeation rates of solutes through the gap junction channel. One of the experimental preparations in which junctional membrane conductance and permeability can be assessed in conjunction with conductance and diffusion in the cytoplasm is the earthworm (*Lumbricus*) median giant septate axon. The septa of these axons contain gap junctions that allow communication between adjacent axon segments (Gunther 1975; Kensler et al. 1979).

RESULTS AND DISCUSSION

Morphology of the Median Giant Septate Axon of the Earthworm

The median giant axon of the earthworm is myelinated, with a periaxonal space (extracellular space between axon and myelin) typically 20 nm in width. Figure 1, a and b, consists of freeze-fracture images that show the nature of the myelin and the axolemma-myelin interface, respectively. Small protuberances of the axolemma (nodes) penetrate through the myelin at intersegmental regions to form what are presumably functional "nodes" (Gunther 1975), whereas septa, spaced about 1 mm apart, are found near segmental branches (Stough 1926; Mulloney 1970). The myelin appears more loosely wrapped than its mammalian counterpart and there are numerous desmosomes in the myelin (Fig. 1a). The septa were first described by Stough (1926) as having a geometrically simple design. He found that the septa traverse the axon perpendicularly or at an obtuse angle of about 140°. They therefore cover a longitudinal distance of 100–200 μm along the axon, unlike the crayfish septate axon in which the septa extend a distance of 500–600 μm. Figure 1c is a transmission electron micrograph of a portion of the septum arising from the axolemma of two adjacent axons. Freeze-fracture of the septum reveals gap junctions that have protruding particles in the P-face and concomitant pits in the E-face (Fig. 1d). The junctions are randomly arranged over the surface of any individual septum (Kensler et al. 1979). The geometry of the septate axon makes it an excellent cell system to study the movement of a solute in the cytosol and through the gap junction.

Gap Junction Permeability in the Septate Axon in H_2O Saline

Carboxyfluorescein (CF) is an excellent probe for monitoring intercellular communication and has been used extensively for that purpose. In all the subsequent fluorometric scans the dye used was CF. In the median giant axon, charged fluorescent probes can be introduced by iontophoresis (Brink 1983); Figure 1e shows an example of an injected cell and an adjacent cell into which dye from the injected cell has diffused. The septum at the opposite end of the adjacent cell is discernible. Determining the spatial distribution of any injected fluorescent probe allows measurement of axoplasmic diffusion and gap junction membrane permeability for the probe used as a tracer (Brink 1983). Quantitative measurements were obtained by moving the injected axon by a photomultiplier at constant velocity while excitation (470–490 nm) was delivered to the axon via epi-illumination. The output of the phototube is acquired by computer through an A/D converter. The data are then stored on disk for later analysis. Figure 2a shows three such scans or spatial distributions of CF from an individual experiment taken at three different times. The solvent was H_2O at 21°C. In all three cases junctional membrane permeability (P_j) remained constant, assuming that P_j = flux/concentration gradient (i.e., Fickian diffusion). The diffusion coefficient of various probes in axoplasm has been measured and found to be suppressed relative to the predicted value in water by a factor of about 5 to 10 (Brink and Dewey 1981), indicating that the dye is being retarded in its "random walk" by binding of the dye to axoplasmic elements, viscosity of the axoplasm, tortuosity factors, such as mitochondria, sequestration, or a combination of all four. Models have been used to compute the permeability of an interface with an infinite volume of either side (see Crank 1975 for a detailed analysis; see also Carslaw and Jaeger 1959). Application of Fickian diffusion ($\partial C/\partial t = D\partial^2 C/\partial x^2$) to the geometry just given yields solutions of the following form:

$C_1(x, t) = C_0/2[1 + erf(x/2\sqrt{Dt}) + exp(hx + h^2Dt)erfc((x/2\sqrt{Dt}) + (h\sqrt{Dt}))]$

for all values of $x > 0$ (dye-containing compartment).

$C_2(x, t) = C_0/2[erfc(x/2\sqrt{Dt}) - exp(hx + h^2Dt)erfc((x/2\sqrt{Dt}) + (h\sqrt{Dt}))]$

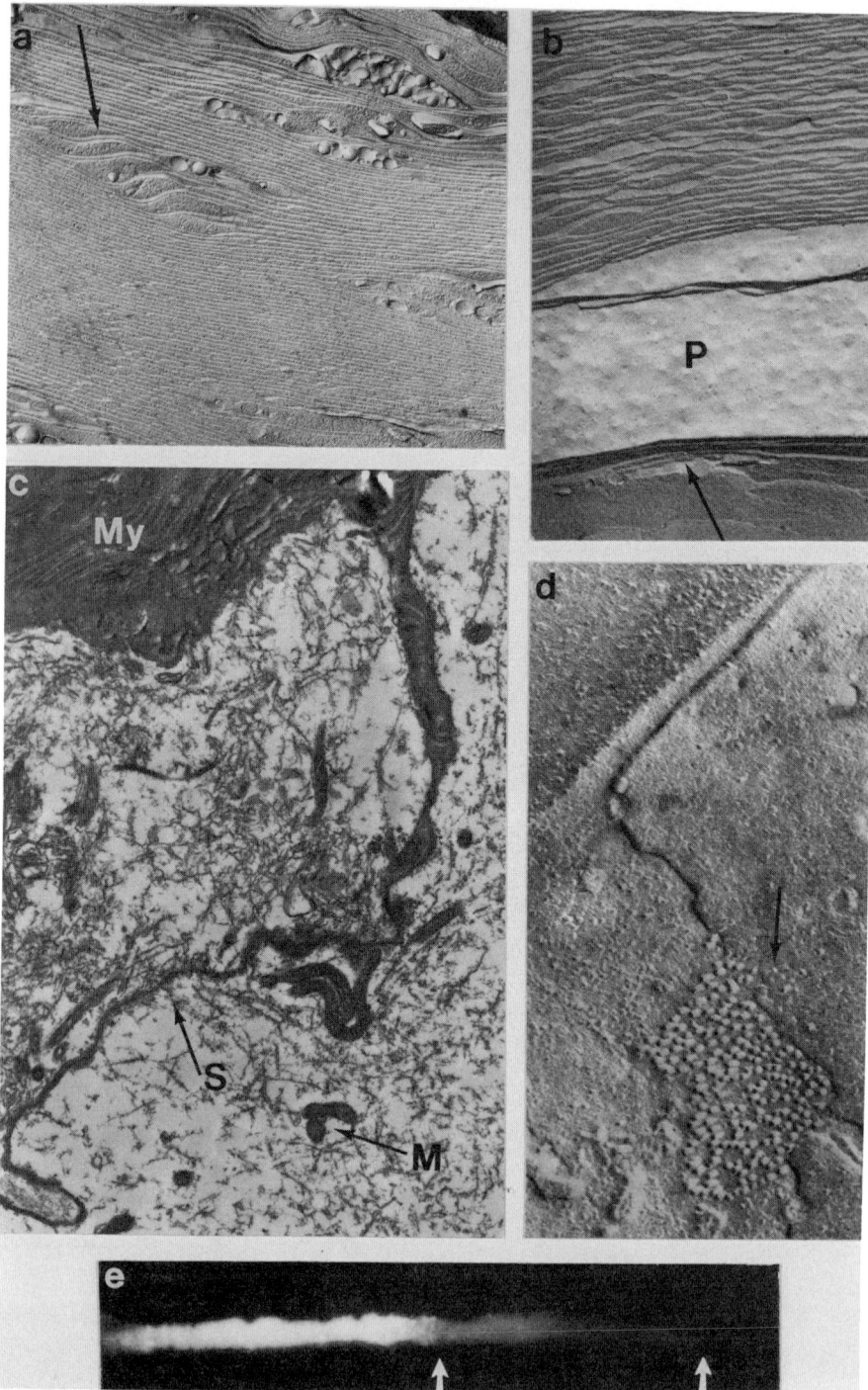

Figure 1 *(See facing page for legend.)*

for all values of $x < 0$. In Equation (1) all values of x appear as absolute values. Symbols: D, diffusion coefficient, assumed to be constant for x and t; $h = 2(P_j/D)$ where P_j = junctional membrane permeability; $D\partial C_1/\partial x + P_j(C_2 - C_1) = 0$ at the interphase taken as $x = 0$; C_0, concentration of the dye at $t = 0$ for all $x > 0$.

Thus at $t = 0$ the cell is initially filled with dye that diffuses longitudinally down the axon. The model assumes that the septum lies perpendicular to the long axis of the axon when in fact it traverses the axon at an angle (Stough 1926; Gunther 1975). The total distance the septum traverses averages about 50–200 μm. For preparations where the septa traverse the axon for distances greater than 200 μm, no analysis was performed. Calculations of the form given here yield information about the permeability of the junctional membrane and the diffusion coefficient of the cytosol for only short time intervals because of the constraining boundary condition that the volumes on either side of a junction are infinite. In the case of the earthworm septate axon at times greater than 15 minutes, the solutions of Equations (1) and (2) are unable to fit the diffusion data. Recently, a new diffusion model for the septate axon was presented which allows computation of diffusional processes for all time in the axon (Brink et al. 1984a). For the dye CF both P_j and D_a were found to be constant in time.

In H_2O saline at temperatures above 6°C, all the probes that have been studied diffuse through the junctions of the septa. But at temperatures at or below 6°C, the diffusion is drastically reduced to a level that is below that detectable by the recording apparatus. The diffusion of CF within an axon at 6°C in H_2O is shown in Figure 2b to demonstrate this phenomenon. There is little or no longitudinal diffusion into the adjacent cell to the left and only slight diffusion into the adjacent cell to the right, even after a rather long time. At temperatures above 6°C diffusion can be observed, as shown in Figure 2a. The temperature dependence of Cl_2F gap junction permeability has been shown to have a Q_{10} of 2.0 in H_2O at temperatures above the 6–4°C range (H_2O) (Brink 1983). The lack of or greatly reduced permeability in H_2O saline at or below 6°C for CF (Fig. 2b) suggests, by itself, that the gap junctional channels are changing their characteristic conductance or permeance state at specific temperatures. In fact for a number of probes, the "block temperature" is in the range of 6°C in H_2O saline (Brink et al. 1984). If the channel is changing diameter in response to temperature as these data suggest, then

Figure 1
Freeze-fracture image of the myelin surrounding the median giant axon of earthworm is shown in a; the arrow points to a series of desmosomes. Freeze-fracture of a myelin-axon interface is shown in b; the arrow indicates the interface. The large plane above the interface is a P-face of a myelin membrane (P). (c) Thin-section image of septum. MY, myelin; M, mitochondrondria; S, septum. (d) Freeze-fracture picture of septal membrane showing a gap junction (nexus). Arrow indicates pits on the E-face. (e) Fluoromicrograph of dye-injected septate axon. The cell adjacent to the one iontophorized with dye has begun to fill with dye and the next septum is visible. The axon is approximately 90 μm in diameter. The septum transverses an area of about 100 μm. The arrows indicate the position of two septa.

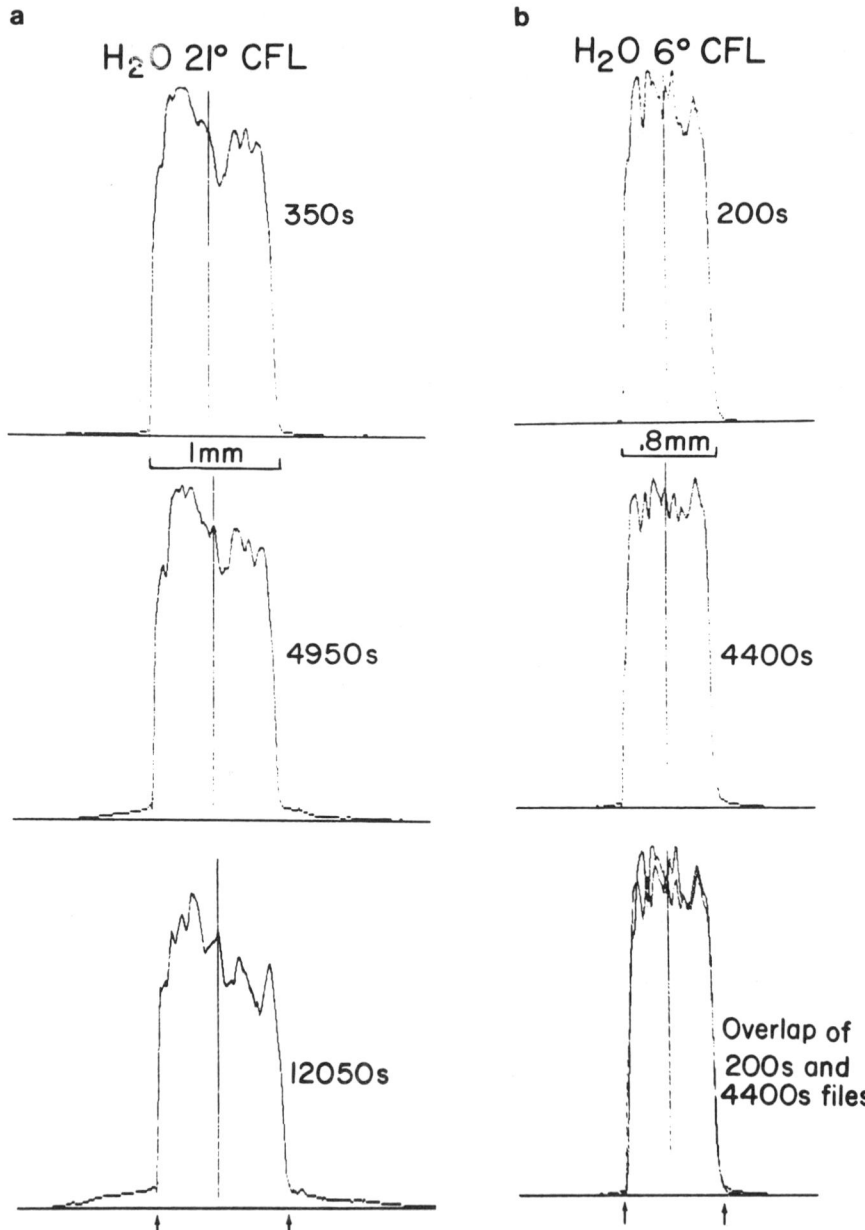

Figure 2
Diffusion profiles for the dye CF at 21°C (a) and 6°C (b). The sharp changes in intensity (arrows) indicate the position of the septa at either end of the cell. Dye diffuses across the septa at 21°C but not at 6°C. The injected cell length is 1 mm for the 21°C preparation and 0.8 mm for the 6°C preparation (distance between arrows).

one might predict that dye molecules such as CF will have a similar block temperature whether in H_2O saline or D_2O saline. The same should also be true for junctional conductance in either solvent if the channel is responding to temperature. If hydration of solutes is an important feature of a molecule's ability to move through the junctional channel, then different block temperatures would be predicted for probes of varied charge and size in a solvent like D_2O because of its ability to increase the hydration state of a solute relative to H_2O. Channel wall solvation will also result in narrowing the intercellular channel. If increased channel hydration with cooling were the dominant feature regulating solute movement through the junction, then all the dyes would not only have similar block temperatures in H_2O but similar block temperatures in D_2O. But different blocking temperatures for the various probes studied and the linearity of electrical conductance with temperature (see below) suggest that it is the hydration state of the probes and their interaction with the channel or channel hydration shell that is most influential.

Gap Junction Permeability in the Septate Axon in D_2O Saline

The solvent isotope effects of D_2O should cause a reduction in P_j if it is influenced by hydration and solvent clustering. If a solute maintains a hydration or deuteration shell as it traverses the gap junction, then the "block temperature" for a probe might be a higher temperature in D_2O than H_2O. Figure 3a shows two scans from an individual experiment of CF diffusion. The diffusion is noticeably less than that of Figure 2a and inspection indicates that P_j is reduced. The reduction in P_j is consistent with solvent isotope effects. The junctional permeability for Cl_2F has been studied more extensively (Brink 1983). In that study, both solvent isotope effects (viscosity) and primary isotope effects (hydrogen-deuterium exchange) were found to be at work in determining the flux of the probe through gap junction channels. The Q_{10} for Cl_2F permeability in D_2O above 6°C is 2.5 (2.0 in H_2O Brink 1983). To make a similar analysis for CF a larger sample is required but the trends are clearly demonstrated in Figures 2 and 3. Figure 3b shows the effect of lowered temperature on CF diffusion through the gap junction in D_2O at 15°C. As in Figure 2b little or no diffusion is observable. Results from experiments like those in Figures 2b and 3b show that the dye CF at some critical temperature in either solvent undergoes a drastic reduction in mobility through the gap junction. A simple explanation for the results revolves around the role of solute hydration. In D_2O at 15°C the majority of solute molecules (dye) are deuterated such that their functional diameters are larger than that of the gap junctional channel. The hydrated state is not a static state but a dynamic one in which a solute oscillates between being hydrated and nonhydrated (Bockris and Reddy 1970). Temperature shifts the length of time a solute stays in one state or another. With cooling the hydrated state dominates and the number of solvent molecules in association with a solute increases. In addition to CF and Cl_2F the block temperature for Lucifer Yellow has also been determined in D_2O and H_2O and found to be 12 and 6°C, respectively (Brink et al. 1984b). The elevated block temperatures (greatly reduced permeability) of CF (19°C in D_2O;

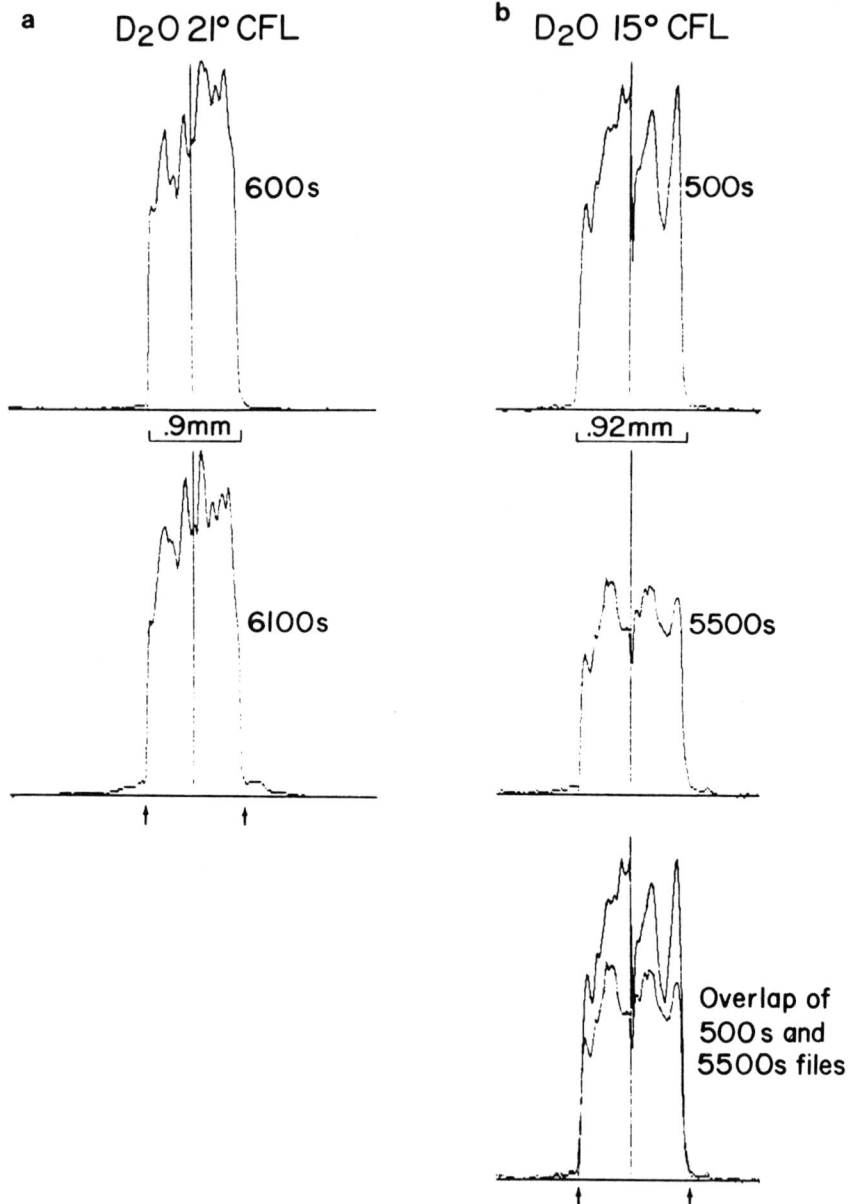

Figure 3
Diffusion profiles for the dye CF at 21°C (*a*) and 15°C (*b*). The arrows indicate the position of the septa. Dye diffuses across the septa at 21°C but not at 15°C. The injected cell length is 0.9 mm for the 21°C preparation and 0.92 mm for the 15°C preparation (distance between arrows).

Brink et al. 1984b) and Lucifer Yellow in D_2O are consistent with the notion that greater charge density on a solute yields a larger hydration shell. The fact that in D_2O saline the block temperature for dyes is much higher than in H_2O supports the hydration hypothesis rather than an iris effect (Unwin and Zampighi 1980) in response to cooling. The probes thus far studied all have dimensions that approximate the channel diameter (Brink and Dewey 1981; Loewenstein 1981; Makowski et al. 1984b) with the exception of Weidmann (1966), Verselis et al. (1985), and Weingart (1974). CF has dimensions of $1.26 \times 12.7 \times 0.85$ nm in the nonhydrated form as determined by Corey-Pauling models. Probes of smaller size, even if charged, should experience no diffusion block with cooling on the basis of the hydration concept (Bockris and Reddy 1970). One such molecule is K^+. To monitor its movement, electrophysiological techniques can be used. This assumes that the principal current carrier within the cell is K^+ ion.

Voltage Clamp Characteristics of the Septum in H_2O and D_2O

The double voltage clamp (Spray et al. 1981) is an excellent method for studying the conductance of gap junctions in both H_2O and D_2O while varying temperature. Figure 4 shows the response of the junctional conductance to an exchange of H_2O saline for D_2O saline at 20°C. Figure 4a shows a clamp record before exchange and Figure 4b shows a record taken 3 minutes after exchange of H_2O with D_2O. The junctional conductance decreased 25%, a decrease that is in the range predicted for solvent isotope effects. After the initial reduction within minutes of exchange there is a slower reduction (by an additional 10%) in junctional conductance which reaches steady state about 15–20 minutes after exchange (Fig. 4c). Solvent isotope effects (viscosity) can only account for a 22–25% reduction at 20°C. The slower reduction can be explained as a primary isotope effect. Primary isotope effects (hydrogen-deuterium exchange) can alter any steric relationship where hydrogen plays a role in the structure and hence function of a cellular component (Schowen 1976). In this case a decrease in junctional conductance (additional 10%) caused by hydrogen-deuterium exchange can be interpreted as causing a decrease in some aspect of the intercellular channel diameter or altering the gating mechanisms that regulate channel patency. Hydration (deuteration) alone would have little or no effect on "flicker time" of a regulatory gate. In addition to demonstrating both solvent and primary isotope effects, the experiment shown in Figure 4 illustrates the speed at which D_2O penetrates the axon.

The effects of temperature on junctional conductance in both solvents are shown in Figure 5. Figure 5a shows the effect of H_2O and D_2O on junctional conductance. This represents the data from one experiment. No correction for axoplasmic resistance was made. The Q_{10} values for the septa in both solvents are 1.4 (H_2O) and 1.55 (D_2O). The activation energy for the intercellular conductance in H_2O was approximately 6.2 kcal/mole while in D_2O it was in the range of 7.0 kcal/mole. The difference in activation energy between D_2O and H_2O is about 800 cal/mole. For the dye Cl_2F the difference in activation energy was approximately 1000 cal/mole. Since the difference between hydrogen and deuterium bonds is 250 cal/mole, these data imply that the dye

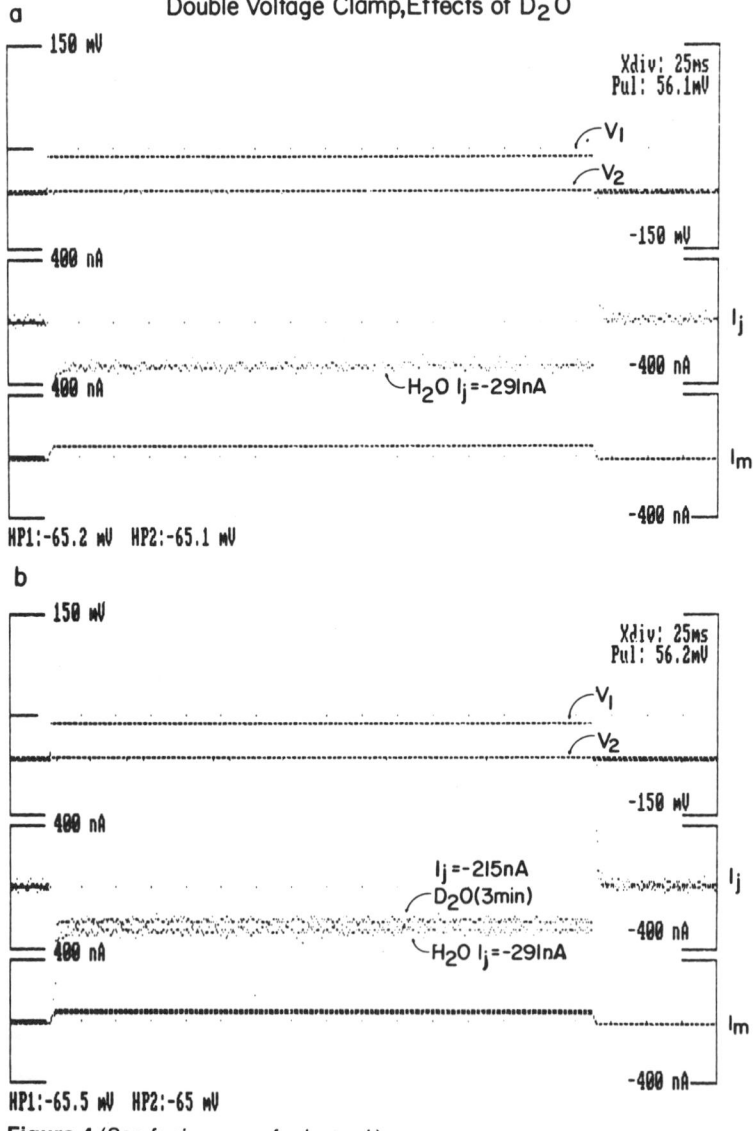

Figure 4 (*See facing page for legend.*)

makes and/or breaks three to four hydrogen (deuterium) bonds in its transit through the junctional channel. The conductance data indicate an interaction between solute (K^+) and channel via the making and/or breaking of two to three hydrogen or deuterium bonds (800 cal/mole). Whether the binding sites are located on the outer surface or within the intercellular channel, or located in one site or many throughout the channel, is impossible to determine from these data. Hill plots for junctional conductance versus pH in amphibian blastomeres suggest that there may be four titratable sites within or on the intercellular channel (Spray et al. 1982). Whether the titratable groups of gap junction channels in the blastomeres and the activation energy differences for intercellular transit in the septa are revealing mechanisic information about the same site(s) remains to be demonstrated. Finally, in Figure 5 there is no break in the slope of the conductance curve in either solvent down to a temperature of 4°C, unlike the behavior of the dye molecules already mentioned.

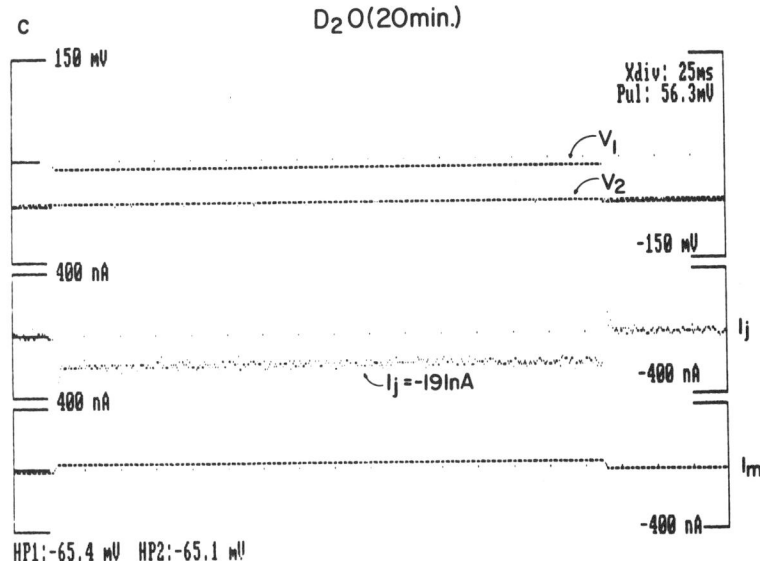

Figure 4
Double voltage clamp of junctional membrane in H_2O saline is shown in *a*. Two superimposed records of junctional conductance before and after (3 min) D_2O exchange as shown in *b*. After 3 min there is an approximately 25% reduction in junctional conductance. These effects are consistent with solvent isotope effects associated with D_2O exchange (i.e., those effects associated with solute solvation and solvent viscosity). After 20 min in D_2O, junctional conductance has finally reached a steady-state level of approximately 65% of the control indicating primary isotope effects (*c*). In all of the records HP1 and HP2 indicate the holding potentials under voltage clamp. The label Pul indicates the amplitude of the voltage pulse. The total time for each record was 500 msec. I_j and I_m are junctional and nonjunctional current, respectively.

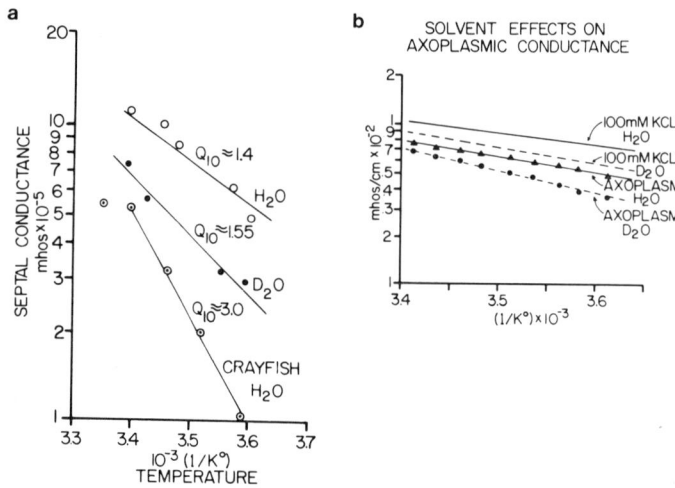

Figure 5
The effect of temperature in H_2O and D_2O saline on septal membrane conductance is shown in *a*. The Q_{10} values are consistent with solvent isotope effects. In addition, data for the lateral axon septum of crayfish are shown. The data are taken from Ramon and Zampighi (1980). The effects of temperature and solvent exchange on axoplasmic conductance of axoplasm and a 100 mM KCl solution are shown in *b*. Once again the effects are consistent with solvent isotope effects.

If an iris effect were influential, then the conductance would also show, if not a block, a change in slope at some temperature. In addition the temperature-dependent conductance behavior of the crayfish septa in H_2O saline is shown (Ramon and Zamphighi 1980). The Q_{10} is much higher than that of the earthworm septa in either solvent. The large Q_{10} of the lateral axon septa of crayfish was first observed and reported by Payton et al. (1969). The reasons for the high temperature dependence were thought to be due to alteration of membrane fluidity or structure which in turn reduced junctional membrane conductance. Cytoplasmic involvement in the form of "plugs" or release of junctional membrane uncouplers is also possible. Whatever mechanisms are at play in the lowering of crayfish junctional membrane conductance in response to cooling, they are not shared by the earthworm septal gap junctions.

The effects of solvent exchange on axoplasmic conductance are also important to ascertain. Figure 5b shows the temperature dependence of axoplasmic conductance in both solvents. These data were collected using the modified cable analysis method devised by Weidmann (1970; see also Brink and Barr 1977; Wojtchak, this volume). The Q_{10} values are 1.3 and 1.4 in H_2O and D_2O, respectively ($n = 5$ at each temperature). Figure 5b also shows the effect of both H_2O and D_2O on the conductance of a 100 mM KCl solution versus temperature. The percent reduction at 20°C is approximately 20% in D_2O versus H_2O. The differences in Q_{10} values are indicative of solvent isotope effects. No primary isotope effects were detectable. In either solvent, junctional conductance showed no voltage-dependent conductance changes, and temperature change also failed to induce voltage-dependent behavior.

Solvent Effects on Gap Junctions 135

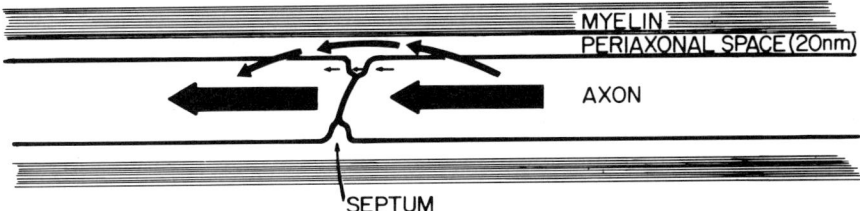

Figure 6
A model for the diffusion pathways within the septate axon of earthworm. One pathway is through the gap junction within the septum; a second pathway is via the periaxonal space (only 20 nm wide). A third could be low-resistance membranes within the septa separated by a 20-nm space.

Alternate Pathways

All of the results given thus far assume that it is the gap junction that is the path between adjacent axons. Another possible pathway is that of the periaxonal space. Figure 6 is a schematic representation of the three possible pathways. A periaxonal space of 20 nm width and 100–200 μm length filled with electrolytes would have a resistance of about 40 MΩ. The input impedance of the axon is 1 MΩ and the septal resistance or composite longitudinal resistance of the septa and periaxonal space is about 40 KΩ (Brink and Barr 1977; Verselis and Brink 1984). Even if the resistance of the plasma membrane were low in regions near the septa, the overwhelming resistance of the restrictive periaxonal space makes it an unlikely as a candidate for the site where most of the ionic current flows along the axon. However, small areas of membrane within the septa with low resistance (1–10 Ω-cm^2) could produce longitudinal resistances in the 1×10^5 range, comparable to that of the septa. In the case of diffusion, if it is assumed that the periaxonal space is freely accessible to the fluorescent probe, then how much would it contribute to the fluorescent signal recorded by the photomultiplier? A typical axon is about 80 μm in diameter. Therefore, in any plane the periaxonal space would contribute about 1 part in 2000 to the fluorescent signal. If the periaxonal space were the path for all the dye transfer rather than the septa, then the transfer rate based on cross-sectional surface area would have to be 1500 times greater than the septa. The presence of gap junctions within the septal membranes of the earthworm giant axon coupled with the physiological data showing the septal membranes to be of low resistance and highly permeable to fluorescent probes make the gap junction the most likely candidate for the longitudinal movement of ions and solutes. However, a contribution from the passage of ions and solutes across the extracellular space of the septa and also along the periaxonal pathway cannot be ruled out.

CONCLUSIONS

The differences in block temperatures for various charged molecules in the solvent D$_2$O lend themselves to the interpretation that molecules are moving through the gap junction channel in a hydrated (deuterated) form. The temperatures at which diffusion is greatly reduced or blocked represent the point at which the majority of the solute molecules being monitored are too large by virtue of hydration of solute and channel to enter and diffuse through the intercellular channel. The conductance data indicate that the current-carrying species, presumably K^+ and Cl^-, are too small even at low temperatures in D_2O to experience a temperature block. Finally, the suppressed diffusion coefficients are best explained by axoplasmic binding, sequestration, tortuosities within cells, and viscosity. The lowered diffusion coefficients point to the role of cell water and its components in retarding solute movement. The data presented here indicate that solutes move throughout the intercellular channel in a hydrated form. When the solvent deuterium oxide is used as a substitute for H_2O, both solvent isotope effects (viscosity) and primary isotope effects (hydrogen-deuterium exchange) are demonstrable with respect to gap junction conductance and permeability.

ACKNOWLEDGMENTS

The author would like to thank Dr. R.W. Kensler for his collaboration on the electron microscopic data presented and for many helpful and insightful suggestions. Dr. V. Verselis was responsible for the collection of much of the voltage and current clamp data of Figure 5. His careful and thoughtful approach is also appreciated. This work was supported by NIH grant GM 24905 and NSF grant BNS 83-14295.

REFERENCES

Arnett, E.M. and D.R. McKelvey. 1969. Solvent isotope effect on thermodynamics of nonreacting solutes. In *Solute-solvent interactions* (ed. J.F. Coetzke and C.D. Ritchie), pp. 314–395. Marcel Dekker, New York.

Asada, Y. and M.V.L. Bennett. 1971. Experimental alteration of coupling resistance at an electrotonic synapse. *J. Cell Biol.* **49:** 159–172.

Bennett, M.V.L. 1977. Electrical transmission: A functional analysis and comparison to chemical transmission. In *The handbook of physiology*, Sec. 1, *The nervous system* (ed. E. Kandel), pp. 357–416. American Physiological Society, Washington.

Bennett, M.V.L. and D.A. Goodenough. 1978. Gap junctions, electronic coupling and intercellular communication. *Neurosci. Res. Program Bull.* **16:** 373–486.

Bockris, J. and A. Reddy, eds. 1970. *Modern electrochemistry*. Plenum Publishing, New York.

Brink, P.R. 1983. Effect of deuterium oxide on junctional membrane channel permeability. *J. Membr. Biol.* **71:** 79–87.

Brink, P.R. and L. Barr. 1977. The resistance of the septum of the median giant axon of earthworm. *J. Gen. Physiol.* **69:** 517–536.

Brink, P.R. and M.M. Dewey. 1981. Diffusion and mobility of substances inside cells. In *Techniques in the life sciences, part 1* (ed. P.F. Baker), p. 1–17. Elsevier/North-Holland, Limerick, Ireland.

Brink, P.R., S.W. Jaslove, and S.V. Ramanan. 1984a. A diffusion model for septate axons. *Soc. Gen. Physiol. Ser.* (Abstr.) (in press).

Brink, P.R., V. Verselis, and L. Barr. 1984b. Solvent-solute interactions within the nexal membrane. *Biophys. J.* **45:** 121–124.

Carslaw, H.S. and J.C. Jaeger, eds. 1959. *Conduction of heat in solids*, 2nd edition. Oxford University Press, Oxford.

Crank, J. 1975. *The mathematics of diffusion*, 2nd edition. p. 15,40. Oxford University Press, Oxford, England.

Gunther, J. 1975. Neuronal syncytia in the giant fibers of earthworm. *J. Neurocytol.* **4:** 55–62.

Hahin, R. and G. Strichartz. 1981. Effects of deuterium oxide on the rate and dissociation constants for saxitoxin and tetrodotoxin action. *J. Gen. Physiol.* **78:** 113–139.

Kensler, R.W., P.R. Brink, and M.M. Dewey. 1979. The septum of the lateral axon of the earthworm: A thin section and freeze fracture study. *J. Neurocytol.* **8:** 565–590.

Loewenstein, W.R. 1976. Permeable junctions. *Cold Spring Harbor Symp. Quant. Biol.* **40:** 49–63.

———. 1981. Junctional intercellular communication: The cell-to-cell membrane channel. *Physiol. Rev.* **61:** 829–913.

Makowski, L., D.L.D. Casper, W.C. Phillips, and D.A. Goodenough. 1984a. Gap junction structure V. Structural chemistry inferred from x-ray diffraction measurements. *J. Mol. Biol.* **174:** 449–482.

Makowski, L., D.L.D. Casper, W.C. Phillips, T.S. Baker, and D.A. Goodenough. 1984b. Gap junction structures VI. Variation and conservation in connexon conformation and packing. *Biophys. J.* **45:** 208–218.

Mulloney, B. 1970. Structure of the giant fibers of earthworms. *Science* **168:** 994–996.

Nemethy, G. and H.A. Scheraga. 1964. Structure of water and hydrophilic bonding in proteins. IV. The thermodynamic properties of liquid deuterium oxide. *J. Chem. Phys.* **41:** 680–689.

Payton, B.W., M.V.L. Bennett, and G.D. Pappas. 1969. Temperature-dependence of resistance at an electrotonic synapse. *Science* **165:** 595–597.

Ramon, F. and G. Zampighi. 1980. On the electronic coupling mechanism of crayfish segmented axons: Temperature dependence of junctional conductance. *J. Membr. Biol.* **54:** 165–171.

Schauf, C.L. and J.O. Bullock. 1979. Modifications of sodium channel gating in *Myxicola* giant axons by deuterium oxide, temperature and internal cations. *Biophys. J.* **27:** 193–208.

Schowen, R.L. 1976. Solvent isotope effects on enzymic reactions. In *Isotope effects on enzyme-catalyzed reactions* (ed. W.W. Copeland et al.), pp. 64–100. University Park Press, Baltimore, Maryland.

Sheridan, J.D. 1974. Electrical coupling of cells and cell communication. In *Cell communication* (ed. R.P. Cox), pp. 30–42. John Wiley, New York.

Spray, D.C., A.L. Harris, and M.V.L. Bennett. 1981. Equilibrium properties of a voltage-dependent junctional conductance. *J. Gen. Physiol.* **77:** 77–94.

———. 1982. Comparison of pH and calcium dependence of gap junctional conductance. In *Intracellular pH: Its measurement and utilization in cellular functions* (ed. R. Nuccitelli and D.W. Deamer), pp. 445–461. A.R. Liss, New York.

Spray, D.C., R.L. White, A. Campos De Carvalho, A.L. Harris, and M.V.L. Bennett. 1984. Gating in gap junction channels. *Biophys. J.* **45:** 219–230.

Stough, H.B. 1926. Giant fibers of the earthworm. *J. Comp. Neurol.* **40:** 409–463.

Turin, L. and A. Warner. 1977. Carbon dioxide reversibly abolishes ionic communication between cells of early amphibian embryo. *Nature* **270:** 56–57.

Unwin, P.N.T. and G. Zampighi. 1980. Structure of the junction between communicating cells. *Nature* **283:** 545–549.

Verselis, V. and P.R. Brink. 1984. Voltage clamp of the earthworm septum. *Biophys. J.* **45:** 147–150.

Verselis, V., R.L. White, D.C. Spray, M.V.L. Bennett. 1985. Correlation of gap junctional conductance and permeability in *Rana* blastomeres. *Biophys. J.* **47:** 504a (Abstr.).

Weidmann, S.J. 1966. The diffusion of radio-potassium across intercalated discs of mammalian cardiac muscle. *J. Physiol.* **187**: 323–342.

———. 1970. Electrical constants of trabecular muscle from mammalian heart. *J. Physiol.* **210**: 1041–1054.

Weingart, R. 1974. The permeability of tetraethylammonium ions of the surface membrane and the intercalated disks of sheep and calf myocardium. *J. Physiol.* **240**: 741–762.

General and Comparative Physiology of Gap Junction Channels

David C. Spray, R.L. White,
V. Verselis, and Michael V.L. Bennett
Department of Neuroscience
Albert Einstein College of Medicine
Bronx, New York 10461

Gap junction channels provide a pathway for flow of electric current and small molecules from a cell to its coupled neighbors. Although current is carried primarily by K^+ in most cells, gap junction channels also allow intercellular exchange of molecules as large as Lucifer Yellow (diameter 1.2 nm; see also Brink, this volume). In excitable tissues, a major purpose of gap junctions is to permit current flow, in many cases for rapid and synchronous activation. Although the purpose of gap junctions between inexcitable cells presumably includes exchange of larger molecules, it is often more convenient to study the channels by measuring electrical parameters than by quantifying intercellular diffusion of dyes or other junction-permeant molecules. Nevertheless, both sorts of measurement depend upon similar cell properties as considered below.

Our approach is to use pairs of cells for both electrophysiological and permeability measurements (Fig. 1). By applying current pulses alternately in each cell and recording voltages, we obtain input and transfer conductances from whih conductances of junctional (g_j) and nonjunctional (g_1, g_2) membranes are unequivocally determined (Bennett 1966; Spray et al. 1981a). The coupling coefficient (k) from cells 1 and 2 is $g_j/(g_2+g_j)$; it follows that the coupling coefficient can be asymmetric in the two directions, can vary from near zero to near unity with constant g_j, and approaches unity when g_2 is low compared with g_j. (When dye coupling is measured with a probe to which nonjunctional membranes are impermeable and there is no loss or binding of dye [a situation approximated by flux measurements of small tetra-alkylammonium ions; Verselis et al. 1985], the analogous dye coupling coefficient is 1; dye

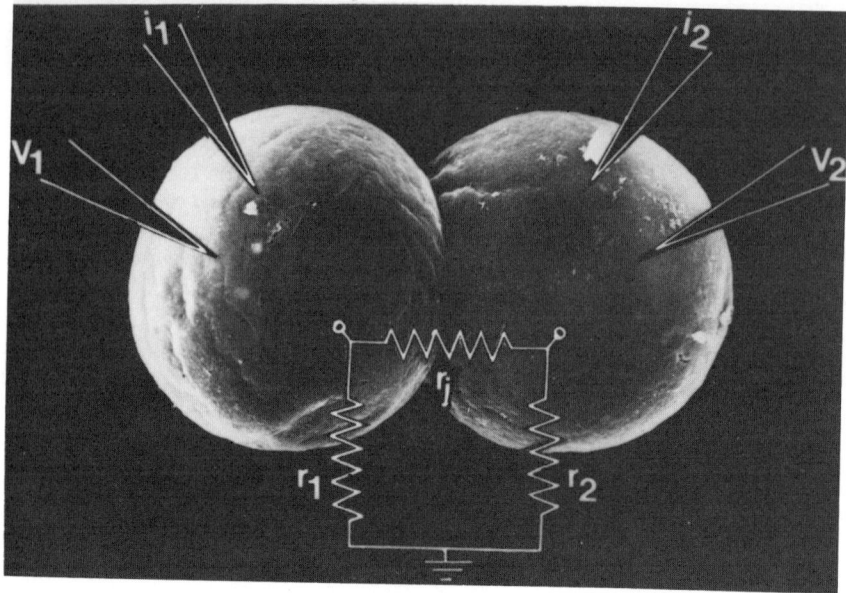

Figure 1
Experimental arrangement used to measure junctional resistance (r_j) and junctional conductance ($g_j = 1/r_j$). Each cell of a pair is impaled with separate electrodes for current delivery and potential measurement in either current or voltage clamp conditions. Under current clamp conditions, pulses are passed alternately in each cell (i_1, i_2) and the equivalent circuit is solved from application of the π-T transform to input and transfer voltages; the result is unambiguous calculation of r_1 and r_2 (the lumped nonjunctional substances of the two cells) and r_j (Bennett 1966). Photo is a scanning electron micrograph by E.A. Morales (Albert Einstein College of Medicine) of a pair of *Fundulus* blastomeres treated with cytochalasin B; the diameter of each cell is about 100 μm.

must eventually equilibrate between the cells if there is a single permeable channel.)

This chapter reviews treatments that alter g_j directly; physiological changes in coupling coefficient without change in g_j are considered elsewhere in this volume (see Bennett et al., this volume). The earliest study of gap junction physiology indicated that junctional conductance can be gated by voltage (the rectifying synapse of crayfish: Furshpan and Potter 1959; see also Giaume and Korn, this volume). Other studies suggested that Ca^{++} and H^+ affect gap junctions, acting on the cytoplasmic face of the channels (Rose and Loewenstein 1975; Turin and Warner 1978). Our work of the past several years has defined the sensitivity of gap junctions in several cell types to these and other stimuli that affect whether gap junction channels are open or closed; in some cases (see Discussion), the sensitivity is high and we argue that these stimuli may operate to gate the channels under physiological conditions. In other cases, the sensitivity is low and may exist only for the convenience of the electrophysiologist or to illustrate how large is the diversity of functional properties within this class of membrane channels. We consider first examples from fish and amphibian embryonic cells that define responses of these junctions to certain gating stimuli.

pH DEPENDENCE AND Ca^{++} INSENSITIVITY

In 1978 Luca Turin and Ann Warner demonstrated that cytoplasmic acidification by CO_2 exposure uncoupled cells in the early amphibian embryo (a phenomenon illustrated in Fig. 2B) and that decreased conductance of the junctional membrane was largely responsible (Turin and Warner 1978, 1980). Using pairs of fish and amphibian embryonic cells, we quantified the relation between g_j and cytoplasmic pH (pH_i; graph, Fig. 2A). With intracellular pH electrodes, we showed that the change in g_j, either during acidification or recovery by rinsing in normal saline, is closely related to intracellular pH (pH_i; Spray et al. 1981b). For these vertebrate embryos, the g_j-pH_i relation is well fit by a Hill plot $G_j = K^n/K^n + H^n$ where G_j is normalized g_j, H is proton concentration, K is the H concentration at which G_j is 0.5, and n, the Hill coefficient, is the effective number of cooperative binding sites. The apparent pK is 7.3 and the Hill coefficient (n) is between 4 and 5 for cells of fish and amphibian embryos (Fig. 2A). Thus a model was proposed in which channel closure is effected by proton binding to four or five highly cooperative sites, although more sites might be involved if cooperativity were lower. The effects are the same for a number of membrane-permeant weak acids (CO_2, lactic, propionic, acetic) and also for treatment with esters that acidify the cytoplasm by esterase-catalyzed hydrolysis (Spray et al. 1982a, 1984a). Strong impermeant acids and buffers do not act when applied to the cell exterior but decrease g_j when injected in concentrations adequate to overwhelm intracellular buffers.

Treatments that elevate cytoplasmic free Ca^{++} decrease intercellular coupling (Rose and Loewenstein 1975), and, since most intracellular buffer systems exchange H^+ and Ca^{++}, it seemed plausible that H^+ ions might act on g_j through a change in the cytoplasmic concentration of ionized Ca^{++}. CO_2 exposure adequate to reduce g_j to very low levels need not affect cytoplasmic free Ca^{++}, as measured by aequorin luminescence or Ca^{++}-sensitive microelectrodes (Bennett et al. 1978; Rink et al. 1980). Ca^{++} sensitivity of gap junctions was quantified in pairs of fish blastomeres with an internal perfusion technique (Spray et al. 1982b). With pH of the perfusate buffered to 7.6, the junctional membrane was insensitive to $[Ca^{++}]_i$ below 100 µM; above 100 µM, g_j decreased, with some conductance remaining even at 1 mM. This relation could be fit by a Hill curve with a pK corresponding to about 0.5 mM Ca^{++} and n of 2-3 (leftmost curve, Fig. 2A). Thus, in this system, Ca^{++} seems to act on g_j independently of H^+ but with an affinity that is lower by a factor of about 10,000. Perhaps, since the Hill coefficient obtained for the divalent cation is about half that obtained for the monovalent one, both ions can react with the same sites on the channel proteins; this question could presumably be resolved with competition experiments or those using reagents that selectively immobilize the pH gate or modify its binding site (see below).

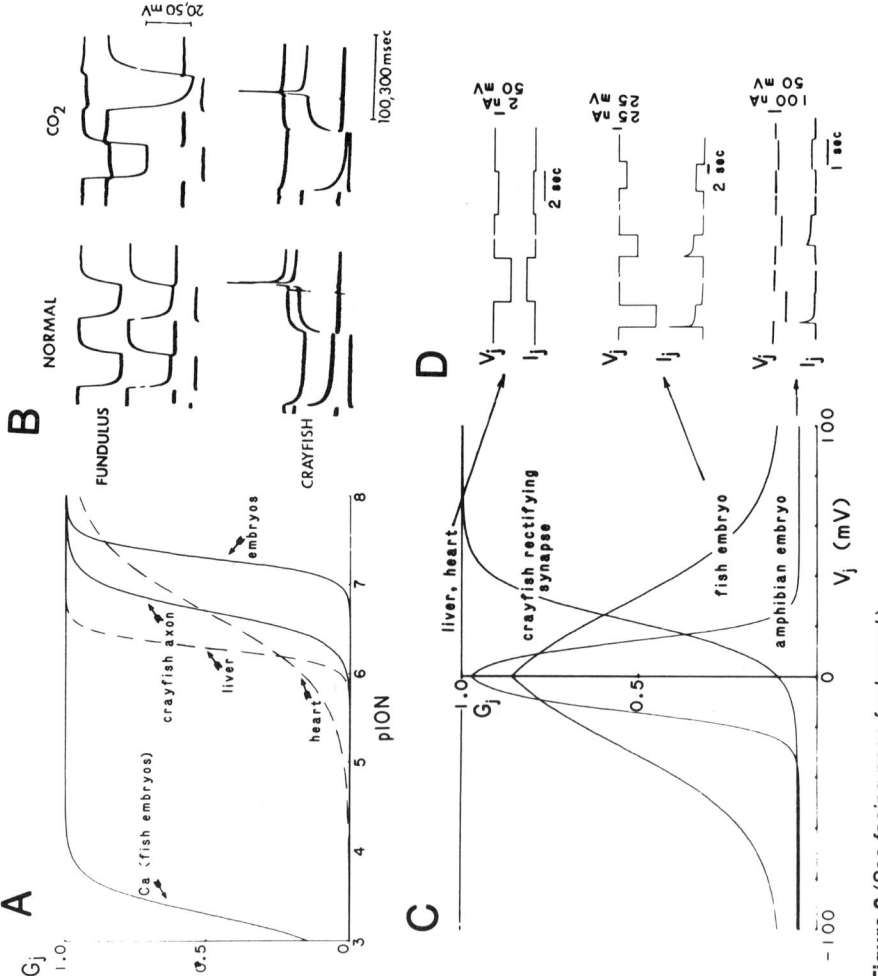

Figure 2 *(See facing page for legend.)*

VOLTAGE DEPENDENCE

Rectification, the easier transjunctional passage of currents in one direction, has been observed at several synapses (Furshpan and Potter 1959; Auerbach and Bennett 1969; Nicholls and Purves 1972). Presumably, in response to transjunctional voltages of appropriate sign, junctional channels are opened or closed: In amphibian embryonic cells the effect of transjunctional voltage is symmetric, voltages of either sign closing the channels. Change in g_j occurs very rapidly in the crayfish rectifying synapse, but the process is slow enough in amphibian embryonic cells that the kinetics can be followed in detail. Under voltage-clamp conditions, junctional current decays exponentially from its initial value when a transjunctional voltage pulse is applied from zero, the decay being more rapid and to a lower junctional conductance for larger voltages (Spray et al. 1979; Harris et al. 1981; Fig. 2C,D). The steady-state conductances attained for various V_j values in each polarity are well fit by Boltzmann relations, as would be expected if the energy difference between open and closed channels were a linear function of the field applied across the junction (Spray et al. 1981a; Fig. 2C). The phenomenon is not current dependent and is not appreciably affected by potential between cytoplasm and exterior (at least over the range ±30 mV). The steady-state voltage dependence is respectable for a membrane channel, being as steep as that of Na^+ activation in nerve membranes, but it is much slower (the time constants are in the range of 0.1 sec, compared with much less than 1 msec for Na^+ channels).

The symmetry in the Boltzmann relation raised the possibility that there might be two gates in series, which was confirmed by detailed examination of the kinetics when transjunctional voltages were reversed (Harris et al. 1981). More recent modelling suggests that this arrangement gives rise to an unexpected if quantitatively minor phenomenon (Fig. 3; Bennett and Spray 1984; Spray et al. 1984b). When one gate is closed by voltage of a given polarity, the other gate sees no voltage and thus will be closed a small fraction of the time (Fig. 3). If it closes, part of the voltage drop occurs across it, decreasing V_j across the first gate and increasing the probability that it will open. If the first gate opens, the second gate sees a voltage that strongly tends to open it, returning the junction to where the first gate sees a voltage that closes it, and

Figure 2
Gating of gap junctional conductance g_j by H^+ and Ca^{++} ions (A,B) and transjunctional voltage (C,D). (B) In most systems, here illustrated for fish blastomeres and crayfish septate axons, electrical coupling is reduced by CO_2 exposure. (A) Using pH-sensitive microelectrodes, G_j(normalized g_j) is related to intracellular pH (pION) by the Hill equation $G_j = K^n/(K^n + H^n)$ where K is the apparant dissociation constant and n, the Hill coefficient, is the equivalent number of highly cooperative binding sites. References are in Table 1. The curve of Ca^{++} effect is from perfusion experiments using well-buffered solutions where pH is maintained at 7.4 (Spray et al. 1982b). (D) Effects of command steps (V_j) under voltage clamp conditions on junctional current (I_j). In ventricular myocytes, I_j is constant during the V_j step; in fish and amphibian embryonic cells, I_j relaxes during the pulse to a steady-state level that depends on V_j. (C) Steady-state G_j-V_j relations are plotted for the several preparations for which data exist. References are in Table 1.

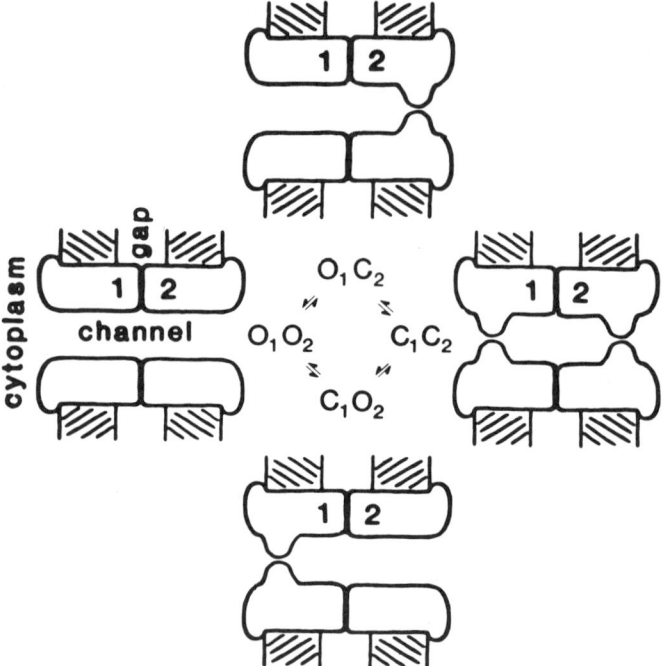

Figure 3
A circular reaction scheme applicable to the gating of voltage-sensitive channels of amphibian gap junctions. Symmetric voltage dependence implies symmetric voltage sensors, which we diagram as gates at the cytoplasmic ends of the channel. This location helps account for independence of g_j from voltage between cytoplasm and extracellular space in the gap. Consider a transjunctional voltage tending to close gate 1. Then most channels will be in state C_1O_2. In this state there will be no voltage across gate 2 and it will occasionally close spontaneously. Now in state C_1C_2 the voltage drop across gate 1 will be reduced and it will open at a greater rate than in state C_1O_2. Thus a few channels may move to state O_1C_2 where now voltage rapidly opens gate 2 and then closes gate 1.

so on. This cycling through a circular reaction scheme occurs because the voltage across each gate depends on the state of the other gate; cycling would be expected wherever channels have series gates or where the voltage seen by the gating mechanism depends on the state of the channel. The energy required for cycling is provided by the applied voltage. A similar scheme has been proposed for Na^+ channels (Finkelstein and Peskin 1984).

The steep voltage dependence in amphibian embryonic cells gives rise to regenerative behavior in the current-voltage curve, as can be illustrated by recording the voltages in a cell pair during application to one cell (cell 1) of a triangle of increasing and decreasing current (Harris et al. 1983; Fig. 4A). As the ramp increases, the voltage of cell 1 increases along a slope that is its coupled input resistance, and the voltage of cell 2 increases along its coupled transfer resistance. At a certain point, the cells' resistances change radically; the voltage in cell 1 rises suddenly to a line of steeper slope and thus its input resistance is higher; simultaneously, transfer to cell 2 is much reduced; the cells change from well to poorly coupled. On the descending limb of the triangle, the cells remain poorly coupled until the voltage difference between them is about what it was when they uncoupled; this occurs at a lower level of current than that which was required to uncouple them, because in the poorly coupled state the higher input and lower transfer resistances result in a larger voltage difference. The current-voltage (I-V) curves thus exhibit a region in which cells are either well or poorly coupled depending on history; in this region they are bistable and any process, even brief, can result in regenerative uncoupling or recoupling if there is a standing current across the junction (as would accompany heterogeneous activation of electrogenic Na^+ pumps) or if there were differences in intrinsic membrane potential (due to receptor activation, for example). An example of regenerative uncoupling in a cell pair that is initiated by a brief event and sustained by a difference in resting potentials is shown in Figure 4B. Although the voltage dependence in embryogenesis has an obvious possible role in establishment of compartmental boundaries between groups of cells with separate developmental fates, direct evidence for or against this function is lacking.

pH AND VOLTAGE ACT ON SEPARATE GATES

In amphibian embryonic cells, pH and voltage independently close gap junction channels. When g_j is reduced by cytoplasmic acidification, sensitivity of residual g_j to voltage is not noticeably altered, either in kinetics of the response or in the proportion of maximal g_j that is obtained in the steady state (Spray et al. 1984b). This implies that the gates involved in channel closure by pH and voltage are distinct. Pharmacological separation of the two gates is even more striking: Certain agents such as glutaraldehyde, ethoxy-n-ethoxydihydroquinoline (EEDQ), and retinoic acid abolish pH dependence but have no detectable effect on voltage dependence (cf. Spray and Bennett 1985).

GATING IS ALL-OR-NONE

In amphibian blastomeres, exponential relaxation of g_j in response to voltage steps is suggestive of a first-order all-or-nothing reaction (Harris et al. 1981). When electrical coupling is reduced, the diffusion of intracellularly injected probe molecules into adjacent cells is also reduced.

In coupled frog blastomeres permeability to symmetric quaternary alkylammonium compounds is proportional to g_j over a wide range when g_j is decreased by long voltage-clamp pulses (Verselis et al. 1985). When fluxes are measured in *Chironomus* salivary gland, reduction by uncoupling treatments is similar for several fluorescent compounds (Zimmerman and Rose 1983). These data do not support an earlier suggestion of the same laboratory of graded closure (Loewenstein 1981).The conclusion from these still somewhat limited data is that closure of the gap junction channel is all-or-none.

DISCUSSION

Comparative Physiology of pH Dependence

Data are now available on responses of gap junctions in various tissues to CO_2 exposure, in several of which the relation between pH_i and g_j has been measured (Table 1). In all these preparations, the action on g_j follows a smooth curve for increasing or decreasing pH values, and the action of H^+ ions can be modeled as titration curves (Fig. 2A) with characteristic pKs and Hill coef-

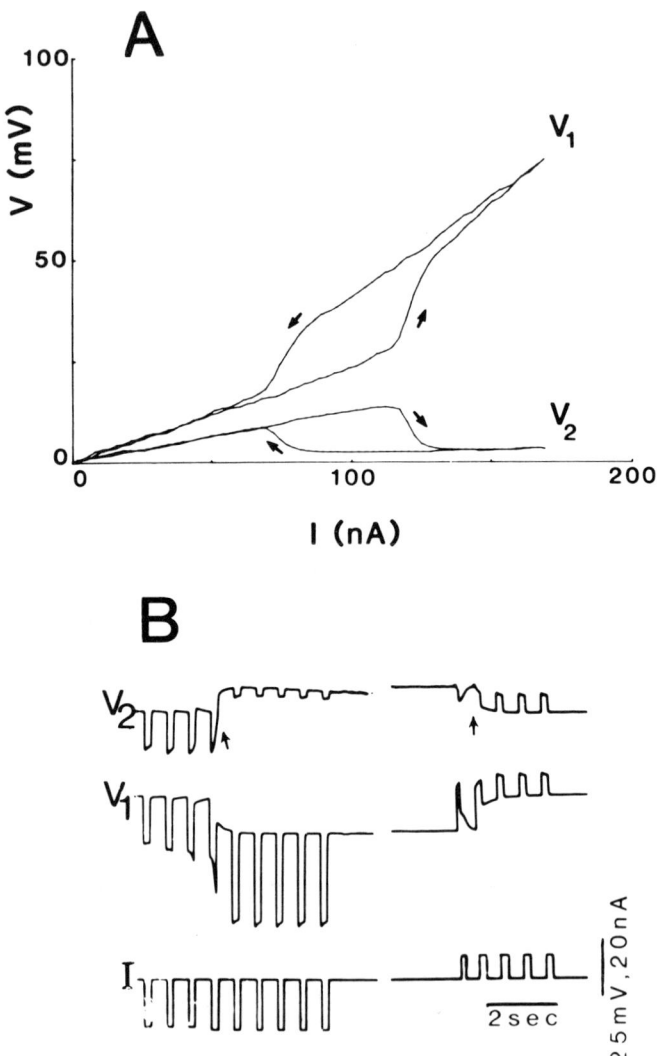

Figure 4 (*See facing page for legend.*)

ficients (n). For crayfish septate axon, reduction in cytoplasmic pH values to low levels or for extended periods results in poor or slow reversibility (Campos de Carvalho et al. 1984); presumably an additional mechanism is responsible. The apparent pK values measured are 7.3 for fish and amphibian blastomeres and about 6.7 for crayfish and earthworm septate axons, *Aplysia* neurons, rat ventricular myocytes and strands of Purkinje fibers; a similar pK has been estimated in preliminary experiments on cultured rat superior sympathetic ganglion neurons. The pK is lower, about 6.4, in pairs of freshly dissociated rat hepatocytes. The Hill coefficient, by which the number of cooperative elements or the extent of cooperativity is estimated, is large in fish and amphibian blastomeres and rat hepatocytes (4.5 and about 8, respectively) and is between 1 and 2 in the other tissues for which data exist.

What do these numbers for apparent pK and n tell us about the gating mechanism of gap junction channels? It seem reasonable that gating by pH involves conformational change in the gap junction protein produced by the binding of protons to specific sites. In six tissues, including some phylogenetically and embryonically diverse, the pKs are very similar, prompting speculation that similar amino acids may comprise the binding region of the gate. On the basis of studies of the amphibian and fish embryonic cells (pK 7.3), we had speculated that histidine residues might be involved in the active site. It seems more likely that if the binding site is homologous among these gap junctions, it is either a more acidic residue or the pK of the basic or neutral residue is modified by adjacent amino acids. Among the amino acid residues that are attractive candidates is lysine, which although intrinsically basic can have in biological tissues an apparent pK as low as 6.1 (Means and Feeney 1971) when in the vicinity of amino acid residues with free carboxyl groups. Support from attempts to modify the residues with group-specific protein reagents is mixed. Glutaraldehyde, which cross-links amino acids with some

Figure 4
Bistability in coupling between amphibian blastomeres resulting from voltage dependence of junctional conductance. (A) In response to a triangle of applied current, input (V_1) and transfer (V_2) voltages increase along a slope (V/I) indicating coupled input and transfer resistances. When $V_1 - V_2$ reaches a certain value (at about 120 nA in this example) the cells uncouple; the input resistance (V_1/I) increases and the transfer resistance (V_2/I) decreases. On the descending limb of the current triangle (arrows pointing toward the origin), input resistance stays high and transfer resistance stays low until $V_1 - V_2$ is small enough for the cell to recouple (corresponding to current of about 70 nA in this case). Over a range of applied currents from about 80 to 120 nA, the cells are either coupled or uncoupled, depending on history. (B) Bistability can occur as the result of endogenous currents. In a pair of cells, coupling is initially strong, as indicated by large V_2 deflections in response to hyperpolarizing current pulses in cell 1. After several successive pulses, the cells uncouple, as indicated by increased input resistance and descreased transfer resistance (first arrow). Thereafter V_2 depolarizes by about 15 mV and V_1 hyperpolarizes by a similar amount and the cells are maintained in a stably uncoupled condition for several minutes (break in record) until depolarizing current pulses (that reduce the transjunctional voltage, $V_1 - V_2$) restore the strong coupling and the initial resting potentials. (Reprinted, with permission, from Harris et al. 1983.)

Table 1
Voltage and pH Gating of Various Gap Junctions

Preparation (reference)	pH_i gating (pKa)	Voltage gating (V_0)
Mammalian liver (Spray and Hertzberg 1985)	yes (~6.4)	no
Mammalian heart (Reber and Weingart 1982; White et al. 1985)	yes (~6.8)	no
Mammalian lens (Schuetze and Goodenough 1982)	maybe not	?
Mammalian SCG neurons (D. Spray et al., unpubl.)	yes (?)	no
Amphibian blastomeres (Spray et al. 1979, 1981a,b)	yes (7.3)	yes (14 mV)
Amphibian Rohon-Beard neurons (Spitzer 1982)	yes (?)	yes (?)
Fish blastomeres (White et al. 1982)	yes (7.3)	yes (28 mV)
Fish horizontal cells (Lasater and Dowling, this volume)	?	no
Hatchetfish rectifying synapse (Auerbach and Bennett 1969)	?	yes (?)
Toadfish pancreatic islet cells (D. Spray and D. Marshak, unpubl.)	yes (?)	no
Squid blastomeres (Spray et al. 1984b)	yes (?)	yes[a]
Aplysia neurons (Bodmer and Spray 1985)	yes (c. 6.8)	no
Chironomus salivary gland (Rose and Rick 1978; Obaid et al. 1983)	yes (?)	yes[a]
Limulus retinular cells (Smith and Baumann 1969)	?	yes (?)
Crayfish septate axon (Campos de Carvalho et al. 1984; Johnston and Ramon 1982)	yes (6.75)	no
Crayfish rectifying synapse (Margiotta and Walcott 1983; Giaume and Korn, this volume)	yes (?)	yes (20 mV)
Earthworm septate axon (Brink et al. 1984; Verselis and Brink 1984)	yes (6.7)	no
Leech rectifying synapse (Nicholls and Purves 1980)	?	yes (?)

[a] In these cases, there is sensitivity to the voltage between inside and outside the gap junction channel (V_{i-o}) as well. For squid V_{i-o} sensitivity is weak compared to V_j sensitivity; for *Chironomus*, V_{i-o} sensitivity is stronger.

specificity for lysine residues, irreversibly blocks g_j, and the small g_j remaining after brief treatment with low concentrations is insensitive to pH_i (Spray et al. 1980). In crayfish septate axon and amphibian blastomeres, EEDQ, a heterofunctional reagent that cross-links carboxylic acid and amino residues, blocks pH dependence but does not affect g_j (Spray et al. 1984b; A.C. Campos de Carvalho et al., in prep.). Other agents expected to react with lysine residues have no effect but in crayfish the sulfhydryl reagents n-ethylmaleimide (NEM) and diamide block g_j (A.C. Campos de Carvalho et al., in prep.). (NEM also uncouples *Chironomus* salivary gland [Politoff et al. 1969] but does not affect pairs of fish embryonic cells; diamide blocks g_j in vertebrate embryonic cells [D.C. Spray et al., unpubl.].) The significance of the Hill coefficient may lie in the tertiary structure of the gap junction protein, and its value may indicate how interactive the subunits are in gap junctions of various tissues. Presumably all gap junctions are dodecamers of subunits, hexamers being contributed by each cell (but see Hanna et al., this volume). If there is one H^+ binding site per subunit, the low n for some tissues (e.g., unity in heart) would represent very low cooperativity among the sites; $n=1$ would also arise if binding of a single H^+ were adequate to close the channel.

Role of Ca^{++} Ions

Ca^{++} sensitivity of g_j has been estimated in several other tissues. In *Chironomus* salivary gland, coupling begins to be reduced at free Ca^{++} concentrations between 40 and 80 μM (Oliviera-Castro and Loewenstein 1971). Since this uncoupling can be reversed by hyperpolarization, Ca^{++} binding is apparently causing a conformational change affecting the dipole moment of the gate (Obaid et al. 1983). In heart, pCa_i has been measured after treatment with agents that are reported in increase Ca_i and to decrease coupling (Dahl and Isenberg 1980). These studies in which the pK for Ca^{++} dependence of uncoupling was estimated at under 5 μM are questionable on several grounds: pH_i was not measured simultaneously and may have changed enough to reduce g_j. Also, a contribution of nonjunctional conductance changes was not excluded. Recent direct measurement of g_j between pairs of ventricular myocytes has shown that g_j does not necessarily decrease when cells undergo irreversible contracture, a process raising myoplasmic free Ca^{++} above 10 μM (White et al. 1985).

That junctional membranes are so insensitive to Ca^{++} (about 10,000 times less than to H^+ in *Fundulus*) suggests that its role in regulating g_j under physiological conditions is minimal. On the other hand, Ca^{++} may play an important role in reducing g_j when a cell is broken open and its cytoplasmic Ca^{++} level elevated to that of extracellular medium. Alternatively, Ca^{++} may be involved in regulation of enzymes that act directly on the junctional membrane, including Ca^{++}-dependent proteases and kinases that may participate in modifying the protein and lipases that may modify the membrane.

Comparative Physiology of Voltage Dependence

Voltage dependence similar to that in amphibians has been seen in only a few other tissues. In fish embryonic cells, transjunctional voltages of either sign reduce g_j but the amphibian g_j-V_j curve is four times as steep as it is in fish and the voltage at which g_j is half-maximal is twice as large in fish as in amphibia. Kinetics in the fish system are apparently more complex, two exponentials being required to fit the decay in current over much of the voltage range (Fig. 2D; White et al. 1982). In the crayfish rectifying synapse, the steepness of the steady-state conductance voltage relation is about half that in amphibia and occurs only for one sign of transjunctional potential (depolarizing if viewed from the giant motor axon; Fig. 2C). Records obtained with current injection in several other systems suggest that g_j is affected by transjunctional potential, including ascidian embryos, *Xenopus* Rohon-Beard neurons, and *Limulus* ommatidium (Table 1). Sensitivity in these sytems seems high but quantitative studies under voltage clamp conditions are needed. Rectification occurs at several electrotonic synapses in leech ganglia (Nicholls and Purves 1972) but in some cases may be due to properties of nonjunctional membrane (Zipser 1979). Some gap junctions are not affected by moderate transjunctional potential, including those of rat superior cervical ganglion neurons in culture, freshly dissociated pairs of rat hepatocyte and ventricular myocytes, toadfish pancreatic islet cells, perch retinal horizontal cells, and septate axons of earthworm and crayfish (Table 1).

Other systems exhibit a dependence of g_j on the potential between inside and outside of the gap junction channel (V_{i-o}). Gap junctional conductance in *Chironomus* salivary gland is high at normal resting potential and decreases when the cells are depolarized (Obaid et al. 1983). In the absence of V_j, g_j as a function of V_{i-o} is well fit by a model with two series gates whose states are independent functions of V_{i-o}. The model does not fit the data when V_j is not small, and evidently g_j depends on both V_{i-o} and V_j. In squid embryonic cells, g_j is normally quite high and is not noticeably affected by V_{i-o} or V_j. When g_j is reduced by cytoplasmic acidification, residual g_j is increased by hyperpolarization of either cell alone or to a lesser extent both cells together, indicating that both V_{i-o} and V_j act on g_j in this system also (Spray et al. 1984b). Emergence of voltage dependence at low pH_i in squid suggests that pH and voltage act on the same gating portion of the channel molecule. Similarly in *Chironomus* the g_j-V_{i-o} curve is shifted by elevated pH_i or free Ca_i. This interaction is distinct from the case in amphibia where the gates are independent.

CONCLUSION: HOMOLOGOUS MEMBRANE CHANNELS CAN HAVE DIVERSE GATES

These studies on voltage and pH dependence of gap junctions in various tissues indicate that although there are common features, regulatory mechanisms are tissue and species specific. It should be admitted that except for rectifying synapses these regulatory mechanisms have not been shown to play a physiological role. Nevertheless, the sensitivity of g_j to voltage or pH is so high in some systems that a physiological role seems highly probable. The bistable region in the current-voltage relation in amphibian blastomeres is certainly broad enough to isolate populations of cells with distinct developmental fates (as can be done experimentally by manipulating the junctional current in a multicellular preparation; Harris et al. 1983). The low degree of voltage sensitivity in *Fundulus* makes a physiological role unlikely; for example, the measured transjunctional voltage at an early compartmental boundary is too small to cause uncoupling (Kimmel et al. 1984).

In crayfish and earthworm septate axon and vertebrate embryos, normal pH_i is just above the point where g_j is sensibly at its maximum (Fig. 2A). Although the cytoplasmic buffering capacity is large, anoxia at least should decrease pH_i into the significant region. In crayfish and *Aplysia* where the role of coupling is presumably electrotonic transmission, there is a decrease in g_{nj} as g_j decreases so that the coupling coefficient is relatively constant over a broad pH range (Giaume and Korn 1982; Campos de Carvalho et al. 1984). In liver, the normal pH_i is so far above the pK (almost a full pH unit) that even though the curve is steep the cells probably never uncouple under physiological conditions (Spray et al. 1984b). In heart the pH_i-g_j relation is not very steep, but the normal pH_i is near the pK of the channel (Spray et al. 1985). In most regions of the heart the safety factor for impulse propagation is very high and small changes in g_j would be of little functional importance. However, where conduction is slowed (e.g., at the interface between Purkinje fibers and ventricular muscle; Overholt et al. 1984) and under pathological conditions in which the safety factor is decreased, even a small decrease in g_j could reduce coupling to below threshold for transmission. From these few examples we conclude that pH control of g_j is likely to have a pathological if not physiological significance.

Work is only beginning on the longer-term regulation of junctional conductance, but questions of interest include whether formation and removal of junctions are affected by gating processes described here.

ACKNOWLEDGMENTS

Work reported here from our laboratory was supported in part by NIH grants NS 16524 and NS 19830 (D.C.S.) and HD 04248 and NS 07512 (M.V.L.B.), a Grant-in-Aid from the American Heart Association, and a McKnight Development Award (D.C.S.).

REFERENCES

Auerbach, A.A. and M.V.L. Bennett. 1969. A rectifying electrotonic synapse in the central nervous system of a vertebrate. *J. Gen. Physiol.* **53:** 211–237.

Bennett, M.V.L. 1966. Physiology of electrotonic junctions. *Ann. N.Y. Acad. Sci.* **37:** 509–539.

Bennett, M.V.L. and D.C. Spray. 1984. Gap junctions: Two voltage dependent gates in series allow voltage induced steady state cycling around a circular reaction scheme. *Biophys. J.* **45:** 60a.

Bennett, M.V.L., J.E Brown, A.L. Harris, and D.C. Spray. 1978. Electrotonic junctions between *Fundulus* blastomeres: Reversible block by low intracellular pH. *Biol. Bull.* **155:** 442.

Bodmer, R. and D.C. Spray. 1985. Permeability and electrophysiological properties of *Aplysia* neurons in situ and in culture. *Biophys. J.* **47:** 504a.

Brink, P.R., V. Verselis, and L. Barr. 1984. Solvent-solute interactions within the nexal membrane. *Biophys. J.* **45:** 121–124.

Campos de Carvalho, A.C., D.C. Spray, and M.V.L. Bennett. 1984. pH dependence of transmission at electrotonic synapses of the crayfish septate axon. *Brain Res.* **321:** 279–286.

Dahl, G. and G. Isenberg. 1980. Decoupling of heart muscle cells: Correlation with increased cytoplasmic calcium activity and with changes of nexus ultrastructure. *J. Membr. Biol.* **53:** 63–75.

Finkelstein, A. and C.S. Peskin. 1984. Some unexpected consequences of a simple physical mechanism for voltage dependent gating in biological membranes. *Biophys. J.* **46:** 549–558.

Furshpan, E.J. and D.D. Potter. 1959. Transmission at the giant motor synapses of the crayfish. *J. Physiol.* **145:** 289–325.

Giaume, C. and H. Korn. 1982. Ammonium sulfate induced uncouplings of crayfish septate axons with and without increased junctional resistance. *Neuroscience* **7:** 1723–1730.

Harris, A.L., D.C. Spray, and M.V.L. Bennett. 1981. Kinetic properties of a voltage-dependent junctional conductance. *J. Gen. Physiol.* **77:** 95–117.

———. 1983. Control of intercellular communication by voltage dependence of gap junctional conductance. *J. Neurosci.* **3:** 79–100.

Johnston, M.F. and F. Ramon. 1982. Voltage independence of an electrotonic synapse. *Biophys. J.* **39:** 115–117.

Kimmel, C.D., D.C. Spray, and M.V.L. Bennett. 1984. Developmental uncoupling between blastoderm and yolk cell in the embryo of the teleost *Fundulus*. *Dev. Biol.* **102:** 483–487.

Loewenstein, W.R. 1981. Junctional intercellular communication: The cell-to-cell membrane channel. *Physiol. Rev.* **61:**829–913.

Margiotta, J.F. and B. Walcott. 1983. Conductance and dye permeability of a rectifying electrical synapse. *Nature* **305:** 52–55.

Means, G.E. and R.E. Feeney. 1971. *Chemical modification of proteins*. Holden-Day, San Francisco.

Nicholls, J.G. and D. Purves. 1972. A comparison of chemical and electrical synaptic transmission between single sensory cells and a motoneurone in the central nervous system of the leech. *J. Physiol.* **225:** 637–656.

Obaid, A.L., S.J. Socolar, and B. Rose. 1983. Cell-to-cell channels with two independently regulated gates in series: Analysis of junctional conductance modulation by membrane potential, calcium, and pH. *J. Membr. Biol.* **73:** 69–89.

Oliviera-Castro, G.M. and W.R. Loewenstein. 1971. Junctional membrane permeability: Effects of divalent cations. *J. Membr. Biol.* **5:** 51–77.

Overholt, E.D., R.W. Joyner, R.D. Veenstra, D. Rawlings, and R. Weidmann. 1984. Unidirectional block between Purkinje and ventricular layers of papillary muscles. *Am. J. Physiol.* **247:** H584–H595.

Politoff, A.L., S.J. Socolar, and W.R. Loewenstein. 1969. Permeability of a cell membrane junction. Dependence on energy metabolism. *J. Gen. Physiol.* **53:** 498–515.

Reber, W.R. and R. Weingart. 1982. Ungulate cardiac Purkinje fibres: The influence of intracellular pH on the electrical cell-to-cell coupling. *J. Physiol.* **328:** 87–104.

Rink, T.J., R.Y. Tsien, and A.E. Warner. 1980. Free calcium in *Xenopus* embryos measured with ion-selective microelectrodes. *Nature* **283:** 658–660.

Rose, B. and W.R. Loewenstein. 1975. Permeability of cell junctions depends on local cytoplasmic calcium activity. *Nature* **254:** 250–252.

Rose, B. and R. Rick. 1978. Intracellular pH, intracellular free Ca, and junctional cell-cell coupling. *J. Membr. Biol.* **44:** 377–415.

Scheutze, S.M. and D.A. Goodenough. 1982. Dye transfer between cells of the embryonic chick lens becomes less sensitive to CO_2 treatment during development. *J. Cell Biol.* **92:** 694–705.

Smith, T.G. and F. Baumann. 1969. The functional organization within the ommatidium of the lateral eye of the Limulus. Prog. Brain Res. **31:** 313-349.
Spitzer, N. 1982. Voltage and stage dependent uncoupling of Rohon-Beard neurones during embryonic development of Xenopus tadpoles. J. Physiol. **330:** 145-162.
Spray, D.C. and M.V.L. Bennett. 1985. Physiology and pharmacology of gap junctions. Annu. Rev. Physiol. **47:** 281-303.
Spray, D.C. and E.L. Hertzberg. 1985. Biophysical properties of rat liver gap junctional channels. Biophys. J. **47:** 505a.
Spray, D.C., A.L. Harris, and M.V.L. Bennett. 1979. Voltage dependence of junctional conductance in early amphibian embryos. Science **204:** 432-434.
―――. 1980. Glutaraldehyde differentially affects gap junctional conductance and its pH and voltage dependence. Biophys. J. **33:** 108a.
―――. 1981a. Equilibrium properties of a voltage-dependent junctional conductance. J. Gen. Physiol. **77:** 75-94.
―――. 1981b. Gap junctional conductance is a simple sensitive function of intracellular pH. Science **211:** 712-715.
―――. 1982a. Comparison of pH and Ca dependence of gap junctional conductance. In Intracellular pH (ed. R. Nuccietelli and D. Deamer), p. 445-461. A.R. Liss, New York.
Spray, D.C., J. Stern, A.L. Harris, and M.V.L. Bennett. 1982b. Gap junctional conductance: Comparison of sensitivities to H and Ca ions. Proc. Natl. Acad. Sci. **79:** 441-445.
Spray, D.C., R.L. White, F. Mazet, and M.V.L. Bennett. 1985. Regulation of gap junctions. Am. J. Physiol. **248:** H753-H764.
Spray, D.C., J.M. Nerbonne, A.C. Carvalho, A.L. Harris, and M.V.L. Bennett. 1984a. Substituted benzyl esters: A new class of compounds that reduce gap junctional conductance by cytoplasmic acidification. J. Cell Biol. **99:** 174-179.
Spray, D.C., R.L. White, A.C. Campos de Carvalho, A.L. Harris, and M.V.L. Bennett. 1984b. Gating of gap junction channels. Biophys. J. **45:** 219-230.
Spray, D.C., R.D. Ginzberg, E.A. Morales, M.V.L. Bennett, M. Babayatsky, Z. Gatmaitin, and I.N. Arias. 1984c. Physiological and pharmacological properties of gap junctions between dissociated pairs of rat hepatocytes. J. Cell Biol. **99:** 344a.
Turin, L. and A.E. Warner. 1978. Carbon dioxide reversibly abolishes ionic communication between cells of early amphibian embryo. Nature **270:** 56-57.
―――. 1980. Intracellular pH in early Xenopus embryos: Its effects on current flow between blastomeres. J. Physiol. **300:** 489-504.
Verselis, V. and P.R. Brink. 1984. Voltage clamp of the earthworm septum. Biophys. J. **45:** 147-150.
Verselis, V., R.L. White, D.C. Spray, and M.V.L. Bennett. 1985. Gap junctions permeability to TEA and conductance are proportional, suggesting all-or-none gating. Soc. Neurosci. Abstr. **9:** (in press).
White, R.L., D.C. Spray, A.C. Carvalho, and M.V.L. Bennett. 1982. Voltage dependent gap junctional conductance between fish embryonic cells. Soc. Neurosci. Abstr. **8:** 944 (Abstr.).
White, R.L., D.C. Spray, A.C. Campos de Carvalho, B.A. Wittenberg, and M.V.L. Bennett. 1985. Some physiological and pharmacological properties of cardiac myocytes dissociated from adult rat. Am. J. Physiol. (in press).
Zimmerman, A.L. and B. Rose. 1983. Analysis of cell-to-cell diffusion kinetics: Changes in junctional permeability without accompanying changes in selectivity. Biophys. J. **41:** 216a.
Zipser, B. 1979. Voltage-modulated membrane resistance in coupled leech neurons. J. Neurophysiol. **42:** 465-475.

Control of Junctional Permeability

Fidel Ramón, Guido A. Zampighi,* and Amelia Rivera

Departmento de Fisiología y Biofísica
Centro de Investigación del IPN
Apartado Postal 14-740, México, 07000, D.F. Mexico

*Department of Anatomy and
Jerry Lewis Neuromuscular Research Center
UCLA School of Medicine, University of California
Los Angeles, California 90024

The lateral giant axons of the crayfish nerve cord meet at the septa in regions called "windows," openings in the septal connective tissue where the membranes of both axons come together to form communicating (gap) junctions. These junctions perform the role attributed to the gap junctions of other tissues, although there are some morphological differences between these and, for example, liver gap junctions (see Table 1 in Zampighi et al. 1978).

In rat liver, the communicating junctions are formed by a membrane protein with a central hydrophilic channel. When isolated, these structures appear in two configurations that can be correlated with the open and closed state of the channels (Unwin and Zampighi 1980). The mechanisms by which these channels are regulated is a subject that has recently received a great deal of attention. In this report we present an analysis of some of the experimental procedures that have resulted in closing the junctional channels of crayfish axons and other preparations, and some considerations on whether or not these maneuvers may also be a means of cellular control under physiological conditions.

EXPERIMENTAL CONTROL

Regulation of the state of the channels was first attributed to the effects of changes in the intracellular Ca^{++} concentration $[Ca^{++}]_i$, with elevated Ca^{++} producing uncoupling (Loewenstein 1966, 1981). More recently Turin and Warner (1977) and Spray et al. (1981a) have proposed that uncoupling may be a consequence of increased internal H^+ concentrations.

At the time that these hypotheses appeared, the response of the junctional channels to either Ca^{++} or H^+ was measured under conditions in which no information was directly available on the internal concentrations of either ion. To circumvent this difficulty in the interpretation of the effects of the ions, it was decided to implement an internal perfusion technique on coupled cells. This way the internal concentrations of either H^+ and Ca^{++} ions at which uncoupling occurred could be accurately measured. The preparation used in these experiments was the septate lateral axon from crayfish, whose large average diameter of 80–100 μm permitted internal perfusion by means of glass cannulas (Johnston and Ramón 1981).

The results from experiments on internally perfused axons demonstrated that neither internal Ca^{++} (up to 1 mM) or internal pH (down to 5.4) affected the coupling parameters of these axons (Fig. 1), whereas similar Ca^{++} con-

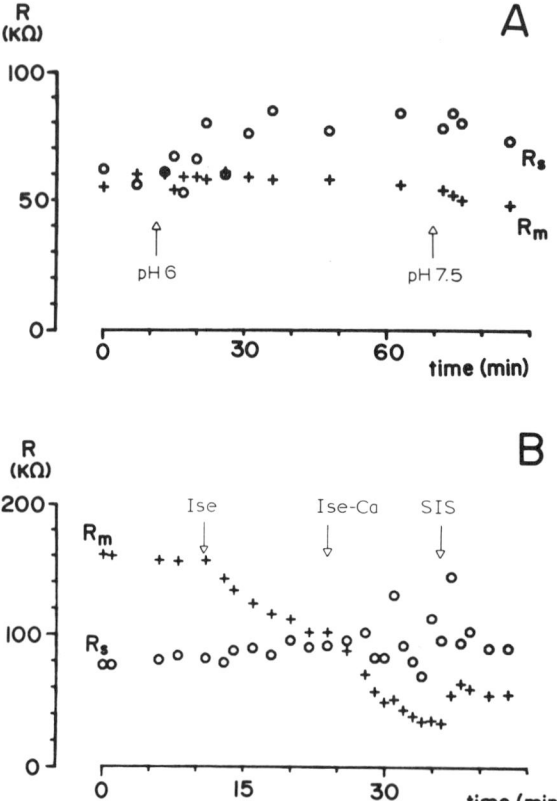

Figure 1
Junctional and membrane resistance of crayfish axons during internal perfusion. (A) Membrane (+) and septal (○) resistances change very little during 60 min perfusion at pH 6.0. (B) During perfusion with 1.25 mM Ca^{++} the septal resistance (○) remains constant, while membrane resistance (+) decreases slightly. (Reprinted, with permission from Johnston and Ramón 1981).

centrations and internal acidification to about pH 6.4 uncoupled axons that were not perfused. Since perfused axons could still be uncoupled with glutaraldehyde (3%), heptanol (3 mM), or octanol (1 mM; Fig. 2) added to the external medium, and since the effect of the alcohols was largely reversible and seemed to be a direct one on the junctional proteins (see below), the possibility that the perfusion maneuvers had somehow damaged the coupling structures was discarded. When octanol, up to a concentration of 10 mM, was applied to the interior of the axons the junctional resistance remained unchanged. These results suggested that the perfusion solution had washed away, or somehow inactivated, a soluble intermediary that was necessary for the junctions to perform their gating function (Johnston and Ramón 1981). More recently, calmodulin has been proposed to be this intermediary (Peracchia et al. 1983).

Similarly perfused axons were used to conduct voltage clamp studies of the junctional region and to obtain its current voltage relationships. These results demonstrated that coupling junctions from crayfish axons are voltage independent in the range ±100 mV (Johnston and Ramón 1982). These results are in contrast to the voltage dependence of the junctions between embryonic cells of amphibia (Spray et al. 1981b).

The effect of some alcohol anesthetics has also been tested on perfused and intact axons. In both, extracellular application of the alcohols 1-heptanol (3 mM) and 1-octanol (1 mM) produced reversible uncoupling within 10–20 minutes of their application to the bath solution. Other anesthetics, such as procaine (5 mM), benzocaine (3 mM), propranolol (1 mM), ketamine (2 mM), 1-hexanol (10 mM), and 1-nonalol (0.3 mM) had no effect on the coupling parameters. Since the effects of heptanol and octanol were similar whether the axons had been perfused or not, this suggested that the alcohols reached a "site" directly on the protein of the junction. Furthermore, the same alcohols applied internally to the perfused axons had no effect on the coupling function, even at concentrations up to 10 mM, suggesting that the site is located on, or near, the external hydrophilic region of the protein (Johnston et al. 1980).

An alternative explanation for the effects of heptanol and octanol could be

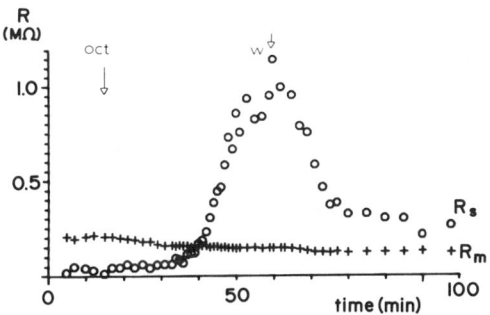

Figure 2
Effect of octanol (1 mM) on the junctional resistance of internally perfused crayfish axons. Septal resistance (O) increases dramatically in response to the alcohol and then decreases after washing. Membrane resistance (+) remains constant.

that the alcohols partition into the membrane lipids influencing their fluidity, as has been suggested to explain the effects of alcohols on the miniature endplate currents recorded from toad sartorius muscle (Gage et al. 1974). However, if this were the case for the junctions of crayfish axons, one would expect the uncoupling effect of the alcohols to increase with the hydrocarbon chain length, such as has been found at the neuromuscular junction (Gage et al. 1975). This is not the case and our studies of the effect of several alcohols on the coupling of crayfish axons demonstrated that ethanol, 1-butanol, 1-hexanol, 1-nonalol, and 1-decanol had no effect and only 1-heptanol and 1-octanol produce uncoupling. Thus, the most reasonable explanation for these results is that the alcohols exert their effect directly on the junctional proteins of crayfish axons. Similar uncoupling effects by heptanol and octanol have now been found in cardiac muscle (Délèze and Hervé 1983), *Xenopus* embryos (Bernardini et al. 1984), and fish embryonic cells (Spray et al. 1984).

PHYSIOLOGICAL CONTROL

There is no doubt that cell uncoupling can be experimentally produced by any of the following mechanisms: (1) high internal $[Ca^{++}]_i$; (2) low internal pH, and; (3) the alcohol anesthetics heptanol and octanol. The question remains whether or not one of these is a control mechanism that can uncouple cells during physiological conditions. Clearly the drugs are not physiological, thus only Ca^{++} and/or H^+ remain.

A physiological phenomenon that has been explained as uncoupling due to a high $[Ca^{++}]_i$ is the process of "healing-over." This occurs when a coupled cell is injured, and involves uncoupling due to the $[Ca^{++}]_i$ being raised because of an increased membrane permeability. In this case the phenomenon has been studied carefully and the experimental results can be explained by the increased $[Ca^{++}]_i$ rather than by an internal acidification (De Mello 1983).

Another physiological phenomenon during which intracellular Ca^{++} and internal pH suffer important changes is muscle contraction. Since both cardiac and smooth muscle are coupled via communicating junctions very similar to those of other tissues, it would seem that this is a phenomenon from which very useful information could be obtained about the physiological control mechanism of junctional permeability.

It has been demonstrated that the $[Ca^{++}]$ that uncouples Purkinje fibers of the sheep heart is 5×10^{-6} M (Dahl and Isenberg 1980), whereas complete uncoupling of amphibian blastomeres can be obtained with changes in the internal pH of about 0.6 pH unit (Spray et al. 1981a). Thus, in view of the fact that during muscle contraction internal Ca^{++} rises to levels probably in the micromolar range, and internal pH drops, it is conceivable that cardiac and smooth muscle cells might uncouple, even though transiently, during contraction.

During tetanic stimulation of frog skeletal muscle, the Ca^{++} released from the sarcoplasmic reticulum increases the myoplasmic concentration from a basal level of about 10^{-7} M to about 10^{-3} M, as measured by electron probe analysis (Somlyo et al. 1981). The $[Ca^{++}]_i$ reached during contraction of vascular smooth muscle cells, even though it does not reach 10^{-5} M (Bond et al. 1984), is also well above threshold for uncoupling cardiac muscle cells.

Simultaneously with muscle contraction, and due to the H^+ released when Ca^{++} binds, the internal pH of skeletal muscle changes by about 0.074 pH unit. Since measured changes are about 0.4 pH unit, the remaining change must be due to the demonstrated acceleration of glycogenolysis by micromolar Ca^{++} concentrations in the presence of troponin. This must be intensive and prolonged enough to lead to lactic acid production in an amount sufficient to explain the observed fall in internal pH (Abercrombie and Ross 1983).

Two facts seem to contribute to the small changes in internal pH that are produced during muscle contraction. These are: (1) The high pH buffering capacity of the intracellular medium. The buffering capacity of both vertebrate and invertebrate muscle is of the order of 40–50 mmol/L pH (with this amount of buffering roughly 40,000 H^+ ions must be injected into the cytoplasm for each free H^+ appearing intracellularly). (2) H^+ are actively and rapidly transported out of the cell. This mechanism accounts for the long-term maintenance of resting pH_i, since H^+ extrusion neutralizes the tendency of the cytoplasm to acidify due to both metabolic and passive H^+ influx. In the case of smooth muscle cells, these regulate their internal pH very closely in the face of loads of CO_2 in the bath solution, and their intracellular buffering power is even higher than that of skeletal or cardiac muscle (Aickin 1984).

It would follow from the above considerations that, since changes in $[Ca^{++}]_i$ of cardiac and smooth muscle cells are considerable, they should uncouple during contraction, even if only for a brief period of time. We would not expect the muscle cells to uncouple in response to the very small changes that occur in internal pH.

To detect a possible uncoupling of cardiac or smooth muscle cells during physiological contraction one can measure the cable parameters related to the internal resistance of a muscle strip during the passage of a propagated action potential. This resistance reflects the myoplasmic and junctional resistances. Since changes in myoplasmic resistance are not expected during the propagation of action potentials, the internal resistance values should yield an indication of the state of the coupling structures. These measurements can be performed by injecting small current pulses during an action potential propagating in a muscle strip and measuring the resulting voltage changes. This method was used by Weidmann (1970) who demonstrated that there were no changes in longitudinal resistance during the cardiac muscle action potential. In spite of possible shortcomings of the method, this result indicates that cardiac muscle cells do not uncouple during the physiological contraction triggered by the depolarizing wave. A similar conclusion in regard to a single twitch has been reached by Weingart (1977).

Figure 3A shows a cardiac muscle action potential and the superimposed voltage changes produced by the injection of small pulses of current. As can be seen the amplitude of the recorded voltage changes is similar during the plateau and recovery phases of the action potential. To improve the chances of measuring junctional resistance, these pulses were injected during the plateau phase of cardiac action potentials, when membrane resistance and intracellular Ca^{++} levels are also high.

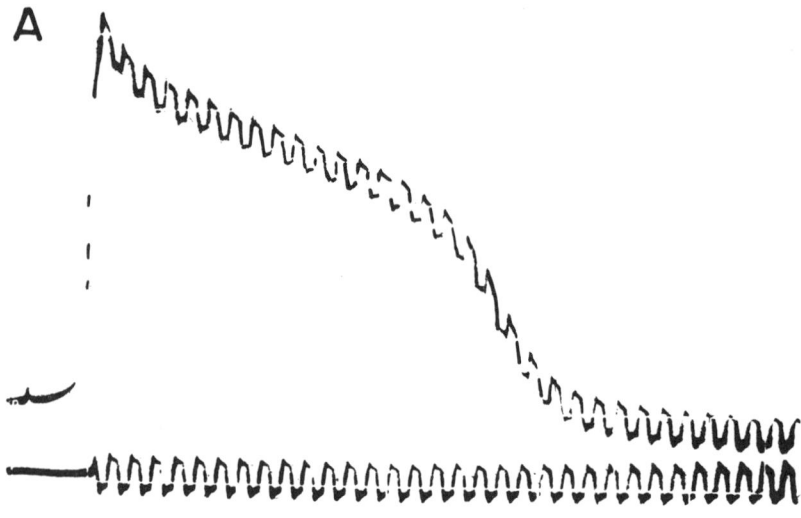

Figure 3
(A) Action potential from a bundle of dog Purkinje fibers. Small voltage changes in response to current pulses are superimposed. The amplitude of the voltage response does not change throughout the plateau phase and is similar to responses in the absence of an action potential (*lower trace*). (B) Action potential from a strip of rat uterine smooth muscle. The voltage responses are of equal magnitude throughout the propagated action potential. Calibration marks are 20 and 30 mV and 50 and 20 msec for panels A and B, respectively.

Figure 3B shows another action potential (from rat uterine smooth muscle) with superimposed voltage changes that are of similar magnitude throughout the entire action potential. However, since in this case there is no plateau phase, membrane resistance might be changing more rapidly than during cardiac action potentials.

The results from the experiments described above indicate that the longitudinal resistance of both cardiac and smooth muscle cells does not change radically during the passage of propagated action potentials. These results suggest that muscle cells do not uncouple during the contraction wave produced by the propagated action potential, and this junctional insensitivity can be explained by one of the following mechanisms: (1) the coupling junctions from cardiac and smooth muscle cells are of a different nature from those of the cells in which sensitivity to Ca^{++} and pH has been shown, or (2) the muscle junctional regions are somehow protected from changes in the internal ionic environment. In either case, it would seem that neither Ca^{++} nor H^+ could be a physiological control mechanism of the state of the junctional channels. A similar conclusion has been reached using intracellular current pulses and recording of electrotonic potentials in cardiac muscle (Pressler 1984).

UNCOUPLING BY A CELL NOT INVOLVED IN COUPLING

Cell uncoupling has been assumed to be due, in most cases, to closure of junctional channels. However, at least in one case, uncoupling seems to be produced by a cell not involved in the coupling, as suggested by the experiments of Pappas et al. (1971). These authors observed that during mechanical trauma of crayfish septate axons, the surrounding Schwann cells invaded the region of close apposition of the axons. Axon uncoupling was simultaneous with the invasion, as measured with standard electrophysiological techniques. Since this seemed an example of uncoupling produced by a mechanism that could have physiological implications, we investigated the effect of a drug that induces similar behavior of the Schwann cells.

Figure 4 shows micrographs of the coupling region of septate lateral axons of crayfish 60 minutes after application of nicotine (1 mM) to the bath solution. At this time action potentials stopped traversing the septal area and junctional resistance had increased from a control value of 65 KΩ to about 820 KΩ. Figure 4a shows the control aspect of the junctional region between the axons. The characteristic rows of vesicles at both sides of the junction are clearly seen. Figure 4, b–e, can be described as a time series, although the photographs were obtained from the same experiment. In a first stage of uncoupling produced by nicotine, the Schwann cells seem to compress the junctional region in the direction of the arrows producing its folding, as shown in Figure 4b. Figure 4c illustrates the disarray of membranes in a junctional area that has been completely disorganized by the presence of Schwann cells. Regions where the junction has been obliquely and transversally sectioned can be seen. Later on, Figure 4d shows the remnant of the junctional region (arrows) with a point area of contact to the neighboring axon, but without the charac-

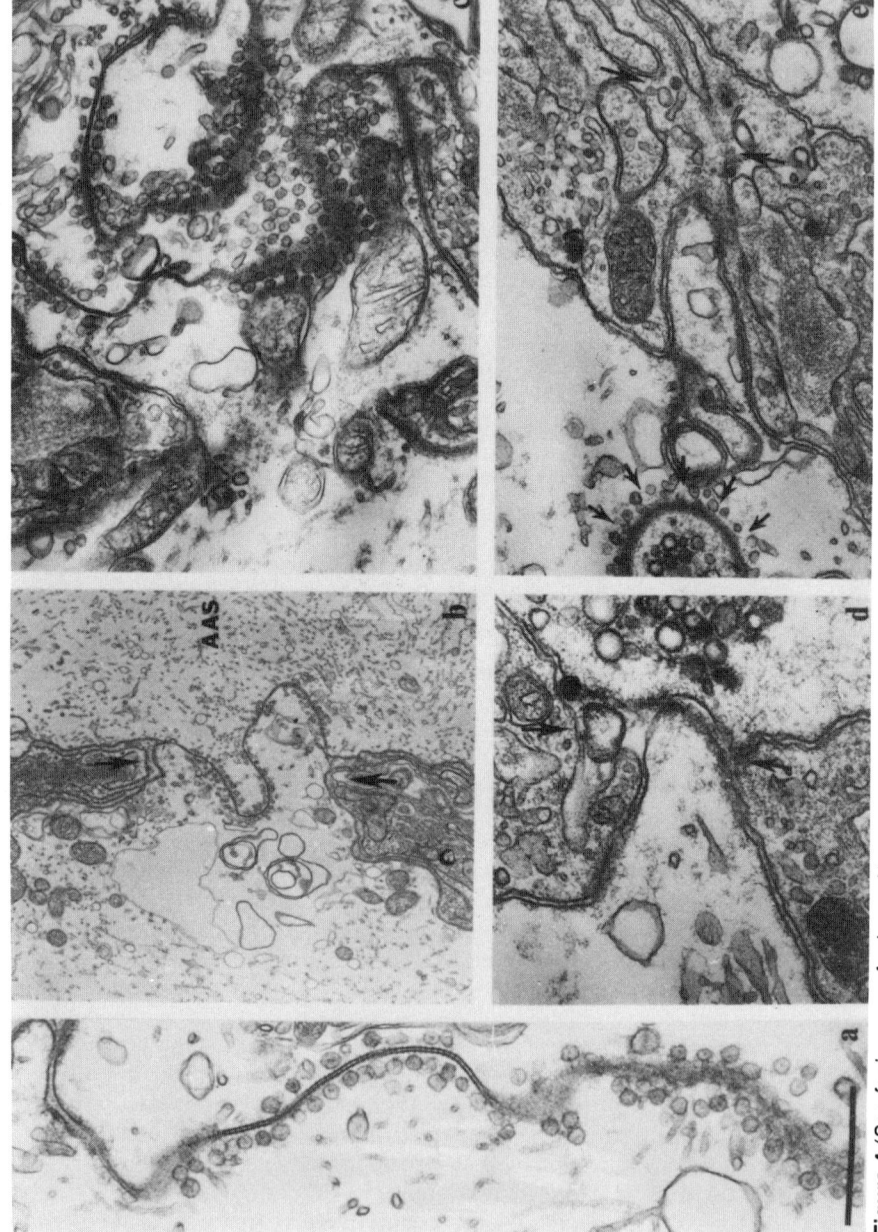

Figure 4 (See facing page for legend.)

teristic vesicular pattern shown in Figure 4a. In a final stage, Figure 4e shows that the area of contact of the axons has disappeared. On the far left side an annular junction (arrows) can now be seen. On washing the nicotine from the bath solution, electrophysiological recoupling appears after approximately 30-45 minutes.

To eliminate the possibility that the annular junction shown in Figure 4e belonged to a transverse section of an interdigitation from one of the axons, we performed a semiserial study. Figure 5 shows micrographs from this semiserial study of the same annular junction shown in Figure 4e. The annular junction is large enough to include a mitochondria, and still maintains the rows of vesicles associated with the membranes. Its cross section appears circular in all photographs, suggesting that it is a spherical structure and not a digitation from the neighboring axon.

Annular junctions have been seen in crayfish axons after injury (Hanna et al. 1984) and also in ovarian granulosa cells (c.f., Larsen, this volume). In general they are found in tissues known or suspected to be targets of peptide hormones and it appears that hormones specifically acting on adenylate cyclase may also stimulate their internalization (Azarnia and Larsen 1977). Presumably these annular junctions also play a role in recycling the junctional areas (Burghardt and Anderson 1979; Larsen 1983). However, in the case of crayfish axons, the annular junctions break the pattern in that they do not seem to be associated to a known hormone or to cAMP, although they might be part of the physiological recycling process.

Thus, it seems that nicotine is able to induce uncoupling of crayfish axons by a mechanism where channel closure, if existent, is only secondary. We do not, at this point, have an explanation for this behavior of Schwann cells; however, reactivity of these cells to nicotine would not be unusual, as it has been shown that they have cholinergic receptors of the nicotinic type (Lieberman et al. 1981) and that they are able to produce release of the neurotransmitter acetylcholine (Dennis and Miledi 1974).

In summary, it would seem that although changes in internal levels of Ca^{++} and H^+ can produce uncoupling, these are useful experimental procedures that might not contribute to physiological uncoupling. It is possible that junctional channels are permanently open and that to uncouple the cells, a mechanism that removes the membrane region containing the channels is required.

Figure 4
Electronmicrographs of the junctional area of crayfish lateral axons 30 min after nicotine (1 mM) was applied to the external solution. (a) Control aspect of the junction. The characteristic rows of vesicles at both sides of the membrane can be seen. (b) Folding of the junction (between arrows) as Schwann cells push the axons apart. (AAS) Anterior axon segment. (c) Oblique and transversal section of the junctional membranes. (d) Between arrows is a remnant of a junctional area without the junctional membranes. (e) The axons have moved apart and the junctional area has disappeared. The space is occupied by Schwann cells. On the left hand side there is now an annular junction delimited by arrows. Calibration bar, 4, 20, 2.5, 7, and 5 nm for panels a–e, respectively.

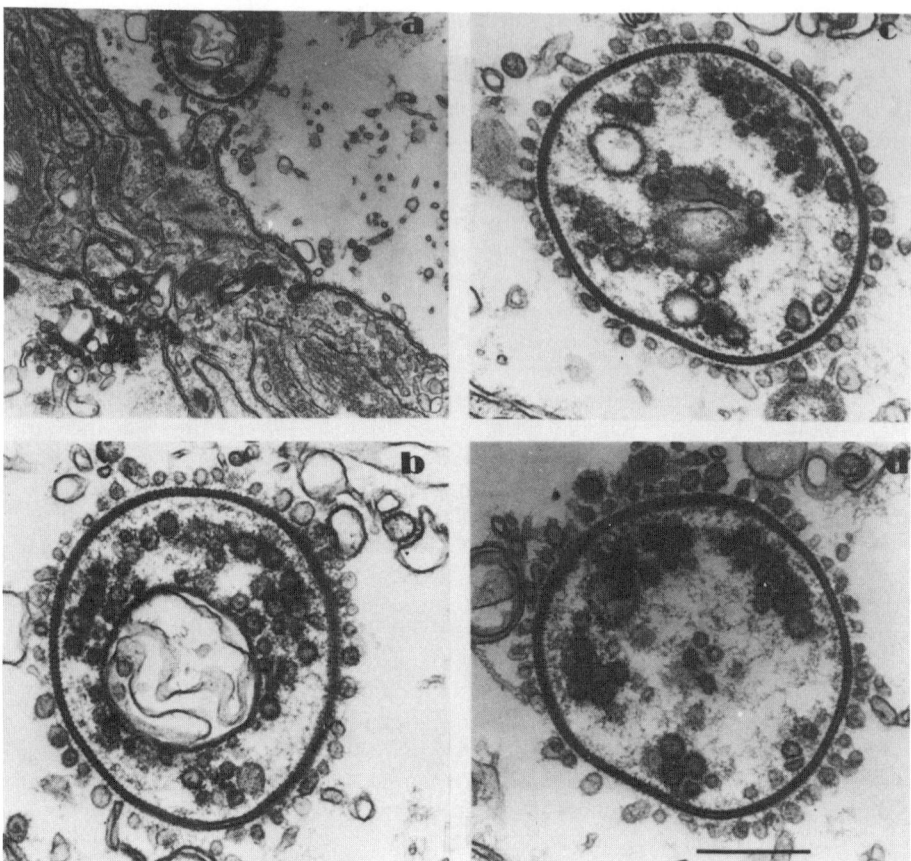

Figure 5
Semiserial sections of an annular junction. (a) Area of the axons where, apparently, there was a communicating junction. An annular junction can be seen in one of the axons. (b, c, and d) Sections of the annular junction seen in a. In all panels the junction is circular and still has the double row of vesicles that characterizes the crayfish junctions. This annular junction is large enough to include a mitochondria. Calibration bar, 7 nm for a and 3 nm for b–d.

ACKNOWLEDGMENTS

This work was partially supported by a grant from CONACyT, México, NIH grant EY04110, and an MDA Jerry Lewis Neuromuscular Research Center Grant.

REFERENCES

Abercrombie, R.F. and A. Ross. 1983. The intracellular pH of frog skeletal muscle: Its regulation in hypertonic solutions. *J. Physiol.* **345:** 189–204.

Aickin, C.C. 1984. Direct measurement of intracellular pH and buffering power in smooth muscle cells of guinea-pig vas deferens. *J. Physiol.* **349:** 571–585.

Azarnia, R. and W.J. Larsen. 1977. Intercellular communication and cancer. In *Intercellular communication* (ed. W.C. De Mello), p. 145–172. Plenum Press, New York.

Bernardini, G., C. Peracchia, and L.L. Peracchia. 1984. Reversible effects of heptanol on gap junction structure and cell-to-cell electrical coupling. *Eur. J. Cell Biol.* **34:** 307–312.

Bond, M., T. Kitazawa, A.P. Somlyo, and A.V. Somlyo. 1984. Release and recycling of calcium by the sarcoplasmic reticulum in guinea-pig portal vein smooth muscle. *J. Physiol.* **355:** 677–695.

Burghardt, R.C. and E. Anderson. 1979. Hormonal modulation of ovarian interstitial cells with particular reference to gap junctions. *J. Cell Biol.* **81:** 104–114.

Dahl, G. and G. Isenberg. 1980. Decoupling of heart muscle cells: Correlation with increased cytoplasmic calcium activity and with changes of nexus ultrastructure. *J. Membr. Biol.* **53:** 63–75.

Délèze, J. and J.C. Hervé. 1983. Effect of several uncouplers of cell communication on gap-junction morphology in mammalian heart. *J. Membr. Biol.* **74:** 203–215.

De Mello, W.C. 1983. The influence of pH on the healing-over of mammalian cardiac muscle. *J. Physiol.* **339:** 299–307.

Dennis, M.J. and R. Miledi. 1974. Electrically induced release of acetylcholine from denervated Schwann cells. *J. Physiol.* **237:** 431–452.

Gage, P.W., R.N. McBurney, and G.T. Schneider. 1975. Effects of some aliphatic alcohols on the conductance change caused by a quantum of acetylcholine at the toad end-plate. *J. Physiol.* **244:** 409–429.

Gage, P.W., R.N. McBurney, and D. Van Helden. 1974. Endplate currents are shortened by octanol: Possible role of membrane lipid. *Life Sci.* **14:** 2277–2283.

Hanna, R.B., G.D. Pappas, and M.V.L. Bennett. 1984. The fine structure of identified electrotonic synapses following increased coupling resistance. *Cell Tissue Res.* **235:** 243–249.

Johnston, M.F. and F. Ramón. 1981. Electrotonic coupling in internally perfused crayfish segmented axons. *J. Physiol.* **317:** 509–518.

———. 1982. Voltage independence of an electrotonic synapse. *Biophys. J.* **39:** 115–117.

Johnston, M.F., S.A. Simon, and F. Ramón. 1980. Interaction of anesthetics with electrical synapses. *Nature* **286:** 498–500.

Larsen, W.J. 1983. Biological implications of gap junction structure, distribution and composition: A review. *Tissue & Cell* **15:** 645–671.

Lieberman, E.M., J. Villegas, and G.M. Villegas. 1981. The nature of the membrane potential of glial cells associated with the median giant axon of the crayfish. *Neuroscience* **6:** 261–271.

Loewenstein, W.R. 1966. Permeability of membrane junctions. *Ann. N.Y. Acad. Sci.* **137:** 441–472.

———. 1981. Junctional intercellular communication: The cell-to-cell membrane channel. *Physiol. Rev.* **61:** 829–913.

Pappas, G.D., Y. Asada, and M.V.L. Bennett. 1971. Morphological correlates of increased coupling resistance at an electrotonic synapse. *J. Cell Biol.* **49:** 173–188.

Peracchia, C., G. Bernardini, and L.L. Peracchia. 1983. Is calmodulin involved in the regulation of gap junction permeability? *Pfluegers Arch. Eur. J. Physiol.* **399:** 152–154.

Pressler, M.L. 1984. Cable analysis in quiescent and active sheep Purkinje fibers. *J. Physiol.* **352:** 739–757.

Somlyo, A.V., H. González-Serratos, H. Shuman, G. McClellan, and A.P. Somlyo. 1981. Calcium release and ionic changes in the sarcoplasmic reticulum of tetanized muscle: An electron-probe study. *J. Cell Biol.* **90:** 577–594.

Spray, D.C., A.L. Harris, and M.V.L. Bennett. 1981a. Gap junctional conductance is a sensitive function of intracellular pH. *Science* **211:** 712–715.

———. 1981b. Equilibrium properties of a voltage dependent junctional conductance. *J. Gen. Physiol.* **77:** 75–94.

Spray, D.C., R.L. White, A.C. Campos de Carvalho, A.L. Harris, and M.V.L. Bennett. 1984. Gating of gap junctional channels. *Biophys. J.* **45:** 219–230.

Turin, L. and A.F. Warner. 1977. Carbon dioxide reversibly abolishes ionic communication between cells of early amphibian embryo. *Nature* **270:** 56–57.

Unwin, P.N.T. and G. Zampighi. 1980. Structure of the junction between communicating cells. *Nature* **283:**545–549.

Weidmann, S. 1970. Electrical constants of trabecular muscle from mammalian heart. *J. Physiol.* **210:** 1041–1054.

Weingart, R. 1977. The actions of ouabain on intercellular coupling and conduction velocity in mammalian ventricular muscle. *J. Physiol.* **264:** 341–365.

Zampighi, G., F. Ramón, and W. Durán. 1978. Fine structure of the electrotonic synapse of the lateral giant axons in a crayfish (*Procambarus clarkii*). *Tissue & Cell* **10:** 413–426.

Electrical Uncoupling Induced by General Anesthetics: A Calcium-independent Process?

Jacek A. Wojtczak
Department of Anesthesiology
Mount Sinai School of Medicine
New York, New York 10029
and Rockefeller University
New York, New York 10021

It was shown that some pathological conditions, such as hypoxia (Wojtczak 1979, 1982) or digitalis intoxication (Weingart 1977), produce an intercellular electrical uncoupling in cardiac muscle. Those studies have shown that an increase in resting tension (contracture) closely correlates with an increase in internal longitudinal resistivity (R_i), a sign of intercellular electrical uncoupling. Contracture is due to the accumulation of cytoplasmic Ca^{++} and its action on the contractile apparatus. It may also be partly due to the formation of the rigor complexes (ATP depletion) between contractile proteins.

Weingart (1977) has shown that long times (3 hr) are required for the development of contracture and uncoupling during digitalis intoxication. This latency can be explained by the persistent digitalis-induced Ca^{++} overloading, finally leading to a disruption of the normal mitochondrial function. It takes much less time for uncoupling and contracture to develop during hypoxia (about 30 min) which primarily impairs the mitochondrial function. Interestingly, in both cases R_i increases only after the resting tension begins to rise.

The cytoplasmic side of connexons is immersed in subsarcolemmal space, which in cardiac cells is probably well protected against the changes in the concentration of Ca^{++} and possibly H^+. The subsarcolemmal space is delimited from the bulk cytoplasm by a layer of mitchondria which may act as a "sink" for Ca^{++}. This may explain why a considerable rise (up to 10^{-5} M) in intracellular Ca^{++} accompanying every twitch does not uncouple cardiac cells. The concentration of Ca^{++} may increase around the contractile proteins, but mitochondria protect against "spilling" of this Ca^{++} to the subsarcolemmal space. In agreement with this suggestion, exposure of cardiac muscle to increased extracellular K^+ produces transient contracture but no electrical uncoupling (J. Wojtczak, unpubl.). However, once the function of the mitochondria is impaired (by hypoxia or long-term Ca^{++} overloading) Ca^{++} begins to spill into the subsarcolemmal space.

It was shown by Turin and Warner (1977) and Spray et al. (1981) that some embryonic cells can be uncoupled by intracellular acidosis caused, for example, by bathing in solutions gassed with high CO_2. Could the hypoxia-induced uncoupling be explained by the marked acidosis known to develop in these conditions? It is possible that H^+ may be involved in producing uncoupling in cardiac muscle, but the role played by H^+ seems to be small. Reber and Weingart (1982) showed that in cardiac Purkinje fibers exposed to 20% CO_2 (intracellular pH decreased from 7.0 to 6.6) R_i increased by 30%. Hypoxia and digitalis intoxication usually produce a severalfold increase in R_i. DeMello (1983) showed that the healing-over process in cardiac cells, which is very sensitive to Ca^{++} concentration, is practically insensitive to pH change. My own experiments with 20% CO_2 or 20 mM NH_4Cl washout in cat capillary muscle showed that R_i is not changed by pH (J. Wojtczak, unpubl.). On the other hand, White et al. (1984) have shown that isolated, reassociated cardiac myocytes are uncoupled by solutions saturated with 100% CO_2. Spray et al. (1982) have shown in mechanically dissociated and reassociated cells from *Fundulus* blastomeres that gap junctional conductance is much more sensitive to H^+ than to Ca^{++}. It is possible that the relative sensitivites to Ca^{++} and H^+ may be changed in chemically isolated and reassociated cardiac cells.

The supposition that Ca^{++} is responsible for uncoupling during hypoxia or digitalis intoxication in cardiac muscle is supported by other studies. Dahl and Isenberg (1980) measured intracellular Ca^{++}, cell–cell coupling, and gap junction ultrastructure in cardiac Purkinje fibers during the inhibition of the oxidative phosphorylation induced by dinitrophenol (DNP) and during digitalis intoxication. They found that intercellular uncoupling in these states is closely related to increasing intracellular Ca^{++} and produces characteristic morphological changes in gap junctions. Moreover, De Mello (1982) demonstrated that electrical uncoupling can be produced in cardiac Purkinje fibers by intracellular injections of Ca^{++} and other divalent ions.

If the uncoupling that develops in cardiac muscle during pathological processes is dependent on the increasing intracellular Ca^{++}, it should be possible to prevent or reverse it by reducing intracellular Ca^{++} or by inhibiting the protein calmodulin, which modulates intracellular Ca^{++} effects. Indeed, De Mello (1979) showed that by injecting EGTA intracellularly to chelate intracellular Ca^{++} it is possible to reverse uncoupling produced by DNP. Uncoupling produced by hypoxia can be prevented by the pretreatment of preparations with high doses of verapamil and other calcium antagonists, but such treatment is connected with a drastic reduction of contractile force (J. Wojtczak, unpubl.). Thus, the possibility that calmodulin inhibition may prevent uncoupling is particularly attractive because potentially it may not be connected with a reduction of contractile force. Peracchia et al. (1983) and Peracchia (1984, and this volume) have reported that calmodulin inhibitors can prevent uncoupling induced in *Xenopus* embryos by CO_2. One of the aims of this study was to investigate the influence of the calmodulin inhibitor trifluoperazine on the electrical coupling in cardiac muscle as measured by the changes in internal longitudinal resistance of ventricular muscle bundles. Since trifluoperazine is also a potent local anesthetic, its effects on coupling are compared with those of lidocaine, a local anesthetic with no inhibitory effects on calmodulin.

EFFECTS OF LOCAL ANESTHETICS ON INTERCELLULAR COUPLING

Initial experiments designed to prevent hypoxia-induced or digitalis-induced uncoupling with trifluoperazine proved to be very difficult, if not impossible. Preparations quickly became inexcitable since the effects of local anesthetics (including trifluoperazine) on Na^+ channels are potentiated in hypoxic or depolarized cells. It was therefore essential to change the protocol and Ca^{++}-loading procedure.

Cat or dog ventricular papillary or trabecular muscles were exposed to Tyrode's solution made hypertonic by the addition of 900 mOsm of sucrose (cumulative tonicity, 1200 mOsm). Hypertonicity has been shown by Barr et al. (1965) to disrupt gap junctions in cardiac muscle. Methods used were similar to those previously described (Wojtczak 1979). Briefly, the ratio of internal (V_i) to external (V_o) potentials for a propagating impulse was measured in the muscle immersed in silicon oil. This ratio is equal to the ratio of internal (r_i) to external longitudinal resistance (r_o). To normalize the data obtained from various preparations, R_i was calculated. As the preparations were mounted between the pins of the force transducer, tension was measured simultaneously with the changes in R_i.

Figure 1a shows the effects of hypertonic solution on the force of contraction and R_i. Hypertonicity quickly produces inexcitability and marked contracture. Resting tension increases even more on washing out and then slowly decreases. When excitability is restored after 20–30 minutes, R_i can be measured. Typically it was measured after 45 minutes of washout, and at that time it was still elevated far above control. It has been shown previously that even a moderate hypertonicity produces Ca^{++} overloading, so an increase in R_i after hypertonic exposure shown here may be due to Ca^{++} overloading or mechanical disruption of gap junctions or both. Figure 1b shows that the Ca^{++}-overloading component is probably the most important because the pretreatment of fibers with Ca^{++}-free solution reduces both hypertonic contracture and increase in R_i. Some contracture (and uncoupling) developing during washout may be due to Ca^{++}-paradox (sudden return from Ca^{++}-free to normal Ca^{++} superfusate). Figure 1c shows the effects of trifluoperazine, which prevents very markedly the development of contracture and increase in R_i. Figure 1d shows that lidocaine also exerts some protective effects but those effects are less marked. Since trifluoperazine was more effective, it may indirectly imply that calmodulin may be involved in the modulation of intercellular coupling.

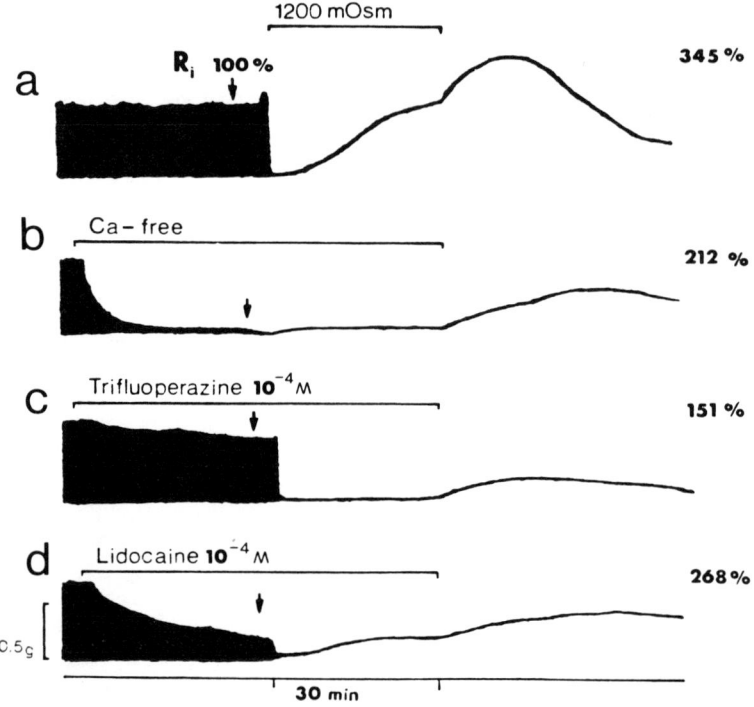

Figure 1
Effects of local anesthetics on tension and R_i in cat papillary muscles exposed to hypertonic Tyrode's solution. Preparations were superfused either with Ca^{++}-free solutions or 10^{-4} M trifluoperazine and 10^{-4} M lidocaine 30 min before and during exposure to hypertonic (1200 mOsm) solutions. An increase in R_i and resting tension (a) induced by hypertonicity is largely prevented by Ca^{++}-free solutions (b), trifluoperazine (c), and lidocaine (d). R_i was measured during control and 45 min after removal of hypertonic Tyrode's solution. Changes in R_i are expressed as a percent change from its value in control.

EFFECTS OF GENERAL ANESTHETICS ON INTERCELLULAR COUPLING

Recently, Johnston et al. (1980) showed that aliphatic alcohols (heptanol, octanol) can electrically uncouple segments of the crayfish septate axon. I have repeated those experiments in cardiac muscle using a modified sucrose gap technique for measuring internal longitudinal resistance (Kleber 1973). In this method, relative changes in internal longitudinal resistance can be evaluated by measuring changes in the amplitude of the trans-gap action potential (TGP). Its amplitude decreases when the total resistance of the gap goes up due to the increase in internal longitudinal resistance of the muscle in the gap produced by intercellular uncoupling. Alcohols were added to the isotonic sucrose solution which superfused the muscle in the gap. Sucrose contained 10^{-4} M $CaCl_2$ to prevent uncoupling developing in Ca^{++}-free isotonic sucrose solutions (Kleber 1973). The contracture that develops in normal muscle superfused with isotonic sucrose was prevented by exposing the muscle to isotonic KCl first (which produces only transient contracture) and then to isotonic sucrose.

Figure 2 shows the effects of octane, octanol, and octanoic acid on intercellular coupling measured as a change of TGP. Data are expressed as a percent change of TGP from values during control (100%). When the intercellular coupling decreases, the amplitude of TGP also decreases, reaching 0 when cells in the gap are totally electrically uncoupled. Octane produces no effect. Octanoic acid (1 mM) induces total electrical uncoupling which is as fast as octanol-induced uncoupling but is only partially reversible. Octane-induced changes in action potential configuration show that this compound is able to penetrate cardiac membranes but exerts no effect on gap junctions. Thus, the presence of the hydrophilic group at the end of the hydrophobic chain seems to be important. The length and configuration of the hydrophobic chain is also of importance, as only medium-chain aliphatic alcohols (C_6–C_8) were able to uncouple cardiac cells, confirming the results of Johnston et al. (1980) on the crayfish septate axon. It is possible that the reversible uncoupling produced by the aliphatic alcohols is due to the participation of the hydroxyl group in weak, reversible hydrogen bonds. Also, one may speculate that octanoic acid, which possesses a terminal carboxyl group, can form a more stable ionic bond, which explains its poor reversibility.

Alcohols are not used clinically as general anesthetics. Do other general anesthetics used in clinical anesthesia exert some influence on cell–cell coupling? I have investigated the influence of halothane, since it often produces arrhythmias in the operating room. Halothane was dissolved in isotonic sucrose solution by vaporization. The vaporizer was calibrated so that different concentrations of halothane could be used to saturate superfusing solutions (1–3% mixtures with oxygen). Halothane produces a dose-dependent decrease in electrical coupling (Wojtczak 1984). Uncoupling is nearly complete when a mixture of 3% halothane is used (Fig. 3). Uncoupling induced by halothane is relatively fast to develop and is fully reversible. As it was shown on Figure 1, trifluoperazine could prevent intercellular uncoupling induced by Ca^{++} overloading. If the mechanisms involved in uncoupling produced by alcohols and halothane are similar to those involving increased intracellular Ca^{++}, then trifluoperazine should at least partly prevent this type of uncoupling. As shown in Figure 3, trifluoperazine is unable to prevent the uncoupling produced by halothane. The fiber was pretreated with trifluoperazine long before it was exposed to halothane for the second time. This unexpected result suggests that the mechanism by which general anesthetics uncouple cardiac cells need not involve Ca^{++} binding to the cytoplasmic site of the gap junction. Indeed, the fact that aliphatic alcohols and halothane exert profound negative inotropic effects on the heart and induce pronounced relaxation may imply that Ca^{++} is pumped out of the cytoplasm under these conditions.

Trifluoperazine and lidocaine can be classified according to Seeman (1972) as positively charged anesthetics, whereas alcohols and halothane can be classified as neutral anesthetics. As shown by Seeman (1972), both types of anesthetics produce membrane expansion in erythrocyte membranes. As both groups of drugs produce opposite effects on the intercellular coupling, it is

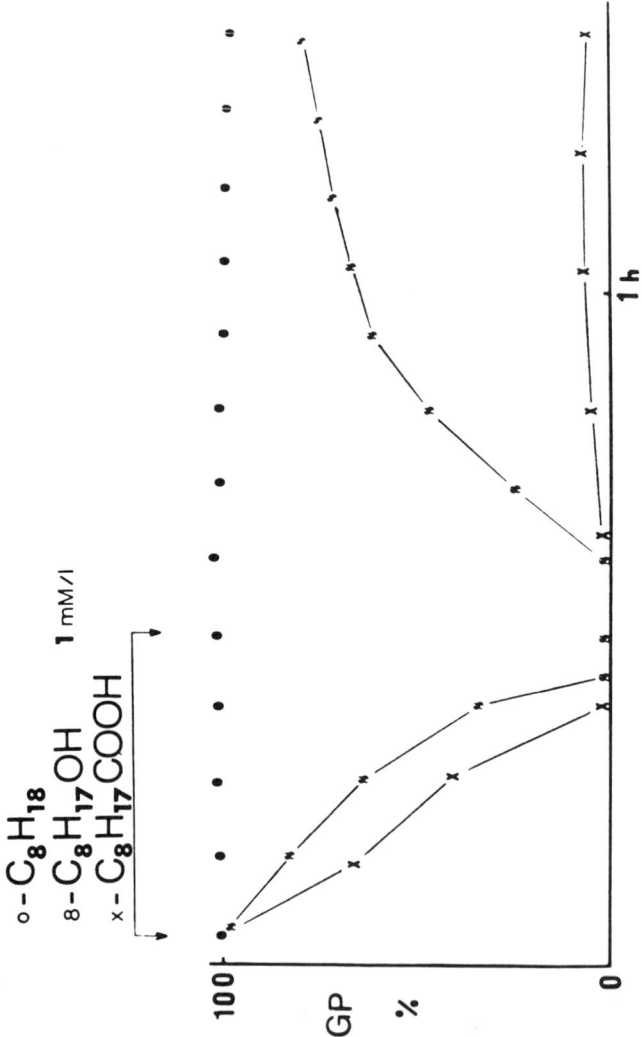

Figure 2
Effects of octane (○, C_8H_{18}), octanol (8, $C_8H_{17}OH$), and octanoic acid (×, $C_8H_{17}COOH$) on TGP measured across the sucrose gap. Changes in TGP reflect changes in intercellular coupling (100%, normal coupling; 0, total electrical uncoupling). They are expressed as a percent change from its value in control.

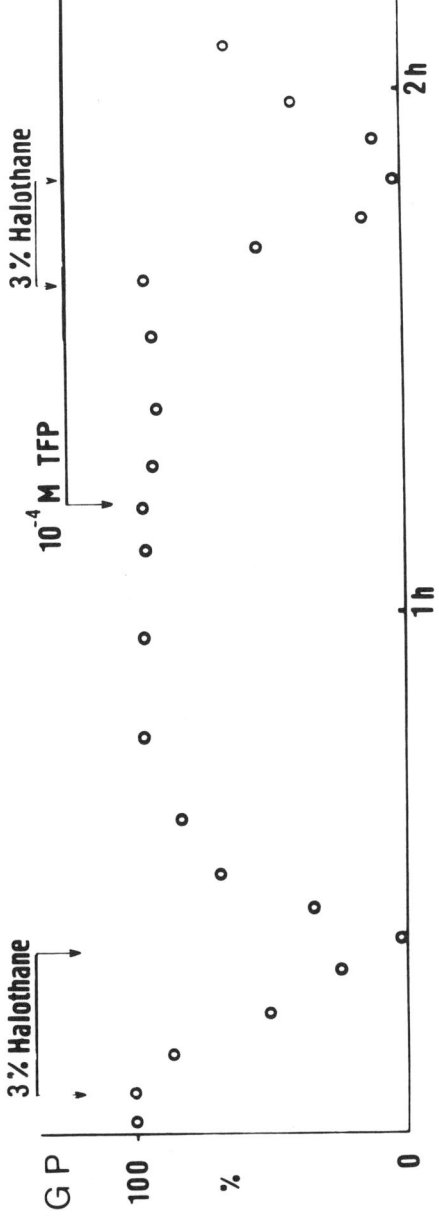

Figure 3
Effects of halothane on TGP which reflect changes in intercellular electrical coupling. Changes in TGP are expressed as a percent change from its value in control. Trifluoperazine (TFP) is unable to prevent uncoupling induced by this volatile anesthetic.

unlikely that membrane expansion is responsible for the effects exerted by those drugs. It can be speculated that neutral anesthetics produce intercellular uncoupling by a direct action on some hydrophobic site of the channel-forming protein, most probably in the outer bilayer. Whether calmodulin is involved in the modulation of intercellular coupling in cardiac muscle needs further clarification. Peracchia et al. (1983) have reported that intercellular uncoupling in *Xenopus* embryos can be prevented with trifluoperazine and also with the more specific inhibitor of calmodulin, calmidazolium (Peracchia 1984). According to my preliminary results, calmidazolium can prevent healing over in cardiac muscle.

SUMMARY

Neutral (general) anesthetics, such as aliphatic alcohols and halothane, induce intercellular electrical uncoupling in cardiac muscle. Some local anesthetics, including the calmodulin inhibitor, trifluoperazine, can prevent intercellular uncoupling produced by Ca^{++} overloading during hypertonic shock. Since trifluoperazine cannot prevent uncoupling produced by alcohols and halothane, it is possible that the mechanism of this uncoupling does not depend on Ca^{++} and may depend on the direct interactions of those compounds with proteins forming gap junction channels.

REFERENCES

Barr, L., M. Dewey, and W. Berger. 1965. Propagation of action potentials and the structure of the nexus in cardiac muscle. *J. Gen. Physiol.* **48:** 797–823.

Dahl, G. and G. Isenberg.1980. Decoupling of heart muscle cells: Correlation with increased cytoplasmic calcium activity and with changes of nexus ultrastructure. *J. Membr. Biol.* **53:** 63–75.

De Mello, W. 1979. Effect of 2-4-dinitrophenol on intercellular communication in mammalian cardiac fibers. *Pfluegers Arch.* **380:** 267–276.

———. 1982. Cell-to-cell communication in heart and other tissues. *Progr. Biophys. Mol. Biol.* **39:** 147–182.

———. 1983. The influence of pH on the healing-over of mammalian cardiac muscle. *J. Physiol.* **339:** 299–307.

Johnston, M., S. Simon, and F. Ramón. 1980. Interaction of anesthetics with electrical synapses. *Nature* **286:** 498–500.

Kleber, A. 1973. Effects of sucrose solution on the longitudinal tissue resistivity of trabecular muscle from mammalian heart. *Pfluegers Arch.* **345:** 195–205.

Peracchia, C. 1984. Communicating junctions and calmodulin: Inhibition of electrical uncoupling in *Xenopus* embryo by calmidazolium. *J. Membr. Biol.* **81:** 49–58.

Peracchia, C., G. Bernardini, and L. Peracchia. 1983. Is calmodulin involved in the regulation of gap junctional permeability? *Pfluegers Arch.* **399:** 152–154.

Reber, W. and R. Weingart. 1982. Ungulate cardiac Purkinje fibers: The influence of intracellular pH on the electrical cell-to-cell coupling. *J. Physiol.* **328:** 87–104.

Seeman, P. 1972. The membrane action of anesthetics and tranquilizers. *Pharmacol. Rev.* **24:** 583–655.

Spray, D., A. Harris, and M. Bennett. 1981. Gap junctional conduction as a simple and sensitive function of intracellular pH. *Science* **211:** 712–715.

Spray, D., J. Stern, A. Harris, and M. Bennett. 1982. Gap junctional conductance: Comparison of sensitivities to H and Ca ions. *Proc. Natl. Acad. Sci.* **79:** 441–445.

Turin, L. and A. Warner. 1977. Carbon dioxide reversibly abolishes ionic communication between cells of early amphibian embryo. *Nature* **270:** 56–57.

Weingart, R. 1977. The action of ouabain on intercellular coupling and conduction velocity in mammalian ventricular muscle. *J. Physiol.* **264:** 341–365.

White, R., D. Spray, B. Schwartz, B. Wittenberg, and M. Bennett. 1984. Some physiological and pharmacological properties of cardiac gap junctions. *Biophys. J.* **45:** 279a.

Wojtczak, J. 1979. Contractures and increase in internal longitudinal resistance of cow ventricular muscle induced by hypoxia. *Circ. Res.* **44:** 88–95.

———. 1982. Influence of cyclic nucleotides on the internal longitudinal resistance and contractures in the normal and hypoxic mammalian cardiac muscle. *J. Mol. Cell Cardiol.* **14:** 259–265.

———. 1984. Effects of general and local anesthetics on intercellular coupling in the heart muscle. *Biophys. J.* **45:** 22a.

Protein from Purified Lens Junctions Induces Channels in Planar Lipid Bilayers

James E. Hall
*Department of Physiology and Biophysics
University of California, Irvine
Irvine, California 92717*

Guido A. Zampighi
*Department of Anatomy and
Jerry Lewis Neuromuscular Research Center
UCLA School of Medicine, University of California
Los Angeles, California 90024*

Gap junctions are (as anyone reading this volume probably knows already) regions where adjacent cells are intimately apposed. Very strong circumstantial evidence suggests that they provide a low-resistance pathway where electrolytes and nonelectrolytes can pass from cell to cell: wherever low resistance and dye passage between cells is seen, gap junctions have also been found; and wherever gap junctions have been found, electrical coupling and dye passage are also seen (Bennett 1973). Numerous electrophysiological studies have defined the macroscopic properties of the electrical coupling in a variety of tissues (Bennett and Goodenough 1978; Loewenstein 1981; Spray et al. 1984). Such studies have shown that gap junctions from different tissues differ in their voltage dependence, size of the dyes that they pass, and sensitivity to changes in hydrogen and calcium concentrations. These studies suggest that all gap junctions provide a channel between cells that probably has a conductance on the order of hundreds of picosiemens. It has not been possible to obtain reliable measurements of single-channel conductance in vivo in most tissues; nor have the molecular species that form the putative channels been conclusively identified.

Reconstitution of junctional function in planar bilayers using defined components provides a method for filling in both these gaps. The data in this paper show that purified lens fiber-junction protein forms channels when incorporated in planar bilayers. The data imply that, although lens junction protein differs markedly from the junctional proteins of liver and heart (Nicholson et al. 1982), it can nevertheless form channels with properties that resemble those of other junctions. Reconstitution of junctional channels makes possible, in principle, the determination of which proteins are responsible for channel formation. In practice the possibility that the few channels observed are due to a minor contaminant must be dealt with. Our data make it possible to estimate the probability that channels we observe are due to a contaminating protein. We find that it is very likely that the channel is formed by the major intrinsic protein of lens fiber junctions.

DISCUSSION

Basic Methods

Junctions between mature lens fibers were isolated without using detergents or extrinsic proteases according to a previously described method (Zampighi et al. 1982). The isolated fraction contained numerous large junctions with a pentalamellar structure 13–14 nm in overall thickness. SDS-PAGE showed this fraction to be composed almost entirely of a single band with an apparent molecular weight of 27,000. This fraction was then solubilized in octylglucoside under conditions that produced mostly monomeric junction protein (G. Zampighi, unpubl.). The solubilized junctions were centrifuged at 100,000g for 90 minutes to remove particulate material and the supernatant was used for all further experiments.

Bilayer experiments were performed using methods previously described. All the membranes used in this study were formed of bacterial phosphatidyl ethanolamine (Avanti Biochemicals, Birmingham, Alabama) and squalane (Atomergic, Plainview, New York) using the double-monolayer technique (Hall et al. 1984).

Junctional Material Increases Bilayer Conductance in Two Ways

We incorporated the junctional material described above into the bilayers in two ways. First lens junctional protein was dissolved in octylglucoside and added directly to the aqueous solutions bathing a planar bilayer. This procedure resulted in a membrane current-voltage (*I-V*) curve like that shown in Figure 1. The resistance of the bilayer was slightly decreased at low voltages. At higher voltages in the region of 150–200 mV, the membrane conductance increased in a noisy, voltage-dependent manner. We did not see uniform conductance steps in this conductance, even at a resolution of 2 pA in 1 msec. Addition of large amounts of octylglucoside with no dissolved protein increased the membrane conductance slightly, but only at low voltages. The noisy voltage-dependent conductance increase seen at higher voltages with protein did not occur when octylglucoside alone was added to the membrane, even in amounts 100-fold greater than the amounts added with dissolved protein.

A second method was to incorporate lens protein first into vesicles by mixing octylglucoside-dissolved protein (identical to that used in the first experiments) together with octylglucoside-dissolved dioleolyl phosphatidyl choline and dialyzing against several liters of buffer to remove the detergent (Mimms

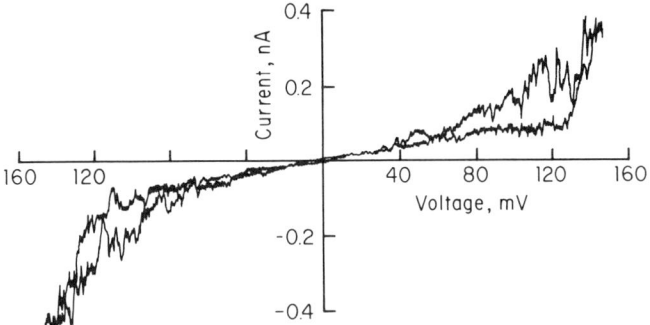

Figure 1
I-V curve of a membrane exposed to octylglucoside-dissolved protein added directly to the aqueous phases bathing both sides of the membrane (1.0 M KCl, 25 mM MES, pH 5.7). Note that the conductance increases at elevated voltages of both signs, but does not appreciably increase above bare membrane conductance near zero voltage. Control experiments in which octylglucoside alone was added in much larger quantities showed that noisy conductance increase at high voltage depends on the presence of protein. The slight increase in zero-voltage conductance above the bare membrane level can be attributed at least in part to the presence of octylglucoside.

et al. 1981). This procedure produced a suspension of vesicles which were characterized by freeze-fracture electron microscopy. A typical low-magnification view of this suspension is shown in Figure 2A, which shows numerous unilamellar vesicles whose apparent diameters in the plane of fracture range from 800 to 3000 Å. The fracture faces of most of the vesicles show distinct intramembrane particles. The outlined area in Figure 2A is shown at higher magnification in Figure 2B. Here convex and concave fracture faces of several unilamellar vesicles are seen. Both types of fracture face show intramembrane particles which cast shadows on the lower-lying surface. A few pits are also seen. (Pits are intrinsically harder to detect than particles for the same reason that telephone poles are easier to see in aerial photographs than 12-inch holes in the ground; i.e., poles cast a longer shadow.) Increasing the lipid-to-protein ratio in the dialysate decreased the number of particles found per unit area of vesicle. The number of particles per vesicle is roughly correlated with the ratio of protein molecules to lipid molecules in the dialysate.

When vesicles containing an average of about two particles per vesicle were added to both sides of a lipid bilayer, *I-V* curves like that shown in Figure 3 were obtained. In contrast to the conductance changes induced by the octylglucoside-solubilized material, the conductance induced by the vesicle-borne material is large at low voltage, shows clear single-channel steps, and turns off as the absolute magnitude of the voltage increases. The voltage dependence of this conductance resembles that of gap junctions between the cells of the amphibian or teleost blastomere (Obaid et al. 1983; Spray et al. 1984). Because the conductance of the vesicle-borne material was reminiscent of gap junction conductances in vivo, we chose to study it in detail.

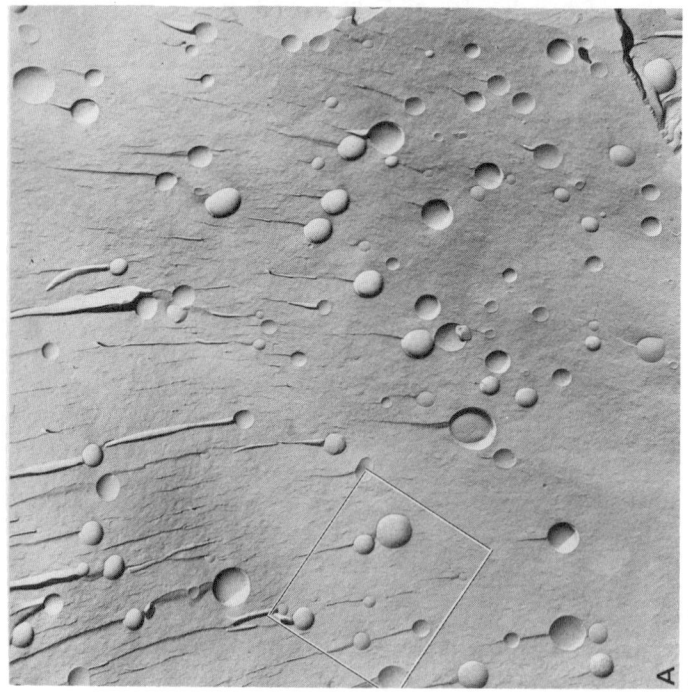
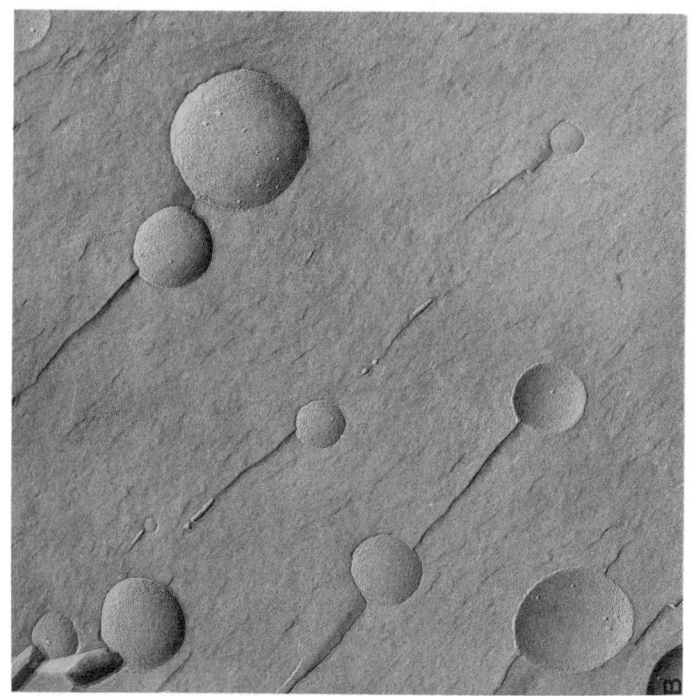

Figure 2
Freeze-fracture electron micrographs of vesicles containing lens fiber junctional protein. (A) Most vesicles contain one or more particles. Magnification, 13,800 ×. (B) Enlargement of the area shown in A. Magnification, 57,500 ×. The field shows both concave and convex faces. Some of the particles, those best situated for favorable shadowing, show dark central deposits of shadowing material. The smaller of the two vesicles at the top right shows one particle and one pit on a convex fracture face.

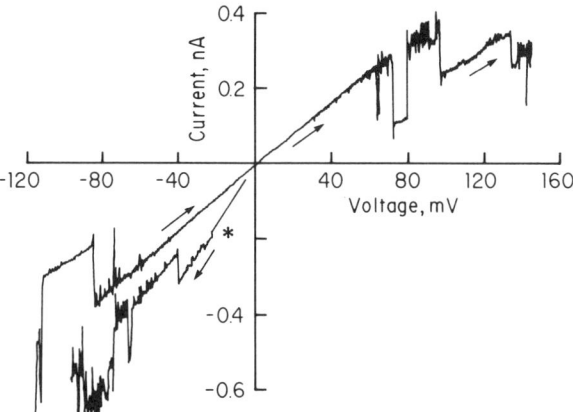

Figure 3
I-V curve of phosphatidyl ethanolamine-squalane membrane exposed to protein-containing vesicles added to both aqueous phases. Voltage was swept first in the negative direction at 5 mV/sec to about 125 mV. The starting point of the sweep is marked by an asterisk (*). The direction of sweep was then reversed, and the voltage swept to about +150 mV (arrows mark the direction of sweep). Note that the conductance is on near 0 V and tends to decrease in a stepwise fashion at higher voltages. Note also that there is increasing noisiness as the magnitude of the voltage increases. This noisiness results from an increasing tendency for the channel to turn off as the voltage increases. The solutions were 1.0 M KCl, 25 mM HEPES (pH 5.7) on both sides of the membrane. Membrane diameter was 0.3 mm. Channels of the type seen in this figure will be called "voltage-dependent channels" because they tend to close with elevated voltage of either sign.

Single-channel Characteristics—Each Channel Has Four Sublevels

The single-channel characteristics of the voltage-dependent conductance induced by the vesicle-carried material are consistent with the properties expected from a junctional channel. The single-channel conductance in 100 mM KCl is about 200 pS. In 1.0 M salt the single-channel conductance ranges from 1500 to 2000 pS. The larger the absolute magnitude of the voltage, the greater the probability that the channel will be off. This is illustrated in Figure 4 by a series of single-channel current records at 35, 50, and 58 mV. The traces show current responses to voltages of both signs. The data were recorded on FM tape with a bandwidth of 5 kHz and played back to a strip chart recorder at the desired speed. The left-hand records in Figure 4 were recorded on the strip chart recorder at low time resolution (1 sec/mm). The right-hand traces are short pieces of the data shown in the low-time resolution strips, but recorded at a much greater time resolution (0.01 sec/mm). As the magnitude of the voltage is increased, the channel spends proportionally more time in the off state. Even at very high voltages (100–200 mV), however, the conductance never drops to the unmodified membrane level. In fact at voltages above about 175 mV, the conductance begins to increase with voltage in the same way that it does when the octylglucoside-solubilized material is added directly to the bilayer. Thus the noisy conductance, the only conductance seen when octyl-glucoside-solubilized material is added to the membrane, is incorporated along with the crisp channel when protein is added to the bilayer in vesicles.

At voltages below 150 mV, the most characteristic feature of the conductance is the single-channel step, a large step of 1500–2000 pS in 1.0 M KCl or NaCl, or 200 pS in 0.1 M KCl. This large step, which is the most frequently observed change, is further divided by two nearly equally spaced relatively unstable levels. The right-hand traces in Figure 4 are time-expanded records

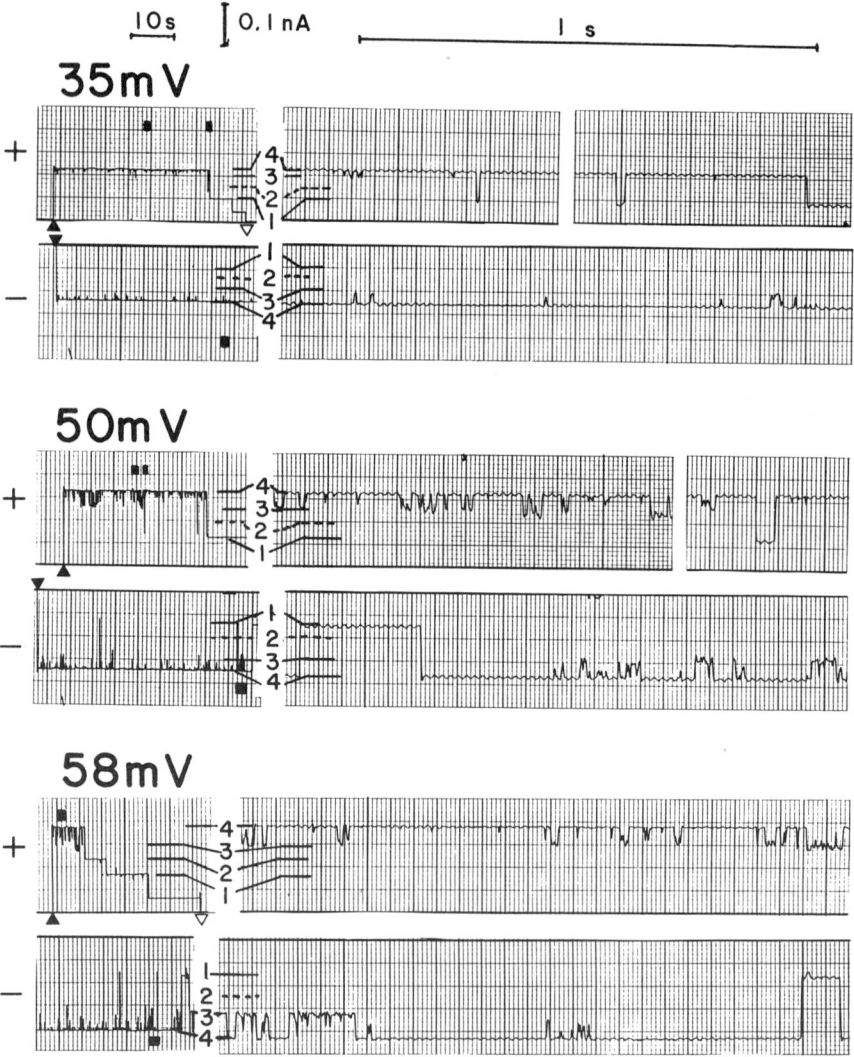

Figure 4 (See facing page for legend.)

of the same channels shown in the left-hand traces. The four levels are marked on each of the traces and labeled 1, 2, 3, and 4, starting from the lowest conductance level. Level 2 is nearly always reached from level 1, an upward transition; and level 3 is nearly always reached from level 4, a downward transition. Levels 2 and 3 are also usually noisier than levels 1 and 4. (In the 58-mV trace shown in Figure 4, level 2 is unusual from two points of view: first it is less noisy than usual, and second it is entered directly from level 4, not from level 1 as is usually the case.) Thus in most cases, levels 2 and 3 are correlated with levels 1 and 4. All four levels must thus be associated with the same membrane structure and cannot arise from independent entities.

The probability of the channel being open decreases as the absolute value of the voltage increases; but even at voltages larger than 150–200 mV, the conductance of the membrane does not decrease to the unmodified membrane conductance level. Furthermore, the maximum conductance (that at 0 V) is not always an integral multiple of the major step (1500–2000 pS in 1.0 M salt, 200 pS in 0.1 M salt). There is thus no well-defined relation between the maximum conductance at low voltage and the minimum conductance at high

Figure 4
Single-channel characteristics of the voltage-dependent channels. Records are shown from a single-channel membrane (phosphatidyl ethanolamine squalene) in 1.0 M KCl and 25 mM MES (pH 5.7). Current recordings were made at positive and negative voltages of magnitudes 35, 50, and 58 mV. The currents were recorded on FM magnetic tape at a bandwidth of 5 kHz for off-line analysis. The traces shown above were made by playing the taped data to a strip chart recorder at two different speeds to reveal both long-time and short-time characteristics of the channels. The strips at the left of the figure are all played back at a speed 1 mm/sec (calibration bar 10 sec long at the top of the figure refers to these left-hand strips). If the record includes the point where the voltage was turned on, that point in marked by a dark triangle (▲). If the record includes the point where the voltage was turned off, that point is marked with an open triangle (△). Plus and minus signs to the left of the traces indicate the sign of the applied voltage. The current scale is the same for all of the traces shown (0.1-nA calibration bar shown at the top of the figure). Positive currents are upward (zero current at the bottom of the trace); negative currents are downward (zero at the top of the trace). The traces at the right of the figure were played back at a speed of 100 mm/sec (1-sec calibration bar at the top of the figure). Each right-hand trace shows a part of the low time resolution trace to its left. Black boxes on the left-hand traces mark the times shown in the right-hand traces. As the magnitude of the voltage is increased for either positive or negative voltages, the current tends to spike more and more often toward zero current. At +35 mV the channel is almost always in the fully on state (what we have called level 4), but at +58 mV, the channel spends much more time in the lower conductance states. The channel has four conductance levels that are numbered in the spaces between the low-time resolution traces and the high-resolution traces. Level 1 is the lowest conductance level, and level 4 the highest. In this set of records, levels 1, 3, and 4 appear in either the positive or negative voltage trace at all the voltages shown. Level 2 is relatively rare and is seen clearly in this set of records only in the +58 mV trace. This particular record represents a rare case in that level 2 is entered from level 4. In most cases level 2 is entered from level 1. Level 2 was also seen (entered from level 1 and usual) in plus and minus 42 mV records taken on this membrane.

voltage. In many-channel membranes, this ratio is around 0.5, but in few-channel membranes, it is much more variable. In fact the value of the off state sometimes changes during the course of an experiment. This is shown in Figure 5, a histogram of the relative probability of a channel being in a given state. The amplitudes of the peaks are proportional to the probability of the channel having the indicated conductance. The solid line shows the relative probabilities at 35 mV and the dashed line those at 58 mV. The highest conductance peak is level 1. There is a very small peak corresponding to level 2, larger ones corresponding to levels 3 and 4, and finally a peak between level 4 and the zero conductance level. This last peak between level 4 and ground corresponds to one of the variable off state levels seen in some membranes. This level is not always seen; it is not clear whether or not it is due to the closing of an independent channel or is a sublevel of the major four-level step. Note that even when there is the additional complication of a variable off level, the major four-level step is still clearly seen.

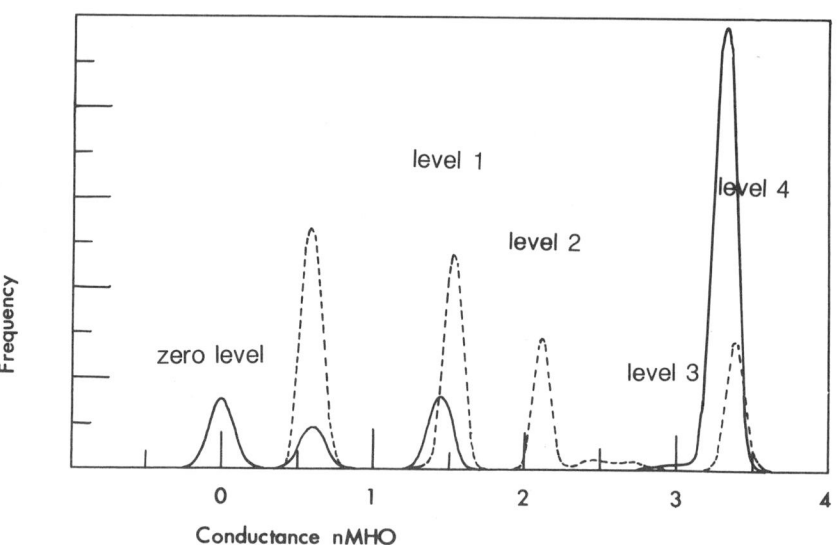

Figure 5
Histogram of channel conductance at 35 and 58 mV. The aqueous solutions were 1.0 M KCl 25 mM MES (pH 5.7). The solid trace shows a histogram at 35 mV. The peak at 3.3 nS has the maximum amplitude and corresponds to level 4. The shoulder on this peak corresponds to level 3, and the peak at 1.4 nS is level 1. Level 2 is not seen at 35 mV. The dotted trace shows the 58-mV conductance histogram. The relative amplitudes of the higher conductance peaks have decreased, and the relative amplitudes of the lower conductance peaks increased. In additon to levels 4, 3, and 1, level 2 is seen at 2.2 nS. The peak at 0.6 nS is an additional level not part of the four-level channel structure. This level is the most commonly seen type of channel which is not associated with the four-level channel structure.

Voltage Dependence

The conductances of both many-channel and few-channel membranes are voltage dependent. The current-voltage curve of a many-channel membrane is shown in Figure 6A. The conductance decreases as the voltage increases from about 50 mV up and then begins to increase again at about 170 mV. This increase in conductance above about 175 mV is similar in character to that induced by octylglucoside-dissolved protein added in the absence of vesicles (compare the *I-V* curve in Fig. 1.). It thus may be that the vesicle-borne material has the capacity to induce both kinds of channels, while the octylglucoside-solubilized junctions can only induce the kind of channel that turns on with increasing voltage.

Conductance versus voltage curves for both many channel and few channel membranes are plotted in Figure 6B. Because the channel spends nearly all of its time in either level 1 or level 4, the data can be fit to a two-level Boltzmann distribution of the form:

$$G = \frac{(G_{on} - G_{off}) \exp[-ne(V - V_o)]}{1 + \exp[-ne(V - V_o)]} + G_{off}$$

where V_o is the voltage at which half of the channels are open, e the electronic charge, V the applied voltage, and n the effective gating charge in units of e. Our data for both single-channel and many-channel membranes show wide variation in V_o and in n. V_o ranges from 60 mV to 100 mV and n from 1 to 2.

Octanol Causes Complete Disappearance of the Channel

Octanol added to both sides of a bilayer containing the voltage-dependent channel causes the channel to close. Figure 7A shows an *I-V* curve before the addition of octanol, and Figure 7B shows an *I-V* curve 30 minutes after the addition of octanol. The solutions are 0.1 M KCl and 25 mM (2 [*N*-morpholino]ethane sulfonic acid) (MES) (pH 5.7). In this solution the single-channel conductance is about 200 pS. Note that the zero-voltage conductance after the addition of octanol has decreased from 1000 pS, about five open channels, to about 100 pS, about half the conductance of an open channel. In the experiment shown in Figure 7, 70 µl of octanol were added to each side of the membrane (chamber volume was 2.3 ml). This is a relatively large amount of octanol and the membrane broke shortly after the curve in Figure 7B was taken. In other experiments where about one-tenth as much octanol was added, the conductance returned completely to the unmodified membrane level before the membrane broke, but not until several hours after the octanol was added. Artifacts may explain the long latency and high concentration for octanol observed to alter channel conductance: Octanol dissolves slowly in water so that much of the time lag may be time of solvation; this would also reduce the effective concentration of octanol at the membrane.

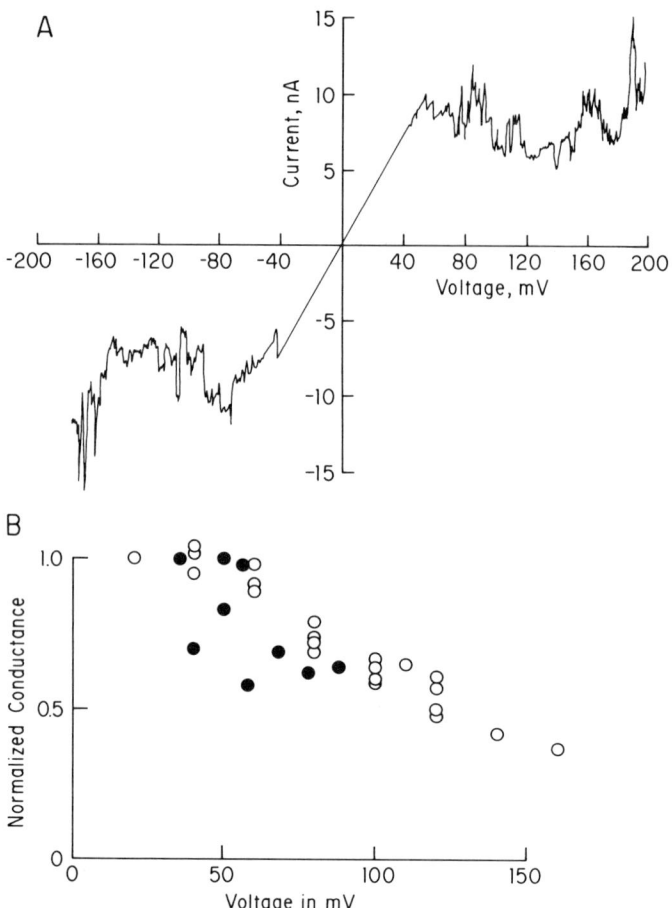

Figure 6
(*A*) Current voltage curve of a many-channel membrane. The aqueous solutions are 1.0 M KCl, 25 mM MES, and 26 mM CaCl$_2$ (pH 5.7). Between 50 and 150 mV, the conductance decreases with increasing voltage, but as the voltage is increased above about 150 mV, the conductance begins to increase with voltage as it does in the *I-V* curve shown in Figure 1. (*B*) Normalized conductance-voltage curves of both single and many-channel membranes. Conductance divided by conductance at 0 V is plotted as a function of voltage. Both single- and many-channel membranes show the same weak voltage dependence. If fitted to a two-state Boltzmann distribution (a reasonable approximation since the channel spends most of its time in either state 4 or state 1), the effective charge ranges from 1 to 2 electronic charges and the voltage at which half the channels are open ranges from 60 to 100 mV. (●) Data from single-channel membranes; (○) data from many-channel membranes.

The Major Lens Junction Protein Probably Forms the Channels

Although the protein we are using to reconstitute channels is very pure by biochemical standards (90–98%), we only observe a few channels in any given

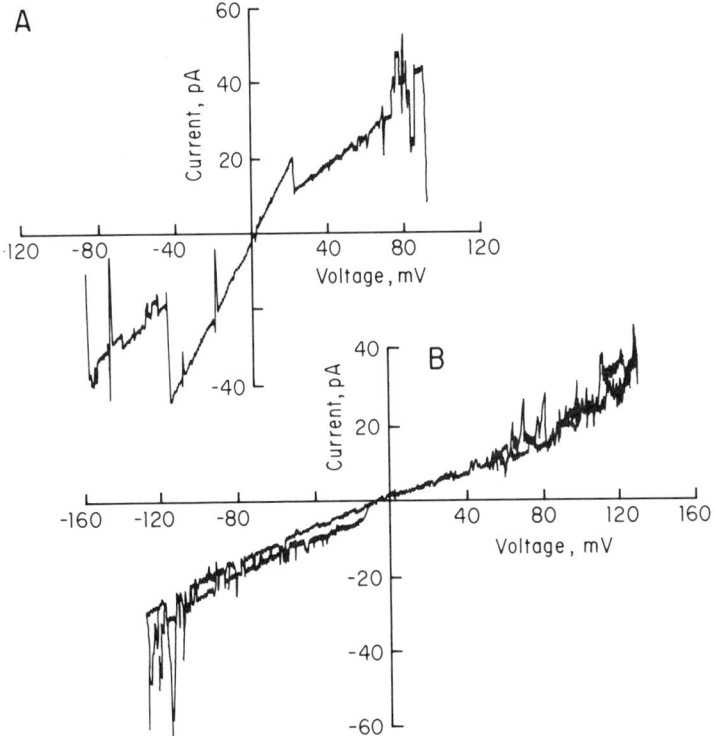

Figure 7
Effect of octanol on channel conductance. Membrane is formed from phosphatidyl ethanolamine squalene in 0.1 M KCl, 25 mM MES (pH 5.7). Curve A is a I-V curve after the insertion of three channels into the membrane. Curve B is 30 min after the addition of 70 µl of octanol to both sides of the membrane. (Chamber volume was 2.3 ml on each side of the membrane.) The membrane broke shortly after curve B was taken. In other experiments where only 5 µl of octanol was added to each side, the channel was also blocked, but several hours after the addition of octanol. The membrane was much more stable with less octanol, however, and was monitored long enough to see the conductance return completely to bare membrane level.

experiment, a few hundred at most in a many-channel membrane. It is thus quite possible that the channels we observe could be formed by some very minor contaminant. We can easily eliminate the possibility that the putative contaminant is water soluble because octylglucoside-solubilized material added to the bilayer chamber at a concentration 50 times higher than that resulting when vesicles are added does not form voltage-dependent channels. In the present case we can also argue fairly strongly that the channels are probably not formed by a membrane-bound contaminant either.

The argument depends on careful characterization of the vesicular material added to the planar bilayers. In our experiments, we add 2 µl of solution containing about 10^{12} vesicles to each side of the membrane. Each vesicle contains on the average about two intramembrane particles. Observation of direct adhesion of vesicles containing a fluorescent dye to the planar membrane (D.J. Woodbury and J.E. Hall, unpubl.) indicates that only about 20–50 vesicles will fuse with the membrane under conditions of our experiment. We have had 38 successful membranes (where at least one voltage-dependent channel has been seen), and so far we have failed to see this kind of channel in only one experiment having the correct lipid-to-protein ratio.

The probability p that a given vesicle contains a channel-forming protein can be estimated from the average number of vesicles per experiment that adhere to the membrane and donate a channel to it, v, and the fraction of experiments in which no channels are observed. If m is the number of experiments in which channels are not observed and n the number of experiments in which channels are observed, p is given by:

$$p = 1 - (m/(m+n))^{1/v}$$

All the numbers on the right-hand side of the equation can be estimated: v is 20–50; n is about 50; m is 1. If v is 20, p must be greater than about 0.17 (that is 17 out of a 100 vesicles are expected to contain the channel-forming protein). The only protein present in sufficiently large concentration to be found in 18 out of 100 vesicles is the major intrinsic protein of the lens. This argument thus eliminates trace contaminants from being the channel-forming protein, and in the worst case, rules out as the channel-former any protein not present in the original material at less than the 1 or 2% level. Proteins present at this level should have been detected on the SDS gel. We thus have a high degree of confidence that the channels we observe are due to the major intrinsic protein of lens.

SUMMARY/CONCLUSION

The reconstitution studies described here show clearly that material isolated from bovine lens can form channels in planar bilayers. These channels have properties similar to those observed in gap junctions of some tissues. Because we have been able to study them in bilayers, we have been able to provide detailed characterization of their single-channel properties. We found that channels that have a conductance of about 200 pS in 0.1 M KCl or NaCl have a much higher probability of being in a high conductance state at low voltages than at high, and that the channels characteristically exhibit four sublevels of conductance.

By carefully characterizing the vesicles used in reconstituting the channels, we have shown it to be very likely that the channels we have studied are formed by the major intrinsic protein of the lens recently sequenced by Gorin et al. (1984).

Our results do not, however, reveal the architecture of the channel in the planar bilayer. A particularly important unanswered question is whether the channels we observe are in a single bilayer or a double one. That we must add vesicles to both sides of the bilayer before seeing channels suggests that the double-bilayer possibility cannot be lightly discarded. This question can be addressed by electron microscopic analysis of bilayers in which channels have been observed.

Finally, the use of antibodies to purified junctional protein that alter channel properties may serve to increase our confidence that the channels observed are formed by the major intrinsic lens protein.

REFERENCES

Bennett, M.V.L. 1973. Function of electrotonic junctions in embryonic and adult tissues. *Fed. Proc.* **32**: 65-74.

Bennett, M.V.L. and D.M. Goodenough. 1978. Gap junctions, electrotonic coupling and intercellular communication. *Neurosci. Res. Program Bull.* **16**: 346-373.

Gorin, M.B., S.B. Yancey, J. Cline, and J.-P. Revel. 1984. The major intrinsic protein (MIP) of the bovine lens fibre membrane: Characterization and structure based upon cDNA cloning. *Cell* **39**: 49-59.

Hall, J.E., I. Vodyanoy, T.M. Balasubramanian, and G.R. Marshall. 1984. Alamethicin: A rich model for channel behavior. *Biophys. J.* **45**: 233-247.

Lowenstein, W. 1981. Junctional intercellular communication and the cell-to-cell membrane channel. *Physiol. Rev.* **61**: 829-913.

Mimms, L.T., G. Zampighi, Y. Nozaki, C. Tanford, and J.A. Reynolds. 1981. Phospholipid formation and trans membrane protein incorporation using octylglucoside. *Biochemistry* **20**: 833-840.

Nicholson, B.J., L.J. Takemoto, M.W. Hunkapiller, L.E. Hood, and J.-P. Revel. 1982. Differences between liver gap junction protein and lens MIP26 from rat: Implications for tissue specificity of gap junctions. *Cell* **32**: 967-978.

Obaid, A.L., S.J. Socolar, and B. Rose. 1983. Cell-to-cell channels with two independent regulated gates in series: Analysis of junctional channel modulation by membrane potential, calcium and pH. *J. Membr. Biol.* **73**: 69-89.

Spray, D.C., R.L. White, A. Campos de Carvalho, A.L. Harris, and M.V.L. Bennett. 1984. Gating of gap junction channels. *Biophys. J.* **45**: 219-230.

Zampighi, G.A., S.A. Simon, J.D. Robertson, T.J. McIntosh, and M.J. Costello. 1982. On the structural organization of isolated bovine lens fibre. *J. Cell Biol.* **91**: 175-189.

An In Vitro Approach to Cell Coupling: Permeability and Gating of Gap Junction Channels Incorporated into Liposomes

Camillo Peracchia and Stephen J. Girsch
Department of Physiology, University of Rochester School of Medicine and Dentistry, Rochester, New York 14642

Over two decades of research on direct cell–cell communication have delineated the principal factors involved in the control of channel permeability (Peracchia 1985). Many physical and chemical changes leading to channel occlusion have been determined and agents mediating the effect of uncoupling treatments on channel gates have been identified. However, the precise site on which the uncouplers act, the possible synergism or competition among uncoupling agents, the likely involvement of soluble intermediates, and the molecular mechanism of channel gating are still poorly understood.

In a continuing effort to simplify the research approach to coupling regulation, our interest has moved from isolated tissues to dimer cells and eventually to internally perfused systems, the result being a progressive improvement in our capacity to dissect and analyze individually the elements controlling cell–cell transport. A further simplification in research approach is the utilization of in vitro reconstituted systems, whose limited number of variables can be more easily and effectively controlled.

Two in vitro systems have recently been employed. One involves the incorporation of purified lens gap junction proteins into liposomes and the evaluation of permeability and gating of the resulting channels (hemichannels) by an osmotic swelling assay (Girsch and Peracchia 1983, 1985a; Nikaido and Rosenberg 1984; Peracchia and Girsch 1984). The other system utilizes planar bilayers to which liposomes containing lens junction protein are incorporated by fusion (Hall and Zampighi, this volume).

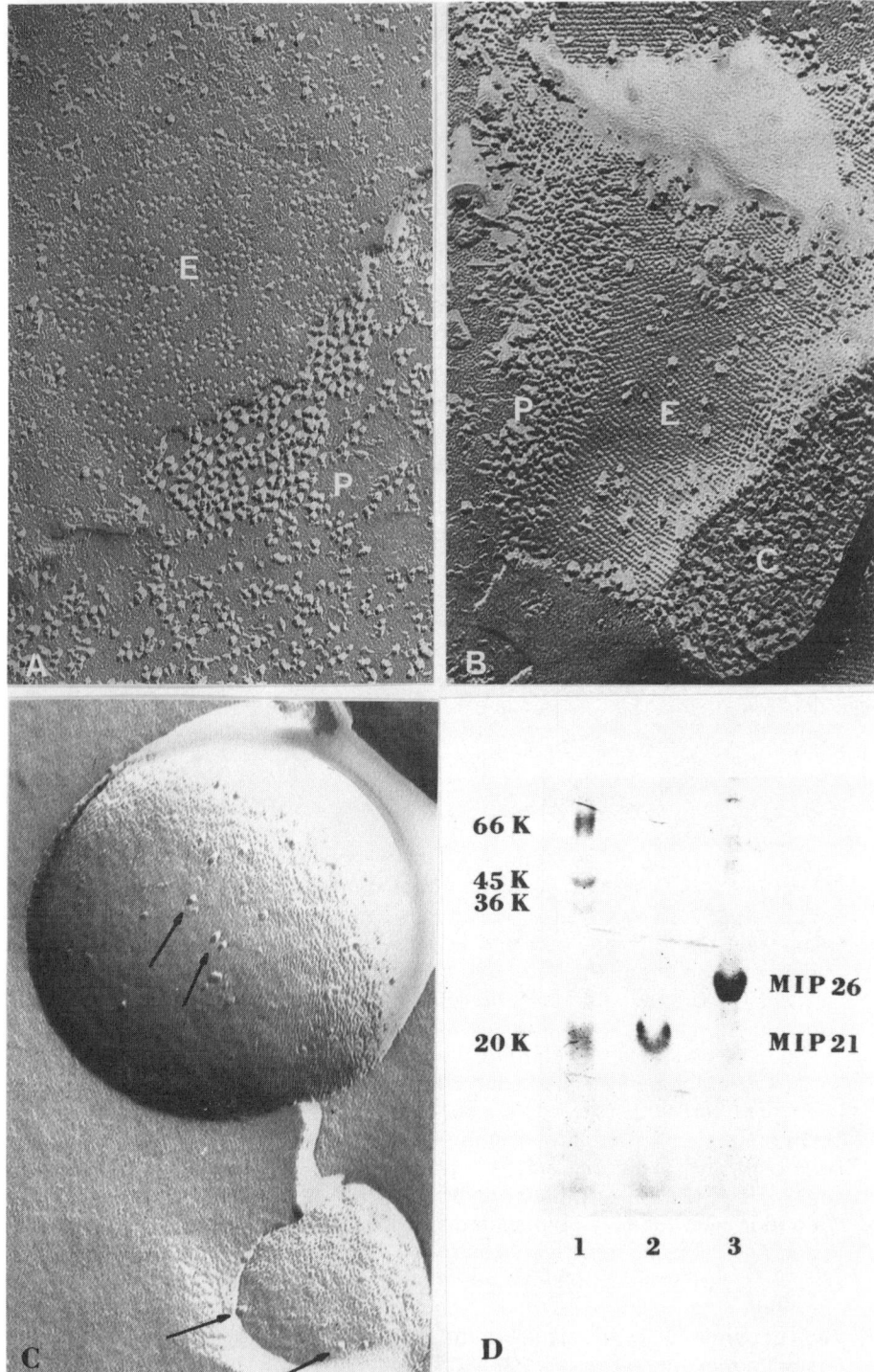

Figure 1 (See facing page for legend.)

DISCUSSION

Permeability of Liposome Incorporated Lens Junction Channels

In the lens, while the surface epithelial cells are coupled by typical gap junctions, the fiber cells are extensively joined (~60% of their surface) by junctions (Fig. 1A) that differ from typical (liver type) gap junctions in many respects. Morphologically, they are thinner and crystallize (Fig. 1B) into pseudohexagonal, rhombic, and orthogonal arrays with a 6.5-nm periodicity (Peracchia and Peracchia 1980a,b; Bernardini and Peracchia 1981; Zampighi et al. 1982). Biochemically, they are made of a 28.2-kD protein (MIP26, Fig. 1D) different in amino acid composition from the 28-kD protein of liver gap junctions (Nicholson et al. 1983). Immunologically, they have only a weak, if any, antigenic cross-reactivity with the liver protein (Hertzberg et al. 1982; Traub and Willecke 1982). Nevertheless, their physiological properties are thought to be basically the same as those of other communicating junctions because they are the only junctions between cells that communicate electrically with each other, pass dyes and metabolites, and can be uncoupled with treatments effective in a variety of cell systems (Bernardini et al. 1981; Rae et al. 1982; Jacob 1983).

For liposome incorporation, MIP26 proved ideal because, unlike other gap junction proteins, it can be purified in large quantities by a simple procedure (Russell et al. 1981; Girsch and Peracchia 1985a). This involves removal of soluble and extrinsically bound proteins from lens fiber junctions by successive washes in Tris buffer, 8 M urea, and 0.1 N NaOH, and either extraction with 2% octylpolyoxyethylene or solubilization of the junctions with 2% SDS and isolation of MIP26 with SDS-polyacrylamide thick-gel electrophoresis (SDS-PAGE) and electroelution. Liposome incorporation is easily obtained by adding MIP26 (final conc., 1 mg/ml) to brain phospholipids (lipid/protein ~250), vortex mixing, and sonicating. For studying channel permeability, both incorporated (Fig. 1C) and protein-free liposomes are loaded with a solution of a channel impermeant, dextran T-10 (10 kD), and suspended into solutions of channel permeants, either KCl, sucrose, or polyethyleneglycol (PEG, mw, 1.450 ± 300), 50% hypertonic to the T-10 solution. All solutions are buffered to pH 7.4 with Tris (Girsch and Peracchia 1985a).

The presence of open channels is determined spectrophotometrically by measuring the decrease in optical density (scattering, $OD_{500\ nm}$) caused by liposome swelling as permeants and water diffuse through the channels while T-10 does not (Lukey and Nikaido 1980). In hypertonic solutions of per-

Figure 1
Incorporation of lens junction protein into liposomes. In freeze-fracture replicas, control fiber cell gap junctions show a loose and disordered array of particles and complementary pits (A). In lenses with increased $[Ca^{++}]_i$ the junctions crystallize into ordered arrays with a 6.5-nm periodicity (B). Liposomes incorporated with MIP26 (C) or MIP21 show ~8-nm particles (arrows) on both the convex and the concave (not shown) fracture face, suggesting a bilateral protein orientation. In D is shown the electrophoretic profile of SDS-extracted lens junction proteins: MIP26 (lane 3) and its trypsin-cleaved product, MIP21 (lane 2). Standards are in lane 1.

meants there is a biphasic change in optical density with open channels: a rapid increase that peaks briefly, followed by a slow decrease (Fig. 2A). The rapid increase in optical density is due to the initial osmotic gradient which causes water efflux and liposome shrinkage; the slow decrease results from a channel-mediated influx of permeants, causing a progressive increase in the internal tonicity and consequently water influx and liposome swelling. In con-

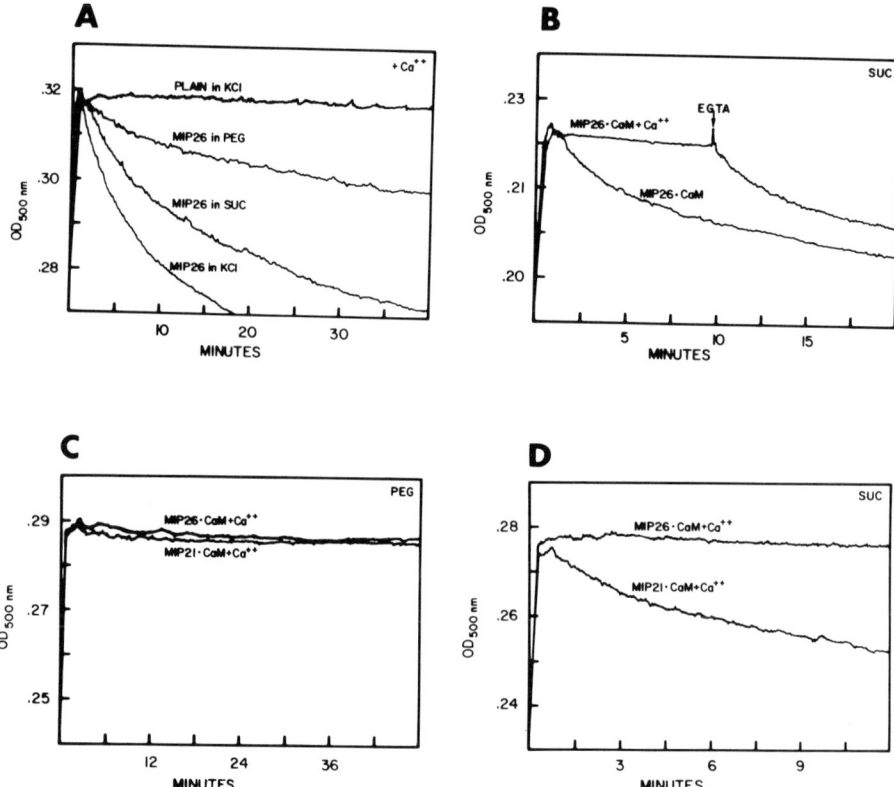

Figure 2
Permeability and gating of liposomes incorporated with MIP26, MIP21, MIP26-CaM, and MIP21-CaM. Channel permeability is determined spectrophotometrically by measuring the decrease in optical density ($OD_{500\ nm}$) as the liposomes, loaded with an impermeant (dextran T-10), swell after a brief initial shrinkage, when suspended in solutions of permeants (KCl, sucrose, or PEG) 50% hyperosmotic to the T-10 solution. MIP-26-incorporated liposomes (A) swell in KCl, sucrose, and PEG with progressively lower rates, whereas protein-free (plain) liposomes do not, indicating the presence of channels permeable to probes as heavy as ~1.5 kD (PEG). MIP26-CaM liposomes (B) swell only in the absence of added Ca^{++} (100 μm). The Ca^{++}-gated channels reopen with addition of EGTA (B). In the presence of Ca^{++} (100 μm) both MIP26-CaM and MIP21-CaM liposomes do not swell in PEG (C) but only MIP26-CaM liposomes do not swell in sucrose (D), indicating that with Ca^{++} the MIP21-CaM channels become impermeable to the larger probe (PEG), but remain permeable to sucrose.

trast, in the absence of channels, as with unincorporated liposomes (Fig. 2A, plain) or with closed channels (Fig. 2B), the initial rapid increase in optical density is not followed by a slow decrease, because only water diffuses through the liposome membrane causing liposome shrinkage up to osmotic equilibrium.

MIP26-incorporated liposomes, but not controls, swell following a brief initial shrinkage in either hypertonic KCl, sucrose, and PEG (Fig. 2A), suggesting the presence of channels permeable to molecules as heavy as $\sim$1.5 kD (Girsch and Peracchia 1983, 1985a). By using the initial rate of change in OD as measure of probe permeance, the log–log plot of OD rate versus molecular weight of probe size is linear (see Fig. 4 below; Peracchia and Girsch 1985a). Recently, Nikaido and Rosenberg (1984) have also succeeded in incorporating MIP26 into liposomes. From the permeability rates of the resulting channels they have estimated a channel diameter of $\sim$1.4 nm.

Preliminary experiments with liposomes incorporated with the gap junction protein isolated from rat liver (28-kD component) show that this protein forms large channels as well. The liver channels may be smaller than those of the lens, because the incorporated liposomes swell in KCl and sucrose but not in PEG (Girsch and Peracchia 1985c).

In an attempt to study channel gating, solutions of permeants varying in [Ca^{++}] and [H^+] were tested. An increase in [Ca^{++}] up to 100 μM (Fig. 2A) or a decrease in pH down to 5.5 has no effect on channel permeability. Since both ions are known to be involved in cell–cell channel gating (Turin and Warner 1977, 1980; Loewenstein 1981; Spray et al. 1981, 1982), these negative results seemed contradictory. A number of interpretations could be proposed: (1) the in vitro channels could be different from intact channels; (2) the in vitro system may lack some regulatory components; (3) the [Ca^{++}] and [H^+] tested may be insufficient for closing a mammalian cell–cell channel (Spray et al. 1982). In view of recent evidence for possible participation of calmodulin (CaM)-like proteins in gap junction function, we chose to test the hypothesis that this protein is a regulatory component of channel gating and that its absence in our in vitro system may explain the apparent lack of channel gating.

CaM-mediated Gating of Incorporated Channels

Johnston and Ramon (1981) have proposed the existence of a soluble intermediate of uncoupling on the basis of evidence for a loss of the Ca^{++} or H^+-induced uncoupling effects in internally perfused septate axons of crayfish. Peracchia and Bernardini (1984) have proposed a CaM-like involvement in uncoupling, on the basis of observed inhibitory effects of CaM inhibitors, trifluoperazine (Peracchia et al. 1981, 1983) and calmidazolium (Peracchia 1984), on CO_2-induced uncoupling of amphibian embryonic cells (Fig. 3). Welsh et al. (1981, 1982) and Hertzberg and Gilula (1982) have demonstrated the binding of CaM to lens and liver gap junction protein in gel overlay experiments. Recently, Wojtczak (1984 and this volume) has shown that trifluoperazine inhibits the increase in longitudinal resistance in papillary muscle fibers caused by Ca^{++} overloading and that calmidazolium prevents the healing over of heart cells. On the other hand, Lees-Miller and Caveney (1982) have re-

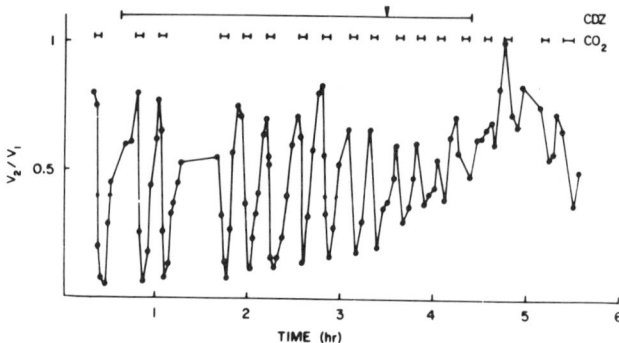

Figure 3
Time course of coupling coefficient between neighboring cells of Xenopus embryo (16- to 32-cell stage). In the absence of calmidazolium (CDZ, a calmodulin inhibitor) exposure to 100% CO_2 causes a large decrease in coupling coefficient. With 5×10^{-8} M calmidazolium (left of arrowhead) the decrease in coupling coefficient with CO_2 is progressively less pronounced. Upon exposure to a higher calmidazolium concentration (1×10^{-7} M, right of arrowhead) the effect of CO_2 on the coupling coefficient is almost completely abolished. This indicates that CaM inhibition prevents the cell–cell channel closure induced by low pH_i (Peracchia 1984).

ported an uncoupling effect of CaM inhibitors in insect cells and Cole and Garfield (1984 and this volume) have shown an inhibition of metabolic coupling with several CaM blockers.

To test a possible CaM involvement in gating in vitro-reconstituted lens junction channels, we have incorporated into liposomes MIP26 in the presence of CaM at equimolar concentration (Girsch and Peracchia 1983, 1985a). Since CaM binds to MIP26 both in the presence and absence of Ca^{++} (Welsh et al. 1981, 1982; Hertzberg and Gilula 1982), we assumed that it remains linked to the incorporated protein thereafter. Thus, both the T-10 loading solution and the solutions of permeants did not contain CaM.

Without added Ca^{++}, MIP26-CaM liposomes swell, following rapid initial shrinkage, when suspended into hypertonic KCl, sucrose (Fig. 2B), or PEG, with an initial swelling rate only slightly lower than that of MIP26 liposomes (Fig. 4). In the presence of Ca^{++} (100 μM) added to the hypertonic solutions from a 10 mM $CaCl_2$ stock solution, MIP26-CaM liposomes do not swell in any of the probes (Fig. 2B), indicating complete channel closure. Addition of 500 μM EGTA to Ca^{++}-treated MIP26-CaM liposomes reinitiates swelling (Fig. 2B), demonstrating the reversibility of the channel gating mechanism. Mg^{++} has no effect on either type of liposome, while preliminary data indicate that lowering the pH from 7.4 to 6.5 sizably reduces the permeability of MIP26-CaM channels and a further reduction to pH 5.5 completely closes these channels. Addition of CaM to the external medium of MIP26 liposomes has only a small effect on channel gating, whereas exposing both inner and outer surfaces of MIP26 liposomes to CaM by loading them with T-10 and CaM restores full gating competency. This indicates that CaM must be on both liposome membrane surfaces to provide effective gating, consistent with the idea that chan-

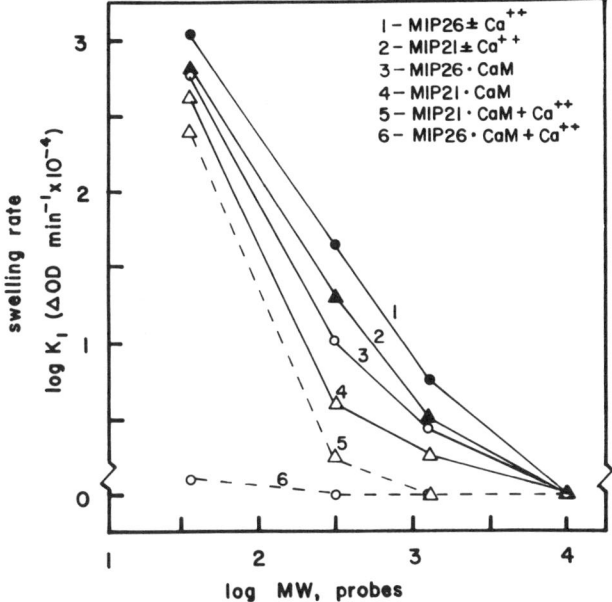

Figure 4
Dependence of liposome swelling rate (channel permeation rate) on the size of the permeant in four liposome types. For MIP26 liposomes (1) the log of swelling rate decreases linearly with the log of probe molecular weight, with or without Ca^{++} (100 μm). MIP21 liposomes (2) behave in a similar way but with slower kinetics. MIP26-CaM (3) and MIP21-CaM (4) liposomes swell, but less rapidly than either type without CaM. Addition of Ca^{++} closes the channels of MIP26-CaM liposomes completely (6), but only partially those of MIP21-CaM liposomes (5). Permeants tested are: K^+ = 40 daltons; sucrose = 342 daltons; PEG = 1.5 kD; dextran T-10 = 10 kD.

nel proteins are incorporated into liposomes from either surface. The bilateral protein incorporation is suggested by freeze-fracture images of intramembrane particles on both liposome fracture surfaces (Girsch and Peracchia 1985a).

These data indicate that the addition of CaM is sufficient to render functional the gating mechanism of reconstituted lens channels. In this system, gating seems to require only MIP26, CaM, and phospholipids, in addition to uncoupling agents (Ca^{++} or H^+), suggesting that all the gating information arises from tertiary and quaternary interactions of the two proteins and the lipid matrix.

In cells, a similar arrangement could exist, but at present it would be hazardous to assume it a priori. The evidence of uncoupling inhibition by CaM blockers supports equally well a direct and/or indirect CaM involvement. Other factors, like protein phosphorylation, for example, could participate in the process by modulating a channel permeability regulation. Data showing a cAMP dependent phosphorylation of MIP26 (in serine only) are available (Johnson and Johnson 1982; Garland and Russell 1984; Johnson; Willecke; both this volume) and cAMP has been shown to affect coupling regulation, in some cases increasing (Hax et al. 1974; DeMello 1984), in others decreasing (Piccolino et al. 1982; Wojtczak 1982; Teranishi et al. 1983; Lasater and Dowling, this volume), coupling. Channel protein phosphorylation could result in either a positive or negative sensitivity modulation of gating. Examples of Ca^{++}-CaM-activated enzymes showing positive and negative sensitivity modulation by phosphorylation are phosphorylase *b* kinase and myosin light-chain kinase, respectively, where phosphorylation shifts the Ca^{++} sensitivity in either direction by approximately one order of magnitude (Rasmussen 1983). Another reason for cautiousness is the fact that the in vitro channels are actually hemichannels rather than whole channels, as they span one membrane only; permeability and gating of whole channels could be different. Moreover, the in vitro channels could be architecturally different and further studies are needed to determine the effect, if any, of different lipid media.

What Structure Makes the Channel Gate?

The gating of a membrane channel could conceivably be attained in various ways. A channel could be closed by a plugging molecule, as tetrodotoxin closes Na^+ channels, by a conformational change of the entire channel framework, resulting in a uniform constriction of the bore diameter, by the door-like movement of a protein side arm, by modified surface charges, etc.

In gap junctions, the molecular events that result in channel closure are still unclear. Recent data from low-dose microscopy and X-ray diffraction (Unwin and Ennis 1983, 1984), showing Ca^{++}-induced changes of subunit orientation in isolated liver gap junctions, may support the model of major protein conformational change. However, in this study the relationship between subunit movement and channel patency was not established and no structural change was detected by lowering pH.

Recent work on the molecular structure of MIP26 and the 28-kD liver protein indicates that these proteins have both the amino- and carboxyterminal arms on their cytoplasmic side, the carboxyterminal arm being the major side-chain (Nicholson et al. 1981, 1983; Gorin et al. 1984). We have tested the possibility that the carboxyterminal arm participates in channel gating by incorporating into liposomes the trypsin cleavage product of MIP26, a 21-kD protein (MIP21) that has lost the 5- to 7-kD carboxyterminal arm (Peracchia and Girsch 1985b). The permeability and gating of MIP21 channels incorporated into T-10-loaded liposomes was studied as previously described.

In the absence of CaM, MIP21 liposomes swell, following a brief initial shrinkage, when suspended in hypertonic KCl, sucrose, or PEG, solutions with or without added Ca^{++} (100 μM), indicating the presence of channels permeable to molecules as heavy as 1.5 kD. The swelling rates of MIP21 channels are lower than those of MIP26 channels (Fig. 4), either because these channels are less permeable or because they are less efficiently incorporated. Without Ca^{++}, the swelling rate of MIP21-CaM liposomes is only slightly lower than that of MIP21 liposomes (Fig. 4). Addition of Ca^{++} (100 μM) to MIP21-CaM liposomes prevents swelling in PEG (Fig. 2C), but not in sucrose (Fig. 2D) or KCl (Fig. 4), indicating that the channels are able to close only partially.

They become impermeable to the larger probe (PEG) and yet remain permeable to sucrose and KCl. A partial occlusion of the MIP21-CaM channels is also suggested by the observation of a decreased swelling rate in sucrose upon Ca^{++} addition (Fig. 4).

These results show that the removal of the carboxyterminal arm has a sizeable effect on the gating competency of the incorporated lens junction channels. A possible interpretation is that the 5- to 7-kD carboxyterminal arm is indeed the major structure of the channel gate and that channel occlusion results from a Ca^{++}-induced conformational change in MIP26, causing the movement of this side-arm toward the channel opening (Fig. 5). Consistent with this hypothesis is recent evidence for Ca^{++}-induced conformational change in isolated MIP26, obtained with fluorescence and circular dichroism spectrophotometry (Girsch and Peracchia 1985b). Alternatively, trypsin cleav-

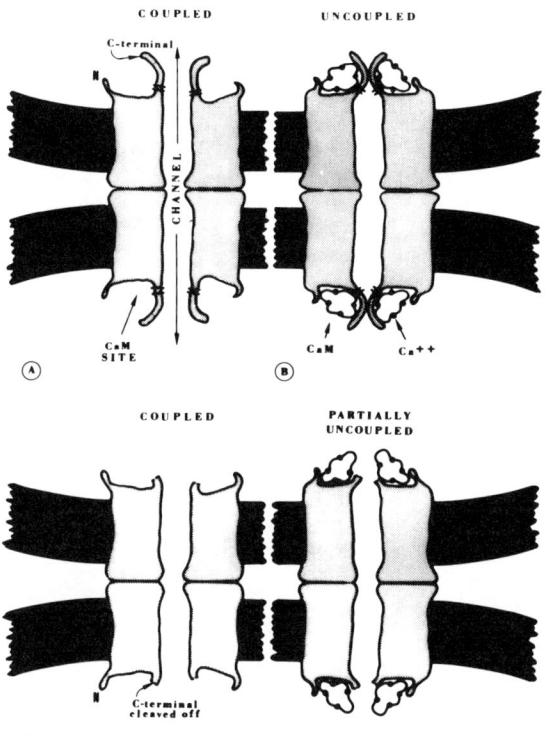

Figure 5
Hypothetical model of channel gating via CaM. The channel protein is believed to have CaM sites (A). CaM would bind to the sites and cause protein conformational changes, when activated, resulting in channel occlusion (uncoupled) near the cytoplasmic end (B). The channels of liposomes incorporated with MIP21 (C), the trypsin-cleaved lens channel protein which has lost the 5- to 7-kD carboxyterminal component, close only partially with Ca^{++}-CaM (D), because they have lost the major portion of the gating structure.

age of the carboxyterminal arm could affect the binding of CaM to channel protein subunits, rather than the channel gate; CaM could bind less strongly and/or inappropriately to MIP21, resulting in less efficient gating.

Data showing an effect of ethoxy-*N*-ethoxydihydroquinoline (EEDQ), a carboxyl reagent, on H^+-induced electrical uncoupling (Spray et al. 1984) and the actual location of the carboxyterminal chain in MIP26 seem consistent with a carboxyterminal involvement in gating. A recent representation of MIP26, derived from cDNA sequence data (Gorin et al. 1984), shows the molecule traversing the lipid bilayer six times and depicts the carboxyterminal arm at the cytoplasmic end of the channel as an extension of the amphiphilic transmembrane segment believed to represent the hydrophilic surface structure of the channel. In addition, the carboxyterminal arm contains serine and threonine residues in a configuration favorable to cAMP-dependent phosphorylation, which might explain the cAMP effect on channel gating mentioned above.

It is interesting that also in the case of Na^+ channels a proteolytic enzyme with a specificity close to that of trypsin has been found to prevent gating (eliminate inactivation) (Armstrong et al. 1973; Rojas and Rudy 1976). In this system, however, the aminoterminal peptide appears to be the cleavable gating structure (Armstrong and Bezanilla 1977). In MIP26, only five amino acid residues are removed by trypsin from the aminoterminus (Nicholson et al. 1983), virtually all the cleavage occurring at the carboxyl terminus. Thus, the possibility that the aminoterminal arm participates in cell–cell channel gating cannot be completely ruled out yet.

In view of the fact that the MIP21/MIP26 ratio physiologically increases with age, one should consider the possibility that the presence of poorly gated channels is needed for the normal function of aging lenses. The presence of MIP21 in older lenses could be a safety device needed for maintaining open ionic and metabolic pathways between cortical fibers and degenerating fibers of nuclear and perinuclear regions.

CONCLUSIONS

For years gap junctions have been assumed to be the structures that mediate direct cell–cell communication, but until recently, only circumstantial evidence has supported this notion (Peracchia 1985). During the past year, data showing the electrical and dye uncoupling effects of intracellularly injected monoclonal antibodies to the liver gap junction protein (Warner et al. 1984) and evidence for the ability of isolated gap junction proteins to create large channels once incorporated into liposomes (Girsch and Peracchia 1983, 1985a; Nikaido and Rosenberg 1984; Peracchia and Girsch 1984) or planar bilayers (Hall and Zampighi, this volume) have supported this notion more directly.

The channel-incorporated liposome system has enabled us to study in vitro channel gating and permeability. The incorporated lens channels are permeable to molecules as heavy as 1.5 kD, whereas the liver channels appear to be slightly smaller. The channels are not gated by Ca^{++} or H^+ alone, but become gating competent in the presence of CaM. Channels made of MIP21, a trypsin cleavage product of MIP26 which has lost the 5- to 7-kD carboxyterminal arm, close only partially with Ca^{++}-CaM, suggesting that the carboxyterminal arm may be the major structure of the channel gate.

The recent developments in gap junction study open a new exciting chapter in cell communication research. The application of monoclonal antibodies will provide the means for determining the biogenesis of cell-cell channel protein and the intimate meaning of gap junctions in cell function—their possible involvement in differentiation and transformation and their role in maintaining the ionic and metabolic balance of cells engaged in cooperative activities. The in vitro approach will bring to focus the molecular basis of channel gating and will enable one to test the suspected participation of protein phosporylation in channel modulation.

REFERENCES

Armstrong, C.M. and F. Bezanilla. 1977. Inactivation of the sodium channel. II. Gating current experiments. *J. Gen. Physiol.* **70:** 567-590.

Armstrong, C.M., F. Bezanilla, and E. Rojas. 1973. Destruction of sodium conductance inactivation in squid axons perfused with pronase. *J. Gen. Physiol.* **62:** 375-391.

Bernardini, G. and C. Peracchia. 1981. Gap junction crystallization in lens fibers after an increase in cell calcium. *Invest. Opthalmol. Visual Sci.* **21:** 291-299.

Bernardini, G., C. Peracchia, and R.A. Venosa. 1981. Healing over in rat crystalline lens. *J. Physiol.* **320:** 187-192.

Cole, W.C. and R.E. Garfield. 1984. A23187 and calmodulin antagonists inhibit metabolic coupling between parturient rat myometrial smooth muscle cells. *Biophys. J.* **45:** 23a.

DeMello, W.C. 1984. Effect of intracellular injection of cAMP on the electrical coupling of mammalian cardiac cells. *Biochem. Biophys. Res. Commun.* **119:** 1001-1007.

Garland, D. and P. Russell. 1984. Phosphorylation of lens fiber cell membrane proteins. *Proc. Natl. Acad. Sci.* **82:** 653-657.

Girsch, S.J. and C. Peracchia. 1983. Lens junction protein (MIP26) self-assembles into liposomes forming large channels regulated by calmodulin (CaM). *J. Cell Biol.* **97:** 83a.

———. 1985a. Lens cell-to-cell channel proteins: I. Self-assembly into liposomes and permeability regulation by calmodulin. *J. Membr. Biol.* **83:** 217-225.

———. 1985b. Lens cell-to-cell channel protein: II. Conformational changes in the presence of calmodulin. *J. Membr. Biol.* **83:** 227-233.

———. 1985c. Liposome-incorporated rat liver gap junction channels are less permeable than lens channels. *Biophys. J.* **47:** 507a(Abstr.).

Gorin, M.B., S.B. Yancey, J. Cline, J.-P. Revel, and J. Horwitz. 1984. The major intrinsic protein (MIP) of the bovine lens fiber membrane: Characterization and structure based upon cDNA cloning. *Cell* **39:** 49-59.

Hax, W.M.A., G.E.P.M. Venrooij, and J.B.J. Vossenberg. 1974. Cell communication: A cyclic AMP mediated phenomenon. *J. Membr. Biol.* **19:** 253-266.

Hertzberg, E.L. and N.B. Gilula. 1982. Liver gap junctions and lens fiber junctions: Comparative analysis and calmodulin interaction. *Cold Spring Harbor Symp. Quant. Biol.* **46:** 639-645.

Hertzberg, E.L., D.J. Anderson, M. Friedlander, and N.B. Gilula. 1982. Comparative analysis of the major polypeptides from liver gap junctions and lens gap junctions. *J. Cell Biol.* **92:** 53-59.

Jacob, T.J. 1983. Raised intracellular free calcium within the lens causes opacification and cellular uncoupling in the frog. *J. Physiol.* **341:** 595-601.

Johnson, K.R. and R. Johnson. 1982. Bovine lens MP26 is phosphorylated in vitro by an endogenous cAMP-dependent protein kinase. *Fed. Proc.* **41:** 755 (Abstr.).

Johnston, M.F. and F. Ramón. 1981. Electrotonic coupling in internally perfused crayfish segmented axons. *J. Physiol.* **317:** 509-518.

Lees-Miller, J.P. and S. Caveney. 1982. Drugs that block calmodulin activity inhibit cell-to-cell coupling in the epidermis of *Tenebrio molitor*. *J. Membr. Biol.* **69:** 233-245.

Loewenstein, W.R. 1981. Junctional intracellular communication: The cell-to-cell membrane channel. *Physiol. Rev.* **61:** 829-913.

Luckey, J. and H. Nikaido. 1980. Specificity of diffusion channels produced by phage receptor protein of *Escherichia coli*. *Proc. Natl. Acad. Sci.* **77:** 167-171.

Nicholson, B.J., M.W. Hunkapiller, L.B. Grim, L.E. Hood, and J.-P. Revel. 1981. Rat liver gap junction protein: Properties and partial sequence. *Proc. Natl. Acad. Sci.* **78:** 7594-7598.

Nicholson, B.J., L.J. Takemoto, M.W. Hunkapiller, L.E. Hood, and J.-P. Revel. 1983. Differences between liver gap junction protein and lens MIP26 from rat: Implications for tissue specificity of gap junctions. *Cell* **32:** 967-978.

Nikaido, H. and E.Y. Rosenberg. 1984. Functional reconstitution of lens junction proteins into proteoliposomes. In *Proceedings of the 42nd Annual Meeting of the Electron Microscopy Society of America* (ed. G.W. Bailey), p. 130-133. San Francisco Press, California.

Peracchia, C. 1984. Communicating junctions and calmodulin: Inhibition of electrical uncoupling in *Xenopus* embryo by calmidazolium. *J. Membr. Biol.* **81:** 49-58.

―――. 1985. Cell coupling. In *The enzymes of biological membranes* (ed. A. Martonosi), vol. 1, p. 81-130. Plenum Publishing, New York.

Peracchia, C. and G. Bernardini. 1984. Gap junction structure and cell-to-cell coupling regulation: Is there a calmodulin involvement? *Fed. Proc.* **43:** 2681-2691.

Peracchia, C. and S.J. Girsch. 1984. Calmodulin-mediated gating of lens gap juncion channels in vesicles. In *Proceedings of the 42nd Annual Meeting of the Electron Microscopy Society of America* (ed. G.W. Bailey), p. 134-137. San Francisco Press, California.

―――. 1985a. Permeability and gating of lens gap junction channels. *Curr. Eye Res.* **49:** (in press).

―――. 1985b. C-terminal arm and cell-to-cell channel gating. *Biophys. J.* (Abstr.) **47:** 506.

Peracchia, C. and L.L. Peracchia. 1980a. Gap junction dynamics: Reversible effects of divalent cations. *J. Cell Biol.* **87:** 708-718.

―――. 1980b. Gap junction dynamics: Reversible effects of hydrogen ions. *J. Cell Biol.* **87:** 719-727.

Peracchia, C., G. Bernardini, and L.L. Peracchia. 1981. A calmodulin inhibitor prevents gap junction crystallization and electrical uncoupling. *J. Cell Biol.* **91:** 124a.

―――. 1983. Is calmodulin involved in the regulation of gap junction permeability? *Pfluegers Arch.* **399:** 152-154.

Piccolino, M., J. Neyton, P. Witkovsky, and H.M. Gerschenfeld. 1982. γ-Aminobutyric acid antagonists decrease junctional communication between L-horizontal cells of the retina. *Proc. Natl. Acad. Sci.* **79:** 3671-3675.

Rae, J.L., R.D. Thompson, and R.S. Eisenberg. 1982. The effect of 2-4 dinitrophenol on cell-to-cell communication in the frog lens. *Exp. Eye Res.* **35:** 598-609.

Rasmussen, H. 1983. Pathways of amplitude and sensitivity modulation in the calcium messenger system. In *Calcium and cell function* (ed. W.Y. Cheung), vol. IV, p. 1-61. Academic Press, New York.

Rojas, E. and B. Rudy. 1976. Destruction of the sodium conductance inactivation by a specific protease in perfused nerve fibers from *Loligo*. *J. Physiol.* **262:** 501-531.

Russell, P., G. Robinson, and J. Kinoshita. 1981. A new method for rapid isolation of the intrinsic membrane proteins of the lens. *Exp. Eye Res.* **32:** 511–516.

Spray, D.C., A.L. Harris, and M.V.L. Bennett. 1981. Gap junction conductance is a simple and sensitive function of intracellular pH. *Science* **211:** 712–715.

Spray, D.C., J.H. Stern, A.L. Harris, and M.V.L. Bennett. 1982. Gap junctional conductance: Comparison of sensitivities to H and Ca ions. *Proc. Natl. Acad. Sci.* **79:** 441–445.

Spray, D.C., R.L. White, A. Campos de Carvalho, A.L. Harris, and M.V.L. Bennett. 1984. Gating of gap junction channels. *Biophys. J.* **45:** 219–230.

Teranishi, T., K. Negishi, and S. Kato. 1983. Dopamine modulates S-potential amplitude and dye-coupling between external horizontal cells in carp retina. *Nature* **301:** 243–246.

Traub, O. and K. Willecke.1982. Cross-reaction of antibodies against liver gap junction protein (26 K) with lens fiber junction protein (MIP) suggests structural homology between these tissue specific gene products. *Biochem. Biophys. Res. Commun.* **109:** 895–901.

Turin, L. and A.E. Warner. 1977. Carbon dioxide reversibly abolishes ionic communication between cells of early amphibian embryo. *Nature* **270:** 56–57.

―――. 1980. Intracellular pH in early *Xenopus* embryos: Its effect on current flow between blastomeres. *J. Physiol.* **300:** 489–504.

Unwin, P.N.T. and P.D. Ennis. 1983. Calcium-mediated changes in gap junction structure: Evidence from the low angle X-ray pattern. *J. Cell Biol.* **97:** 1459–1466.

―――. 1984. Two configurations of a channel-forming membrane protein. *Nature* **307:** 609–613.

Warner, A.E., S.C. Guthrie, and N.B. Gilula. 1984. Antibodies to gap-junctional protein selectively disrupt junctional communication in the early amphibian embryo. *Nature* **311:** 127–131.

Welsh, M.J., J. Aster, M. Ireland, J. Alcala, and H. Maisel. 1981. Calmodulin and gap junctions: Localization of calmodulin and calmodulin binding sites in chick lens cells. *J. Cell Biol.* **91:** 123a.

―――. 1982. Calmodulin binds to chick lens gap junction protein in a calcium-independent manner. *Science* **216:** 642–644.

Wojtczak, J. 1982. Influence of cyclic nucleotides on the internal longitudinal resistance and contractures in the normal and hypoxic mammalian cardiac muscle. *J. Mol. Cell Cardiol.* **14:** 259–265.

―――. 1984. Effect of general and local anesthetics on intercellular coupling in the heart muscle. *Biophys. J.* **45:** 22a.

Zampighi, G., S.A. Simon, J.D. Robertson, T.J. McIntosh, and M.J. Costello. 1982. On the structural organization of isolated bovine lens fiber junctions. *J. Cell Biol.* **93:** 175–189.

Reduced Junctional Permeability in Cells Transformed by Different Viral Oncogenes

Michael M. Atkinson and Judson D. Sheridan
*Department of Anatomy
University of Minnesota
Minneapolis, Minnesota 55455*

The possibility that cancer cells have impaired intercellular communication via permeable junctions was raised more than a decade ago (Loewenstein and Kanno 1967; Furshpan and Potter 1968), but the evidence has remained mixed. Some cancer cells apparently lack such junctions (Loewenstein and Kanno 1967; Borek et al. 1969), whereas many others have junctions qualitatively similar to those between normal cells (Furshpan and Potter 1968; Sheridan 1970). However, there have been few quantitative studies, particularly of cell systems in which the cancerous, or "transformed," state can be easily manipulated, or the transformed cells directly compared with their nontransformed counterparts. Thus, it has remained possible that, in general, the junctional permeance of cancer cells is reduced, although not necessarily to zero (Sheridan and Johnson 1975; Loewenstein 1979).

As an initial test of this possibility, we have reported that dye transfer between cells infected with a temperature-sensitive Rous sarcoma virus (RSV) mutant is reduced, but not eliminated, shortly after shifting the cells to the transformation-permissive temperature (Atkinson et al. 1981). We have quantitated this response more precisely with computer-assisted video analysis of dye spread, and have expanded these studies to cells infected with other retroviruses having different types of oncogenes. Reduction of junctional dye permeability is again correlated with the transformed state, even though the different oncogenes are believed to act via different primary mechanisms. Our freeze-fracture studies of RSV-infected cells indicate that the decreased transfer is most likely due to reduced junctional permeability rather than to reduced junctional area. Thus, our results provide new support for the general hypothesis that reduced junctional communication and cell transformation are associated.

ASV-INFECTED CELLS

We began our studies on avian sarcoma virus-infected cells because it had been previously suggested that the transforming gene product of these viruses, pp60^{src}, appears to accumulate in the vicinity of cell junctions (Willingham et al. 1979). For our studies we used NRK cells infected with the temperature-sensitive mutant LA25, which is active at 33–35°C ("permissive" temperatures) and inactive at 39–40.5°C ("restrictive" temperatures) (Wyke 1973). That is, the cells have a transformed phenotype when grown at the permissive temperatures and a nontransformed phenotype when grown at restrictive temperatures (Wang and Goldberg 1979).

Initially, we tested the ability of the cells to transfer dye after being grown for long times (generally more than a day) at the two different temperatures. For these studies we injected Lucifer Yellow CH (MW 457) into single cells and determined the minimum time elapsed from the begining of the injection until dye first appeared in an adjacent cell (transfer time interval). As we have reported (Atkinson et al. 1981), dye transfer occurred regardless of the growth temperature, but the rate of transfer was significantly slower (i.e., the transfer time interval was longer) for the transformed cells when compared with either the nontransformed, infected, or the uninfected cells (Table 1).

Thus, the results were consistent with an association between decreased junctional transfer and cell transformation. To characterize this association further, we determined the kinetics of the junctional change after short-term temperature shifts between the permissive and restrictive temperatures. Whereas it was possible to use transfer time intervals for these kinetic experiments (Table 2), a more direct measure of junctional permeance was desirable. Consequently, we applied a method that we had recently developed for estimating junctional permeability times area for pairs of cultured cells (M.M. Atkinson and J.D. Sheridan, in prep.). The method involves injecting one cell of an isolated pair with the fluorescent dye, Lucifer Yellow CH, and videotaping dye movement with the use of a high-sensitivity video camera and tape recorder. A digital representation of the fluorescence intensity is obtained with

Table 1
Average Transfer Time Intervals of Cells Maintained at 35°C or 40.5°C

Cell line	Growth temperature	Average transfer time interval ± S.E.M. (sec)	Number of penetrations
LA25-NRK	35°C	26.1 ± 3.4	35
LA25-NRK	40.5°C	6.6 ± 0.4	35
Uninfected NRK	35°C	5.8 ± 0.5	14
Uninfected NRK	40.5°C	5.7 ± 0.8	11

Data for the average transfer time intervals were obtained from three to six separate experiments. (Reprinted, with permission, from Atkinson et al. 1981.)

Table 2
Transfer Time Intervals from LA25-NRK Subjected to Reciprocal Temperature Shifts

Direction of shift	Time of incubation (min)	Average transfer time interval ± S.E.M. (sec)	Number of penetrations
40.5°C → 35°C	0	6.6 ± 0.4	35
	15	13.9 ± 1.5	19
	30	18.0 ± 1.6	23
	60	24.9 ± 3.6	15
35°C → 40.5°C	0	26.1 ± 3.4	35
	15	14.9 ± 1.8	24
	30	6.6 ± 0.4	20
	60	6.2 ± 0.5	10

Data for the average transfer time intervals were obtained from three to six separate experiments. (Reprinted, with permission, from Atkinson et al. 1981.)

a Digisector board and Apple computer. Control experiments confirm that the video signal is directly proportional to fluorescence intensity, which in turn is proportional to dye concentration within a limited, but usable, range. Given these relationships, diffusion theory predicts that the following formula should hold true:

$$\ln(I_1 - I_2) = -PA(1/V_1 + 1/V_2)t + \ln I_0,$$

where I_1 and I_2 are the instantaneous digitized intensities at time t, I_0 is the initial intensity in the injected cell, P is the permeability coefficient, A is junctional area, and V_1 and V_2 are the volumes of the two cells. When $\ln(I_1 - I_2)$ is plotted against time, a straight line should result. The slope of the line $-PA(1/V_1 + 1/V_2)$ can be used to compare junctional permeance (PA), provided the volumes change rather little. If the cells are spherical, the volumes can be estimated and PA actually calculated. The reliability of the slope and derived PA values can be assessed by inspecting the fit of the values to the calculated straight line and by determining a correlation coefficient R. Typically, the visual fit of the data to the straight line is quite good (Fig. 1) with $R > 0.95$. When the fit is good, potential complications such as binding of dye, differential cell thicknesses, etc., that tend to produce nonlinear functions, can be excluded. When present, these factors usually can be corrected for as will be described elsewhere.

When the computer-assisted video analysis was applied to the LA25-infected cells in temperature-shift experiments, we saw that nearly complete changes in junctional permeability (as reflected in the slopes) had occurred after only 15 minutes at the new temperature, irrespective of the direction of the temperature shifts (Table 3). It is particularly interesting that the junctional changes occur with nearly the same time course as the changes in the activity of the virus gene product (A. Goldberg, pers. comm.; see Radke and Martin 1979 for data on a related temperature-sensitive mutant). To the best of our knowledge, this makes the junctional changes the earliest alteration of phenotype occurring as LA25-NRK cells assume or leave the transformed state,

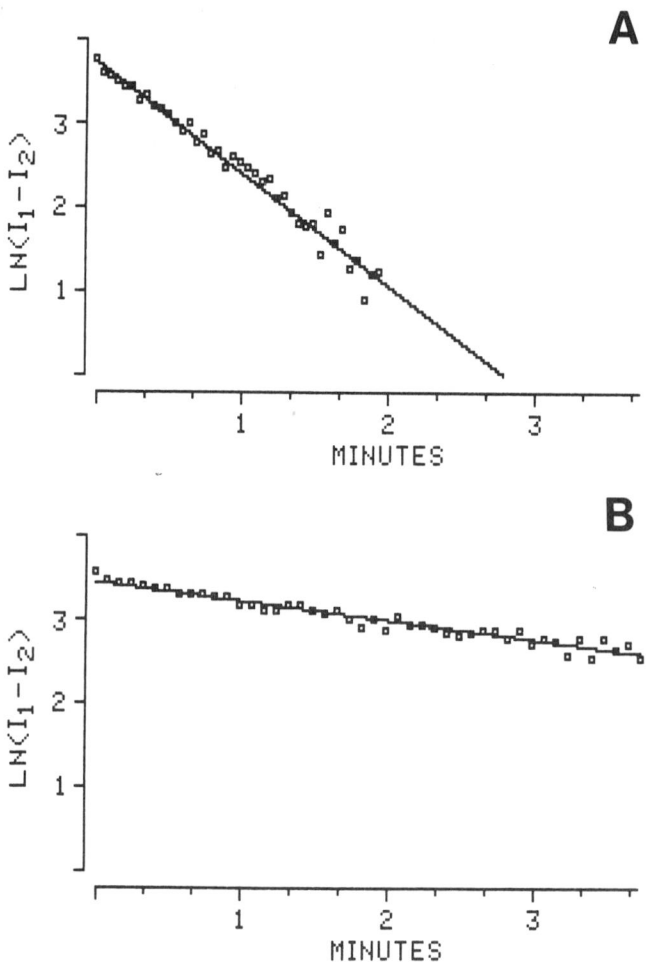

Figure 1
Computer printout of ln $(I_1 - I_2)$ for a pair of LA25 cells grown for more than 24 hr at the restrictive, nontransforming temperature, 39°C (A), and at the permissive, transforming temperature, 33°C (B). The slope, $PA (1/V_1 + 1/V_2)$ for A is 0.022 with $R = -0.99$; the slope for B is 0.004 with $R = -0.98$.

thus increasing the possibility that the junctional defect somehow contributes to the transformation process.

There are two ways that dye transfer as measured by either transit time interval or slope might be altered, both of which would lead to fewer open junctional channels. First, the total junctional area (and therefore the number of channels) might be decreased. Second, the proportion of open versus closed channels and/or channel diameter might be decreased while the junctional area remained constant. By studying freeze-fracture replicas of unin-

Table 3
Permeability of Junctions from LA25-NRK Subjected to Reciprocal Temperature Shifts

Direction of shift	Time of incubation (min)	$PA(1/V_1 + 1/V_2)$	Number of penetrations
39°C → 33°C	0	0.026 ± 0.003	21
	15	0.010 ± 0.001	9
	30	0.007 ± 0.001	10
33°C → 39°C	0	0.008 ± 0.001	14
	15	0.026 ± 0.006	7
	30	0.023 ± 0.005	7

PA is the permeance term for the cell pair having volumes V_1 and V_2, where P is the permeability constant and A is the total junctional area.

fected cells and LA25-infected cells at the permissive and restrictive temperatures, we have found little difference in the gap junctional area per interface between infected or uninfected cells grown for short (1 hr) or long (>24 hr) times at either temperature (M.M. Atkinson, Anderson, and J.D. Sheridan, in prep.). Thus, there do not appear to be sufficient changes in junctional area to explain the differences in dye transer, implying that the existing channels are being altered. This conclusion is further supported by our observation that the junctional particles undergo a change in packing arrangement that correlates temporally with the change in junctional permeability following temperature shifts. Most of the large junctions (i.e., those that can be scored reliably) in cells at the restrictive temperature (and in uninfected cells at either temperature) have an ordered pattern (Fig. 2A), whereas those in infected cells at the permissive temperature have a looser, more random pattern (Figure 2B). As shown in Table 4, the manifestation of the change is different for the two temperature shifts. That is, the onset of transformation induced by a temperature downshift is accompanied by a decrease in the number of interfaces that contain junctions with ordered particle packing; reversal to the nontransformed phenotype by switching the cells to the nonpermissive temperature is accompanied by an increase in the area of junctions displaying the ordered particle arrangement, as well as an increase per interface in interfaces with ordered packing. Thus, ordered packing is apparently associated with high junctional permeance and loose packing associated with low permeance. These relationships agree with those seen in the heart (Page et al. 1983; Green and Severs 1984), but are opposite those seen in certain other preparations (see Peracchia 1980). These correlations could reflect a real difference in the in situ packing of the junctional subunits or merely a change in their respective sensitivities to fixative-induced rearrangements. Whichever the explanation, the correlation is interesting since, in either case, the results indicate a modification of the particles (or associated lipids) during the transformation process.

The mechanism underlying the change in junctional permeability is unknown, but an interesting possibility is that the gene product, pp60^{src}, which

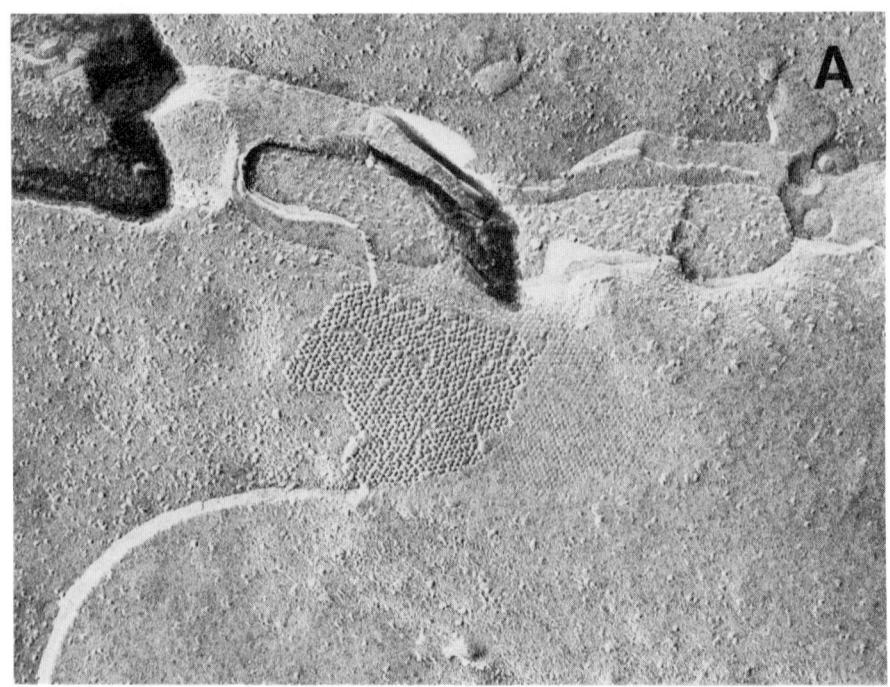

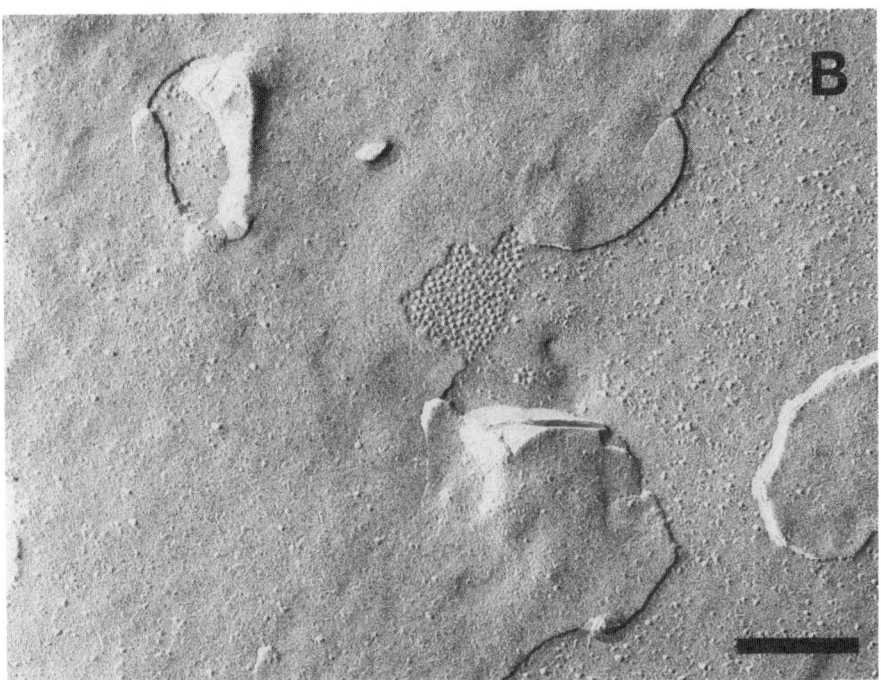

Figure 2 *(See facing page for legend.)*

Table 4
Number and Area of Ordered Junctions per Scored Interface

Temperature (°C)	Percent of scored interfaces with ordered junctions	Area of ordered junctions per scored interface ($\mu m^2 \pm$ S.D.)	Number of interfaces scored
40.5	70%[a]	5.7 ± 7.4	23
40.5 → 35	20%	5.8 ± 7.0	34
35	25%	1.6 ± 1.2[b]	16
35 → 40.5	52%	6.3 ± 11.7	27

[a] $p < 0.001$ that the frequency for 40.5°C cells is not different from the frequency for downshifted cells (chi-square).

[b] $p < 0.05$ that this sample population and the sample from the downshifted cells are from the same population (Mann-Whitney two-tailed).

is a tyrosine-specific protein kinase (Collett and Erikson 1978; Hunter and Sefton 1980), acts directly on junctional protein. To date, there is no evidence that junctional proteins contain phosphotyrosine, but further studies are warranted. Of course, less direct mechanisms, e.g., via phosphorylation of inositol, decrease in pH, or increase in cytoplasmic calmodulin or calcium are also possible and should be explored.

OTHER VIRALLY TRANSFORMED CELLS

Our results with ASV-infected cells immediately prompt the question of whether viral transformation involving other primary mechanisms, i.e., oncogene products with properties distinct from pp60^{src}, might also decrease junctional permeance.

Studies on an NRK cell clone (6m2), which is infected with temperature-sensitive mutant (ts110) of the Moloney murine sarcoma virus (Blair et al. 1979), have given results similar to those seen with the LA25-infected cells. These cells have permissive and restrictive temperatures similar to those for the LA25 cells. They differ, however, in taking much longer after temperature shift to express the transformed phenotype or to revert to normal (Brown et al. 1981). This delay may in part relate to the fact that the temperature-sensitive step is at the transcriptional level (Horn et al. 1981). As can be seen from Table 5, 6m2 cells, grown at the permissive temperatures, also show decreased slopes. At present, we do not have kinetic data on the effects of temperature shift except to note that no change in junctional permeability is apparent by 6 hours after downshift.

Figure 2
Sample micrographs of freeze-fracture replicas taken from LA25 cells grown for more than 24 hr at restrictive (A) or permissive (B) temperatures. Note the regular, nearly crystalline particle domains in A (ordered) and the dispersed pattern of particles in B (nonordered). Calibration bar, 0.2 μm.

Table 5
Mean Junctional Permeability of Normal and Transformed ts110-NRK Cells and Uninfected NRK Cells

Cell line	Growth temperature	PA($1/V_1 + 1/V_2$)	Number of penetrations
ts110-NRK	39°C	0.058 ± 0.01	10
ts110-NRK	33°C	0.009 ± 0.004	10
Uninfected NRK	39°C	0.032 ± 0.008	14
Uninfected NRK	33°C	0.025 ± 0.005	14

PA is the permeance term for the cell pair having volumes V_1 and V_2, where P is the permeability constant and A is the total junctional area.

Besides the more complete analysis of the LA25 and 6m2 cells, we have preliminary information that junctions between NRK cells infected with a Kirsten murine sarcoma virus have reduced dye permeability.

Thus, our studies indicate that transformation of NRK cells by different retroviruses is accompanied by quantitative reductions in junctional permeance. These observations strengthen the idea that there may be some general connection between impaired junctional permeance and cancerous transformation and indicate that the deficit need not involve the complete loss of junctional communication. Future experiments with other RNA as well as DNA viruses and with chemically induced transformation will be needed before the connection can be generalized. Even with the few systems we now have, however, we may be able to begin unravelling the possible association between junctional defects and abnormal regulation of cell proliferation.

ACKNOWLEDGMENTS

Susan Anderson provided excellent technical contributions to the freeze-fracture studies. We are grateful to Ron Furnival and Janet Balson for maintaining the tissue culture stock, and to Chris Frethem and Jennifer Steinert for their help in preparing the manuscript.

REFERENCES

Atkinson, M.M., A.S. Menko, R.G. Johnson, J.R. Sheppard, and J.D. Sheridan. 1981. Rapid and reversible reduction of junctional permeability in cells infected with a temperature-sensitive mutant of avian sarcoma virus. *J. Cell Biol.* **91:** 573–578.

Blair, D.G., M.A. Hull, and E.A. Finch. 1979. The isolation and preliminary characterization of temperature-sensitive tranformation mutants of Moloney sarcoma virus. *Virology* **95:** 303–316.

Borek, C., S. Higashino, and W. Loewenstein. 1969. Intercellular communication and tissue growth. IV. Conductance of membrane junctions of normal and cancerous cells in culture. *J. Membr. Biol.* **222:** 78–86.

Brown, R., J.P. Horn, L. Wible, R.B. Arlinghaus, and B.R. Brinkley. 1981. Analysis of the sequence of events in the transformation process in cells infected with a ts transformation mutant of Moloney murine sarcoma virus. *Proc. Natl. Acad. Sci.* **78:** 5593–5597.

Collett, M.S. and R.L. Erikson. 1978. Protein kinase activity associated with the avian sarcoma virus src gene product. *Proc. Natl. Acad. Sci.* **75:** 2021-2024.

Furshpan, E.J. and D.D. Potter. 1968. Low resistance junctions between cells in embryos and tissue culture. *Curr. Top. Dev. Biol.* **3:** 95-127.

Green, C.R. and N.J. Severs. 1984. Gap junction connexon configuration in rapidly frozen myocardium and isolated intercalated disks. *J. Cell Biol.* **99:** 453-463.

Horn, J.P., T.G. Wood, E.C. Murphy, Jr., D.G. Blair, and R.B. Arlinghaus. 1981. A selective temperature sensitive defect in viral RNA expression in cells infected with a ts transformation mutant of murine sarcoma virus. *Cell* **25:** 37-46.

Hunter, T. and B.M. Sefton. 1980. The tranforming gene product of Rous sarcoma virus phosphorylates tyrosine. *Proc. Natl. Acad. Sci.* **77:** 1311-1315.

Loewenstein, W.R. 1979. Junctional intercellular communication and the control of growth. *Biochim. Biophys. Acta* **560:** 1-65.

Loewenstein, W.R. and Y. Kanno. 1967. Intercellular communication and tissue growth. I. Cancerous growth. *J. Cell Biol.* **33:** 225-234.

Page, E., T. Karrison, and J. Upshaw-Earley. 1983. Freeze-fractured cardiac gap junctions: Structural analysis by three methods. *Am. J. Physiol.* **244:** H525-H539.

Peracchia, C. 1980. Structural correlates of gap junction permeation. *Int. Rev. Cytol.* **66:** 81-146.

Radke, K. and G.S. Martin. 1979. Transformation by Rous sarcoma virus: Effects of src gene expression on the synthesis and phosphorylation of cellular polypeptides. *Proc. Natl. Acad. Sci.* **76:** 5212-5216.

Sheridan, J.D. 1970. Low-resistance junctions between cancer cells in various solid tumors. *J. Cell Biol.* **45:** 91-99.

Sheridan, J.D. and R.G. Johnson. 1975. Cell junctions and neoplasia. In *Molecular pathology* (ed. R.A. Good et al.), p. 354-378. C.C. Thomas, Springfield, Illinois.

Wang, E. and A.R. Goldberg. 1979. Effects of the src gene product of microfilament and microtubule organization in avian and mammalian cells infected with the same temperature-sensitive mutant of Rous sarcoma virus.*Virology* **92:** 201-210.

Willingham, M.C., G. Jay, and I. Pastan. 1979. Localization of the ASV src gene product to the plasma membrane of transformed cells by electron microscopic immunocytochemistry. *Cell* **18:** 125-134.

Wyke, J.A. 1973. The selective isolation of temperature-sensitive mutants of Rous sarcoma virus. *Virology* **52:** 587-590.

Alterations in Coupling in Uterine Smooth Muscle

W.C. Cole and R.E. Garfield
*Department of Neurosciences
McMaster University Health Sciences Center
Hamilton, Ontario, Canada L8N 3Z5*

Parturition in animals and humans is marked by the development of intense, synchronous, and coordinated contractile activity in the smooth muscle layers (myometrium) of the uterus from a relatively inactive state during most of pregnancy (Csapo 1981). In all species studied, the onset of this activity during term or preterm labor is invariably associated with the precipitous development of large numbers of gap junctions between the myometrial cells (Garfield et al. 1977, 1978). Moreover, improved electrical (Sims et al. 1982) and metabolic (Cole et al. 1985) communication between uterine smooth muscle cells is observed concomitant with the formation of the junctions, supporting the hypothesis that the gap junctions permit the myometrium to behave as a functional syncytium during parturition. These alterations in the extent of structural and functional coupling are significant, therefore, in that the presence of gap junctions and cell–cell communication probably represents the biophysical basis for synchronous and effective uterine contractile activity during labor.

Evidence from our group at McMaster University and that of others suggests the presence of specific physiological mechanisms for regulating and producing alterations in structural and functional coupling in the myometrium during pregnancy and parturition. This paper will describe the possible mechanisms involved in the control of coupling in uterine smooth muscle. In particular, it will indicate the hormonal mechanisms thought to regulate: (1) the appearance of the gap junctions, hence, the extent of structural coupling, and (2) the permeability of the gap junctions, hence, the extent of functional coupling in the myometrium. The integrated function of these control mechanisms prob-

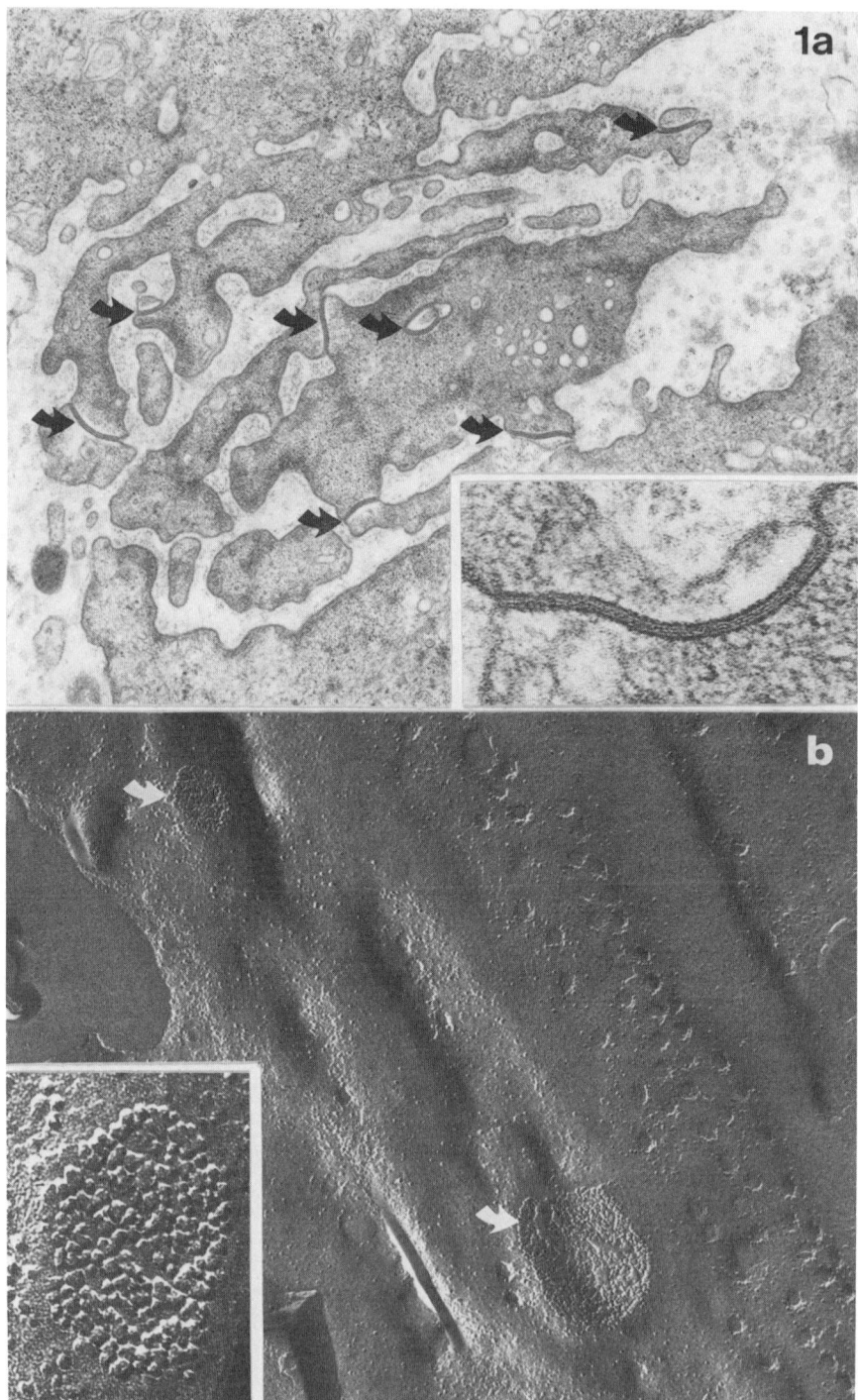

Figure 1 (See facing page for legend.)

ably operates to ensure appropriate activation and maintenance of synchronous activity in the myometrium and effective delivery of the fetus(es).

ALTERATIONS IN STRUCTURAL COUPLING

It is now established that gap junctions (Fig. 1) occupy a significant percentage of the area of the uterine smooth muscle cell plasma membrane (ca. 0.1–0.4%) only during term or preterm labor, that is, when the muscle is functionally active. Quantitative transmission electron microscopic morphometric analysis was employed to document alterations in structural coupling in the uterine muscle of a variety of mammalian species throughout pregnancy (Fig. 2). Gap junctions are consistently absent, or present in low frequency and small size in nonpregnant, as well as preterm and postpartum, animals (Garfield et al. 1977, 1978, 1979). In pregnant animals, the junctions begin to form immediately (0–24 hr) prior to the onset of labor (Fig. 2). Gap junctions are always present in large numbers ($\sim$1000 per cell) and increased size ($\sim$250 nm) during normal delivery of the fetuses, but disappear within 24 hours after parturition (Garfield et al. 1977). This pattern of altered structural coupling in the myometrium is particulary prominent in rats and rabbits (Fig. 2) (Garfield et al. 1978; Demianczuk et al. 1984). Guinea pigs and sheep (Garfield et al. 1978) appear to differ slightly in that they demonstrate higher numbers of gap junctions prior to term (Fig. 2). The gap junction profile in women during pregnancy and normal, spontaneous vaginal delivery is not known, but junctions are present in greater numbers in tissues from women undergoing caesarean section and in labor compared with those not in labor (Fig. 2) (Garfield and Hayashi 1982). It is also significant that gap junctions are invariably present in myometrial tissues from animals undergoing premature labor, either as a result of experimental manipulation or pathology (Garfield et al. 1977). Moreover, if the appearance of gap junctions is delayed, then pregnancy is prolonged. Thus, myometrial gap junctions are dynamic and ephemeral structures whose presence is inextricably associated with the conversion of the uterus into an active organ just prior to parturition. There is no known exception to this phenomenon and, for this reason, gap junctions appear to be necessary for effective labor.

REGULATION OF STRUCTURAL COUPLING

A variety of studies show that the synthesis and/or breakdown of several circulating and local hormones change immediately prior to term or preterm labor (Thorburn and Challis 1979). In particular, in species such as rats and sheep, estrogens are thought to increase, whereas progesterone decreases

Figure 1
(a) An electron micrograph of parturient rat myometrium with gap junctions (arrows). Magnification, 34,000×. (*Inset*) High magnification (75,000×) of one gap junction. (b) Freeze-fracture micrograph of parturient rat myometrium with gap junctions (arrows). Magnification, 50,000×. (*Inset*) High magnification (175,000×) of a gap junction.

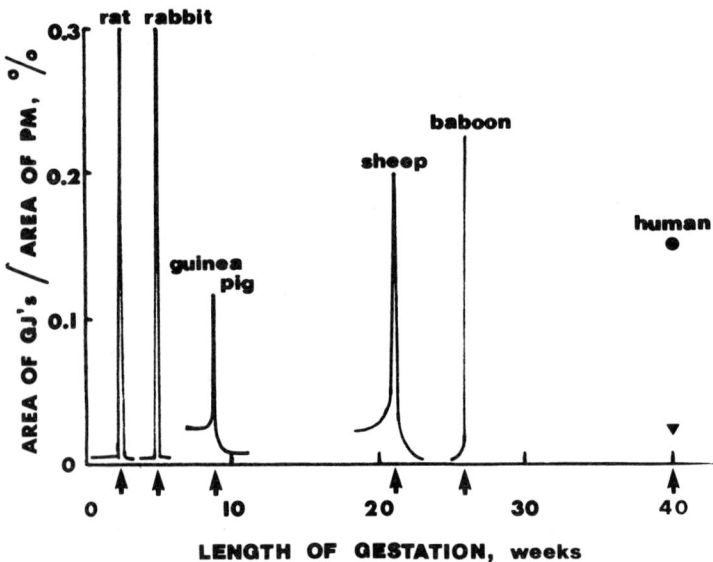

Figure 2
Area of myometrial gap junction membrane as a percentage of plasma membrane during late pregnancy and at parturition in various species. Arrows indicate date of parturition in various species. Note: Human data are for tissues from women undergoing caesarean section either prior to labor (▼) or during labor (●).

(progesterone withdrawal), at term (Fig. 3). Some of the cycloxygenase breakdown products of arachidonic acid, such as prostaglandin F_2 (Fig. 3), E_1, and E_2 are believed to increase whereas others, such as prostacyclin (PGI_2) may decrease immediately prior to parturition (Thorburn and Challis 1979). The role(s) of these hormones in regulating myometrial gap junctions has been studied both in vivo and in vitro. It would appear that estrogen stimulates, whereas progesterone inhibits, gap junction formation in the myometrium. The role of the prostaglandins remains unclear at this time in that there is evidence for both stimulatory and inhibitory prostanoids.

Steroids

Ovariectomy of pregnant rats subsequent to day 15 following conception leads to premature formation of gap junctions and labor within 24–48 hours (Garfield et al. 1977). The fact that this premature alteration in gap junctions results from an experimentally induced progesterone withdrawal is suggested by a drop in progesterone levels and the ability of progesterone therapy to prevent this drop precludes the formation of gap junctions and prolongs pregnancy for as long as treatment is continued. Administration of progesterone to intact rats over the last few days of pregnancy (i.e., day 19 onward) similarly will prevent normal progesterone withdrawal, the appearance of gap junctions, and parturition. This hormone will also inhibit gap junction development in

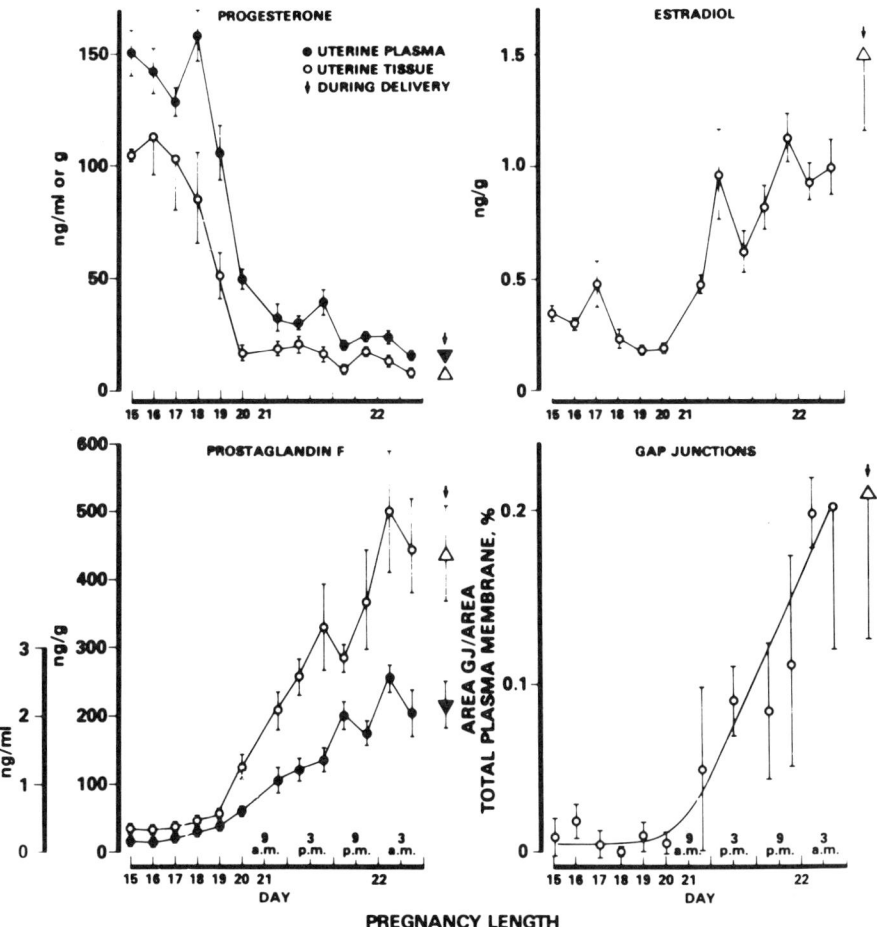

Figure 3
Changes (mean ±S.E.M.) in progesterone, estradiol, prostaglandin F in uterine vein plasma (●) and tissues (○), and gap junctions during the latter days of pregnancy and parturition (△, ▼) in the rat. The axis of the time scale is expanded at day 21 onward. (Reprinted, with permission, from Puri and Garfield 1982.)

tissues in vivo (Table 1), or incubated in vitro, and treated with estrogen (Garfield et al. 1978). These observations prompted Garfield et al. (1977, 1978) to suggest that progesterone inhibits labor and maintains pregnancy by repressing the formation of gap junctions and limiting electrical cell-cell coupling in the myometrium. Moreover, it was proposed that the development of gap junctions and their control by steroids may be the basis of the progesterone block hypothesis advanced by Csapo (1981).

It is not entirely evident, however, how progesterone manifests its influence on myometrial gap junction formation. Puri and Garfield (1982) contend that this steroid could act at one or more of the following: inhibition of protein (con-

Table 1
Gap Junctions in Myometria of Immature Rats Treated with Estradiol, Progesterone, Indomethacin, or Sodium Meclofenamate

Treatment[a]	Gap junction frequency per 1000 μM membrane	Size (nM)	Fractional area as a percent of plasma membrane
1. Immature[b]			
Control	0.103 ± 0.103	36	0.001 ± 0.001
E_2 (50 μg/day)	0.325 ± 0.325	52 ± 14	0.003 ± 0.003
E_2 (500 μg/day)	3.300 ± 1.260*	135 ± 14*	0.091 ± 0.038*
E_2 (500 μg/day) + P_4 (2.5 mg × 2/day)	0.336 ± 0.221**	69 ± 5**	0.004 ± 0.003**
E_2 (50 μg/day) + IDM (0.2 μg × 2/day)	3.170 ± 0.767	93 ± 11	0.061 ± 0.024
E_2 (500 μg/day) + IDM (0.2 μg × 2/day)	8.185 ± 3.707	114 ± 8	0.180 ± 0.080
P_4	0	0	0
IDM	0	0	0
2. Immature			
Control	0	0	0
E_2 (50 μg/day)	0.650 ± 0.450*	107 ± 38*	0.013 ± 0.007*
E_2 (500 μg/day)	0.363 ± 0.238*	132 ± 30*	0.009 ± 0.006*
E_2 (50 μg/day) + IDM	3.425 ± 0.912**	121 ± 13	0.081 ± 0.036
E_2 (50 μg/day) + P_4	0.725 ± 0.912**	107 ± 35	0.015 ± 0.012
E_2 (500 μg/day) + IDM	5.425 ± 1.220**	106 ± 53	0.121 ± 0.030**
E_2 (500 μg/day) + P_4	0.240 ± 0.147**	104 ± 15	0.005 ± 0.003**
E_2 (500 μg/day) + MFA (0.7 mg/day)	3.620 ± 0.596**	91 ± 10	0.067 ± 0.018**
IDM	0	0	0
P_4	0	0	0
MFA[c]	0	0	0

Similar data were obtained for mature, ovariectomized rats.
[a]E_2, Estradiol; P_4, progesterone; IDM, indomethacin; MFA, sodium meclofenamate.
[b]Following 3 or 4 days treatment.
[c]Data from MacKenzie et al. (1983) and MacKenzie and Garfield (1985).
*Different from control tissues ($p < 0.05$).
**Different from E_2-treated tissues ($p < 0.05$).

nexon) synthesis either directly or indirectly through inhibition of the estrogen nuclear receptor-genome interaction, a direct effect on the plasma membrane, and/or an indirect effect through manipulation of the synthesis of prostaglandins (see Fig. 6 below).

Injection of relatively high doses of estrogen (either 17β-estradiol or diethylstilbesterol, a synthetic estrogen) into immature (Merk et al. 1980; MacKenzie et al. 1983), mature ovariectomized nonpregnant (MacKenzie and Garfield 1985) (Table 1), and intact, day 18-20 pregnant (L.W. MacKenzie and R.E. Garfield, unpubl.) rats will induce the formation of myometrial gap junctions and, in the latter animals, it will provoke premature labor. Similarly, estrogen will potentiate the formation of junctions in tissues incubated in vitro (Garfield et al. 1978). This stimulatory influence of estrogen on gap junction synthesis is not unique to the myometrium and has been observed in other reproductive tissues, such as the ovary (Merk et al. 1972), where steroid receptors are present.

It is thought that the estrogens give rise to the development of gap junctions by stimulating the synthesis of connexon proteins (Garfield et al. 1978) (see Fig. 6 below). The ability of this steroid to stimulate uterine protein synthesis is well recognized and is believed to result from a direct stimulation of transcription through interaction of estrogen-nuclear receptor complexes with the genome (Gorski and Gannon 1976). The ability of an antiestrogen, tamoxifen (MacKenzie and Garfield 1985), as well as actinomyocin D and cyclohexamide (Garfield et al. 1980b), to inhibit gap junction development in response to estrogens is consistent with a nuclear control of connexon synthesis. Moreover, mRNA obtained from the myometria of estrogen-primed rats is thought to induce the formation of gap junctions when incorporated into mutant cells that normally lack the capability to form junctions (Dahl et al. 1980).

Prostaglandins

The ability of metabolites of arachidonic acid liberated from the plasma membrane of smooth muscle cells, endometrial cells, or the fetal membranes to influence the development of gap junctions in the myometrium is suggested by several lines of experimentation. However, the role and mechanism by which they operate remain to be identified.

Garfield et al. (1977) demonstrated that gap junctions were significantly less numerous in the nondistended horns of unilaterally ovariectomized, parturient rats. Similarly, gap junction frequency in the myometrium of postpartum rats following estrogen injections was the greatest, and more closely approached that observed in parturient animals, in uterine horns distended by intrauterine balloons (Wathes and Porter 1982). Distention of the myometrium is a well-documented stimulus of uterine prostaglandin synthesis (Thorburn and Challis 1979), and on the basis of these observations it seems probable that prostanoids and stretch are required to attain the extent of structural coupling observed in the distended horns of parturient animals.

Evidence for stimulation and inhibition of myometrial gap junction development by prostaglandins has been obtained through the use of two inhibitors of cycloxygenase (the initial enzyme in the prostaglandin pathway), indomethacin, and sodium meclofenamate. When myometrial tissues from midterm pregnant rats are incubated in vitro in the presence of indomethacin, the normal development of gap junctions is suppressed dramatically (Garfield et al. 1980a,b). This would suggest that a stimulatory prostaglandin is involved in the in vitro formation of gap junctions. In contrast, however, treatment of immature or ovariectomized mature rats (with nondistended uteri) with a combination of estrogen and either indomethacin or sodium meclofenamate potentiates the stimulatory effect of the steroid (Table 1) (MacKenzie and Garfield 1985). Neither cycloxygenase inhibitor exerts an effect when administered alone.

In vitro experiments show that some prostaglandins may stimulate (thromboxanes and endoperoxides), and others inhibit (PG I_2), the development of gap junctions (Garfield et al. 1980a,b). These data exemplify the complex nature of the prostaglandin influence on myometrial gap junctions. The in vivo actions of the cycloxygenase inhibitors have been interpreted to indicate that some product of the cycloxygenase pathway, perhaps prostacyclin, inhibits the stimulus provided by estrogen treatment or, alternatively, that a product of the lipoxygenase pathway of arachidonic acid metabolism may potentiate the estrogenic stimulation of gap junctions (MacKenzie et al. 1983). In the presence of the cycloxygenase inhibitors, the activity of lipoxygenase may be enhanced and the effect of estrogen potentiated. It should be noted, however, that the effects of the lipoxygenase products or its inhibitors on gap junction formation are as yet untested because of the unavailability of these compounds.

How the arachidonic acid metabolites may influence gap junction synthesis remains to be clarified. They could influence the synthesis of connexons directly, either at the level of the steroid receptors or translation (MacKenzie et al. 1983) (see Fig. 6 below). Alternatively, it was previously suggested that they may be involved in the regulation of connexon aggregation by either altering membrane fluidity or influencing protein–protein cross-linking in the plasma membrane leading to a stabilization of developing gap junctions (Garfield et al. 1980a,b) (Fig. 6). Prostaglandins can influence the interaction between steroids and their specific receptors and can inhibit the binding of these hormones by uterine tissues (Sanfilippo et al. 1983). Thus, an interaction between, and downregulation of, the estrogen receptor by some inhibitory prostanoid is a distinct possibility. Clearly, considerably more experimentation is required before we fully understand the role of the prostaglandins in the regulation of myometrial gap junction development.

It should be noted that many of our studies have implied that changes in the synthesis of gap junctions are important determinants in regulating their presence in the myometrium. However, the destruction or degradation of the gap junctions may also play a significant role in this process. Gap junction degradation previously was suggested to involve an aggregation of small gap junctions into very large contacts, prior to the formation of annular junctions and a withdrawal of these into one of the adjacent cells by an endocytotic mechanism (Garfield et al. 1980b). It remains to be shown what regulates gap junction degradation; perhaps prostaglandins or other substances are responsible for either stimulating or inhibiting their destruction (Fig. 6).

ALTERATIONS IN FUNCTIONAL COUPLING

Garfield et al. (1977) proposed that the role of the gap junctions is to couple uterine smooth muscle cells into a functional syncytium, facilitating synchronized electrical and contractile activity in the uterine wall. That the junctions should function in this capacity is consistent with their role as sites for cell–cell communication in other tissues. However, attempts to evaluate whether there is a change in electrical coupling in the myometrium at term have produced conflicting results. This is most likely due to the inherent difficulties in the study of electrical coupling between smooth muscle cells. In smooth muscles, such as the myometrium, it is very difficult to use a double-microelectrode technique to evaluate the extent of coupling directly because the cells are coupled into a complex three-dimensional syncytium; the density of injected current falls so rapidly with distance from the source electrode that it is difficult to record any voltage deflection in an adjacent electrode at more than 20–30 μm of separation (Holman and Neild 1979). Thus, regardless of the area of gap junctions present, the cells appear to be poorly coupled. This difficulty has led to the development of techniques that are thought to evaluate the extent of coupling indirectly. Four studies on coupling in the myometrium during pregnancy and parturition have used the so-called Abe-Tomita method (see Sims et al. 1982). The technique employs extracellular polarization to reduce the complex geometry of small strips of muscle to that of a simple, one-dimensional cable. The distance required for the steady-state amplitude of propagating potential changes to fall to $1/e$ of their value at the source is referred to as the length or space constant λ, and this value is taken as a measure of the degree of cell–cell coupling. Sims et al. (1982) observed a significant increase in λ in myometrial tissues from parturient (3.68 ± 1.0 mm) compared with day 17–22 pregnant, nondelivering (2.59 ± 0.84 mm) rats, suggestive of improved cell–cell flow of current during labor. However, the inability of three previous studies (Daniel and Lodge 1973; Kuriyama and Suzuki 1976; Zelcer and Daniel 1979) to identify similar changes in λ at term and the consistently large magnitude of this parameter in nondelivering tissues, which were thought to possess very few junctions, remain to be adequately explained. However, they may be the result of in vitro gap junction formation during recording, a failure to document the extent of structural coupling in the tissues employed for the functional measurements, and/or that the technique does not accurately reflect the extent of cell–cell coupling. In addition, it may be that the level of coupling prior to term as measured in vitro, overestimates that in vivo, because of the lack of possible physiological regulation of functional coupling by circulating hormones (see below).

In light of these problems, we chose to evaluate alterations in coupling by comparing cell–cell diffusion of a radiolabeled metabolite, [^{3}H]2-deoxyglucose (2DG) in small strips of longitudinal myometrium from day 17–20 pregnant (few gap junctions) and parturient (many gap junctions) rats (Cole et al. 1985). A two-compartment bathing chamber technique similar to that used by Weidmann (1966) was employed. One portion of strips of longitudinal myometrium was exposed to the tracer and its longitudinal distribution determined following 5 hours for diffusion. 2DG enters the cells by substituting for glucose on the facilitative carrier; once inside it is phosphorylated and in this form is not metabolized further nor is it able to recross the plasma membrane. Thus, the tracer remains in a diffusible pool in the cytoplasmic compartment of the muscle cells and, given appropriate cell–cell pathways, it may diffuse through the muscle strip.

The spatial distribution of 2DG was found to be considerably greater in strips from parturient compared with the nonparturient rats (Fig. 4). This implies an increased diffusivity of 2DG in the myometrium of delivering rats which is reflected, in quantitative terms, by an almost 10-fold change in the apparent diffusion coefficient (D_a) for the tracer at term (Fig. 5). Similar experiments using tritiated sucrose and mannitol indicated that diffusion through the extracellular space could not account for the distribution of 2DG in the different tissues nor its altered diffusivity in those from parturient animals. We concluded that the increased rate of tracer redistribution in the parturient tissues was the result of the larger area of gap junctions between the smooth muscle cells (Fig. 5). The results of these diffusion experiments suggest that there is a dramatic increase in the extent of functional coupling in the myometrium at term concomitant with the formation of gap junctions. Thus, the data are consistent with the hypothesis that the development of the gap junctions serves to convert the myometrium into a functional syncytium during parturition.

Regulation of Functional Coupling

Alterations in the extent of functional coupling are thought to occur in the absence of a change in structural coupling in a wide variety of cell types (e.g.,

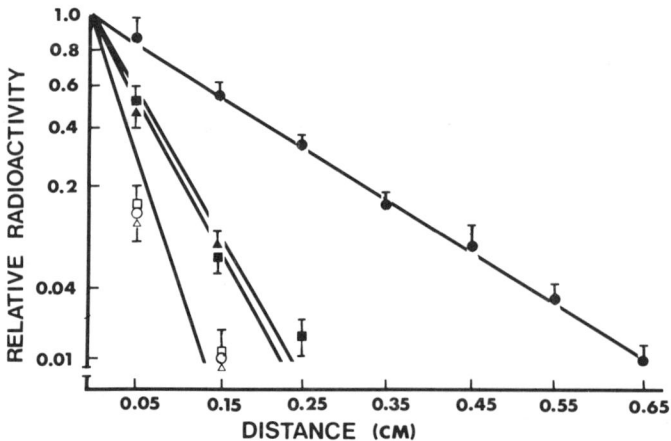

Figure 4
The longitudinal distribution of [^{3}H]2-deoxyglucose-P (▲, ●, ■) and sucrose (△, ○, □) in myometrial strips from day 17–20 pregnant (▲, △), parturient (●, ○), and day 2–3 postpartum (■, □) rats following a 5-hr diffusion experiment. Data are plotted on arithmetic probability graph paper and all values are means ± S.E.M. of at least 14 strips. (Data are from Cole et al. 1985.)

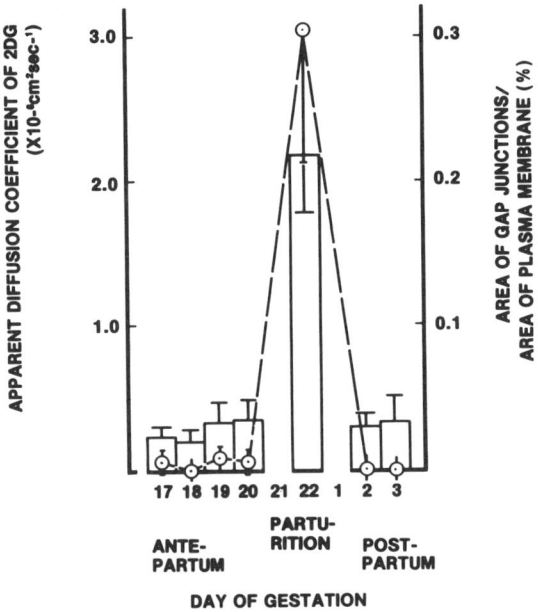

Figure 5
The apparent diffusion coefficient (C_a) of [^{3}H]2-deoxyglucose-P (histogram) and the area of gap junctions as a percent of the plasma membrane (○) versus day of gestation, parturition, and postpartum. Values for D_a and gap junction area are means ± S.E.M. (Reprinted, with permission, from Cole et al. 1985.)

Peracchia 1980; Spray et al. 1984). These observations suggest that a change in junctional permeability may be obtained through alterations in the gap junctional connexons, such as an all-or-none closure or dilation of the cell–cell channel, leading to a state of either decreased or enhanced coupling, respectively. Instances of a modulation of junctional permeability by hormones (Hermsmeyer 1982) or neurotransmitters (Laufer and Salas 1981; Piccolino et al. 1982) have been described in the literature. That the permeability of the junctions and, therefore, the extent of functional coupling in the myometrium, may be regulated by endogenous mechanisms is an intriguing possibility. Improved functional coupling would be expected to promote greater electrical and contractile synchrony in the uterine wall and lead to an enhanced rate of intrauterine pressure development and more effective labor. Alternatively, gap junction closure would reduce synchrony and lead to ineffective labor and a prolongation of pregnancy.

We have used the 2DG diffusion technique described above to study the influence of several agents on coupling in the myometrium of parturient rats. Preliminary evidence suggests that the permeability of the junctions in the parturient myometrium is influenced by intracellular Ca^{++} (Cole and Garfield 1984), pH, and cAMP (Cole and Garfield 1985) (Table 2; see also Fig. 6). Elevated intracellular Ca^{++} and lowered pH produced by the calcium ionophore

Table 2
Apparent Diffusion Coefficient (D_a) for [^{3}H]2-deoxyglucose-P in Parturient Rat Myometrium Following Various Treatments

	Drug	Concentration	n	Apparent diffusion coefficient (D_a in cm^2 sec^{-1} + S.E.M.)
1.	Control	—	15	1.91 ± 0.10*
	A23187	1 µM	8	0.16 ± 0.03
	A23187 and CAM-ANT[a]	1 µM, 0.1 mM	15	1.39 ± 0.17*
	CAM-ANT			
	chlorpromazine	1 µM	7	0.84 ± 0.14
	calmidazolium	1 µM	7	0.44 ± 0.08
	O-nitrobenzyl acetate	1 mM	7	1.14 ± 0.15
2.	Control	—	17	1.78 ± 0.11*
	Dibutyryl cAMP	1 mM	8	0.33 ± 0.09
	8-Bromo cAMP	0.1 mM	6	0.66 ± 0.03
	Theophylline	1 mM	4	0.89 ± 0.09
	Forskolin	1 µM	6	0.55 ± 0.07
	Theophylline and forskolin	1 mM, 1 µM	6	0.3 ± 0.08
	Relaxin	0.1 µg/ml	5	0.46 ± 0.09
	Carbacylin	1.0 µM	4	0.54 ± 0.08
	Isoproterenol	1.0 µM	4	0.62 ± 0.12

[a]CAM-ANT, Calmodulin antagonist; data are for chlorpromazine.
*Not significantly different ($p < 0.05$).

A23187 (1 µM) and the substituted benzyl ester (see Spray et al. 1984) o-nitrobenzyl acetate, respectively, reduced the longitudinal distribution and apparent diffusion coefficient of 2DG in the parturient myometrium; however the latter drug is rather ineffective compared with A23187 (Table 2). Smooth muscle cells apparently possess a very effective mechanism for regulating intracellular pH (Aiken 1984) and this probably explains the inability of the substituted ester to reduce coupling in the myometrium as markedly as that seen in other cell types (Spray et al. 1984). It appears that calmodulin may be required to confer calcium sensitivity to the junctions because a normal distribution of 2DG is observed in tissues treated with A23187 in the presence of calmodulin antagonists, such as chloropromazine and calmidazolium (Table 2). However, the role played by calmodulin in the regulation of junctional permeability is complex in that the calmodulin antagonists are able to produce a dose-dependent uncoupling when administered alone (Table 2). This latter result would seem to suggest that calmodulin may be involved indirectly in the regulation of coupling as well, perhaps by participating in the regulation of intracellular cAMP. Morphometric analysis of the tissues used in these uncoupling experiments failed to show any differences in the extent of structural coupling in the treatment groups, consistent with uncoupling by a modulation of channel permeability.

Elevated cAMP produced by treatment with dibutyryl- or 8-bromo-cAMP also reduces cell–cell diffusion of 2DG in the myometrium, and this can be mimicked by inhibiting phosphodiesterase activity with theophylline or by stimulating adenylate cyclase with forskolin (Table 2). That cAMP may play a role in regulating functional coupling in the myometrium is significant in that relaxin, prostacyclin (PG I_2), and β_2-adrenoceptor agonists appear to influence labor and parturition (Thorburn and Challis 1979) and exert inhibitory effects on the myometrium by elevating intracellular cAMP (Vesin et al. 1979; Sanborn et al. 1980). Moreover, in preliminary experiments with porcine relaxin, carbacyclin (a stable PG I_2 analog), and isoproterenol (a nonspecific β-adrenoceptor agonist) the diffusivity of 2DG was found to be significantly lower in treated compared with control tissues (Table 2). Evidently, there may be specific, receptor and secondary messenger-mediated physiological mechanisms for controlling cell–cell communication in the myometrium independent of the systems controlling structural coupling.

It is tempting to speculate that a cAMP-mediated uncoupling mechanism may be involved in maintaining pregnancy in instances of premature junction formation and/or in species such as the guinea pig, sheep, and possibly human, in which low but significant numbers of gap junctions are present throughout pregnancy. Perhaps the high levels of relaxin and prostacyclin observed in preterm pregnant animals act to elevate intracellular cAMP and prevent synchronous activity in the myometrium. A decline in these hormones or their receptors, or an antagonism of their action by oxytocin or stimulatory prostaglandins (e.g., PGF_2 or PGE_2), may facilitate a shift to patent gap junction channels and the development of syncytial behavior.

Whether junctional permeability in the myometrium can be enhanced by circulating hormones is unresolved at this time. It may be that oxytocin or a stimulatory prostaglandin, such as PGF_2, E_1, or E_2, may increase cell–cell communication between uterine smooth muscle cells either through a direct interaction with the gap junctions or, as noted above, by reversing the inhibition of coupling produced by relaxin and/or prostacyclin. These agonists all promote labor and there is evidence that their action is in part to increase the rate of intrauterine pressure development (Csapo 1981). As noted previously, the rate of pressure development is dependent on the rate of activation of individual cells in the myometrium and, hence, influenced by the extent of functional coupling.

SUMMARY

This paper describes alterations in, and the possible endogenous regulatory mechanisms of, structural and functional coupling between uterine smooth muscle cells. At the end of pregnancy gap junctions develop in large numbers between uterine smooth muscle cells to facilitate increased intercellular exchange of small ions and metabolites and permit syncytial electrical and contractile behavior in the myometrium. Since the development of normal, effective labor is dependent on the acquisition of this improved cell–cell communication, the physiological mechanisms regulating the appearance of gap junctions in the myometrium and their permeability must be identified to understand the regulation of parturition in animals and humans.

A schematic representation of the possible physiological mechanisms controlling the extent of structural and functional coupling in the myometrium is shown in Figure 6. We believe the appearance of the junctions to be controlled

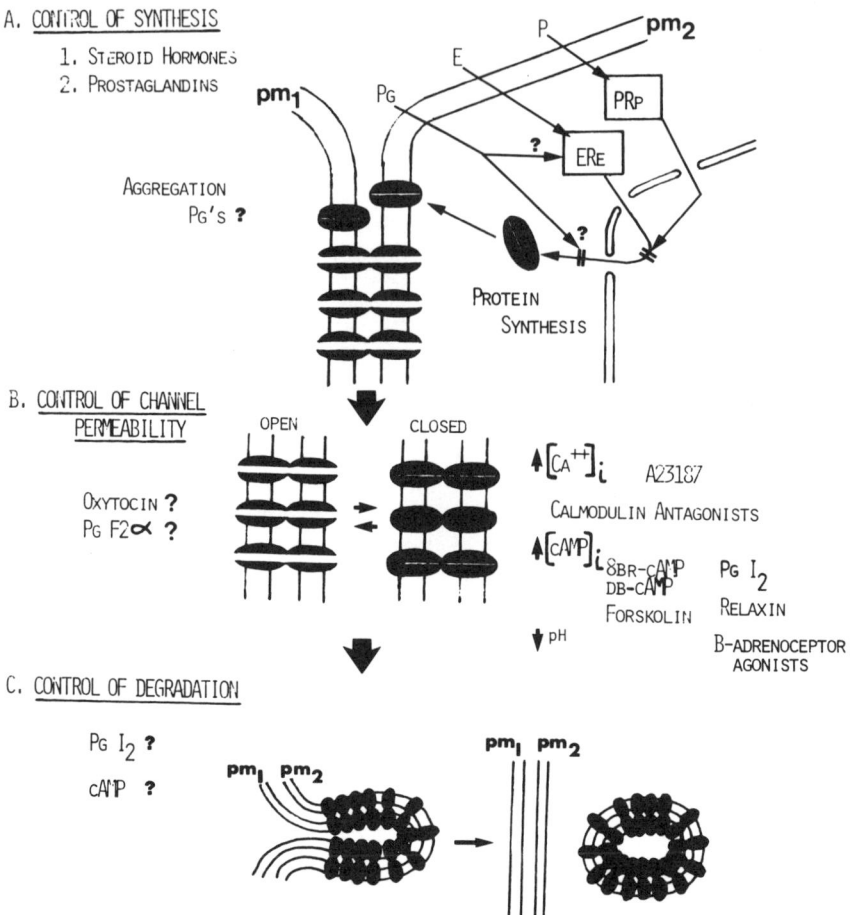

Figure 6
A schematic representation of the possible mechanisms regulating alterations in structural and functional coupling in the myometrium. Possible roles for: (1) estrogen (E) and progesterone (P) interacting with their receptors (R_E and R_p) to regulate connexon protein synthesis; (2) prostaglandins in controlling either the interaction with the steroids or aggregation of connexons; (3) Ca^{++}, cAMP (8Br-cAMP, DB-cAMP), and pH in regulating functional coupling; (4) the possible roles for endogenous hormones (prostacyclin, relaxin) or neurotransmitters (β_2-adrenoceptor agonists) in regulating functional coupling; (5) an irreversible endocytotic pathway for gap junction degradation.

by the changes in the steroid hormones, estrogen and progesterone, and prostaglandins, which may serve to regulate the synthesis, insertion, and degradation of the junctions. Preliminary results identify several intracellular factors, such as Ca^{++}, pH, and cAMP, which appear to influence functional coupling in the absence of an alteration in structural coupling. Moreover, we have observed an inhibition of functional coupling produced by physiologically relevant hormones, such as relaxin and prostacyclin. We feel that these data are evidence for the presence of specific receptor- and second messenger-mediated endogenous mechanisms for regulating direct intercellular communication between uterine smooth muscle cells. The integrated function of these regulatory mechanisms represents an important control of the onset and maintenance of parturition in animals and humans.

ACKNOWLEDGMENTS

This work was supported by grants from the Medical Research Council of Canada to R.E. Garfield. W.C. Cole and R.E. Garfield are grateful to the Canadian and Ontario Heart Foundation for fellowships. The authors also wish to thank Dr. O.D. Sherwood for his generous gift of porcine relaxin.

REFERENCES

Aiken, C. 1984. Direct measurement of intracellular pH and buffering power in smooth muscle cells of guinea pig vas deferens. *J. Physiol.* **349:** 571–585.

Cole, W.C. and R.E. Garfield. 1984. A23187 and calmodulin antagonists inhibit metabolic coupling between parturient rat myometrial smooth muscle cells. *Biophys. J.* **45:** 23a.

―――. 1985. Elevated intracellular cAMP reduces intercellular communication between parturient rat uterine smooth muscle cells. *Biophys. J.* **47:** 448a.

Cole, W.C., R.E. Garfield, and J.S. Kirkaldy. 1985. Gap junctions and direct intercellular communication between rat uterine smooth muscle cells. *Am. J. Physiol.* (in press).

Csapo, A.I. 1981. Force of labor. In *Principles and practice of obstetrics and perinatology* (ed. L. Iffy and H.A. Kaminetzky), p. 761–802. Wiley, New York.

Dahl, G., R. Azarnia, and R. Werner. 1980. De novo construction of cell-to-cell channels. *In vitro* **16:** 1068–1075.

Daniel, E.E. and S. Lodge. 1973. Electrophysiology of myometrium. In *Uterine contraction* (ed. J.B. Josimovich), p. 19. Wiley, New York.

Demianczuk, N., M.E. Towell, and R.E. Garfield. 1984. Myometrial electrophysiologic activity and gap junctions in the pregnant rabbit. *Am. J. Obstet. Gynecol.* **149:** 485–491.

Garfield, R.E. and R.H. Hayashi. 1982. Appearance of gap junctions in the myometrium in women in labor. *Am. J. Obstet. Gynecol.* **140:** 254–260.

Garfield, R.E., M.S. Kannan, and E.E. Daniel. 1980a. Gap junction formation in myometrium: Control of estrogens, progesterone and prostaglandins. *Am. J. Physiol.* **238:** C81–C89.

Garfield, R.E., D. Merrett, and A.K. Grover. 1980b. Gap junction formation and regulation in myometrium. *Am. J. Physiol.* **239:** C217–C228.

Garfield, R.E., S. Sims, and E.E. Daniel. 1977. Gap junctions: Their presence and necessity in myometrium during parturition. *Science* **198:** 958–960.

Garfield, R.E., S. Sims, M.S. Kannan, and E.E. Daniel. 1978. Possible role of gap junctions in activation of myometrium during parturition. *Am. J. Physiol.* **235:** C168–C179.

Gorski, J. and F. Gannon. 1976. Current models of steroid hormone action: A critique. *Annu. Rev. Physiol.* **38:** 425–450.

Hermsmeyer, K. 1982. Angiotensin II increases electrical coupling mammalian ventricular myocardium. *Circ. Res.* **47:** 524–529.

Holman, M.E. and T.O. Neild. 1979. Membrane properties. *Br. Med. Bull.* **35:** 235–241.

Kuriyama, H. and H. Suzuki. 1976. Changes in electrical properties of rat myometrium during gestation and following hormonal treatments. *J. Physiol.* **261**: 315–333.

Laufer, M. and R. Salas. 1981. Intercellular coupling and retinal horizontal cell receptive field. *Neurosci. Lett.* **7**: 5339a.

MacKenzie, L.W. and R.E. Garfield. 1985. Hormonal control of gap junctions in the myometrium. *Am. J. Physiol.* **248**: C296–C308.

MacKenzie, L.W., C.P. Puri, and R.E. Garfield. 1983. Effects of estradiol-17β and prostaglandins on rat myometrial gap junctions. *Prostaglandins* **26**: 925–944.

Merk, F.B., C.R. Botticelli, and J.T. Albright. 1972. An intercellular response to estrogen by granulosa cells in the rat ovary: An electron microscopic study. *Endocrinology* **90**: 992–1007.

Merk, F.B., P.W.L. Kwan, and I. Leav. 1980. Gap junctions in the myometrium of hypophysectomized estrogen-treated rats. *Cell Biol. Int. Rep.* **4**: 287–294.

Peracchia, C. 1980. Structural correlates of gap junction permeation. *Int. Rev. Cytol.* **66**: 81–146.

Piccolino, M., M.J. Neyton, P. Witkovsky, and H.M. Gerschenfeld. 1982. γ-aminobutyric acid antagonists decreases junctional communication between L-horizontal cells of the retina. *Proc. Natl. Acad. Sci.* **79**: 3671–3675.

Puri, C.P. and R.E. Garfield. 1982. Changes in hormone levels and gap junctions in the rat uterus during pregnancy and parturition. *Biol. Reprod.* **27**: 967–975.

Sandborn, B.M., H.S. Kuo, N.W. Weisbrodt, and O.D. Sherwood. 1980. The interaction of relaxin with the rat uterus. I. Effects on cyclic nucleotide levels and spontaneous contractile activity. *Endocrinology* **106**: 1210–1215.

Sanfilippo, J.S., J. Teichman, T.R. Melvin, C.O. Osyamkpe, and J.L. Wittliff. 1983. Influence of certain prostaglandin synthetase inhibitors on cytoplasmic estrogen receptors in the uterus. *Am. J. Obstet. Gynecol.* **145**: 100–104.

Sims, S.M., E.E. Daniel, and E.E. Garfield. 1982. Improved electrical coupling is associated with increased numbers of gap junctions in uterine smooth muscle at parturition. *J. Gen. Physiol.* **80**: 353–375.

Spray, D.C., R.L. White, A. Campos de Carvalho, A.L. Harris, and M.V.L. Bennett. 1984. Gating of gap junction channels. *Biophys. J.* **45**: 219–230.

Thorburn, G.D. and J.R.G. Challis. 1979. Endocrine control of parturition. *Physiol. Rev.* **59**: 863–918.

Vesin, M.F., L. Dokhac, and S. Harbon. 1979. Prostacyclin as an endogenous modulator of adenosine cyclic 3′-5′ monophosphate in the myometrium and endometrium. *Mol. Pharmacol.* **16**: 823–840.

Wathes, D.C. and D.G. Porter. 1982. Effect of uterine distension and estrogen treatment on gap junction formation in the myometrium of the rat. *J. Reprod. Fertil.* **65**: 497–505.

Weidmann, S. 1966. Diffusion of radiopotassium across intercalated discs of mammalian cardiac muscle. *J. Physiol.* **187**: 323–342.

Zelcer, E. and E.E. Daniel. 1979. Electrical coupling in rat myometrium during pregnancy. *Can. J. Physiol. Pharmacol.* **57**: 590–495.

Development and Regulation of Electrotonic Coupling between Cultured Sympathetic Neurons

John A. Kessler,*† David C. Spray,†
Juan C. Saez,† and Michael V.L. Bennett†
Departments of *Neurology and †Neuroscience
Rose F. Kennedy Center for Mental Retardation
Albert Einstein College of Medicine
Bronx, New York 10461

Plasticity of neuronal phenotype and formation and remodeling of chemical synapses are well documented. For example, neurons may alter the transmitters they synthesize and secrete, and thus their phenotypes, under appropriate circumstances (Patterson and Chun 1977; LeDouarin 1980; Kessler 1984). Indeed, emerging evidence suggests that continuing neuronal change is the rule, not the exception, and that transmitter and synaptic modifiability may be a common crucial feature of neurons and neuronal circuits. Is the formation of electrotonic synapses similarly influenced by the environment? Sympathetic neurons cultured in the presence of serum do not exhibit electrotonic coupling, and the structural correlate of this mode of synaptic transmission (the gap junction) has not been observed in sympathetic ganglia in situ (cf. Kondo et al. 1980). However, coupling is common when sympathetic neurons are cultured in a defined serum-free medium (Higgins and Burton 1982; Kessler et al. 1984). Moreover, these neurons continue to express noradrenergic characteristics and remain excitable and responsive to exogenous neurotransmitters (Higgins and Burton 1982; Iacovitti et al. 1982), suggesting that the emergence of electrical coupling represents a highly specific change in neuronal differentiation. Consequently, rat sympathetic neurons in culture provide a unique model system for studying mechanisms governing formation of electrotonic synapses by mammalian neurons.

ELECTROTONIC COUPLING IN DEFINED MEDIUM

Why do sympathetic neurons form electrotonic synapses when they are cultured in the defined serum-free medium (DM) but not when they are in serum-containing medium (SM)? The emergence of coupling could result either from the addition of one or more of the ingredients of DM or from the deletion of serum. In fact, addition of the five primary additives in DM (insulin, progesterone, selenium, transferrin, putrescine) to SM resulted in coupling between neurons obtained from the rat sympathetic superior cervical ganglion (SCG), suggesting that one or more of these chemicals promoted coupling. However, only 20% of neuron pairs were coupled under these conditions, while 40% were coupled in DM. This result suggests that serum inhibits electrotonic synapse formation, and that the development of coupling in DM resulted both from the addition of ingredients that promote coupling, and from deletion of serum that inhibits electrotonic synapse formation.

Which of the five chemical additives in DM promoted coupling? Addition of insulin alone to cultures in SM resulted in coupling, as did addition of selenium, whereas progesterone, transferrin, or putrescine alone had no effect (Fig. 1). Consequently, insulin and selenium each apparently cause sympathetic neurons to couple electrically. The effects of insulin were first detectable within 24 hours after treatment (8% of pairs coupled), but maximal incidence of cou-

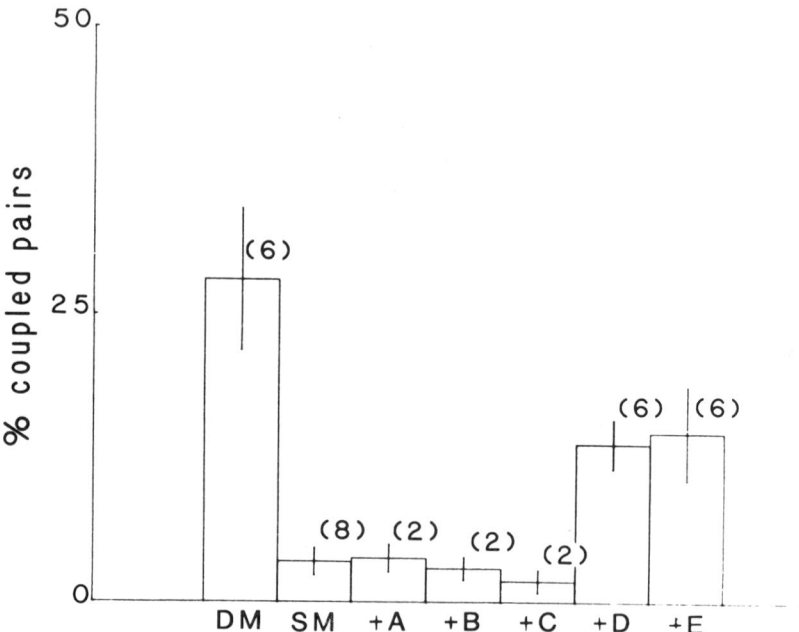

Figure 1
Effects of constituents of defined medium on incidence of electrical coupling. Cells were cultured in defined medium (DM) or serum-containing medium (SM) for 2 weeks. SM with individual constituents of DNA added was applied to cells after they had been cultured in SM for 2 weeks, and coupling was assayed 2 days later. Numbers in parentheses represent the number of dishes examined; at least 40 pairs of neurons were assayed in each dish. Results represent mean ± S.E.M. A, Progesterone; B, putrescine; C, transferrin; D, selenium; E, insulin.

pling (22%) occurred after 3–4 days. Insulin promoted coupling even at concentrations as low as 0.01 mg/ml. Maximal incidence of coupling occurred at 1–10 mg/ml, and the incidence of coupling decreased at higher doses.

cAMP AND COUPLING

These observations indicate that specific chemical signals may govern the extent to which neurons are electrically coupled. What intracellular mechanisms mediate formation of electrotonic synapses? Since cAMP promotes coupling between certain types of cultured nonneuronal cells (Hax et al. 1974; Flagg-Newton et al. 1981), we examined its effects on sympathetic neurons. Treatment with any one of several membrane-permeable cAMP derivatives rapidly resulted in coupling between neurons (Fig. 2). Coupling appeared within 4 hours of treatment, and maximal incidence (34% of neuron pairs) occurred at 12 hours. Addition of a phosphodiesterase inhibitor (caffeine or isobutyl methylxanthine) prolonged the period of coupling after treatment with

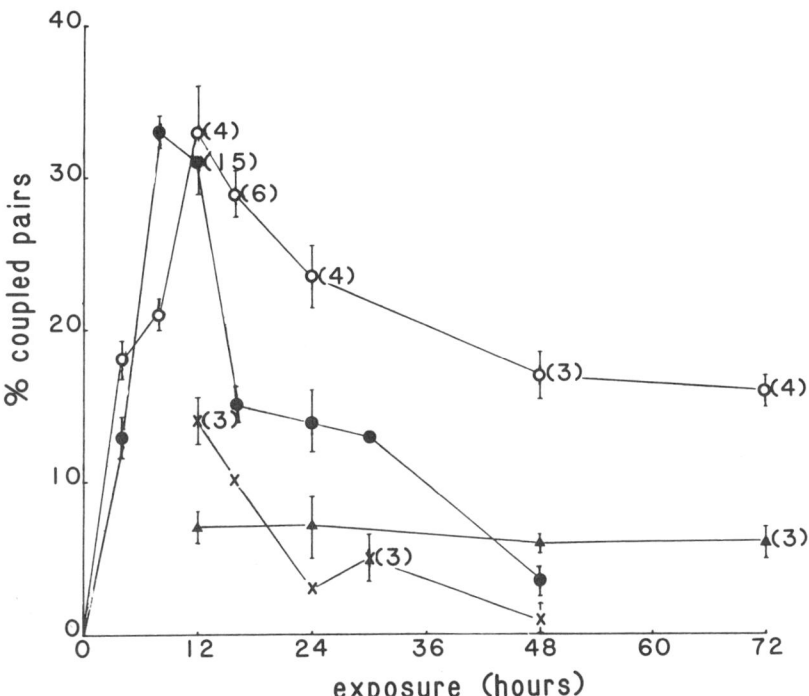

Figure 2
Effects of cAMP and phosphodiesterase inhibitors on incidence of electrical coupling. Incidence of coupling increased in SM containing dibutyryl-cAMP (db-cAMP), and addition of caffeine prolonged the effects of db-cAMP. All compounds were added at a final concentration of 1.0 mM. (○) db-cAMP; (●) db-cAMP + caffeine; (△) caffeine; (X) butyrate.

a cAMP derivative, presumably by inhibiting metabolism of the cyclic nucleotide. Dose-response relationships were examined for the dibutyryl derivative of cAMP (db-cAMP); the incidence of coupling was increased with doses as low as 0.01 mM, and coupling incidence increased with increasing dose of the cyclic nucleotide. Finally, treatment with forskolin, a stimulator of adenylate cyclase, also promoted coupling between sympathetic neurons. These observations suggest that intracellular cAMP stimulates electrotonic synapse formation by sympathetic neurons.

Endogenous levels of cAMP in sympathetic neurons were examined after treatment with insulin and selenium to determine whether the development of coupling after treatment might be mediated through an increase in cytoplasmic cyclic nucleotide (Fig. 3). cAMP levels were also examined in control cultures grown in SM and DM. Treatment with insulin or selenium in SM did not increase neuronal levels of cAMP above those in SM alone, suggesting that cAMP does not mediate the effects of insulin or selenium. However, neurons cultured in DM contained significantly more cAMP than those in SM, suggesting that serum supresses cAMP levels. Consequently, coupling between neurons cultured in DM apparently results both from an increase in cAMP (due to deletion of serum) and from the addition of insulin and selenium.

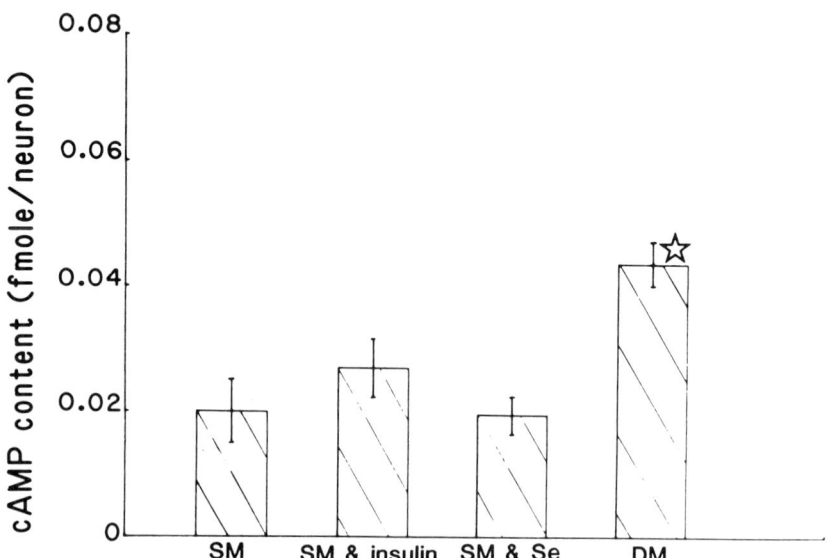

Figure 3
Endogenous levels of cAMP are similar in neurons cultured in SM with or without insulin or selenium (Se) but are increased in neurons cultured in DM. Cultures were grown for 7 days in either DM or SM. At that time, insulin (5 μg/ml) or selenium (30 nM) was added to some of the cultures in SM, and 7 days later all cultures were harvested for cAMP assay. cAMP levels are expressed as mean fmoles per neuron ± S.E.M. (n = 8). Star indicates that group differs from SM control at $p < 0.01$ (analysis of variance). (Reprinted, with permission, from Kessler et al. 1984.)

INCREASED INCIDENCE OF COUPLING REPRESENTS FORMATION OF ELECTROTONIC SYNAPSES

Although morphological data are lacking, other data suggest that electrotonic coupling in these cultures is mediated by gap junctions that function as electrotonic synapses. First, transfer of dyes to which nonjunctional membranes are impermeable was observed in every pair of electrotonically coupled cells tested, including cells from DM, insulin, and cAMP cultures (Fig. 4). Moreover, in no case tested was dye coupling seen where electrotonic coupling was absent. Finally, treatment with insulin or db-cAMP did not appreciably alter nonjunctional conductance compared with the value for noncoupled cells, indicating that coupling was due to increased junctional conductance between treated neurons.

ROLE OF RNA AND PROTEIN SYNTHESIS IN ELECTROTONIC SYNAPSE FORMATION

Does the formation of electrotonic synapses in response to insulin or cAMP represent utilization of preexisting proteins or are ongoing protein and/or RNA synthesis required? To approach this question, neurons exposed to insulin or db-cAMP were treated with cycloheximide (an inhibitor of protein synthesis), camptothecin or actinomycin D (RNA synthesis inhibitors), or cytosine arabinoside (a DNA synthesis blocker) (Fig. 5). Camptothecin, actinomycin D, and cycloheximide each markedly reduced formation of electrotonic synapses; cytosine arabinoside, by contrast, had no effect on coupling (not illustrated). Neuronal resting membrane potentials and action potentials remained normal during the relatively short exposure (12–24 hr) to the metabolic inhibitors. These observations suggest that RNA and protein synthesis but not DNA synthesis are required for the formation of electrotonic synapses.

Effects of Membrane Depolarization

Depolarization of sympathetic neurons increases endogenous levels of cAMP. Since cAMP stimulated electrical coupling, we asked whether membrane depolarization similarly promoted electrotonic synapse formation. Treatment with 35 mM K^+ or with veratridine increased electrotonic coupling within 24 hours. Tetrodotoxin (TTX), which blocks the Na^+ channel effects of veratridine, prevented the effects of veratridine on coupling, but TTX alone had no effect. These observations suggest that membrane depolarization and transmembrane Na^+ flux stimulate formation of electrotonic synapses, possibly by increasing cAMP levels.

Effects of Neurotransmitters

Since membrane depolarization promoted coupling, the effects of neurotransmitters that normally influence SCG neurons were examined. The innervation of the SCG is primarily cholinergic; consequently the effects of carbachol, an

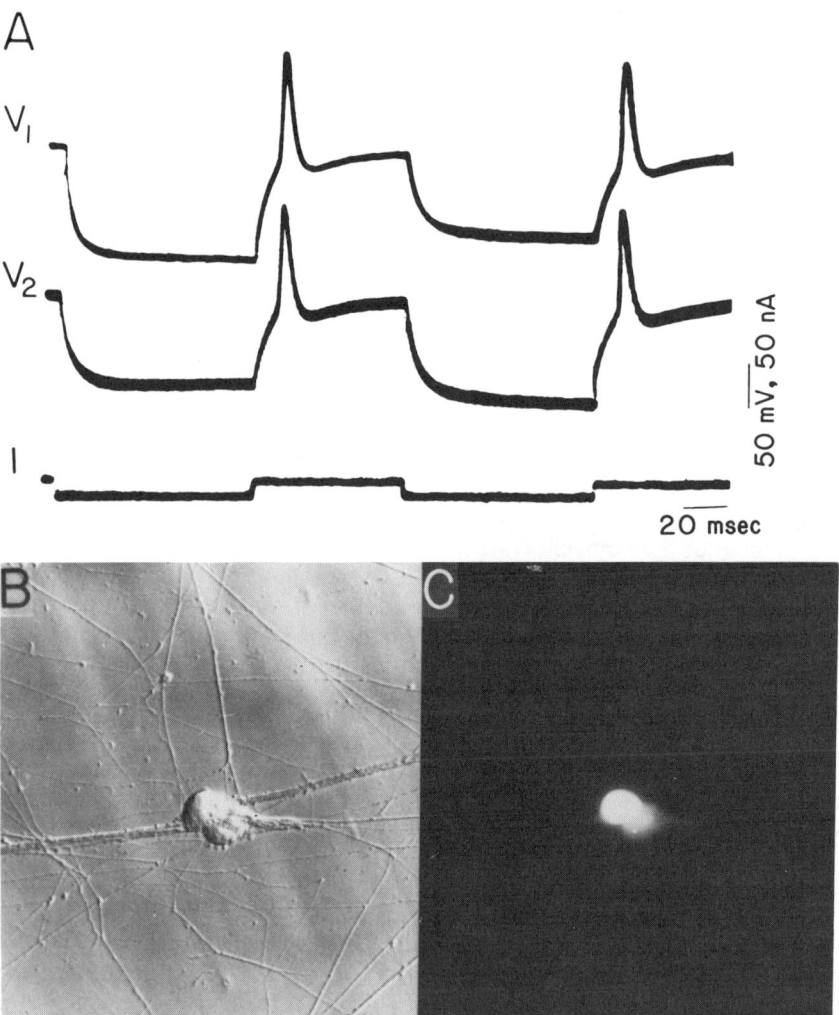

Figure 4
Exposure of cell pairs to SM with insulin (5 µg/ml) stimulates formation of electrotonic and dye coupling between SCG neurons. (A) Currents (I, bottom trace) injected into cell 1 (first pulse) or cell 2 (second pulse) produced almost equal voltages (V_1 and V_2) in the two cells, which were excitable as evidenced by the generation of impulses at the end of the hyperpolarizations. Application of the π-T transform to these data gave conductances of the nonjunctional membrane of 79 nS and 78 nS and junctional conductance of 330 nS (B and C). In another well-coupled cell pair, Lucifer Yellow injected into one cell spread to the other cell within a few minutes, as recorded in the fluorescence micrograph in C. (Reprinted, with permission, from Kessler et al. 1984.)

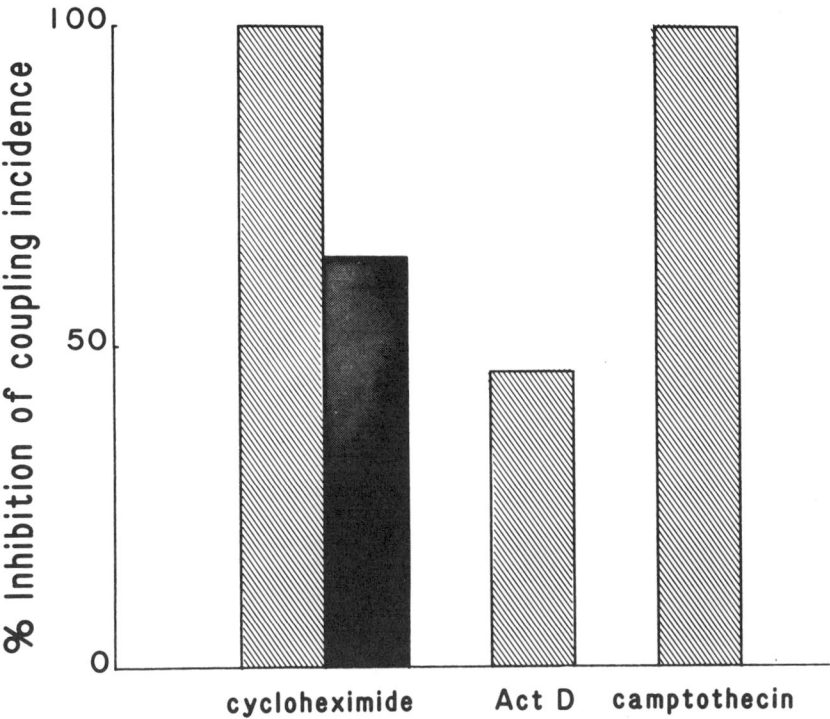

Figure 5
Effects of metabolic inhibitors on incidence of coupling. Coupling between cell pairs was assayed 12 hr after the addition of cycloheximide (2 µg/ml), actinomycin D (1 µg/ml), or camptothecin (2 µg/ml) and either db-cAMP (light bars) or insulin (dark bar).

acetylcholine agonist, were examined in culture. Exposure to carbachol (0.1 µM to 0.1 mM) resulted in a dose-dependent increase in the number of coupled neurons (Fig. 6). The percentages of neurons coupled after carbachol treatment and after db-cAMP treatment were similar. Addition of both carbachol and db-cAMP to the medium did not result in a further increase in the percentage of neurons coupled, but did significantly increase the strength of the coupling (i.e., increased junctional conductance) compared with treatment with either drug alone (Fig. 6). These observations suggest that carbachol and cAMP promote coupling in the same subpopulation of sympathetic neurons.

SCG sympathetic neurons also receive dopaminergic innervation from ganglion interneurons (SIF cells). This input is believed to be inhibitory, in contrast to the stimulatory cholinergic innervation. Consequently, the effects of dopamine on electrotonic coupling were examined in culture (Fig. 6). Pargyline (a monoamine oxidase inhibitor) was added to the medium to inhibit degradation of dopamine. Pargyline alone had no effect on electrical coupling. However, addition of dopamine rapidly promoted formation of electrotonic coupling. The

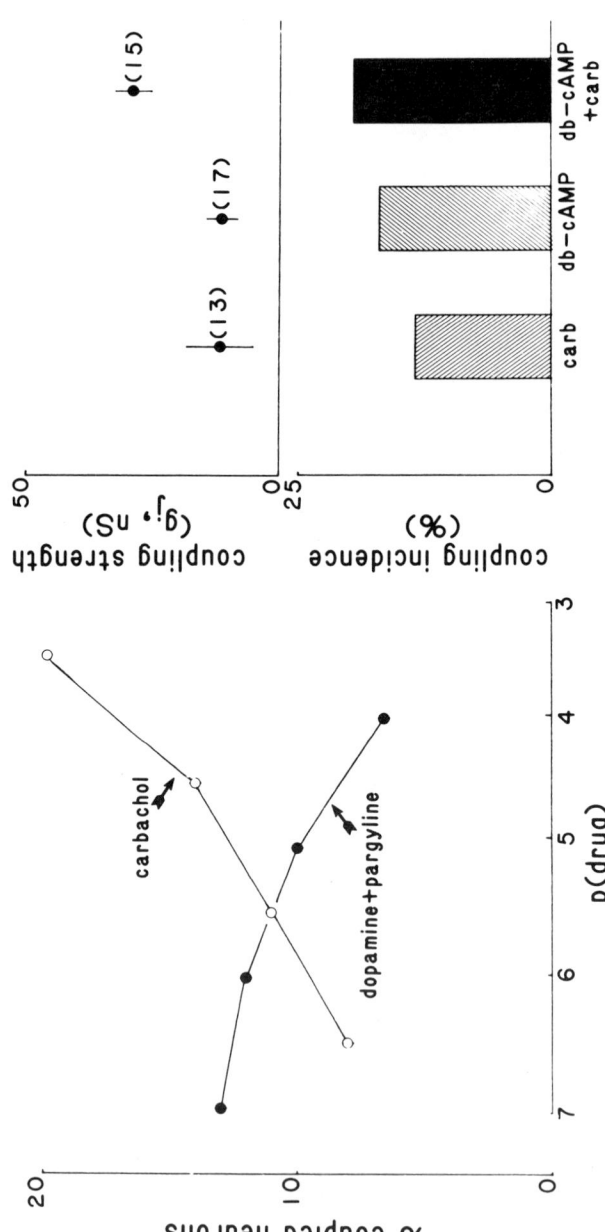

Figure 6
Effects of neurotransmitters on electrical coupling. Coupling between cell pairs was assayed 12 hr after the addition of carbachol or dopamine and pargyline (10^{-5} M). db-cAMP was used at 5 mM. Carbachol was used a 10^{-5} M in the additivity experiments (*right side*) with db-cAMP.

maximal increase in the percentage of coupled cells occurred at 0.1 µM. Diminishing the concentration below 0.1 µM reduced the stimulatory action on coupling. Similarly, increasing concentration above 0.1 µM also reduced the effect on coupling with loss of action at 0.1 mM. Submaximal responses at concentrations above 0.1 µM may reflect a variety of effects, including receptor desensitization or downregulation, as described for other actions of dopamine.

Effects of Carbon Dioxide

Previous studies demonstrated that lowering cytoplasmic pH greatly reduces conductance of gap junctions in many tissues (cf. Spray et al. 1984). To begin to determine whether pH might exert a long-term effect on coupling between sympathetic neurons, the effects of increasing CO_2 concentration in the incubator were examined. Since the culture system uses an HCO_3 buffering system, increased CO_2 leads to a substantial decrease in pH in the medium (and presumably intracellularly). Transient exposure of sympathetic neurons to elevated CO_2 concentrations (8–12% instead of the normal 5%) rapidly promoted formation of electrical coupling. Only brief exposure (2 hr) to elevated CO_2 was required for the development (24 hr later) of electrotonic synapses. Increasing the CO_2 concentration from 8% to 12% only marginally increased the percentage of coupled neurons. It has not yet been established whether these effects are due to changes in pH; measurement of intracellular pH and correlation of pH with coupling should help resolve this question. However, the observation that butyrate (which promotes intracellular acidification) also promotes electrical coupling between sympathetic neurons supports a role for pH, which would be slower and opposite in direction to the fast block of junctional conductance produced in most tissues by 100% CO_2.

These observations indicate that the formation of electrotonic synapses depends upon the environment of the neuron, and that specific chemical signals may govern the extent to which neurons are electrically coupled. What relevance do these studies have to neuronal function in vivo? Electrotonic synapses have not been demonstrated in sympathetic ganglia in vivo (cf. Kondo et al. 1980), and it is possible that the formation of coupling in vitro represents altered differentiation related to tissue culture conditions. However, even in culture sympathetic neurons did not form electrotonic synapses except in response to insulin, selenium, or cAMP; consequently, failure to find such coupling in normal intact ganglia is not surprising. Alteration of the neuronal milieu in vivo, however, might lead to the formation of electrotonic synapses and such a phenomenon could play a role in disease processes. Consequently, it will be important to determine whether stimuli that promote electrotonic coupling in vitro also occur and are effective in the intact organism.

ACKNOWLEDGMENTS

We are grateful to Kathryn Sweeney and Joe Zavilowitz for superb technical assistance. This work was supported by National Institutes of Health Grants

NS 16524, NS 07512, NS 14830, NS 20013, and NS 20778 and was aided by grants from the American Heart Association and the Dysautonomia Foundation. D.C.S. is the recipient of a McKnight Development Award. J.A.K. is the recipient of the George Cotzias Award of the American Parkinson Disease Association.

REFERENCES

Flagg-Newton, J.L., G. Dahl, and W.R. Loewenstein. 1981. Cell junction and cyclic AMP. I. Upregulation of junctional membrane permeability and junctional membrane particles by cyclic nucleotide treatments. *J. Membr. Biol.* **63:** 105-121.

Hax, W.M.A., G.E. VanVenrooij, and J.B. Vossenberg. 1974. Cell communication: A cyclic-AMP mediated phenomenon. *J. Membr. Biol.* **19:** 253-266.

Higgins, D. and H. Burton. 1982. Electrotonic synapses are formed by fetal-rat sympathetic neurons maintained in a chemically-defined culture medium. *Neuroscience* **7:** 2241-2253.

Iacovitti, L., M.I. Johnson, T.H. Joh, and R.P. Bunge. 1982. Biochemical and morphological characterization of sympathetic neurons grown in chemically-defined medium. *Neuroscience* **7:** 2225-2239.

Kessler, J.A. 1984. Non-neuronal cell conditioned medium stimulates peptidergic expression in sympathetic and sensory neurons in vitro. *Dev. Biol.* **106:** 61-69.

Kessler, J.A., D.L. Spray, J.C. Saez, and M.V.L. Bennett. 1984. Determination of synaptic phenotype: Insulin and cAMP independently initiate development of electrotonic coupling between cultured sympathetic neurons. *Proc. Natl. Acad. Sci.* **81:** 6235-6239.

Kondo, H., N.J. Dunn, and G.D. Pappas. 1980. A light- and electron-microscopic study of rat superior cervical ganglion cells by intracellular HRP-labeling. *Brain Res.* **197:** 193-199.

LeDouarin, N.M., M.A. Teillet, C. Ziller, and J. Smith. 1978. Adrenergic differentiation of cells of the cholinergic ciliary and Remaks ganglion in avian embryos after in vivo transplantation. *Proc. Natl. Acad. Sci.* **75:** 2030-2034.

Patterson, P.H. and L.L.Y. Chun. 1977. The induction of acetylcholine synthesis in primary cultures of dissociated rat sympathetic neurons. I. Effect of conditioned medium. *Dev. Biol.* **56:** 263-280.

Spray, D.C., R.L. White, A. Campos de Carvalho, A.L. Harris, and M.V.L. Bennett. 1984. Gating of gap junction channels. *Biophys. J.* **45:** 219-230.

Determinants of Specificity of Electrical Synapses: The Making and Breaking of Connections in *Helisoma*

S.B. Kater, C.S. Cohan, and P.G. Haydon
Department of Biology, University of Iowa
Iowa City, Iowa 52242

During the last decade the electrical synapse has joined the chemical synapse as a recognized form of intercellular communication in the nervous system. Despite considerable attention to the elucidation of rules regulating the formation of chemical synapses, there has been little or no progress toward understanding parallel processes for electrical synapses. Here, we report our findings on a small network of electrically coupled neurons in which we have begun to test and perhaps understand some of the mechanisms underlying the formation of neuronal connections.

Much of our present understanding of electrical synapses comes from studies on invertebrate nervous systems, where electrical synapses are commonly found between identified neurons, and from developing blastomeres and other nonneuronal cells where electrical connections often occur early in development. Additionally, electrical synapses have been found between particular classes of neurons in vertebrate central nervous tissue and in the retina (Bennett 1977; Neyton et al.; Lasater and Dowling; Llinás; all (this volume). Recent data indicate that the formation of some electrical synapses is precisely timed during development of the nervous system. Moreover, once such connections are formed, they may not be maintained indefinitely, but may be broken at a later time (see Bennett et al. 1981). Such developmental observations raise important questions about the rules governing the formation, stabilization, and subsequent modification of electrical synapses and their relation to chemical synapses.

The specificity of neuronal connections can be studied perhaps in greatest detail when the component neurons of a circuit can be identified as individuals. The buccal ganglia of the snail *Helisoma* are composed of large, visually identifiable neurons, many of which have been characterized previously (e.g., Kater 1974). Defined chemical and electrical synapses between particular neurons form the neuronal substrate of the feeding behavior of this animal. These identified neurons have provided a useful system in which to study the rules underlying the formation and maintenance of synaptic connections. We have found that the normal pattern of interconnections between these neurons can be dramatically but predictably altered when these neurons are axotomized. In *Helisoma*, axotomy close to the soma causes a rapid growth of neurites from the damaged neurons and is followed by a sequence of events in which electrical connections among the axotomized neurons become reorganized (Bulloch et al. 1980; Bulloch and Kater 1981). These changes in connections between specific identified neurons occur consistently to produce the same final pattern of connectivity in every preparation. These observations have allowed us to ask what rules might regulate the specificity of synaptic interactions in these ganglia.

By studying the relationship between axotomy, neurite outgrowth, and changes in neuronal connections, we have uncovered some of the regulating conditions that govern the formation and specificity of electrical connections in *Helisoma*. The essential ingredients, which will be detailed in the remainder of this communication, center around the fact that there is: (1) an initial screening amongst the large population of neurons whereby selected neurons interconnect, and (2) a refinement of the initially formed connections such that only specific interconnections are ultimately maintained (Fig. 1).

To summarize:

1. The initial screening is based on the outgrowth status of a given neuron. In order for any pair of neurons to form electrical synapses in *Helisoma*, both neurons must have growing, overlapping neurites (Hadley and Kater 1983; Hadley et al. 1983, 1985; Haydon et al. 1984; Hadley et al. 1985). This initial set of conditions allows a subset of neurons from amongst the larger population to form an electrically coupled network.

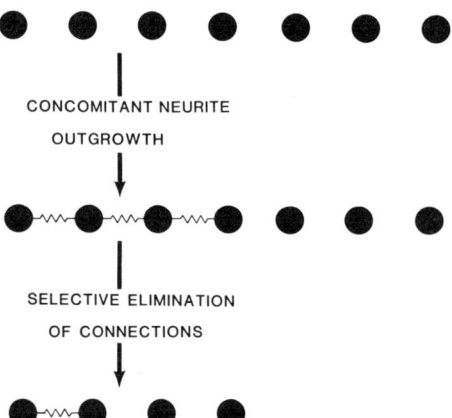

Figure 1
Summary diagram showing mechanisms that lead to specific electrical connections between *Helisoma* neurons. The formation of electrical connections between a subset of neurons within the population depends on the state of neurite outgrowth. Only neurons with concomitant neurite outgrowth form electrical connections with one another. Further refinement of these connections occurs as specific connections are eliminated while others are retained.

2. Further refinement of connections within the subset of electrically coupled neurons occurs. We have found in *Helisoma* that although many neurons form electrical connections with one another, most of these connections are transient. After the initial stage of synaptogenesis, specific electrical synaptic connections are subsequently broken and the final pattern of connectivity is achieved (Bulloch and Kater 1981, 1982; Hadley and Kater 1983).

Taken together, our results demonstrate that when outgrowth is initiated concomitantly among a set of unconnected neurons they inevitably become interconnected. Specificity of connections is achieved by: (1) timing the outgrowth of potential partner neurons; and (2) stabilization of specific connections while others are broken.

Hypothesis: Electrical Synapse Formation Depends on the Interaction of Mutually Growing Neurites

This hypothesis was based initially on experiments that were performed in situ. In these experiments neurite outgrowth within the buccal ganglia was evoked by two distinct methods of axotomy. Whole nerve trunks were crushed to sever all axons within a nerve (Hadley and Kater 1983). Alternatively, individual axons were selectively axotomized by filling their parent neuron with fluorescent dye and irradiating the axon to cause localized photo-dynamic damage, a technique called zapaxotomy (Cohan et al. 1983). Both methods caused axotomized neurons to respond similarly. Whenever pairs of neurons, which were not previously connected with one another, were induced to grow simultaneously such that newly generated neurites from each neuron overlapped one another, new electrical synapses always formed (Fig. 2). Conversely, when only one member of a potential pair within the ganglion was induced to grow, electrical connections never formed even though the newly growing neurites overlapped the nongrowing neuron.

A major test of this hypothesis relied upon our ability to remove individual identified neurons from their ganglionic environment and place them into cell culture where both spatial and temporal aspects of neurite outgrowth could be observed directly (Fig. 3). In cell culture, *Helisoma* neurons grow neurites for a finite period of time and then reach a stable morphological state. Thus, by staggering the time during which neurons were added to the culture dish, it was possible to pair potential partner neurons so that the growth of their neurites occurred either simultaneously, or asynchronously. The results of these experiments (Hadley et al. 1983, 1985) unequivocally supported our hypothesis: Electrical connections formed *only* between pairs of simultaneously growing neurons. Thus, in the widely disparate environments of the buccal ganglia and cell culture, the ability to form electrical connections depends on neurite outgrowth. These findings suggest that some unique property of growing neurites may be responsible for the formation of these connections.

The most obvious feature of growing neurites is their specialized terminal organelle, the neuronal growth cone. In *Helisoma*, actively growing neurons and neurons that spontaneously achieve a stable state show marked differ-

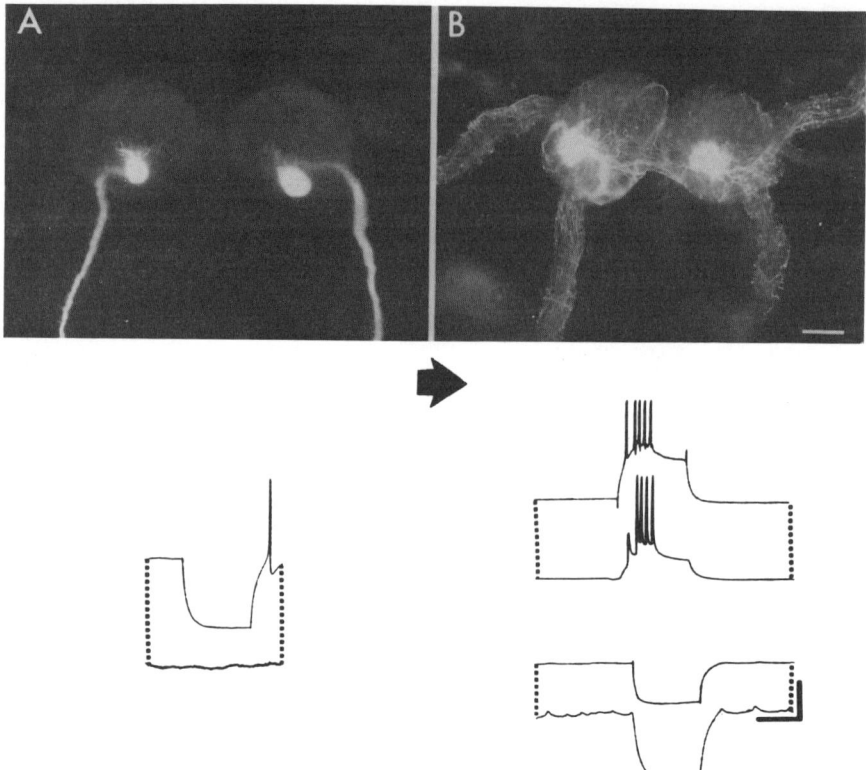

Figure 2
The formation of novel electrical connections between actively growing neurons in situ. (*Upper*) The morphology of neurons 5L and 5R within the buccal ganglia after 5 days in organ culture. (*A*) Isolated control buccal ganglia; (*B*) isolated preparation in which both neurons 5 were selectively axotomized close to the buccal ganglia by focal irradiation of their axons (zapaxotomy) prior to culture. The profuse growth from both neurons has filled the buccal commissure with overlapping neurites. Neurons were filled with the fluorescent dye Lucifer Yellow by intracellular injection. (*Lower*) Electrical recordings of changes in connectivity of neurons 5 evoked by neurite outgrowth. (*A*) Intact, nongrowing preparation has not formed electrical synaptic connections. (*B*) Neurons 5 in which outgrowth was evoked by zapaxotomy formed strong electrical connections with one another. Calibration: Ganglia, 100 μm; electrophysiology, (*A*) 20 mV (*upper*), 5 mV (*lower*); (*B*) 20 mV (*upper pair*), 20 mV (*lower pair, top*), 5 mV (*lower pair, bottom*).

ences in growth cone morphology (Cohan et al. 1985; Hadley et al. 1985; Haydon et al. 1985). The production and cessation of neurite outgrowth from *Helisoma* neurons in cell culture and the corresponding growth cone changes are time-dependent processes that occur over a period of several days. We have recently discovered that the transition between growing and stable states can be controlled experimentally (Fig. 4). Specifically, we have found that the neurotransmitter serotonin (5-hydroxytryptamine, 5-HT) can alter neurite out-

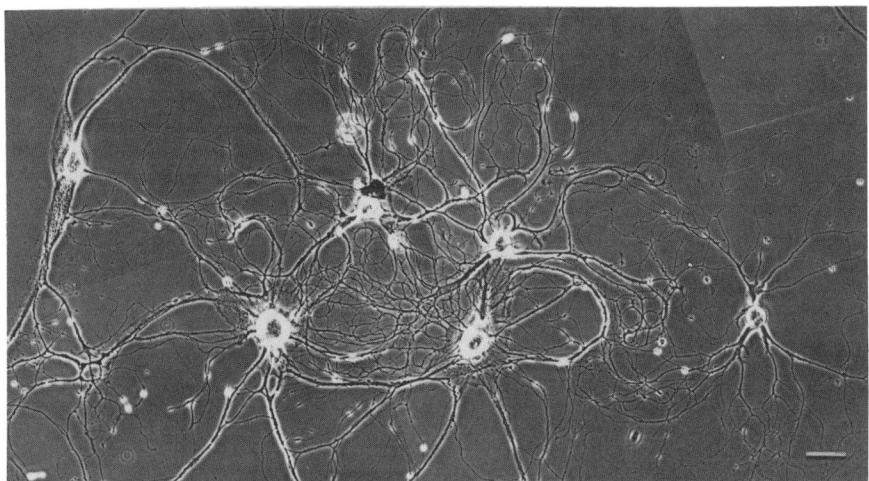

Figure 3
Phase-contrast photomicrograph showing a network of identified snail neurons in cell culture. Six identified neurons (neurons 5 and 19) were individually removed from multiple buccal ganglia and plated near each other in a culture dish. Outgrowth from each neuron resulted in a complex network of overlapping neurites. When assayed electrophysiologically, each of these neurons was electrically coupled to its neighbors. This is in contrast to coupling between identified neurons within the normal ganglionic environment which is limited to specific partners. Calibration, 100 μm.

growth and the motile behavior of growth cones in a neuron-specific manner (Haydon et al. 1984). When exposed to serotonin, growth cones of the identified neuron 19 are dramatically and abruptly inhibited from further motile activity and neurite outgrowth is halted. This is accompanied by retraction of growth cone lamellipodia and filopodia. In contrast, the growth cones of identified neuron 5 are unaffected by serotonin. This degree of experimental control over growth cone behavior and neurite outgrowth allowed us to test whether our previous findings on the formation of electrical connections were simply time-dependent processes or events that were directly related to growth cone activity.

When neurons 5 and 19 were placed into cell culture, electrical connections routinely formed. To test the role of growth cone activity on electrical connections, serotonin was added to cultures of neurons 5 and 19 during their initial stages of neurite outgrowth, before their neurites had overlapped (Fig. 5). Under these conditions electrical connections never formed between 5 and 19 even though the growing neurites of neuron 5 overlapped the nongrowing neurites of neuron 19 (Haydon et al. 1984). However, connections did form between pairs of neurons 5 that were unaffected by the presence of serotonin. These data further support our hypothesis and, moreover, suggest that growth cone activity is important for the formation of electrical connections in this organism.

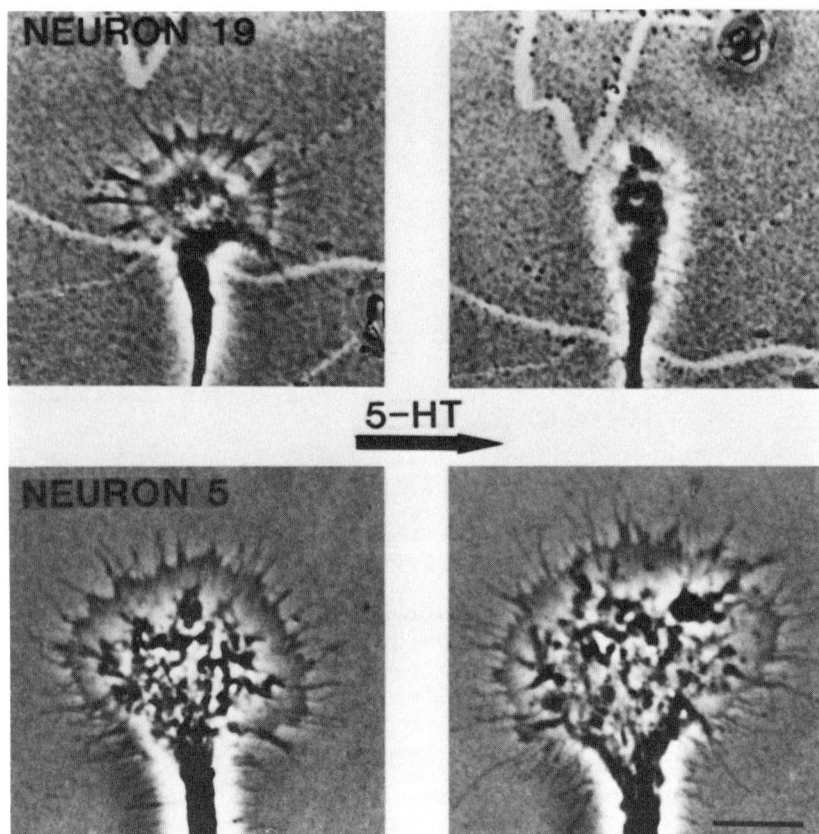

Figure 4
Serotonin causes a neuron-specific inhibition of growth cone motility and outgrowth. Exposure of the growth cones of neuron 19 to serotonin (5-HT, threshold concentration 10^{-7} M) causes a retraction of growth cone filopodia and an inhibition of growth cone motility (*upper*). In contrast, the growth cones of neuron 5 are unaffected by serotonin at concentrations as high as 5×10^{-5} M; these growth cones retain their filopodia and motile activity. Calibration, 20 μm. (Reprinted, with permission, from Haydon et al. 1985.)

We have asked whether these results obtained with identified neurons in vitro have significance in situ. Recent experiments in our laboratory have demonstrated that within the buccal ganglia, outgrowth from neuron 5 is totally unaffected by the presence of serotonin, whereas outgrowth from neuron 19 is always significantly retarded (A.D. Murphy et al., in prep.). When these findings are considered with the additional fact that the buccal ganglia receive a major functional serotonergic input (Granzow and Kater 1977; Murphy et al. 1985), they suggest that the release of a neural transmitter in situ has the opportunity to regulate the formation of electrical connections. Neuron C1 in the cerebral ganglion is a plausible candidate for such a regulatory function.

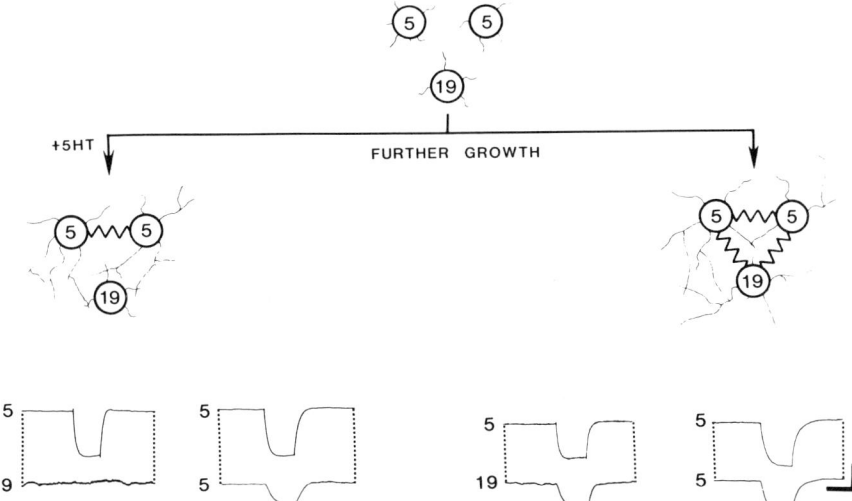

Figure 5
Serotonin's inhibition of neurite outgrowth prevents the formation of electrical synaptic connections between specific identified neurons. Neurons 5 and 19 were plated in culture as isolated spherical cell bodies, whereupon outgrowth was initiated. Serotonin (10^{-6} M) was added to the culture medium prior to neurite overlap (*left*). Consequently, outgrowth of neuron 19 was inhibited whereas outgrowth of neuron 5 was unaffected. Neurites from neurons 5 continued to grow and overlapped one another as well as the nongrowing neurites of neuron 19. As a result, neurons 5 formed strong electrical connections with one another but *never* formed electrical connections with neuron 19. Parallel control cultures (*right*) showed that electrical synaptic connections always formed between *all* neurons when serotonin was *absent* from the medium. Calibration: All upper traces, 20 mV; lower traces, 5 mV (neuron 19) and 10 mV (neuron 5).

This neuron releases serotonin into the buccal ganglia and thereby activates the feeding motor pattern. Might this neuron also regulate the formation of electrical connections in this system by releasing serotonin?

Maintenance and Breaking of Electrical Connections

Although the data presented above demonstrate that all neurons in active stages of outgrowth can potentially connect with one another, further refinement of synaptic connections can occur. Some electrical connections in *Helisoma* buccal ganglia are transient (Fig. 6; Bulloch and Kater 1981, 1982; Hadley and Kater 1983). Each buccal ganglion of *Helisoma* contains approximately 260 neurons. In our survey of a subset containing 28 representative identified neurons, 9 particular neurons were always found to form electrical connections with neuron 5 after outgrowth was evoked by axotomy. Interestingly, however, 3 days after the formation of these new connections, only one connection remained. All coupling with neuron 5, except to its contralateral homolog, was transient and was not observed a few days after initial formation

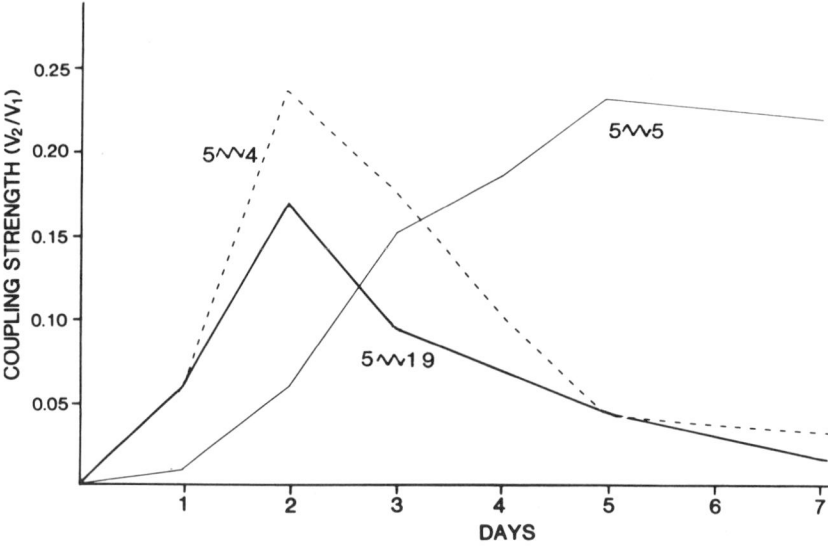

Figure 6
The formation, stabilization, and elimination of specific electrical connections between identified neurons in the buccal ganglia after sprouting was evoked by axotomy. Over a 7-day period, connections between neurons 5 and 19 (heavy solid line) and neurons 5 and 4 (dashed line) are transient and largely eliminated, whereas connections between neurons 5 and 5 (light solid line) are stable and maintained.

(Fig. 6). Thus, specificity of connectivity is attained in *Helisoma's* nervous system by the selection of specific connections from among a subset of interconnected neurons.

Tuning of Synaptic Strength

Although the preceding sections have emphasized the qualitative nature of the dynamics of electrical connections in this organism, it should also be emphasized that quantitative aspects of electrical coupling may also be regulatable. We have previously demonstrated the importance of electrical coupling in the generation of bursting activity within a small group of neurons in the buccal ganglia. Theoretical simulations show that changes in the strength of coupling within this neuronal network can dramatically affect its bursting properties (Merikel et al. 1978). The strength of existing electrical connections (as well as dye coupling) between identified neurons can be changed experimentally by procedures that evoke neurite outgrowth, i.e., axotomy. Neurons 4R and 4L, which innervate the salivary glands of *Helisoma*, are electrically coupled in situ. Following axotomy, the strength of electrical coupling increases from an average coupling coefficient of 0.54 to a new value of 0.72 within a few days (Murphy et al. 1983). Thus, neurons that are already electrically connected can modulate the quantitative strength of their interactions under certain circumstances.

SUMMARY AND PROSPECTUS

The specificity of electrical connectivity is of major importance to the integrative action of the nervous system. We have examined how specificity is attained in *Helisoma* by studying the processes responsible for (1) the formation of electrical synapses and (2) the maintenance or breaking of these initially formed connections. The formation of electrical connections depends on the outgrowth status of potential partner neurons. *The initial formation of an electrical connection relies critically on both a spatial and temporal coincidence of active neurite outgrowth from both potential partner neurons.* Once such electrical connections form, further specificity is conveyed by the stabilization of some connections and the breaking of others. Thus, some subset of neurons is first selected to form connections based on their outgrowth status and then specific connections are again selected to be maintained while others are broken. The combined actions of these two selection processes result in the final pattern of electrical connections that is observed.

Perhaps the major question remaining with respect to electrical connections in the nervous system concerns the extent to which normal physiological processes can modulate electrical coupling between neurons. It is clear that neural transmitters can influence the efficacy of communication between two neurons by shunting synaptic currents through nonjunctional membrane (e.g., Spira et al. 1980). It is much less clear whether similar physiological stimuli can modulate junctional resistance directly (but see Neyton et al.; Lasater and Dowling, both this volume). Even more intriguing is the possibility that neurons not previously connected can be made to connect by certain behavioral and physiological conditions and the correlate of this question, whether neurons already connecting can be made to break their electrical synaptic interactions by similar kinds of conditions. It is of considerable importance to determine to what extent the electrical synapse, which has been regarded historically as a rather fixed relay between neurons, may be a highly malleable element that responds in predictable and precise ways to environmental conditions.

ACKNOWLEDGMENTS

We thank Denise Dehnbostel for preparing the figures and D. McCobb for comments on the manuscript. Supported by Public Health Service grants NS15350, NS18819, and HD18577.

REFERENCES

Bennett, M.V.L. 1977. Electrical transmission: A functional analysis and comparison with chemical transmission. In *Cellular biology of neurons. Handbook of physiology. The nervous system* (ed. E.R. Kandel), vol. 1, p. 357–416. Williams and Wilkins, Baltimore, Maryland.

Bennett, M.V.L., D.C. Spray, and A.L. Harris. 1981. Electrical coupling in development. *Am. Zool.* **21:** 413–427.

Bulloch, A.G.M. and S.B. Kater. 1981. Selection of a novel connection by adult molluscan neurons. *Science* **212**: 79–81.

———. 1982. Neurite outgrowth and selection of new electrical connections by adult *Helisoma* neurons. *J. Neurophysiol.* **48**: 569–583.

Bulloch, A.G.M., S.B. Kater, and A.D. Murphy. 1980. Connectivity changes in an isolated molluscan ganglion during *in vivo* culture. *J. Neurobiol.* **11**: 531–546.

Cohan, C.S., R.D. Hadley, and S.B. Kater. 1983. "Zap Axotomy": Localized fluorescent excitation of single dye-filled neurons induces growth by selective axotomy. *Brain Res.* **270**: 93–101.

Cohan, C.S., P.G. Haydon, and S.B. Kater. 1985. Single channel activity differs in growing and non-growing growth cones of isolated identified neurons of *Helisoma*. *J. Neurosci. Res.* **13**: 285–300.

Granzow, B. and S.B. Kater. 1977. Identified higher-order neurons controlling the feeding motor program of *Helisoma*. *J. Neurosci.* **2**: 1049–1063.

Hadley, R.D. and S.B. Kater. 1983. Competence to form electrical connections is restricted to growing neurites in the snail, *Helisoma*. *J. Neurosci.* **5**: 924–932.

Hadley, R.D., D.A. Bodnar, and S.B. Kater. 1985. Formation of electrical synapses between isolated, cultured *Helisoma* neurons requires mutual neurite elongation. *J. Neurosci.* (in press).

Hadley, R.D., C.S. Cohan, and S.B. Kater. 1983. Electrical synapse formation depends on interaction of mutually growing neurites. *Science* **221**: 466–468.

Haydon, P.G., D.P. McCobb, and S.B. Kater. 1984. Serotonin selectively inhibits growth cone motility and synaptogenesis of specific identified neurons. *Science* **226**: 561–564.

Haydon, P.G., C.S. Cohan, D.P. McCobb, H.R. Miller, and S.B. Kater. 1985. Neuron-specific growth cone properties as seen in identified neurons of *Helisoma*. *J. Neurosci. Res.* **13**: 135–147.

Kater, S.B. 1974. Feeding in *Helisoma trivolvis:* The morphological and physiological bases of a fixed action pattern. *Am. Zool.* **14**: 1017–1036.

Merickel, M., S.B. Kater, and E.D. Eyman. 1978. Burst generation by electrically coupled network in the snail *Helisoma*: Analysis using computer simulation. *Brain Res.* **159**: 331–349.

Murphy, A.D., R.D. Hadley, and S.B. Kater. 1983. Axotomy-induced parallel changes in electrical and dye coupling between identified neurons of *Helisoma*. *J. Neurosci.* **3**: 1422–1429.

Murphy, A.D., D.L. Barker, J.F. Loring, and S.B. Kater. 1985. Sprouting and functional regeneration of an identified serotonergic neuron following axotomy. *J. Neurobiol.* **16**: 137–151.

Spira, M.E., D.C. Spray, and M.V.L. Bennett. 1980. Synaptic organization of expansion motoneurons of *Navanax inermis*. *Brain Res.* **195**: 241–269.

Communication Compartmentation and Pattern Formation in Development

Cecilia W. Lo
Biology Department, University of Pennsylvania
Philadelphia, Pennsylvania 19105

One of the most intriguing problems in the study of development is the question of how pattern formation is regulated. In many developmental systems, this patterning appears to be encoded via gradients; in some cases, this involves maternal gradients laid down during oogenesis and, in other cases, gradients generated de novo in the developing embryo or tissue. Given that gap junctional communication is potentially able to mediate the formation of diffusible chemical gradients (Michalke 1977), it is interesting to consider the possibility that in the developing embryo or tissue, "morphogenetic" gradients may be generated via gap junction-mediated cell–cell exchange (Wolpert 1978). To examine this possibility, we have characterized the gap junctional communication properties of several developmental systems. If gap junctional exchange were indeed involved in the patterning process, we would expect that cell–cell communication may undergo change in a temporally and spatially significant manner in parallel with ongoing developmental events.

DISCUSSION

For these studies, cell–cell communication was examined using microelectrode impalements to monitor either the exchange of ions (electrical coupling) or the intercellular spread of injected fluorescent dye (dye coupling). Using these methods, we probed the pattern of cell–cell communication in the early mouse embryo (Lo and Gilula 1979a,b) and the *Drosophila* wing imaginal disk (Weir and Lo 1982, 1984, 1985).

Mouse Embryo

The mouse embryo is a highly regulative embryo and up until and including the eight-cell stage, all the blastomeres of the embryo are totipotent. The first determination/differentiation event occurs at the late eight-cell stage, at the time of compaction (Gardner and Rossant 1976). This determination event appears to be dictated entirely on the basis of the blastomere's position in the embryo (the "inside-outside" hypothesis; Tarkowski and Wroblewska 1969). Thus blastomeres on the outside of the embryo give rise to the trophoblast cells and the cells on the inside (inner cell mass, ICM) form the embryo proper and the extraembryonic endoderm/ectoderm.

We examined the ionic and dye-coupling properties in the early mouse embryo from the two-cell stage and onwards. Using electrical and dye-coupling measurements, we found that no gap junction mediated cell–cell exchange until the late eight-cell stage. At that time, all eight blastomeres become linked to one another via gap junctional channels. As the late eight-cell embryo also constitutes the stage of development when the trophoblast/ICM cells become determined, this timing is consistent with the possibility that cell–cell communication plays a role in these determination events. In light of the inside-outside hypothesis, we suggest that gap junctional communication might mediate the formation of an intracellular inside-outside gradient and that this gradient may provide the signaling required for specifying these determination events. Note that as tight junctions are also formed de novo at this time (Dulcibella et al. 1975), it is possible that the permeability seal provided by the tight junctional complexes at the apical surface of the embryo may further serve to facilitate the formation of an inside-outside intracellular gradient.

As the compacted embryo develops further, it eventually forms a blastocyst that consists of a hollow vesicle containing two distinct cell types: (1) a single layer of trophoblast cells that form the wall of the vesicle and (2) a knob of undifferentiated cells located at one pole constituting the cells of the ICM. Ionic and dye coupling were observed throughout the entire embryo from the compacted eight-cell morula to the blastocyst stage. Thus once communication is turned on, all the cells of the mouse embryo are essentially linked together as a syncytial unit. Shortly after the formation of the blastocyst, the embryo normally undergoes implantation in utero; when placed in vitro, it will undergo an implantation-like process whereby it becomes attached and spread out on the substratum of the culture vessel. As such in vitro-implanted embryos will undergo further development through to at least the egg cylinder stage of embryogenesis (day 8 of gestation), we were also able to characterize the communication properties of mouse embryos in the early postimplantation stages of development.

The results of our experiment revealed that there are specific changes in the pattern of cell–cell coupling occurring in conjunction with postimplantation development. In the early postimplantation mouse embryo, there are two major differentiation events: first the terminal differentiation of the trophoblast cells to form trophoblast giant cells, and subsequently the differentiation of the ICM cells to form extraembryonic endoderm and embryonic/extraembryonic ectoderm. We found that initially there is complete ionic and dye coupling throughout the entire in vitro-implanted embryo as in the preimplantation blastocyst but that as the embryo undergoes development, this coupling begins to break down. Our experiments revealed that at the time when the trophoblast cells undergo giant cell transformation, they become dye-uncoupled to each other and to the cells of the ICM. However, during this time the cells of the ICM remain well dye-coupled to each other (Fig. 1). We refer to the domain of well-coupled ICM cells as a communication compartment. Surprisingly, when we monitored ionic coupling simultaneously with the dye injections, we found that the trophoblast cells remain ionically coupled to the ICM cells. This coupling was observed even when the trophoblast cell layer was disrupted, thereby indicating that the observed coupling cannot be accounted for by the presence of a tight junctional seal in the trophectoderm. Thus these results suggest that the communication compartment is only partially restricted with respect to cell–cell communication. As such embryos undergo further differentiation, the ICM cells differentiate into two morphologically distinct cell types, the extraembryonic endoderm and the embryonic/extraembryonic ectoderm. During this time, we observed that the ICM cells become further subdivided into a number of additional communication compartment domains.

It is interesting to consider why cell–cell communication would break down between the trophoblast cells and the ICM. The trophoblast cells are known to form gap junctional contacts with the surrounding uterine epithelium (Tachi et al. 1970). Hence should gap junctional communication be involved in generating gradients for specifying pattern formation in mouse development, perhaps it is necessary to isolate the ICM cells so that a critical concentration threshold may be generated for the components of such putative gradients. These gradients can be envisioned to specify differentiation of the ICM cells into ectoderm and endoderm, and, once these cell types are formed, their further segregation into additional communication compartment domains may facilitate the formation of other gradients that may be required for the further development of the embryo, i.e., the differentiaton of the ectodermal cells toward the formation of the embryo proper. The low level of communication that has been observed to persist between cells on opposite sides of a communication compartment border may provide a means for coordinating and cross-regulating the metabolic and differentiation activities of cells in different compartments. This may impart a plasticity to the patterning scheme that would be of great adaptive value for the developing embryo and would account for the highly regulative nature of early mouse embryonic development.

These findings demonstrate a close temporal and spatial correlation between changes in the pattern of cell–cell communication with ongoing developmental events in the mouse embryo, and this correlation is consistent with junctional communication playing an important role in regulating mouse embryonic development. To examine further the role of gap junctional communication in development, we turned to another system, the wing imaginal disk of *Drosophila*. Our aim in these studies was to determine if communication compartments also exist in this developmental system and whether they might be organized in a developmentally signficant manner.

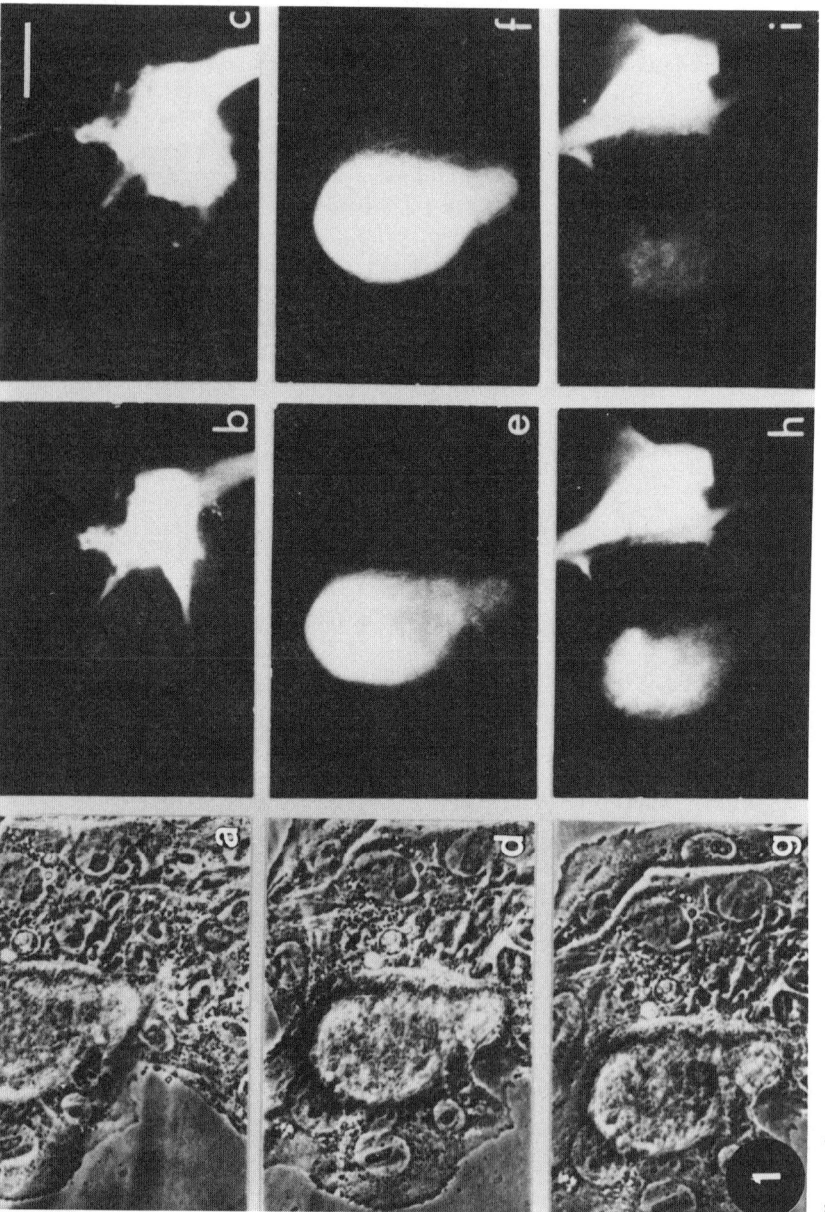

Figure 1
Fluorescein dye transfer in a postimplantation mouse embryo in which the trophoblast cells have undergone giant cell transformation (*a,d,e*). The phase image of an embryo that was successively impaled and injected with fluorescein at three different regions: (*a*) in a trophoblast cell, (*d*) in a cell of the ICM, and (*g*) in another trophoblast cell. (*b–c, e–f, h–i*) The fluorescence dye-spread images from each of the three impalements at various times after the start of injection: (*b*) 4 min, (*c*) 22 min, (*e*) 4 min, (*f*) 28 min, (*h*) 4 min, and (*i*) 19 min. The scale bar in *c* represents 67 μm. (Reprinted, with permission, from Lo and Gilula 1979b.)

Wing Imaginal Disk

In holometabolous insects such as *Drosophila*, most of the adult cuticular structures are formed from groups of cells in the larvae called imaginal disks. For example, the adult wing is formed from the wing imaginal disk, haltere from the haltere disk, etc. These disks consist of hollow sacks of epithelium which at the time of metamorphosis undergo overt differentiation and eversion to form the appropriate adult cuticular structures. The development of the imaginal disk appears to be organized via a number of discrete multicellular domains that are detectable by lineage analysis (Garcia-Bellido et al. 1973, 1976). These domains are referred to as developmental compartments or lineage compartments, and they are thought to play a fundamental role in organizing pattern formation since they appear to constitute the realm within which the pattern-regulating homeotic genes are differentially expressed (Crick and Lawrence 1975; Garcia-Bellido 1975; Morata and Lawrence 1975; Wilcox et al. 1981). Thus, for example, initially an anterior/posterior lineage border is established such that cells on either side give rise to only structures in the anterior or posterior portion of the adult cuticle, and it is only in the posterior compartment that the homeotic engrailed gene appears to be expressed (Kornberg et al. 1985).

Interestingly, these compartments are known to have a polyclonal origin, that is, they are always formed from a group of cells, never just one. This is consistent with the possibility that multicellular domains defined by communication restriction borders (i.e., communication compartments) might play a role in insect pattern formation by providing the basis for the formation or maintenance of these lineage compartments. To examine this possibility, we carried out a detailed analysis of the dye-coupling properties of the *Drosophila* wing imaginal disk. If lineage compartments are functionally related to communication compartments, one might expect to find not only the presence of communication compartments in the wing disk epithelium but also a coincidence between the position of these two sets of compartment borders.

Our experiments were carried out predominantly in the columnar epithelium of the wing disk, the part of the disk epithelium that eventually participates in the formation of the adult cuticle. Using microelectrode impalements we found that the imaginal disk cells indeed are ionically and dye-coupled. Moreover, using dye injections we observed the presence of communication restriction borders that delineate communication compartments (Figs. 2, 3, and 4). These restriction borders were reproducibly observed at the same position as examined by multiple impalements into the same disk and also by numerous impalements into a large number of disks (Table 1). In the latter instance, two criteria were used for orienting the position of a communication restriction border: the overall disk outline and the standard folding pattern of the wing disk epithelium. In some cases, this assignment of position was facilitated when a commmunication restriction border was observed to coincide with folds in the wing disk epithelium. For these latter communication restrictions, the further examination of dye-injected disks by thick-section histology demonstrated that the dye restrictions observed in the intact disk indeed corresponded to sharp discontinuities in fluorescence intensity in the disk epithelium. This latter result indicates that these dye restrictions are indeed real

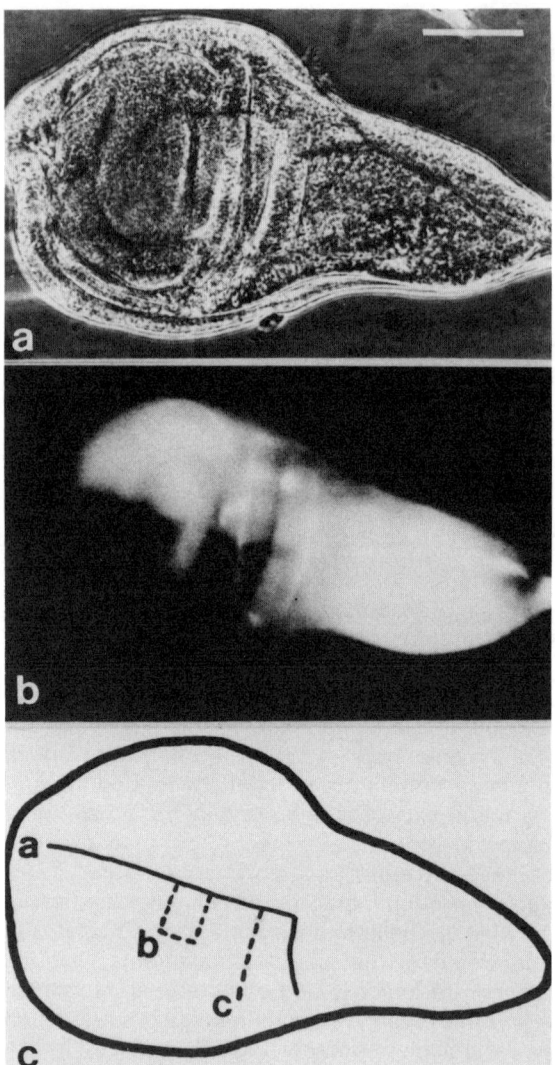

Figure 2
Lucifer Yellow dye transfer in *Drosophila* wing imaginal disk. (*a*) Phase image of a mid-third instar wing disk impaled at the notal (N) region. The site of impalement is marked with an arrow. (*b*) Fluorescence image at 52 min after the start of injection of Lucifer Yellow. (*c*) Line drawing interpretation of the fluorescence image in *b*. The injected dye has moved from the dorsal (D) point of impalement into the anterior (A) half of the disk, but dye movement into the posterior (P) half is restricted at several boundaries. Note that the restriction line a appears to be coincident with the known anterior/posterior lineage border. Also note that the injected dye has spread beyond line a at several positions (compare b and c). The scale bar in *a* represents 100 μm. (Reprinted, with permission, from Weir and Lo 1982.)

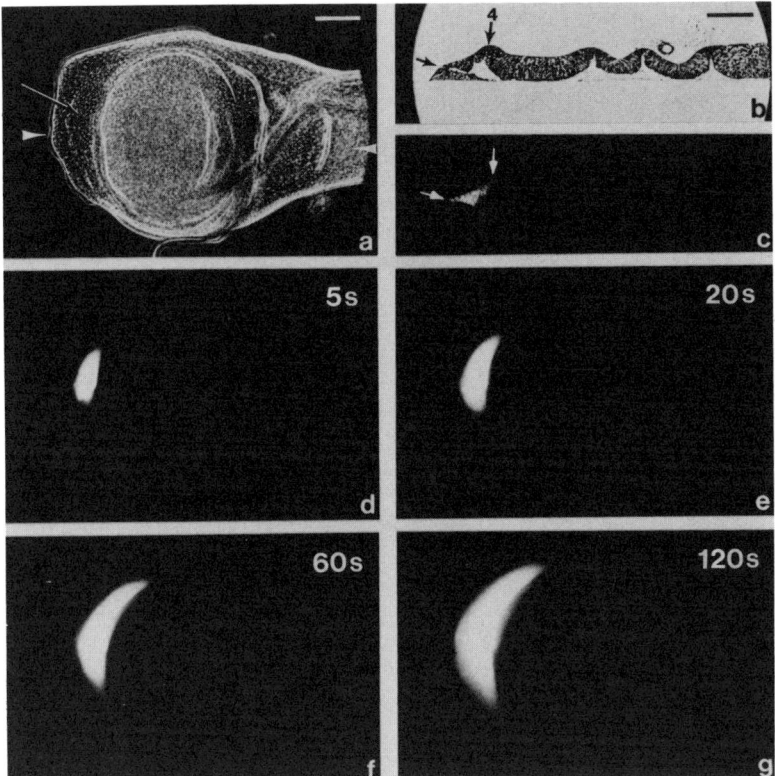

Figure 3
Lucifer Yellow dye spread in a mid- to late-third instar *Drosophila* wing disk. The pattern of dye spread was recorded at various times after the start of injection. At the end of the experiment, the disk was fixed, embedded, and sectioned. (*a*) Phase image of injected disk. The dark arrow indicates the site of impalement. The two white arrows illustrate the cutting plane used to generate the thick sections of the disk as illustrated in *b* and *c*. (*b,c*) Stained and fluorescent image of a thick section obtained from this dye-injected disk. (*d–g*) Fluorescence image of the dye-injected disk at timed intervals. The scale bars in *a* and *b* represent 50 μm. (Reprinted, with permission, from Weir and Lo 1984.)

restrictions in cell–cell communication and not merely optical artifacts arising from the folding pattern of the wing disk epithelium.

After extensive analysis, we detected the presence of nine communication restriction boundaries (Table 1; Fig. 5). These communication restriction boundaries are observed from imaginal disks of the early to late third instar period. Importantly, five of the communication compartment boundaries appear to coincide with the borders of lineage compartments. For example, in Figure 2, communication restriction line a appears to coincide with the position of the A/P lineage restriction and, in Figure 3, the communication restriction

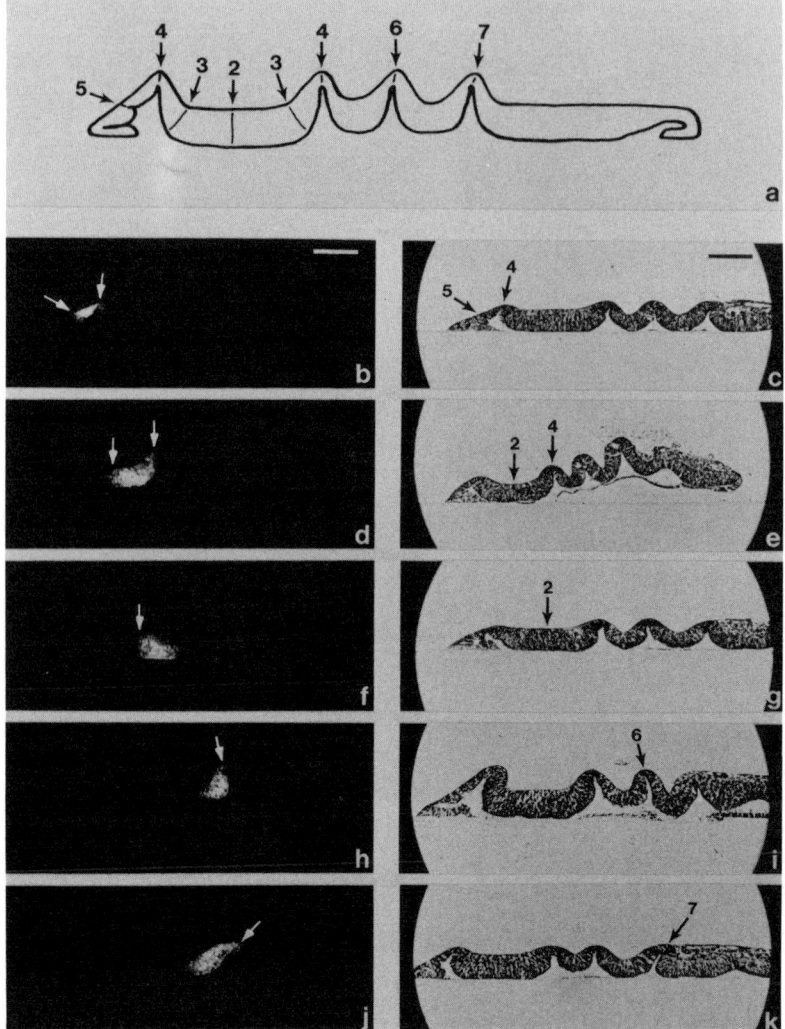

Figure 4
Summary of the positions of communication restriction lines 2–7. (*a*) Line drawing of a disk illustrating the positions of communication restriction lines 2–7 as observed from thick sections of dye-injected disks. (*b–k*) Examples of dye-injected disks that illustrate the position of some of these restriction boundaries. The fluorescence images are illustrated on the left and the corresponding stained sections are illustrated on the right. The scale bars in *b* and *c* represent 50 μm. (Reprinted, with permission, from Weir and Lo 1984.)

border line 4 might correspond to the proximal/distal lineage restriction. All of these restriction boundaries were observed to be only partially restrictive, such that ionic and dye coupling can be detected across the communication compartment border. Thus, in any dye injection experiment, if a large amount

Table 1
Summary of Injections into Wild-type Disks

Line	Impalements			Total	Thick sections
	1	2	≥3		
1	51 (3)	18 (8)	7 (7)	76	
2	50	25 (17)	11 (9)	86	10
3	34	6 (2)	8 (1)	48	8
4	53	18 (1)	15 (5)	86	14
5	7	2		9	3
6	23	2 (1)	1 (1)	26	2
7	17	2 (1)		19	2
8	7			7	
9	3			3	

This table summarizes the results of 540 impalements into 305 third-instar wing disks. The experiments are tabulated according to whether the dye-restriction lines (Fig. 5) were observed with 1, 2, or ≥3 impalements into the same disk. The numbers indicate the number of individual disks in which each line was observed, with dye approaching the border from one side or from two sides. The numbers within the parentheses are a subset of the above and indicate only the number of cases when the borders were observed from two sides. The final column describes the numbers of sectioned disks in which each border was observed in thick sections. (Reprinted, with permission, from Weir and Lo 1984).

of dye is injected, it is possible to observe dye passing across a dye restriction boundary (Fig. 2). These communication restriction borders were observed to be delineated by bands of cells (two to three cells wide) with a low level of cell–cell communication. Such cells presumably provide the basis for the formation and maintenance of communication compartments.

Comparable to our observations in the *Drosophila* wing imaginal disk, the segmental borders of the integument of *Oncopeltus* and *Calliphora* have also been observed to correspond to partial communication restriction borders (Warner and Lawrence 1982; Blennerhassett and Caveney 1984). Moreover at the segmental border of *Oncopeltus*, bands of low-communicating cells have also been observed (Blennerhasset and Caveney 1984). As in the imaginal disk epithelium, these communication restriction boundaries were observed to coincide with the boundaries of lineage restrictions (Lawrence 1973; Wright and Lawrence 1981). Overall these striking results seem to indicate that there may be a functional relationship between lineage and communication compartments. However, the questions remain as to how they are interrelated and how might they participate in organizing patterning events in the wing imaginal disk.

To examine the above question further, we turned to the engrailed homeotic mutant. The wing of this mutant contains a mirror symmetric duplication so that the posterior portion of the wing is transformed into an anterior-like character. In such mutants, the normal A/P lineage border is not observed (Morata and Lawrence 1975). If lineage compartments are responsible for the formation of communication compartments, then wild-type A/P communication compartment border would not be expected to be present. However, if communication compartments are responsible for mediating the formation of lineage

A. Lineage Lines

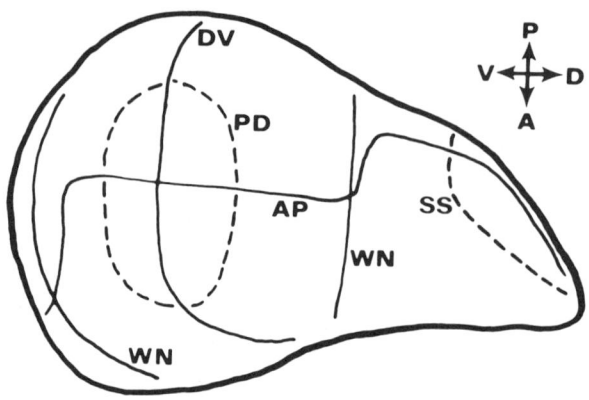

B. Communication Lines

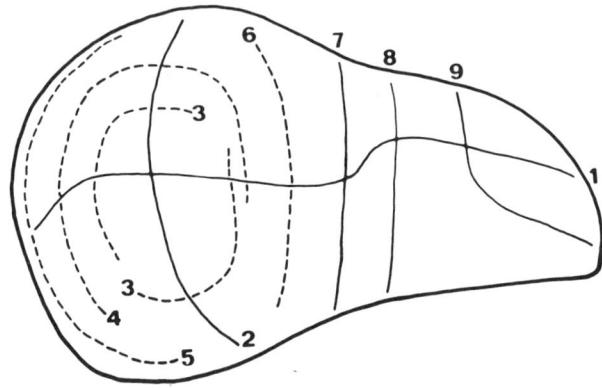

Figure 5
The positions of the major communication restriction borders and the lineage restriction borders in the wing imaginal disk. (a) Summary of all the known lineage restriction lines. (b) Summary of all the major communication restriction borders. (Reprinted, with permission, from Weir and Lo 1984.)

compartments, then one would predict that the A/P communication restriction border (line 1) might remain unperturbed in such mutants. The results of our analysis revealed that the engrailed mutant has an A/P communication restriction border and it is indistinguishable from that of the wild type. This result indicates that lineage compartments are not responsible for the formation of communication compartments. Rather we suggest that perhaps communication compartments may somehow provide the information for specifying the segregation of cells into lineage compartment domains. In essence, we are proposing that the segregation of cells into communication compartment domains is what is fundamentally important in organizing pattern formation in the wing disk epithelium, not the segregation of cells into lineage compartments.

It is interesting to consider how communication compartments might dictate the formation of lineage compartments and specify the compartment-specific differentiation of cells (anterior versus posterior, dorsal versus ventral, etc.). Perhaps communication compartments may serve as the domains within which gradients may be established for coding positional information. This information may regulate the expression of homeotic genes, the products of which may dictate cell-surface changes that can provide the basis for maintaining a lineage separation between cells from different communication compartments. Moreover, these same gene products may also specify the compartment and segment identity of cells (such as anterior versus posterior and wing versus haltere, etc.) so that the appropriate pattern is formed when overt differentiation occurs during metamorphosis. According to this scheme, positional information in the engrailed mutant is normal, as the A/P communication compartments observed are completely wild type in character. What may be disrupted in such mutants is the mechanism required for interpreting that positional information, and for this process to proceed normally, presumably the engrailed gene is required. Consistent with the above interpretation is the fact that the posterior-to-anterior transformation observed in engrailed wings occurs in an entirely position-appropriate manner; for example, dorsal/posterior structures are transformed into dorsal/anterior structures, distal/posterior structures are transformed into distal/anterior structures, etc.

Cell–Cell Communication and Pattern Regulation

It is interesting that in both of the systems we have examined, extensive cell–cell coupling was observed. This is perhaps not surprising as the development in both systems is known to be highly regulative. Thus, a priori, we would have predicted that if cell–cell communication and communication compartmentation were involved in dictating pattern formation, that it would provide a patterning mechanism that would be highly plastic or capable of undergoing regulation. This plasticity would result from the fact that communication compartments are only partially restrictive and hence should have the potential to detect and adapt to environmental perturbations. Moreover, this adaptation or regulation would be further facilitated if gradients were to encode positional information, as such gradients have been hypothesized to undergo smoothing out of newly juxtaposed nonadjacent positional values during pattern regulation (Lawrence 1966; Locke 1967; French et al. 1976). In contrast and as an extension to this logic, we suggest that for systems in which development is dictated by preformed maternal gradients or cytoplasmic determinants, cell–cell communication probably would not play a major role in the patterning process. In such mosaic developmental systems, we predict that there will be either no change in the pattern of coupling during development or little or no coupling at all. Such "negative" correlations would also be consistent with our overall hypothesis on how cell-cell communication might participate in the patterning process.

SUMMARY

Our current hypothesis is that communication compartments may regulate patterning events by mediating the formation of intracellular gradients and such gradients can encode the positional information required for dictating appropriate cell determination/differentiation. We predict that most such communication compartments will only be partially restrictive, as was observed here for the mouse embryo and the insect imaginal disk. This property may provide a plasticity to the patterning mechanism so that the developing tissue can undergo pattern regulation in response to environmental perturbations. In light of this hypothesis, we further suggest that gap junctional communication compartments may only play an important role in organizing pattern formation in tissues or embryos that exhibit highly regulative development.

ACKNOWLEDGMENT

This work was supported by NIH grant GM 30461.

REFERENCES

Blennerhassett, M.G. and S. Caveney. 1984. Separation of developmental compartments by a cell type with reduced junctional permeability. *Nature* **309**: 361-364.

Crick, F.H.C. and P.A. Lawrence. 1975. Compartment and polyclones in insect development. *Science* **189**: 340-347.

Dulcibella, J., D. Albertini, E. Anderson, and J.D. Biggers. 1975. The preimplantation mammalian embryo: Characterization of intercellular junctions and their appearance during development. *Dev. Biol.* **47**: 231-250.

French, V., P.J. Bryant, and S.V. Bryant. 1976. Pattern regulation in epimorphic fields. *Science* **193**: 969-981.

Garcia-Bellido, A. 1975. Genetic control of wing disc development in *Drosophila*. *Ciba Found. Symp.* **29**: 161-182.

Garcia-Bellido, A., P. Ripoll, and G. Morata. 1973. Developmental compartmentalization of the wing disk of *Drosophila*. *Nature New Biol.* **245**: 251-253.

———. 1976. Developmental compartmentalization of the dorsal mesothoracic disc of *Drosophila*. *Dev. Biol.* **48**: 132-147.

Gardner, R.L. and J. Rossant. 1976. Determination during embryogenesis. *Ciba Found. Symp.* **40**: 5-25.

Kornberg, T., I. Siden, P. O'Farrell, and M. Simon. 1985. The *engrailed* locus of *Drosophila*: In situ localization of transcripts reveals compartment-specific expression. *Cell* **40**: 45-53.

Lawrence, P.A. 1966. Gradients in the insect segment: The orientation of hairs in the milkweed bug *Oncopeltus fasciatus*. *J. Exp. Biol.* **44**: 607-620.

———. 1973. A clonal analysis of segment development in *Oncopeltus* (Hemiptera). *J. Embryol. Exp. Morphol.* **30**: 681-699.

Lo, C.W. and N.B. Gilula. 1979a. Gap junctional communication in the preimplantation mouse embryo. *Cell* **18**: 399-409.

———. 1979b. Gap junctional communication in the postimplantation stage mouse embryo. *Cell* **18:** 411–422.

Locke, M. 1967. The development of patterns in the integument of insects. *Adv. Morphog.* **6:** 33–38.

Michalke, W. 1977. A gradient of diffusible substance in a monolayer of culture cell. *J. Membr. Biol.* **33:** 1–20.

Morata, G. and P.A. Lawrence. 1975. Control of compartment development by the *engrailed* gene in *Drosophila*. *Nature* **255:** 614–617.

Tachi, S., C. Tachi, and H.R. Linder. 1970. Ultrastructural features of blastocyst attachment and trophoblastic invasion in the rat. *J. Reprod. Fertil.* **21:** 37–56.

Warner, A.E. and P.A. Lawrence. 1982. Permeability of gap junctions at the segmental border in insect epidermis. *Cell* **28:** 243–252.

Weir, M.P. and C.W. Lo. 1982. Gap junctional communication compartments in *Drosophila* wing imaginal disk. *Proc. Natl. Acad. Sci.* **79:** 3232–3235.

———. 1984. Gap junctional communication compartments in *Drosophila* wing imaginal disk. *Dev. Biol.* **102:** 130–146.

———. 1985. An anterior/posterior communication compartment border in *engrailed* wing disks: Possible implications for *Drosophila* pattern formation. *Dev. Biol.* (in press).

Wilcox, M., D.L. Brower, and R.J. Smith. 1981. A position specific cell surface antigen in the *Drosophila* wing imaginal disk. *Cell* **25:** 159–164.

Wolpert, L. 1978. Gap junctions: Channels for communication in development. In *Intercellular junctions and synapses* (ed. J. Feldman et al.), p. 83–94. Chapman and Hall, London.

Wright, D.A. and P.A. Lawrence. 1981. Regeneration of the segment boundaries in *Oncopeltus:* Cell lineage. *Dev. Biol.* **85:** 238–333.

Control of Molecular Movement within a Developmental Compartment

Stanley Caveney and Richard Safranyos
*Department of Zoology, University of Western Ontario
London, Ontario, Canada N6A 5B7*

The compartment hypothesis states that the fate of cells during insect embryonic development becomes progressively restricted by the formation of compartment boundaries. Once a boundary is drawn, cells on either side of it are confined to their respective compartment and have developmental fates different from the cells in other compartments. Clonal analysis has shown that the epidermis of each body segment in the growing insect is a developmental compartment (Lawrence 1981). Within each segment, the epidermal cells are strongly coupled by membrane channels that allow the rapid movement of inorganic ions and organic tracers from cell to cell, whereas the cells at the segment (compartment) border retard selectively the passage of organic molecules from segment to segment (Warner and Lawrence 1982; Blennerhassett and Caveney 1984). This restriction in junctional communication between segments results from the presence of a discrete population of border cells at the anterior edge of each segment (Blennerhassett and Caveney 1984) that may allow the adjacent segments some degree of developmental autonomy.

The permeability of the junctional channels of the border cells may be developmentally regulated, however, since they are able to pass organic tracers in certain culture media and in response to 20-hydroxyecdysone in vitro (Blennerhassett and Caveney 1984 and unpubl.).

Here we report two ways in which the epidermis controls its level of junctional coupling within the segmental compartment, namely the ability of the epidermal cells to raise their basal level of ionic and dye coupling as the cells increase in number (i.e., the segment grows in size) and as they respond to an insect developmental hormone.

The preparation used in these studies is the larval epidermis of the beetle *Tenebrio molitor*. Each ventral plate of the abdominal wall (sternite) contains up to 60,000 epidermal cells that secrete, but remain attached to, a transparent cuticle through which the cells can be seen in phase-fluorescence microscopy. The cuticle acts also as a mechanical support for the cells during microelectrode penetration. The cells, uniform in height (10 μm in the midinstar epidermis) and in their arrangement (about 11,000 cells/mm^2 packed near-hexagonally), can be maintained in culture medium indefinitely.

The simple monolayer geometry of the epidermis allows the resistance of the cell–cell pathway to ion movement (Caveney and Blennerhassett 1980) to be compared with its resistance to the diffusion of organic tracers (Safranyos and Caveney 1985). In comparing these two independent estimates of junctional permeability, i.e., the effective intercellular resistance and effective diffusion coefficient, we are able to determine the way in which the permeability of the junctional channels is modulated.

Our results suggest that the individual channels in the epidermis open and close in an all-or-none, rather than graded, manner during development of the segmental compartment.

DISCUSSION

Junctional Communication Is Related to Compartment Size

The segment grows in a stepwise and almost isotropic fashion as the larva passes through successive molt cycles in its development. Short bursts of proliferative activity in the epidermal cells are separated in time by long periods when the cells are active in other events, such as cuticle synthesis. Segment growth is due solely to the net increase in cell number in the segment during each molt cycle; epidermal cell density remains constant. Since epidermal cells are added uniformly to the segment throughout its area, and not just at its edges, the detailed proportions of the segment pattern, as well as cell density, are preserved during larval development. The segment pattern is "scale-invariant" (Othmer and Pate 1980).

Organic tracers microinjected into the epidermis pass rapidly from the injected cell into the surrounding cells, as shown for carboxyfluorescein (CF) injection in Figure 1. Within 10 seconds of injection, CF has spread to the third order of cells from the source cell, and by 80 seconds to at least the seventh order. Although this dye diffuses smoothly from cell to cell, it does localize slightly in the nuclei, allowing the individual cells and their packing order to be seen (Fig. 1). The rate at which tracers move in the epidermis is not constant during segment growth, however, but increases in proportion to segment size. Tracers injected into the segmental epidermis excised from large larvae pass considerably faster from cell to cell than when injected into epidermis from smaller (younger) larvae. This is shown diagramatically in Figure 2, a–c. In this experiment, only larvae between molts ("intermolt" or "midinstar" larvae) were used. Between two and six dye "spreads" were recorded from inside the area enclosed by four sets of bristles (Fig. 2e) on each segment.

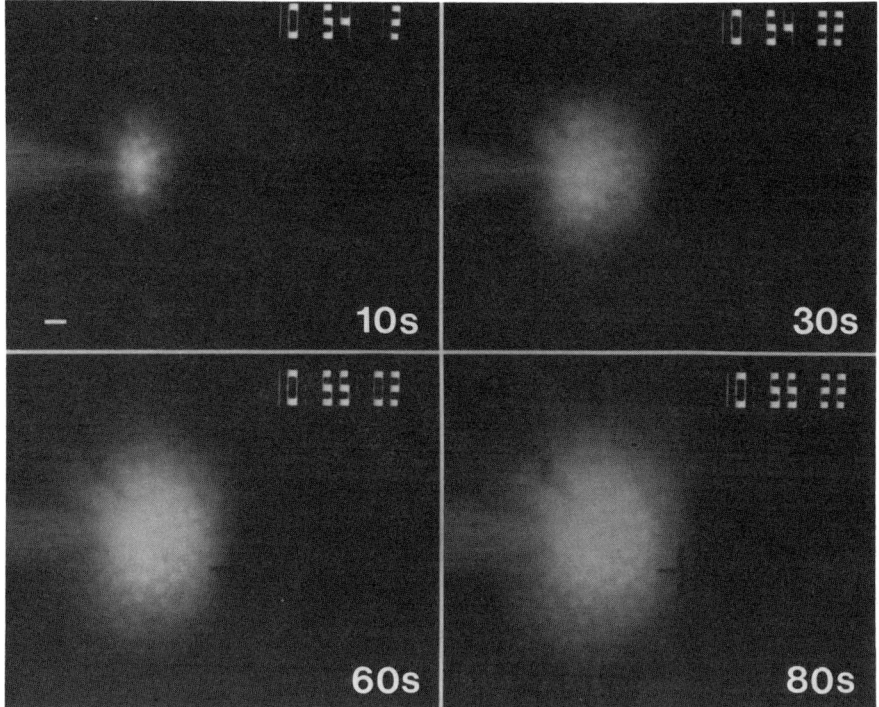

Figure 1
Diffusion of CF in the midinstar epidermis. The range of dye spread is shown after 10 sec, 30 sec, 60 sec, and 80 sec of continuous injection into the cell sheet (200 msec pulse/sec, 2 nA iontophoretic current). The average cell diameter is 9 μm. Scale bar, 10 μm.

Figure 2, a–c, plots the build-up in fluorescence intensity around the source cell during continuous injection of CF into three different segments. The steepness of the fluorescence intensity profiles is a rough indication of the rate of tracer movement from the source. In Figure 2a, in a small segment taken from a young larva, tracer passage is slow compared with Figure 2, b and c, which gives the fluorescence profiles for dye spreads in an intermediate-sized (Fig. 2b) and large (Fig. 2c) segment from successively older larvae.

The rate at which injected CF spreads in the epidermis is given by its effective diffusion coefficient (D_e). The data in Figure 2, a–c, are replotted in Figure 2d to determine D_e; full details of the method are found in Safranyos and Caveney (1985).

The results are summarized in Figure 2e. The index of segment size used is the average dimension of the area between the bristles used for the dye spreads. Segment size strongly influences the level of junctional coupling in the epidermis. Although the reason for this relationship is not known, it is worth mentioning that models predicting scale invariance during tissue growth include as a central feature the need for a mechanism that increases the rate of movement of diffusible morphogens as the tissue increases in size (Othmer and Pate 1980). Our findings in the growing epidermis fit this theoretical prediction. The ability of the epidermis to regulate gap junctional coupling as it grows may also be relevant to dilution models of growth control (Loewenstein 1979).

Junctional Permeability to Organic Tracers Is Raised by a Hormone

The steroid hormone 20-hydroxyecdysone raises junctional conductance to ions in the epidermis by 66% (Caveney and Blennerhassett 1980). This response, complete within 12–18 hours, does not require the synthesis of new gap junctional proteins or junction growth (Caveney et al. 1980), nor does it involve gross structural rearrangements within gap junctional plaques (Berdan and Caveney 1985). Instead, it appears to result from changes in the nature of the individual channels already present in the junctional membrane before hormone stimulation by a gating mechanism that either widens the aperture of already-open channels or opens previously closed channels. Which

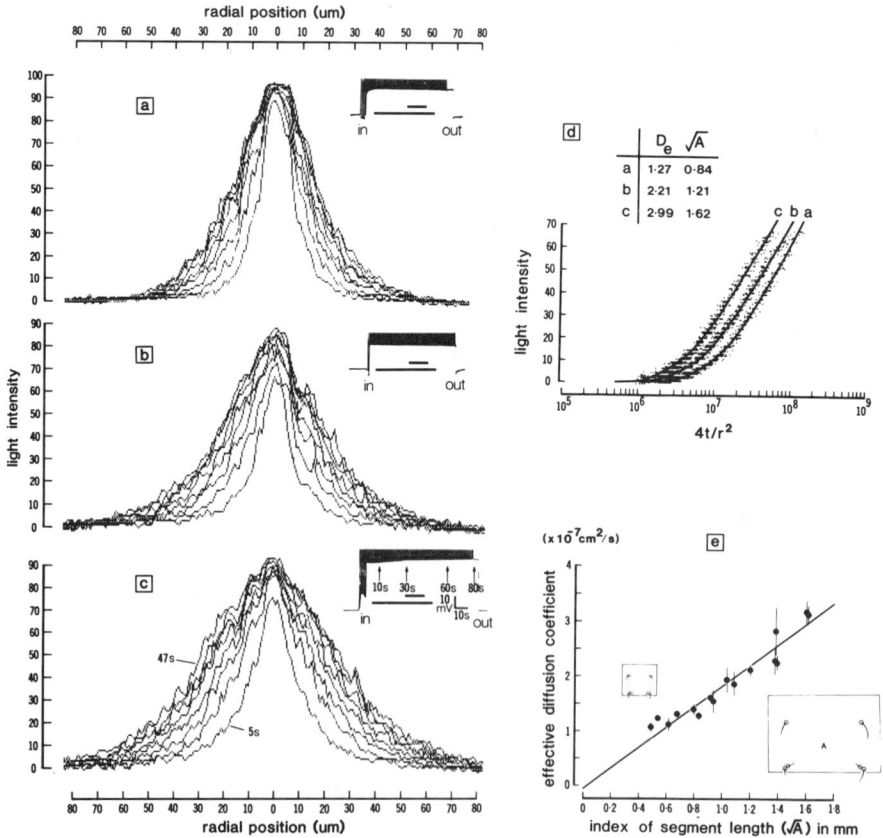

Figure 2 (*See facing page for legend.*)

of these two alternatives—graded versus all-or-none channel opening—is likely to be correct can only be resolved by physiological methods that can detect slight changes in the relative flux rates of differently sized molecules through the junctional membrane at various levels of coupling. If the individual channels have only two states, open or closed, then raising junctional conductance would not influence the size range of molecules capable of passing from cell to cell, but rather their rate of movement. This would apply equally to inorganic ions and to organic tracers of different sizes. On the other hand, if the individual channels open in a graded fashion (i.e., have more than two states), then the rate at which large organic tracers move from cell to cell may be enhanced selectively. The same argument may be made for molecular movement under conditions that reduce junctional permeability, in that channel closure by the two mechanisms would have different effects. Graded channel closure should block the passage of organic tracers before ionic coupling ceased to be detectable, whereas all-or-none channel closure should restrict the movement of all molecules in a coordinate fashion. To examine how 20-hydroxyecdysone acted on the individual channels, CF (m.w. = 376) was coinjected with a larger tracer, lissamine rhodamine B (m.w. = 559) into control and hormone-stimulated epidermis dissected from large larvae. Our premise was that all-or-none channel opening in response to hormone would raise the effective diffusion coefficients for both dyes, but would not affect their relative diffusion rates. The results are shown graphically in Figure 3 and summarized in Table 1. Exposure to hormone for 16–18 hours in vitro raised the D_e of both dyes by 33%, although the ratio of their diffusion coefficients re-

Figure 2
Dye coupling among epidermal cells rises as the segment grows. (a, b, c) Spatial distribution in the fluorescence intensity seen when CF is injected continuously into segmental epidermis from three midinstar larvae of different sizes. Fluorescence intensity profiles, plotted here stepping out radially from the site of injection, are shown at 6-sec intervals between 5 sec and 47 sec of injection in all three cases. Also shown for each spread is a record of the time of dye electrode entry and removal from the source cell and the stability of its membrane potential during dye injection (*inset, top right*). Underneath each membrane potential trace is a long bar marking the period for which the spatial distribution in light intensity is plotted and a short bar indicating the data used in the analysis shown in d. The scale of these three insets is shown in c. The data of the topographic profile shown in c are from the dye spread shown in the micrographs of Fig. 1. (d) Calculation of the effective diffusion coefficient (D_e). D_e is derived from fluorescence values (linearly related to dye concentration) at selected distances and times of injection (here between 20 sec and 43 sec). When plotted as a function of time and area ($4t/r^2$), all values for one side of a dye spread fall on a single curve predicted by an equation describing diffusion from a continuous line source (Safranyos and Caveney 1985). Nonlinear regression through each set of points gives the values of D_e for the right-hand sides of a and c, and the left-hand side of b listed in the table above. Also listed is an indicator of segment size in the three preparations ($\sqrt{A}$, see below). (e) The effective diffusion coefficient for CF is a function of segment size. The values plotted are the mean D_e (± S.D.) for between two and six dye spreads from 16 segments, each from a separate midinstar larva. The relative size of the smallest and largest segment examined is shown. The index of segment size used is the square root of the large area of the segment enclosed within four sets of marker bristles (*A*); the cells used in this analysis lie at the center of this area.

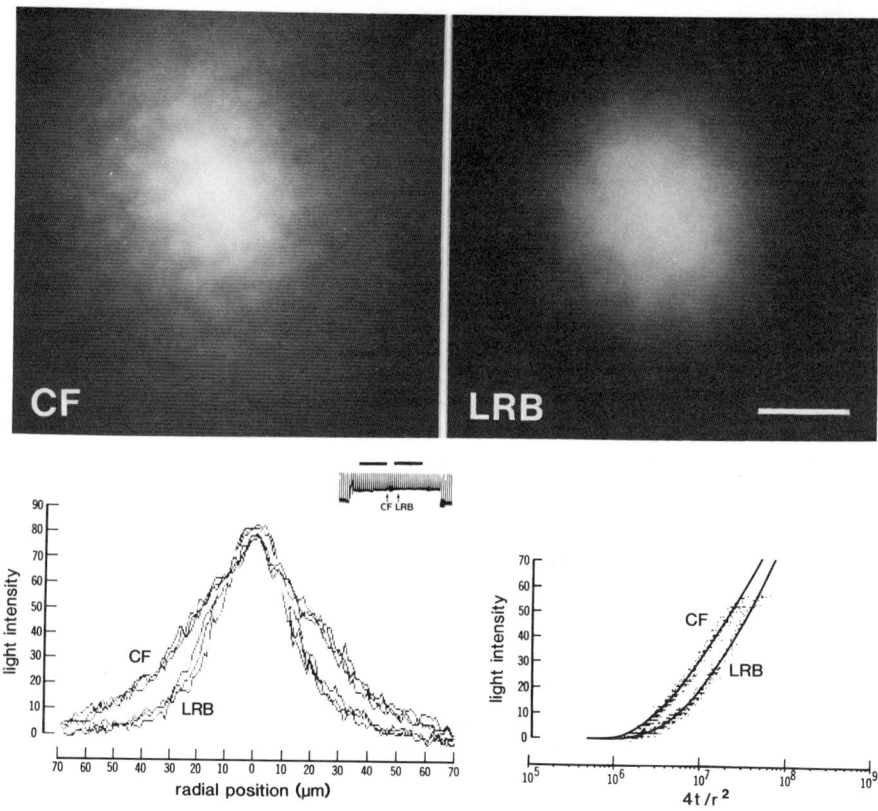

Figure 3
Relative rates of diffusion of two dyes coinjected into the epidermis. Two fluorescent tracers (CF and lissamine rhodamine B [LRB]) were coinjected into the epidermis during exposure to 20-hydroxyecdysone, and the ratio in their rates of diffusion was determined. The record of membrane potential in the source cell during dye coinjection is seen at the top. The arrows mark the time when the two micrographs were taken, the two bars the period during which the light-intensity profiles to the spread of the two dyes are plotted (*bottom left*). The curves fitted to the data (*bottom right*) give the D_e for LRB as 2.43×10^{-7} cm²/sec and that for CF as 4.63×10^{-7} cm²/sec for this dye spread. Scale bar for the micrographs, 20 μm.

mained constant. No disproportionate increase in the movement of the larger tracer was detected.

Although this result is most readily interpreted in terms of an all-or-none model for channel opening, it is difficult to predict how great a change in the diffusion ratio might be expected by a graded mechanism in response to a 40% increase in junctional permeability. The two dyes are rather similar in size and their relative rates of passage may not change in response to graded opening. Other data support the all-or-none mechanism, however. First, when junctional permeability is reduced by chlorpromazine, a calmodulin antagonist (Lees-Miller and Caveney 1982), the rates of diffusion of the two organic tracers

Table 1
Increased Rates of Tracer Movement after the Epidermis Is Exposed to 20-Hydroxyecdysone

Molecule	Effective diffusion coefficient (D_e, $\times 10^{-7}$ cm^2 s^{-1})	
	control (n = 15)	hormone-treated (n = 14)
Carboxyfluorescein	3.72 ± 0.95[a]	4.85 ± 1.00
Lissamine rhodamine B	1.50 ± 0.37	2.00 ± 0.48
CF/LRB ratio	2.54 ± 0.68	2.52 ± 0.61

[a]Mean ± S.D.

drop over a two- to threefold range in a proportional fashion (R. Safranyos, unpubl.). Second, the relationship between ionic and dye coupling in the epidermis remains constant over a threefold range during normal epidermal development, as discussed below.

Channel Gating Appears to Be All-or-None

The mechanism by which individual channels open and close may also be inferred from the precise relationship between ionic and dye coupling in the epidermis. For this reason, directly after several dye spreads had been recorded, ionic coupling was measured in the same region of the segment. This was repeated on epidermis from segments of a wide range of sizes. For example, Figure 4 (left) shows the curves of electrotonic decay for three of these preparations; the resistance to ion movement in the epidermis can be obtained from their slopes (Caveney and Blennerhassett 1980). As the segment grows in size, the intercellular resistance drops.

The relationship between these two indices of junctional permeability is shown in Figure 4 (right). The effective diffusion coefficient for CF is inversely proportional to the effective intercellular resistance. For example, a twofold drop in D_e is matched by a twofold increase in r_i (i.e., a "slope" of -1). This log-scale relationship suggests that the contribution of the junctional resistance to the total pathway resistance is similar in both cases. Previous work showed that for large segments the cytoplasmic resistance to inorganic ions represents about 15% of the total resistance with the remaining 85% of the resistance due to the junctions (Caveney and Blennerhassett 1980). Taking the same ratio in the resistance to dye movement in the epidermis, then for a D_e of 2×10^{-7} to 3×10^{-7} cm^2 s^{-1}, the corresponding cytoplasmic diffusion coefficient is 1×10^{-6} to 2×10^{-6} cm^2 s^{-1} (about right for a molecule the size of CF in cytoplasm; Mastro and Keith 1984). In smaller segments the contribution of the junctions to total pathway resistance is greater, and provided the properties of the cytoplasm remain constant, a slope of -1 means that the rate of ion and tracer movement in the tissue is affected equally. A slope of less than -1 might imply that the cytoplasm provides a greater fraction of the resistance to dye than to ion movement, whereas a slope of greater than -1 might mean the opposite, or that the junctions offer proportionally greater resistance to dye than to ion movement. That is, at low levels of junctional communication, the passage of dye would be more severely retarded,

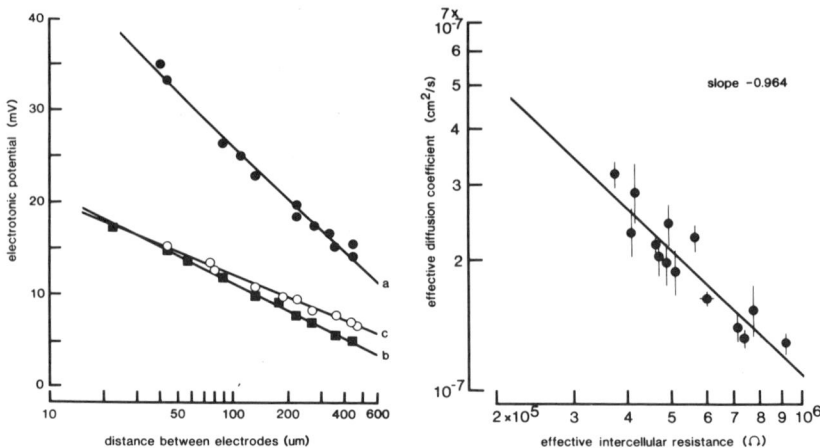

Figure 4
Relationship between ionic and dye coupling in the midinstar epidermis. After a series of CF injections were done on a preparation and video-recorded, ionic coupling in the same area of cells was determined by current injection (60 nA). Sample electrotonic decay curves, in this case from the three preparations analyzed in Fig. 2, are shown on the left. The slope of these curves is directly proportional to the cell–cell resistance to ion movement. The curves fitted to the three sets of data give the intercellular resistances and space constants for the three preparations as: (a) 9.08×10^5 Ω and 1792 μm; (b) 4.42×10^5 Ω and 1300 μm; and (c) 3.79×10^5 Ω and 2473 μm, respectively. On the right, the results from 14 segments, each from a different larva, are given. The effective diffusion coefficient for CF drops off linearly with increasing resistance to ion movement (slope = -0.964).

possibly due to a mechanism involving graded channel closure. Our findings suggest, however, that as the segment increases in size more channels are inserted into the junctional membrane, and that they individually open and close in an all-or-none manner.

These results also support our interpretation of the junctional response to hormone stimulation. The diffusion coefficients for CF in both control and hormone-treated epidermis (Table 1) are higher than those measured immediately after dissection (Fig. 4, right) and the intercellular resistance values are correspondingly lower (R. Safranyos, unpubl.). Preliminary work suggests that these two sets of data fall on the upper end of the slope drawn through the points in Figure 4 (right).

CONCLUSIONS

The insect segment is a developmental compartment with complex spatial and temporal patterns in junctional communication during its growth. Although the epidermal cells appear to have only a single population of junctional channels, the number of channels connecting each cell to its neighbors increases as this tissue grows in size, and the fraction of these channels that are open can be altered by a developmental hormone at any stage during segment development, but each channel opens or closes in an all-or-none fashion. We suspect that these changes in junctional communication are important to normal growth and pattern formation in the segment.

REFERENCES

Berdan, R.C. and S. Caveney. 1985. Gap junction ultrastructure in three states of conductance. *Cell Tissue Res.* **239:** 111–122.

Blennerhassett, M.G. and S. Caveney. 1984. Separation of developmental compartments by a cell type with reduced junctional permeability. *Nature* **309:** 361–364.

Caveney, S. and M.G. Blennerhassett. 1980. Elevation of ionic conductance between insect epidermal cells by β-ecdysone *in vitro*. *J. Insect. Physiol.* **26:** 13–25.

Caveney, S., R.C. Berdan, and S. McLean. 1980. Cell-to-cell ionic communication stimulated by 20-hydroxyecdysone occurs in the absence of protein synthesis and gap junction growth. *J. Insect. Physiol.* **26:** 557–567.

Lawrence, P.A. 1981. The cellular basis of segmentation in insects. *Cell* **26:** 3–10.

Lees-Miller, J.P. and S. Caveney. 1982. Drugs that block calmodulin activity inhibit cell-to-cell coupling in the epidermis of *Tenebrio molitor*. *J. Membr. Biol.* **69:** 233–245.

Loewenstein, W.R. 1979. Junctional intercellular communication and the control of growth. *Biochim. Biophys. Acta* **560:** 1–65.

Mastro, A.M. and A.D. Keith. 1984. Diffusion in the aqueous compartment. *J. Cell Biol.* **99:** 180s–187s.

Othmer, H.G. and E. Pate. 1980. Scale-invariance in reaction-diffusion models of spatial pattern formation. *Proc. Natl. Acad. Sci.* **77:** 4180–4184.

Safranyos, R.G.A. and S. Caveney. 1985. Rates of diffusion of fluorescent molecules via cell-to-cell membrane channels in a developing tissue. *J. Cell Biol.* **100:** 736–747.

Warner, A.E. and P.A. Lawrence. 1982. Permeability of gap junctions at the segmental border in insect epidermis. *Cell* **28:** 243–252.

Antibodies to Gap Junction Protein: Probes for Studying Cell Interactions during Development

Anne E. Warner
Department of Anatomy and Embryology
University College London, London WC1E 6BT, England

Cells in early embryos are directly connected to each other by a low electrical resistance pathway, mediated by the intercellular structure, the gap junction. The presence of this pathway was first described in the squid embryo by Potter et al. (1966), who noted that communication between cells was unbiquitous regardless of eventual developmental fate. Since then the presence of gap junctional communication between embryonic cells has been observed in a wide variety of species, both vertebrate and invertebrate. Gap junctions can allow the direct transfer from cell to cell of both small ions and larger molecules up to a molecular weight of about 1000 daltons (e.g., Simpson et al. 1977). The existence of a direct pathway for cell-cell communication between embryonic cells immediately raised the possibility that this pathway might provide a channel for the exchange of information used to direct development (Potter et al. 1966). Since the original observations, many attempts have been made to test this hypothesis, however it has proved extremely difficult to obtain unequivocal evidence for a functional role for communication through the gap junction during development. In this paper I shall briefly review the experimental approaches used to explore the role of gap junctions in embryogenesis and then describe recent experiments using antibodies raised against gap junction protein to block cell-cell communication during development and observe the developmental consequences (Warner et al. 1984). Since this study has been published very recently, it will be briefly reviewed here and interested readers should refer to the original article.

The Time Course of Elimination of Gap Junctions

One way of examining whether gap junctions are likely to play a role during development is to determine when gap junctional communication disappears between groups of cells with different developmental fates. Both electrophysiological and ultrastructural techniques have been used to see whether gap junctions between groups of cells are lost as they become committed to diverge along separate developmental pathways. In several systems gap junctional communication is lost at about the time of commitment, but the relationship between the loss of the direct cell–cell communication pathway and cell commitment is not sufficiently close to be compelling. In the amphibian embryo electrical coupling is lost between the developing neural tube and the lateral ectoderm when the neural tube closes (Warner 1973), although both anatomical and physiological commitment of neural plate cells probably takes place earlier, during the midneural fold stages (Jacobson 1964; Messenger and Warner 1979). Gap junctions disappear in the developing retina between central retinal cells, close to the time when the polarity of the retina is established (Dixon and Cronley-Dillon 1972, 1974). Apical ridge ectoderm cells in the human limb bud possess gap junctions during induction of underlying mesenchyme to form the various elements of the limb bud; as the apical ridge disappears so do the gap junctions (Kelley and Fallon 1976). In the mouse embryo, dye transfer between inner cell mass and trophectoderm fails close to the time of implantation (Lo and Gilula 1979). A particularly interesting sequence of changes in gap junctional communication occurs during the early development of myotomal muscle in the early *Xenopus laevis* embryo. These are summarized in Figure 1. Cells in the unsegmented mesoderm, which will give rise to the myotomes, are electrically coupled to each other before segmentation. When a new myotome is to be formed, electrical coupling disappears within the unsegmented mesoderm at the site where a new border is to be formed (Blackshaw and Warner 1976a). Once rotation of the myotomal cells is complete, the newly formed myotome again establishes gap junctional communication with the previously formed myotomes. The appearance of electrical communication between formed myotomal cells may be a particular feature of the *X. laevis* embryo because innervation of myotomal muscle begins very shortly after the neural tube closes (Blackshaw and Warner 1976b). In the axolotl embryo, where the somite contains both dermatomal and myotomal elements, electrical coupling between somites is not immediately established after segmentation, although cells are coupled to each other within each somite (Blackshaw and Warner 1976a). However, gap junctions appear both within and between myotomes close to the time of innervation (Keeter et al. 1975). Thus it appears that in both amphibian species gap junctions and electrical coupling are present between developing muscle cells during the early stages of innervation, despite the rather different mechanisms used to form the myotomes. The functional significance of cell–cell communication during innervation is not known. Coupling between striated muscle cells is not a permanent feature since it is well known that adult striated muscle is one of the few adult organs where cells are well insulated from each other. When does gap junctional communication between developing muscle cells disappear? In myotomal muscle cells of *X. laevis* the disappearance of cell–cell commu-

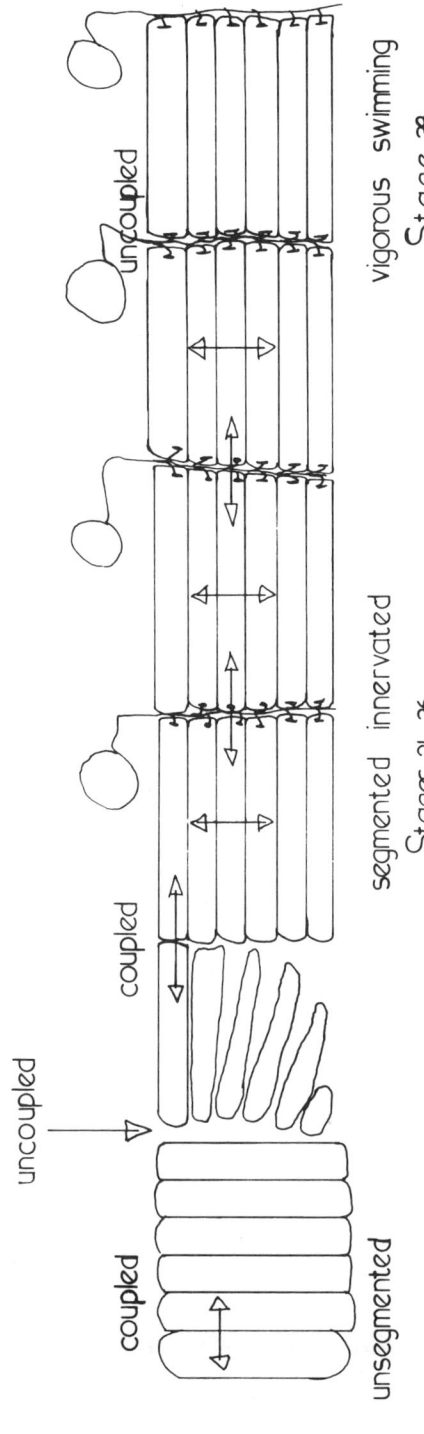

Figure 1
Diagram illustrating the changes in electrical coupling that accompany the formation of myotomes in *X. laevis*. (*Left to right*): Cells of the unsegmented mesoderm (which contains myotomal cells only in *Xenopus*) are electrically coupled, indicated by the linked arrows. Electrical coupling is lost (arrow) at the site of formation of each new myotome. The newly formed myotome immediately makes gap junctions with the adjacent myotome (horizontal linked arrows) and each myotomal cell is electrically coupled to its neighbors in the same myotome (vertical linked arrows). Innervation begins at stages 20–21, and up to stage 36 the myotomal cells are functionally innervated, electrically excitable, contractile, and linked by gap junctions (indicated by horizontal and vertical linked arrows). The right-hand side of the diagram illustrates the situation at stage 38, by which time electrical coupling has been lost between the myotomal cells, triggered by extensive activation of the acetylcholine receptor during the vigorous and sustained swimming which begins at about stage 36.

nication seems to be related to activation of the acetylcholine receptor (AChR) (Armstrong et al. 1983). Tadpoles first begin sustained swimming at about Nieuwkoop & Faber stage 36, and electrical coupling between myotomal cells is gradually lost from this stage and has almost completely disappeared by stage 38. If electrical activity is blocked by treatment with the anesthetic Tricaine, then electrical coupling is retained. Blocking the AChR directly with bungarotoxin is equally effective, suggesting that some event subsequent to the activation of the AChR brings about the elimination of gap junctions. It is not a simple consequence of mechanical stress due to muscle contraction, since tadpoles that lack excitation-contraction coupling, and therefore do not contract when electrically stimulated, nevertheless lose electrical coupling between myotomal cells with the same time course as normal tadpoles (Armstrong et al. 1983).

These various experiments are consistent with a functional role for gap junctional communication during development, but they cannot distinguish between a causal and coincidental role.

The Relation between Gap Junctional Communication and Position within a Developmental Field

In insects such as *Rhodnius* and *Oncopeltus*, studies on the pattern of the cuticular structures have suggested that there is a gradient of positional information running from one segmental boundary to the next, which is repeated in each segment (e.g., Locke 1959). The segment boundary is also a compartment border, as defined by lineage studies (Lawrence 1973). If gap junctions are involved in the control of pattern, one might expect gap junctions to be absent, or have different properties, at the segment boundary. Structurally there seems to be no difference between the size or number of gap junctions linking cells that lie on either side of the segment boundary and those between cells that lie in the same segment (Lawrence and Green 1975) and there is no difference in the degree of electrical coupling within and between segments (Warner and Lawrence 1973). However gap junctions can allow the transfer from cell to cell of larger molecules, and the degree of discrimination against molecules of varying size might be sufficient to define a developmental boundary. Accordingly Warner and Lawrence (1982) tested the transfer of the fluorescent dye Lucifer Yellow (m.w. = 450) in *Oncopeltus* epidermis between cells lying in the same segment and those lying on either side of the segment boundary. They found that Lucifer Yellow spread rapidly and uniformly from cell to cell when injected some distance away from the segmental boundary (Fig. 2). However when injected very close to the boundary (within two cells) dye spread stopped at the segment border in the majority of cases (Fig. 3), although occasionally transfer of Lucifer Yellow was observed between segments. Following these studies, Blennerhassett and Caveney (1984) further examined dye transfer in *Oncopeltus* epidermis and also found both restricted dye transfer at the segment border, which they attributed to a single row of cells at the border with decreased coupling ability and some plasticity in the degree of restriction of dye transfer. This variability in the degree of restriction at a known segmental border makes it difficult to conclude unequivocally that

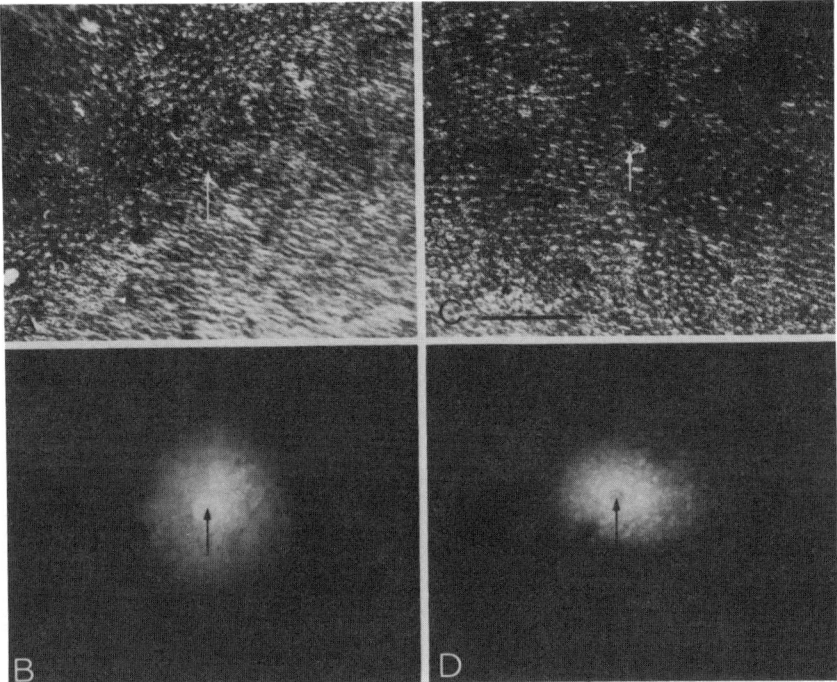

Figure 2
Transfer of Lucifer Yellow between epidermal cells lying in the center of the segment of *Oncopeltus*. Two preparations. (*Upper pair, A* and *C*): Bright-field photographs of the epidermal sheet. (*Lower pair, B* and *D*): Fluorescence photographs to show spread of Lucifer Yellow. Note even spread of dye to neighbors of the injected cell (indicated by arrow).

a difference in the properties of gap junctions underlies the separation of the segments into developmentally autonomous units; nevertheless they show that such a role is plausible.

A further conclusion that can be drawn from such experiments is that position within a developing field must be taken into account when assessing the properties of gap junctions in embryonic systems. In most embryos the dividing line between one developmental field and another cannot be recognized, because there are no clear structural markers. However gap junction permeability may still be related to position and, therefore, developmental fate. Recent experiments by Guthrie (1984), who examined the transfer of Lucifer Yellow between positionally identified cells in the early *X. laevis* embryo, have shown such a correlation. At the 32-cell stage, cells in the animal pole, destined to form dorsal structures, transfer the dye Lucifer Yellow frequently, whereas those in the future ventral region rarely transfer dye. Dye transfer between cells, even in the dorsal region, becomes poor once the embryo moves on to the 64- to 128-cell stage, suggesting that both age and position enter into determining the ability of embryonic cells to exchange molecules

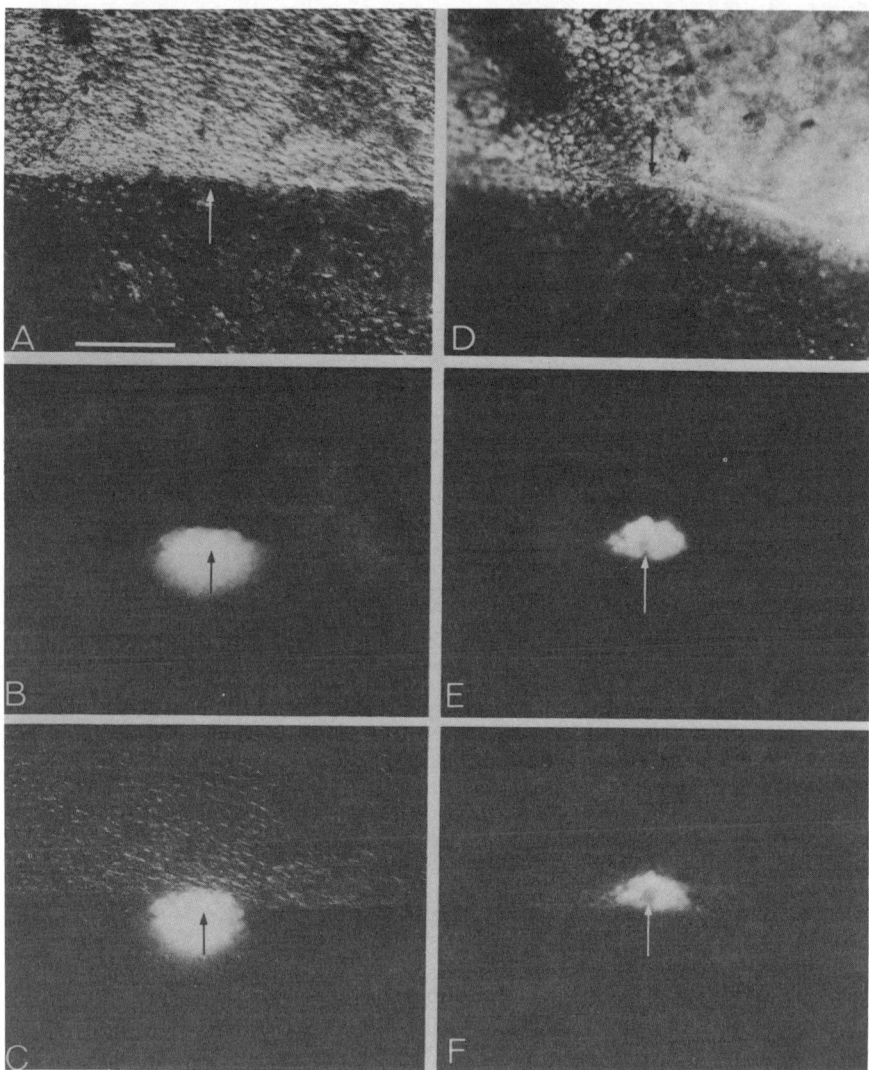

Figure 3
Failure of Lucifer Yellow transfer across the segmental border in two preparations of *Oncopeltus* epidermis. (*Left-hand panels, A, B,* and *C*): Injections made posterior to the segment border. (*Right-hand panels, D, E,* and *F*): Injections made anterior to the segment border. (*Top to bottom*): First pair of photographs *(A,D)* show the bright-field view. Segment border marked by the abrupt change in pigmentation. At the edges of the microscope field the preparation goes out of focus because of buckling of the sheet. The second pair *(B,E)* show the fluorescent images and the bottom pair *(C,F)* simultaneous bright-field and fluorescence photographs. Note asymmetric spread of Lucifer Yellow away from the injected cell (indicated by arrows), with dye failing to cross the segment border. (Reprinted, with permission, from Warner and Lawrence 1982.)

larger than small ions. Such findings might explain some of the different observations in the literature showing both failure of dye transfer between early embryonic cells (e.g., Slack and Palmer 1969; Bennett et al. 1972) and the ability to transfer small dyes (Bennett et al. 1978).

The Consequences for Development of Blocking Cell–Cell Communication

The ideal way to determine the role of gap junctional communication during development would be to block cell–cell communication and then observe the developmental consequences. Until recently only three reagents were known to reduce the permeability of gap junctions: a rise in intracellular free calcium (Rose and Loewenstein 1976), a fall in intracellular pH (Turin and Warner 1977, 1980; Spray et al. 1981), and octanol and heptanol (Johnston et al. 1980). Unfortunately all these reagents perturb cell metabolism also, making it difficult to assign any devlopmental effect they might have to a block of cell–cell communication.

Progress in this field has been made recently with the production of antibodies against the gap junction protein, which can be used to examine the functional role of gap junctions (Warner et al. 1984). Rabbit polyclonal antibodies were raised against the major, 27-kD protein electrophoretically eluted from isolated rat liver gap junctions (prepared according to the procedure of Hertzberg 1984) and twice affinity-purified against the 27-kD protein. The antibodies bind specifically to the 27-kD protein on immunoblots and also recognize a 47-kD protein and a 54-kD protein in homogenates of rat liver and early *X. laevis* embryos. Absorption against the 27-kD protein removes binding to both the 47-kD and 54-kD bands. The 47-kD protein probably represents a "dimeric" form. The precise relation between the 27-kD and 54-kD proteins has yet to be determined, but there are a number of possibilities. The 54-kD protein could be another aggregated form of the 27-kD protein (H. Evans, pers. comm.), a biosynthetic precursor of the gap junction (see Epstein et al. 1977), or it may represent the undegraded form of the gap junction protein, as elimination of endogenous protease activity when preparing plasma membranes is notoriously difficult (see D. Paul, this volume). The affinity-purified antibodies bind to the cytoplasmic surface of intact rat liver gap junctions, as shown by electron microscope immunolocalization both on isolated gap junctions and on intact rat liver. It has not been possible so far to demonstrate immunolocalization in *X. laevis* embryos, probably because the gap junctions are very small and occupy a small fraction of the cytoplasmic membrane (C. Peracchia, pers. comm.).

To test whether either of the two affinity-purified antibody preparations was able to block cell–cell communication, the antibodies were injected into *X. laevis* embryos at the eight-cell stage of development, into the dorsal blastomere destined to give rise to head ectoderm and mesoderm on the right-hand side of the tadpole (see Fig. 4). The following control reagents were injected: two buffers in which the antibodies were suspended, the preimmune sera from the rabbits used to raise the gap junction antibodies, and an antibody raised against an extracellular matrix glycoprotein.

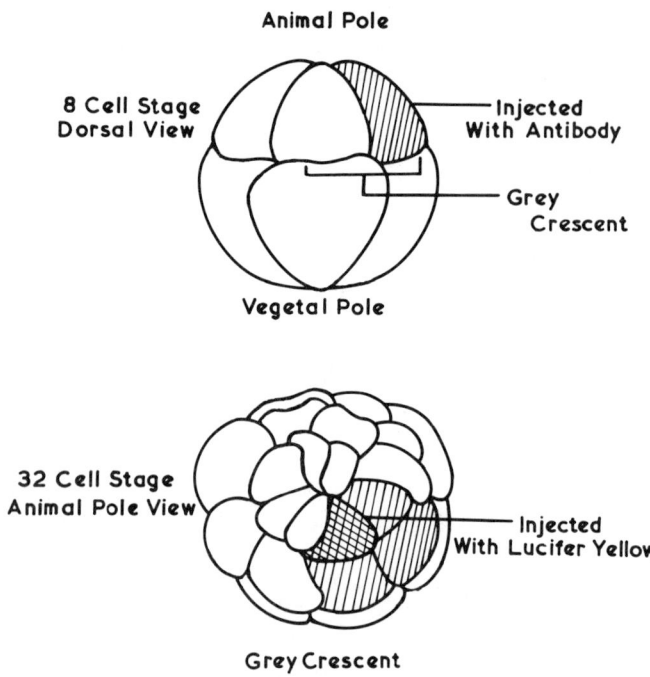

Figure 4
Diagram illustrating the position of injections of gap junction antibody or control reagents (*upper drawing*) at the eight-cell stage. The lower drawing indicates the position of the dorsal animal pole cell used to test for transfer of Lucifer Yellow in control and antibody-injected embryos. (Reprinted, with permission, from Warner et al. 1984.)

This extracellular matrix glycoprotein is normally secreted by rat liver and *X. laevis* embryos. It has a molecular weight of about 54 kD and is the β-chain of fibrinogen (Eastman and N.B. Gilula, in prep.). Although both β-fibrinogen and the largest species recognized by both gap junction antibody preparations are very similar in size (54 kD), the 54-kD protein recognized by the gap junction antibodies cannot be β-fibrinogen because there is no cross-reactivity between the β-fibrinogen antibody and any of the species recognized by the gap junction antibodies. Conversely the gap junction antibodies do not recognize β-fibrinogen. The 54-kD protein recognized by the gap junction antibodies cannot, therefore, be a β-fibrinogen contaminant.

The embryos were left to complete two rounds of cell division after microinjection with one of the reagents and then tested for transfer of Lucifer Yellow out of the progeny of the antibody-injected cell, a dorsal blastomere that normally transfers dye frequently to its lateral neighbors (see Guthrie 1984 and Fig. 4). After prior injection of all control reagents, this cell transferred dye to its lateral neighbors 70–80% of the time, closely similar to normal, uninjected embryos (cf. Guthrie 1984 and Warner et al. 1984). However in embryos that had previously been injected with either gap junction antibody preparation, the incidence of dye transfer fell to 18% (antibody A) and 26% (antibody B). This reduction in ability to transfer Lucifer Yellow is significantly different from controls (Chi-square test, $p \leq 0.005$). The presence or absence of electrical coupling was also tested in embryos previously scored for failure to transfer dye. For these experiments electrical coupling between lateral neighbors in the antibody-free region of the embryo served as a control. Eighty percent of embryos that failed to transfer dye also showed complete block of electrical coupling in the antibody containing region, showing that block of cell–cell communication was complete; in the remaining 20%, electrical coupling was just detectable. Good electrical communication was found in the antibody-free region of these embryos. Therefore, both gap junction antibody preparations are able to prevent the transfer from cell to cell of both small ions and larger molecules. To confirm that gap junction antibodies were inhibiting transfer through already formed junctions, and to obtain some indication of the speed with which the block becomes effective, on some occasions the antibodies were injected at the 32-cell stage, and followed immediately by assay for the ability to transfer Lucifer Yellow. Failure of dye transfer was observed in 83% of the embryos tested within 5 minutes of injection of gap junction antibodies. Thus, block is occurring at previously formed gap junctions. The speed of the block makes it unlikely that the observed effects of injecting gap junction antibodies is the consequence of some other action of the antibody preparations, unrelated to binding to gap junctions.

Cells that had been injected with the gap junction antibodies continued to cleave at a rate indistinguishable from their uninjected neighbors. They also maintained normal resting potentials. This implies that it is unlikely that any fall in intracellular pH follows antibody injection, since a fall in intracellular pH sufficient to block gap junctions inhibits cell division (Turin and Warner 1977, 1980). A rise in intracellular free calcium is also unlikely because direct injection of calcium does not block cell–cell communication in *X. laevis* embryos (Rink et al. 1980).

To determine whether the block of cell–cell communication achieved by injection of gap junction antibodies was the consequence of cross-linking of multivalent IgG on the cytoplasmic surface of the junction, monovalent fragments prepared from the affinity-purified antibody preparations were also tested for an effect on cell–cell communication. Nonimmune Fab fragments served as controls. With the nonimmune fragments, dye transfer failed on 20% of occasions tested, whereas with the immune fragments, dye transfer failed on 52% of occasions tested. The incidence of failure to transfer is significantly different after injection of the immune fragments ($p = 0.03$), suggesting that disruption of gap junction channel permeability arises from a direct interaction of antibody with the channel protein.

Although both gap junction antibodies and monovalent immune fragments greatly reduce the incidence of cell–cell communication, complete block was not observed on 100% of occasions. Why should there be such variability? One reason for the occasional failure of the antibody preparations to block gap junctional communication is almost certainly related to the presence of a store of gap junction protein, or a precursor molecule. Immunoblots carried out on fertilized or unfertilized eggs revealed the presence of a substantial amount of antigenic protein, of the same size as found in rat liver homogenates, which was recognized by both antibodies. One feature of the early *X. laevis* embryo is its ability to go through many rounds of cell division in the absence of substantial protein or RNA synthesis. Zygotic RNA and protein synthesis does not begin until the "midblastula transition," recently described by Newport and Kirschner (1982). All material necessary for progression through development to the midblastula stage must, therefore, be available in the unfertilized egg. That stores of products required for future use, which would include the production of gap junctions, should be present in the egg is, therefore, not surprising. For example fibronectin has been found in the fertilized egg (Lee et al. 1984) and large quantities of sodium are sequestered in the egg for subsequent transfer to the blastocoel fluid (Slack and Warner 1973).

The progeny of the cell injected with Lucifer Yellow to test for dye transfer after antibody injection were clearly visible in the living tadpole for at least 3 days. If the injected cell was deliberately damaged by overirradiation after injection of Lucifer Yellow, cleavage gradually failed and was followed by cytolysis of the progeny of the injected cell; these effects were never seen after injection of either gap junction antibody preparation. How do such embryos develop? Embryos injected with gap junction antibodies generated a high proportion of tadpoles with defects that were related to the developmental fate of the dorsal blastomere injected at the eight-cell stage. The most frequently occurring defect in the antibody-injected tadpoles was varying degrees of right/left asymmetry (63%). In severe cases the tadpoles failed to form eyes and trigeminal ganglia on the right-hand side, along with some mesodermal defect in the notochord and somites and underdevelopment of the brain on the affected side. Figure 5 shows sections through a control and antibody-injected tadpole. In milder cases the notochord was shifted from its central position to the left, uninjected side and the brain and eyes on the right were then smaller than on the left and shifted in position along the rostro-caudal axis. In a few cases (19%), the brain and eyes failed to form altogether; this may be related to the extensive mixing that takes place across the midline in the developing nervous system (Hirose and Jacobson 1979; A.E. Warner, unpubl.). The level of spontaneous abnormalities observed in normal, uninjected tadpoles was low; such abnormalities were highly variable and rarely asymmetric. Embryos injected with all reagents other than gap junction antibodies generated abnormalities of the same kind and equivalent incidence to those seen in uninjected controls.

Thus, blocking cell-cell communication has serious consequences for embryonic development. At present it is difficult to assign a precise role for gap junctional communication, since we do not yet know how long the antibody remains effective or how the embryo responds to a large body of communication-incompetent cells. One possible explanation for the effects observed in the experiments described here is that we are interfering with neural induction, since the antibody is present both in cells destined to induce nervous tissue (mesoderm) and cells destined to respond to the inducing signal (overlying ectoderm).

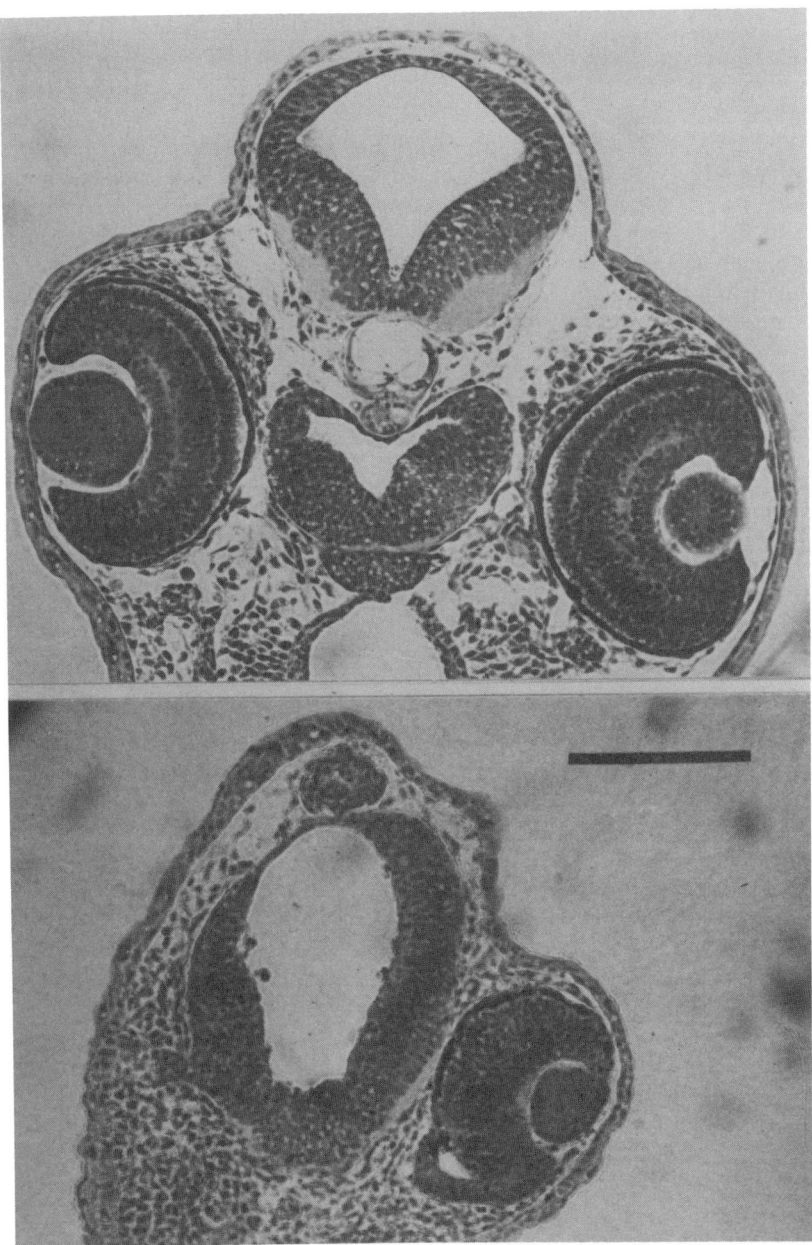

Figure 5
Sections taken through a control embryo (*upper photograph*) and an embryo injected with gap junction antibody at the eight-cell stage (*lower photograph*) at the level of the eyes. Embryos fixed at stage 36. Note absence of the notochord and the eye on the right-hand side of the antibody-injected embryo and the asymmetric development of the brain. Bar, 500 μm. (Reprinted, with permission, from Warner et al. 1984.)

SUMMARY

Since gap junctional communication was first recognized to be a characteristic feature of early embryonic development, it has proved remarkably difficult to determine its developmental role. The changes in direct communication that take place as development proceeds are compatible with the view that developmentally important information is transferred from cell to cell through gap junctions. The position of a cell within a developing system can be reflected in the degree to which it communicates directly with its neighbors, implying that gap junctions may be involved in pattern formation. The first direct evidence that gap junctional communication plays a part in spatial and temporal patterning during development comes from experiments that use antibodies to the major, 27-kD protein extracted from gap junctions to block cell–cell communication. Experiments with these reagents have shown that pronounced patterning defects develop within the region containing antibodies to gap junctions. These experiments provide direct evidence that the 27-kD protein is an integral component of the functional channel for cell–cell communication and make it now possible to identify molecular determinants that control the functional properties of gap junctions. Furthermore, by analyzing in detail the developmental consequences of blocking cell–cell communication during embryogenesis, it should be possible to determine both when and where communication through gap junctions is essential.

ACKNOWLEDGMENTS

This work was supported by the Medical Research Council and the Wellcome Trust. I thank Sarah Guthrie for comments on the manuscript.

REFERENCES

Armstrong, D., L. Turin, and A.E. Warner. 1983. Muscle activity and the loss of electrical coupling between striated muscle cells in *Xenopus laevis* embryos. *J. Neurosci.* **3:** 1414–1421.

Bennett, M.V.L., M.E. Spira, and G. Pappas. 1972. Properties of electrotonic junctions between embryonic cells of *Fundulus. Dev. Biol.* **29:** 419–435.

Bennett, M.V.L., M.E. Spira, and D.C. Spray. 1978. Permeability of gap junctions between embryonic cells of *Fundulus:* A re-evaluation. *Dev. Biol.* **65:** 114–125.

Blackshaw, S.E. and A.E. Warner. 1976a. Low resistance junctions between mesoderm cells during formation of trunk muscles. *J. Physiol.* **255:** 209–230.

Blackshaw, S.E. and A.E. Warner. 1976b. Onset of acetylcholine sensitivity and endplate activity in developing myotome muscles of *Xenopus laevis. Nature* **262:** 217–218.

Blennerhassett, M. and S. Caveney. 1984. Separation of developmental compartments by a cell type with reduced junctional permeability. *Nature* **309:** 361–364.

Dixon, J.S. and J.R. Cronley-Dillon. 1972. The fine structure of the developing retina in *Xenopus laevis. J. Embryol. Exp. Morphol.* **28:** 659–666.

———. 1974. Intercellular gap junctions in pigment epithelium during retinal specification in *Xenopus laevis. Nature* **251:** 505.

Epstein, M.L., J.D. Sheridan, and R.G. Johnson. 1977. Formation of low resistance junctions *in vitro* in the absence of protein synthesis and ATP production. *Exp. Cell Res.* **104:** 25–30.

Guthrie, S.G. 1984. Patterns of junctional communication in the early amphibian embryo. *Nature* **311:** 149–151.
Hertzberg, E.L. 1984. A detergent independent procedure for the isolation of gap junctions from rat liver. *J. Biol. Chem.* **259:** 9936–9943.
Hirose, G. and M. Jacobson. 1979. Clonal organization of the central nervous system of the frog. I. Clones stemming from individual blastomeres of the 16-cell and earlier stages. *Dev. Biol.* **71:** 191–210.
Jacobson, C.-O. 1964. Motor nuclei, cranial nerve roots and fibre pattern in the medulla oblongata after reversal experiments on the neural plate of axolotl larvae. *Zool. Bidr. Upps.* **36:** 73–160.
Johnston, M.F., S.A. Simon, and F. Ramon. 1980. Interaction of anaesthetics with electrical synapses. *Nature* **286:** 498–500.
Keeter, J., G. Pappas, and P. Model. 1975. Inter- and intra-myotomal gap junctions in the *Axolotl* embryo. *Dev. Biol.* **45:** 21–34.
Kelley, R.O. and J. Fallon. 1976. Ultrastructural analysis of the apical ectoderm ridge during vertebrate limb morphogenesis. *Dev. Biol.* **51:** 241–256.
Lawrence, P.A. 1973. A clonal analysis of segment development in *Oncopeltus*. *J. Embryol. Exp. Morphol.* **30:** 681–699.
Lawrence, P.A. and S. Green. 1975. The anatomy of a compartment border: The intersegmental boundary in *Oncopeltus*. *J. Cell Biol.* **65:** 373–382.
Lee, G., R. Hynes, and M. Kirschner. 1984. Temporal and spatial regulation of fibronectin in early *Xenopus* development. *Cell* **36:** 729–740.
Lo, C.W. and N.B. Gilula. 1979. Gap junctional communication in the post-implantation mouse embryo. *Cell* **18:** 411–422.
Locke, M. 1959. The cuticular pattern in an insect *Rhodnius prolixus*. *J. Exp. Biol.* **36:** 459–477.
Messenger, E.A. and A.E. Warner.1979. The function of the sodium pump during differentiation of amphibian embryonic neurons. *J. Physiol.* **292:** 85–105.
Newport, J. and M. Kirschner. 1982. A major developmental transition in early *Xenopus* embryos: 1. Characterization and timing of cellular changes at the mid-blastula stage. *Cell* **30:** 675–686.
Potter, D.D., E.J. Furshpan, and E.S. Lennox. 1966. Connections between cells of the developing squid as revealed by electrophysiological methods. *Proc. Natl. Acad. Sci.* **55:** 328–336.
Rink, T.J., R.Y. Tsien, and A.E. Warner. 1980. Free calcium in *Xenopus* embryos measured with ion-sensitive micro-electrodes. *Nature* **283:** 658–660.
Rose, B. and W.R. Loewenstein. 1976. Permeability of a cell junction and the local cytoplasmic free ionized calcium concentration: A study with aequorin. *J. Membr. Biol.* **28:** 87–119.
Simpson, I., B. Rose, and W.R. Loewenstein. 1977. Size limit of molecules permeating the junctional membrane channels. *Science* **195:** 294–296.
Slack, C. and J.F. Palmer. 1979. The permeability of intercellular junctions in early embryos of *Xenopus laevis* studied with a fluorescent tracer. *Exp. Cell Res.* **55:** 416–419.
Slack, C. and A.E. Warner. 1973. Intracellular and intercellular potentials in the early amphibian embryo. *J. Physiol.* **232:** 313–330.
Spray, D.C., A.L. Harris, and M.V.L. Bennett. 1981. Gap junctional conductance is a simple and sensitive function of intracellular pH. *Science* **211:** 712–715.
Turin, L. and A.E. Warner. 1977. Carbon dioxide reversibly abolishes ionic communication between cells of the early amphibian embryo. *Nature* **270:** 56–57.
―――. 1980. Intracellular pH in *Xenopus* embryos: Its effect on current flow between blastomeres. *J. Physiol.* **300:** 489–504.
Warner. A.E. 1973. The electrical properties of the ectoderm during induction and early development of the nervous system. *J. Physiol.* **235:** 267–286.

Warner, A.E. and P.A. Lawrence. 1973. Electrical coupling across developmental boundaries in insect epidermis. *Nature* **245:** 47-49.

———. 1982. Permeability of gap junctions at the segmental border in insect epidermis. *Cell* **28:** 243-252.

Warner, A.E., S.C. Guthrie, and N.B. Gilula. 1984. Antibodies to gap junctional protein selectively disrupt junctional communication in the early amphibian embryo. *Nature* **311:** 127-131.

Relating the Population Dynamics of Gap Junctions to Cellular Function

W.J. Larsen
*Department of Anatomy and Cell Biology
University of Cincinnati College of Medicine
Cincinnati, Ohio 45267-0521*

As biochemical and physiological probes of gap junction composition and function improve in resolution, the identity of gap junctions in different tissues (Gros et al. 1983; Nicholson et al. 1983; Dermietzel et al. 1984; Gorin et al. 1984; Hertzberg 1984; Hertzberg and Skibbens 1984) and phyla (Epstein and Gilula 1977), or even within different regions of a particular tissue (Warner and Lawrence 1982; Blennerhassett and Caveney 1983), is being questioned. In addition, morphological techniques have also raised questions of functional identity based on dramatic diversity in form (Larsen 1977a, 1983; Peracchia 1980) and, as a consequence, it has been suggested that gap junctions in inexcitable cells may represent a family of structures with somewhat diverse but yet unknown specific functions. Several recent studies that utilize newly developed structural and functional probes (Loewenstein et al. 1978; Unwin and Zampighi 1980; Girsch and Peracchia 1983; Safranyos and Caveney 1983; Makowski et al. 1984; Peracchia and Bernardini 1984) have even begun to suggest mechanisms that could allow the independent modulation of gap junction subunits within an individual junction. Overall, this recent evidence points in the direction of diversity of gap junction structure and function on many levels of comparison and it is becoming apparent that the possibilities for structural and functional variability among gap junctions and their subunits are less restricted than recently believed. This possible diversity should not be surprising, however, when one considers the well-established diversity of structure and function of other membrane organelles such as mitochondria, endoplasmic reticulum, and Golgi apparatus.

Recent work in several laboratories, including our own, supports the idea that some of the postulated structural and functional diversity within the gap junction population in individual cells or in specific tissues may arise as a consequence of developmental alterations of junctional subunits or aggregates of subunits as they form, grow, mature, and undergo degradation (Larsen 1983; Larsen and Risinger 1985). Our more recent studies of normal granulosa cells and cultured tumor cells provide additional support for some of our earlier hypotheses related to junctional turnover and have provided new details related to modulatory mechanisms. These studies have also begun to suggest relationships of junctional modulation to cellular activities underlying follicular development, ovulation, meiotic maturation, cell proliferation, and the modulation of physical cell–cell relationships.

DISCUSSION

When Do Gap Junction Subunits Acquire Function during Their Synthesis and Assembly?

We are almost completely in the dark at this point with regard to the trafficking of gap junction proteins within the cell itself. Very little information is presently at hand, for example, regarding synthetic routes, enzymes, or membrane systems that play a role in gap junction formation. Some evidence does support the idea however that the plasma membrane may contain large amounts of morphologically unrecognizable gap junction precursors since gap junctions have been demonstrated to assemble rapidly in the absence of ATP or protein synthesis (Epstein et al. 1977; Tadvalkar and Pinto da Silva 1983) or with low-temperature treatment (Preus et al. 1983). In addition, it seems likely that these precursors are prevented from aggregation by a microtubule-dependent system since treatments that decrease microtubule stability also cause the rapid appearance of small gap junctions in at least one tissue (Tadvalkar and Pinto da Silva 1983).

The appearance in the plasma membrane of disperse aggregates of 10-nm to 11-nm particles in so-called formation plaque regions is generally considered to be the first observable sign of junction formation. Their aggregation into more densely packed primary plaques appears to be an early but subsequent step that occurs while the apposed cell membranes in the area of the formation plaque approach within approximately 20 nm of each other (Johnson et al. 1974). These small primary plaques in granulosa cells usually consist of only about 10–50 particles but may be related to larger and possibly more mature gap junction plaques through the sequence depicted in Figure 1. The small primary plaques initially formed (Fig. 1a) apparently first aggregate loosely (Fib. 1b) and then may pack more tightly as suggested by the similarity in particle numbers in the more densely packed aggregates making up larger plaques, as shown in Figure 1c (see also Larsen and Tung 1978 and Larsen and Risinger 1985). There are obvious questions that arise from these observations. When do the junctional subunits become functional? Are they "functional" as single 11-nm particles, or is function more restricted? Does the 11-nm particle represent a terminal synthetic step as might be suggested from their apparent conversion to 8-nm particles as they form denser aggregates? It is also open to question as to whether or not the observed change in size represents a simple shape change or actual alteration of composition in each subunit.

In addition, recent observations in our laboratory suggest that the compositional or physical nature of the membrane itself may change as the junctional formation plaque matures into large gap junction aggregates. Filipin, a polyene antibiotic that binds 3-β-hydroxy sterols, appears to deform the membrane within gap junction formation plaques of SW-13 cells where 11-nm particles are relatively sparse (Fig. 2a) but not when these particles are densely aggregated as typical gap junction aggregates (Fig. 2b). We have argued that the absence of apparent filipin sterol complexes in typical gap junction aggregates does not necessarily indicate an absence of 3-β-hydroxy sterols (Risinger and Larsen 1983) but instead may indicate possible physical changes in these membranes during junctional development. It seems possible that these changes in subunit size, organization, and packing as well as membrane properties could parallel the functional maturation of the junctional particles and their membrane lipid substrate. Since many studies have attempted to relate the total junctional area between cell pairs to such physiological parameters as coupling coefficients (Sheridan et al. 1978; Meyer et al. 1981) or to the intercellular transfer of labeled tracers, the functional implications of the observed structural modifications will be of significant interest. Many of these studies have traditionally assumed that all observable gap junction particles within the cells under investigation were functionally equivalent.

When Do Gap Junction Subunits Lose Function during Their Removal from the Plasma Membrane?

The functional equivalence of all gap junction plaques within a given cell's membrane can also be questioned by several of our recent observations in rabbit (Larsen et al. 1981) and rat granulosa cells and in rat gastric mucous epithelium. We have now measured several thousand gap junctions in each of these systems and have discovered that their area distributions are highly skewed (Fig. 3a–c). In a survey of 4911 surface mucous epithelial cell gap junctions, for example, we have found that roughly 48% of the total gap junction area is accounted for by a small number of the largest gap junctions (0.1–1.38 μm^2) which comprise only 5% of all gap junction plaques measured. Nearly identical proportions (albeit different size ranges) are apparent in both rat membrana (5% of largest junctions account for 47% of area) and cumulus (5% of largest junctions account for 45% of area) gap junction populations. Conversely, a very large number of smaller plaques (0–0.01 μm^2), or about 53% of all gap junctions measured, account for only 17% of the total gap junction area measured in the gastric mucous cells with similar proportions apparent in the two granulosa cell populations.

These figures raise several questions of functional significance. The same amount of junctional area, for example, broken up into smaller junctions could theoretically provide more potential interactions between a large number of cell pairs than if, for example, that same area of junctional membrane existed

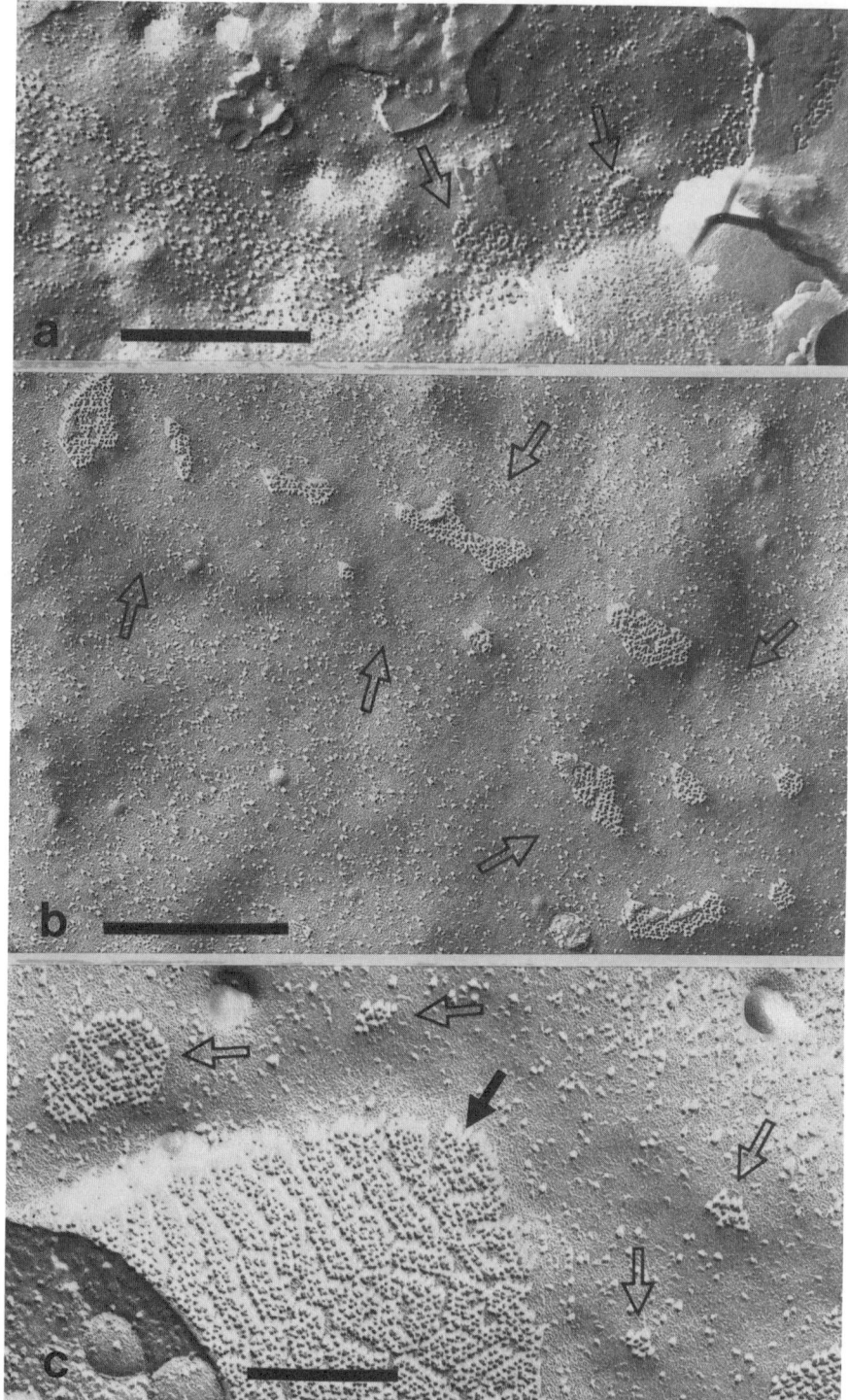

Figure 1 *(See facing page for legend.)*

as a single gap junction plaque. If the smaller junctions were nonfunctional and the larger junctions were functional, one might expect to see a small number of highly interactive cell pairs whereas if the converse held true, one might expect to see a large number of relatively poorly interacting cell pairs. The functional significance of extremely large junctions observed in diverse tissues has been considered previously (Albertini and Anderson 1975; Goodenough 1979; Peracchia 1980; Finbow and Yancey 1981; Larsen 1983). Some junctions in rabbit granulosa cells, for example, may be as large as 40 μm^2, which means that they constitute approximately 20% of the surface of a single cell (Larsen et al. 1981), and we have found that granulosa cells that possess such large gap junctions are usually devoid of smaller junctional aggregates. This suggested to us that smaller plaques may have aggregated to form these excessively large gap junctional caps (Larsen and Tung 1978). More recent quantitative studies of rat cumulus and membrana granulosa cells, respectively, have demonstrated that gap junctions larger than 2.5 μm^2 are lost from these cells within 2.5 hours after an ovulatory stimulus but that before their sudden disappearance we find about one of these junctions per cell.

This calculation and our observations of clathrin-coated blebs on these large gap junctions in rabbit granulosa cells support our earlier hypothesis that these large caps may represent a stage of gap junction removal from the membrane. Additional support for the hypothesis is provided by the observation that although the calculated surface areas of cytoplasmic gap junctional vesicles in different tissues may vary widely, they correspond rather closely to the areas of the large junctional caps in the same tissue (Larsen and Risinger 1985).

Since we have developed a wide range of data supporting the idea that the apparent gap junction vesicles observed in these tissues are internalized for degradation (Larsen and Tung 1978; Larsen et al. 1979, 1981; Murray et al. 1981; Larsen 1983; Risinger and Larsen 1983; Larsen and Risinger 1985), it seems likely that the large junctional caps represent a stage of junctional development just penultimate to destruction. If this is true, it is plausible that these large caps may not represent functional gap junction membrane at all, and it is therefore conceivable that functional gap junction area may be significantly overestimated in cases where a large proportion of gap junction membrane is present in large pre-endocytotic caps.

Figure 1
Formation plaques and developing gap junctions. (a) This formation plaque in a reaggregating SW-13 cell membrane is typical of those observed in this malignant line and in normal granulosa cells. A large band of P-fracture-face 10-nm to 11-nm particles is interspersed with small junctional aggregates (arrows). Scale bar, 0.5 μm. (Reprinted, with permission, from Murray et al. 1981.) (b) A group of primary aggregates in membrana granulosa cell membrane of a developing rabbit follicle. The band of slightly elevated membrane (enclosed by arrows) that includes these aggregates is relatively "particle poor." Scale bar, 0.5 μm. (Reprinted, with permission, from Larsen 1977a.) (c) Small primary aggregates (open arrows) surround a larger junctional plaque in rabbit membrana granulosa cell membrane which is composed of an assemblage of smaller particle plaques (solid arrow) similar in size to those surrounding the junction. Alternatively, we were not able to exclude the possibility that the pattern observed in the larger more mature junctions is the consequence of fixation. Scale bar, 0.25 μm.

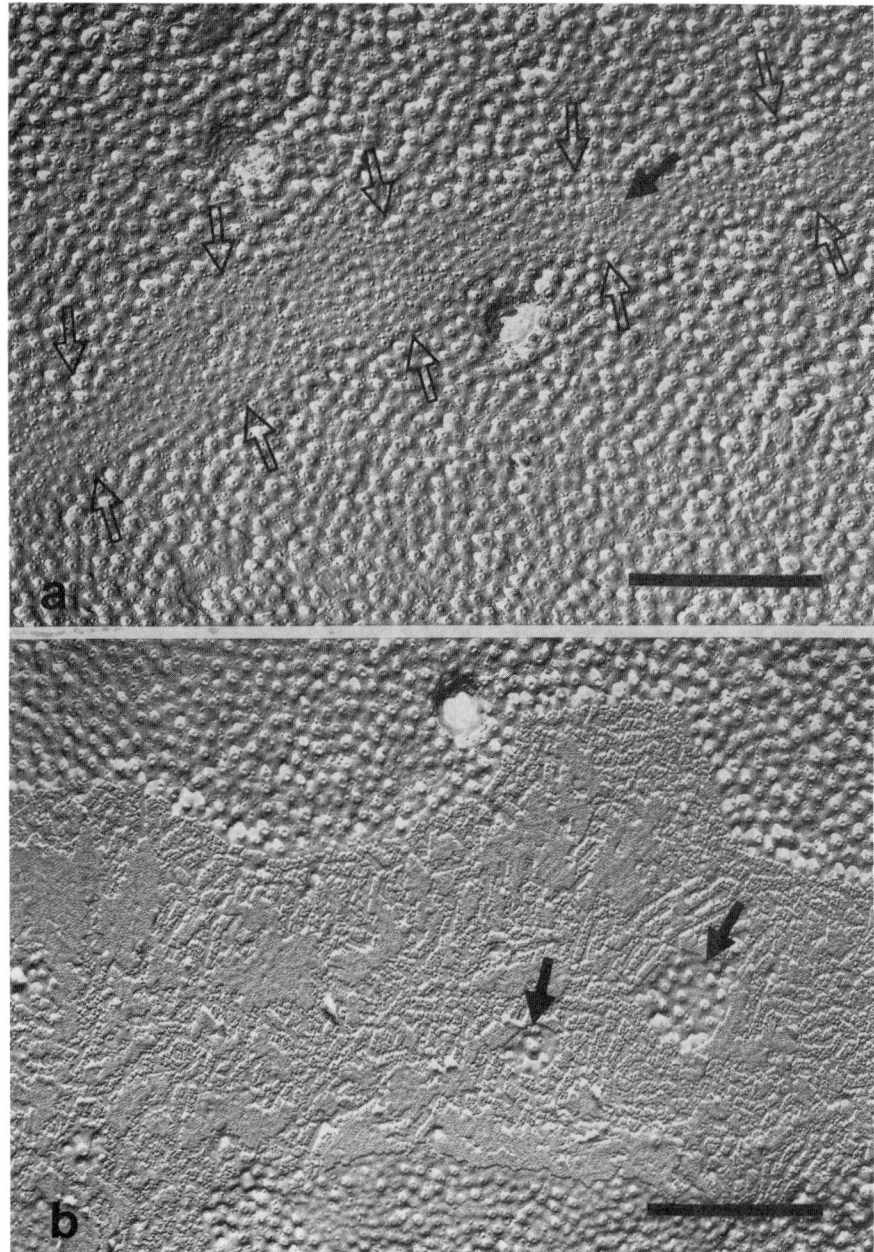

Figure 2 (See facing page for legend.)

To complicate further our earlier discussion related to the functional significance of the smaller junctions in these tissues, we have also made recent observations in the preovulatory cumulus and membrana cell system just mentioned to suggest that the endocytosis of some of the larger junctions may be piecemeal. The consequence of such piecemeal uptake may result in the images depicted in Figure 4, a and b. Indeed, in both cumulus and membrana cell systems, a significant and rapid loss of gap junctions from the surface (as reflected in the significant reduction of gap junction fractional area) is correlated directly with a significant increase in the appearance of smaller gap junctional plaques. In the cumulus cell system particularly, we find that an initial 60% reduction in total gap junction surface area is correlated with an increase in the total number of individual gap junction plaques per cell from about 20 to 50. Similarly, a 25% drop in the fractional area of membrana gap junctions is accompanied by an increase in individual gap junction plaques from about 27 to 58 per cell. These observations support the possibility that some small junctional plaques may represent leftover fragments or terminal surface membrane aggregates rather than newly formed aggregates, and again raise questions related to the functional identity of all junctions within a given population.

Once Utilized, Are Gap Junction Subunits Recycled for Future Use?

The movement of gap junction protein from the surface into the cytoplasm for apparent degradation has been studied in relative detail by our laboratory over the last few years as mentioned. Briefly, we have accumulated broad-based evidence to suggest that gap junction membrane in some cell systems is endocytosed through an actin-based mechanism (Fig. 5a–c) (Larsen et al. 1979). The vesicles (Fig. 5d,e) produced by this process may then fuse with lysosomes, resulting in the degradation of contents of the cytoplasmic gap junction vesicle matrix and elements of the membrane as well (Larsen and Tung 1978; Murray et al. 1981; Larsen 1983; Risinger and Larsen 1983; Larsen and Risinger 1985). One would expect that by the time the acidic environment inferred from acid phosphatase localization studies (Larsen and Tung 1978) is established, and degradation is initiated, that the junctional subunits have lost typical function. This hypothesis seems particularly plausible in view of the finding that low pH uncouples paired cells in several systems (Turin and Warner 1980; Peracchia and Peracchia 1980; Spray et al. 1981, 1982). It also seems unlikely that any functional advantage could result from communication between the vesicle interior and the host cytoplasm at this stage since

Figure 2
(a) Freeze-fracture replica of SW-13 cell membrane treated with filipin. Note the P-face elevations in nonjunctional membrane indicative of filipin-sterol complexes but note also the lesser deformations within the formation plaque area enclosed within the open arrows. A small junctional plaque is denoted by the solid arrow. Scale bar, 0.5 μm. (b) This larger SW-13 cell gap junction is not deformed by filipin as is the surrounding nonjunctional membrane. This junction does include areas of apparent nonjunctional membrane containing filipin sterol complexes (solid arrows). Scale bar, 0.5 μm. (Figs. 2a, b reprinted, with permission, from Risinger and Larsen 1983.)

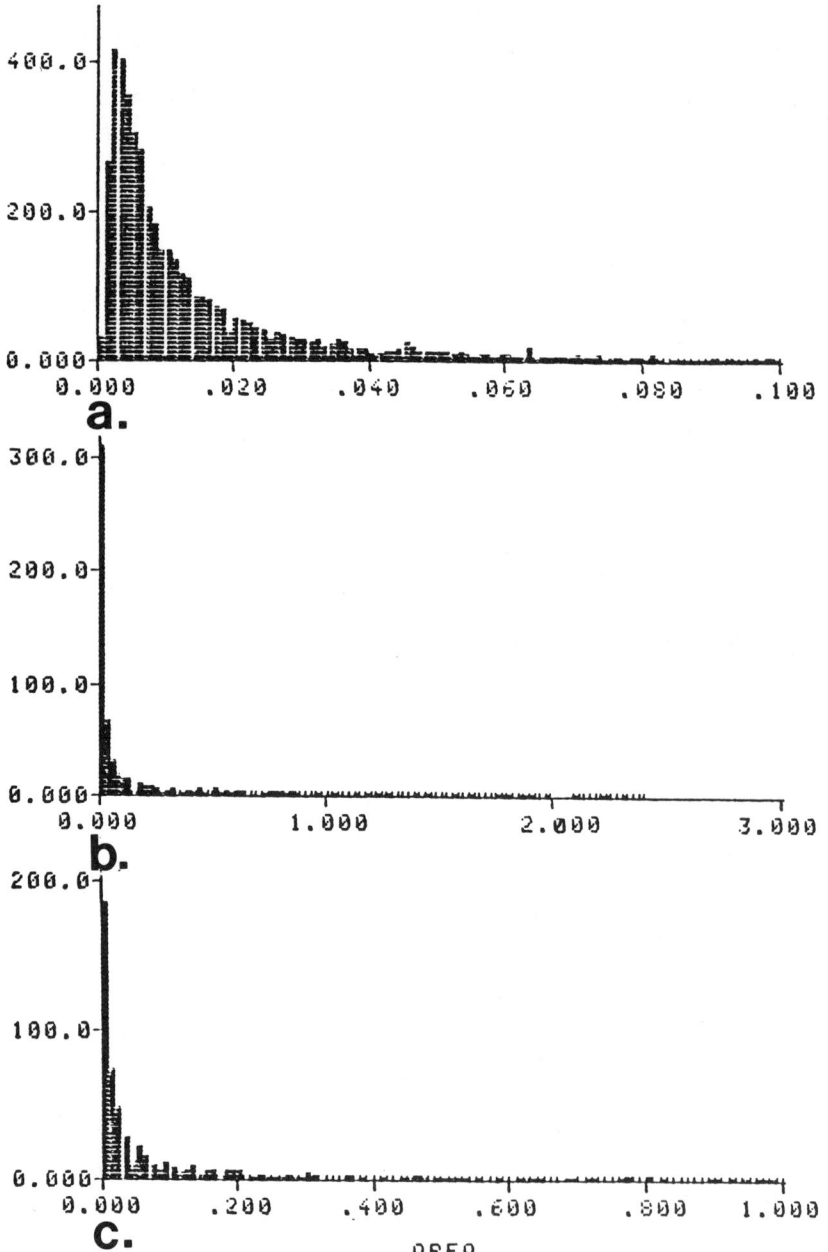

Figure 3
Area distribution of gap junctions in rat gastric mucous cells (*a*), rat cumulus cells (*b*), and rat membrana granulosa cells (*c*). For illustrative purposes each distribution does not include the largest junctions, which comprise 5% of the population (see text).

the lysosomal environment of the vesicle would probably be harmful to the cell. Clearly, however, there is at present no method to determine when during endocytosis or movement into the cytoplasm this postulated curtailment of communication function actually occurs.

Beyond the structural degradation observed in many of these vesicles (Larsen and Tung 1978; Murray et al. 1981; Risinger and Larsen 1983), it is not at all clear as to what the ultimate fate of these endocytosed junctional elements may be. Are they degraded completely by the cell's digestive apparatus or are junctional subunits or junctional proteins recycled after transport back to the plasma membrane and utilized again? Although it has been suggested recently that junctional elements within these vesicles in adrenal cortical cells may cycle back to the surface (Decker 1984), it is also possible that cells may use other mechanisms to remove gap junctions temporarily when rapid recycling is desirable. In regenerating rat liver (Yancey et al. 1981), insect neuroglial cells at metamorphosis (Lane and Swales 1980), or injured crayfish lateral axons (Hanna et al. 1984), it is possible that the disappearance of gap junctions observed during these processes may result from dispersion of junctional plaques into individual particles which are then not recognizable as gap junctions in freeze-fracture replicas. Upon the reestablishment of cell function subsequent to regeneration or metamorphosis or injury, it is possible that the gap junctions could simply reassemble. Since coupling coefficients during liver regeneration decrease by a factor of 10 whereas recognizable junctional area decreases by a factor of 100, it is conceivable that small groups of particles or particle pairs could still maintain function during this hypothesized dispersal. In contrast to these systems, however, it has been demonstrated that gap junctions that occur in the precursors to hair cells in the developing chick otocyst are irreversibly lost from the plasma membrane during development (Ginzberg and Gilula 1979) and in this system internalized junctions are apparently destroyed within typical autophagic vacuoles. The specific mechanism of junctional removal may, therefore, be related to the potential functional role of gap junctions during future cellular development in a given tissue.

Functional Implications of Junctional Modulation

Suffice it to say that the functional ramifications of the relatively subtle alterations in area distributions and junctional sizes just discussed are yet inadequately understood and will probably be difficult to investigate. Our laboratory has been working on two cell systems, however, where a dramatic and significant net loss of gap junctions from the cell surface appears to be correlated with cellular properties and activities of general interest in cell biology: namely mitosis, meiosis, cell cohesion, tumor cell "piling up" in culture, and release of the mammalian ovum from the follicle at ovulation. Our studies relevant to these points have utilized a cultured SW-13 adrenal cortical adenocarcinoma and the preovulatory rabbit and rat ovarian follicle.

Piling Up and Mitosis in SW-13 Cells

SW-13 cells, after numerous passages in culture, exhibit three-dimensional growth or the typical piling up behavior demonstrated by many tumor cell lines.

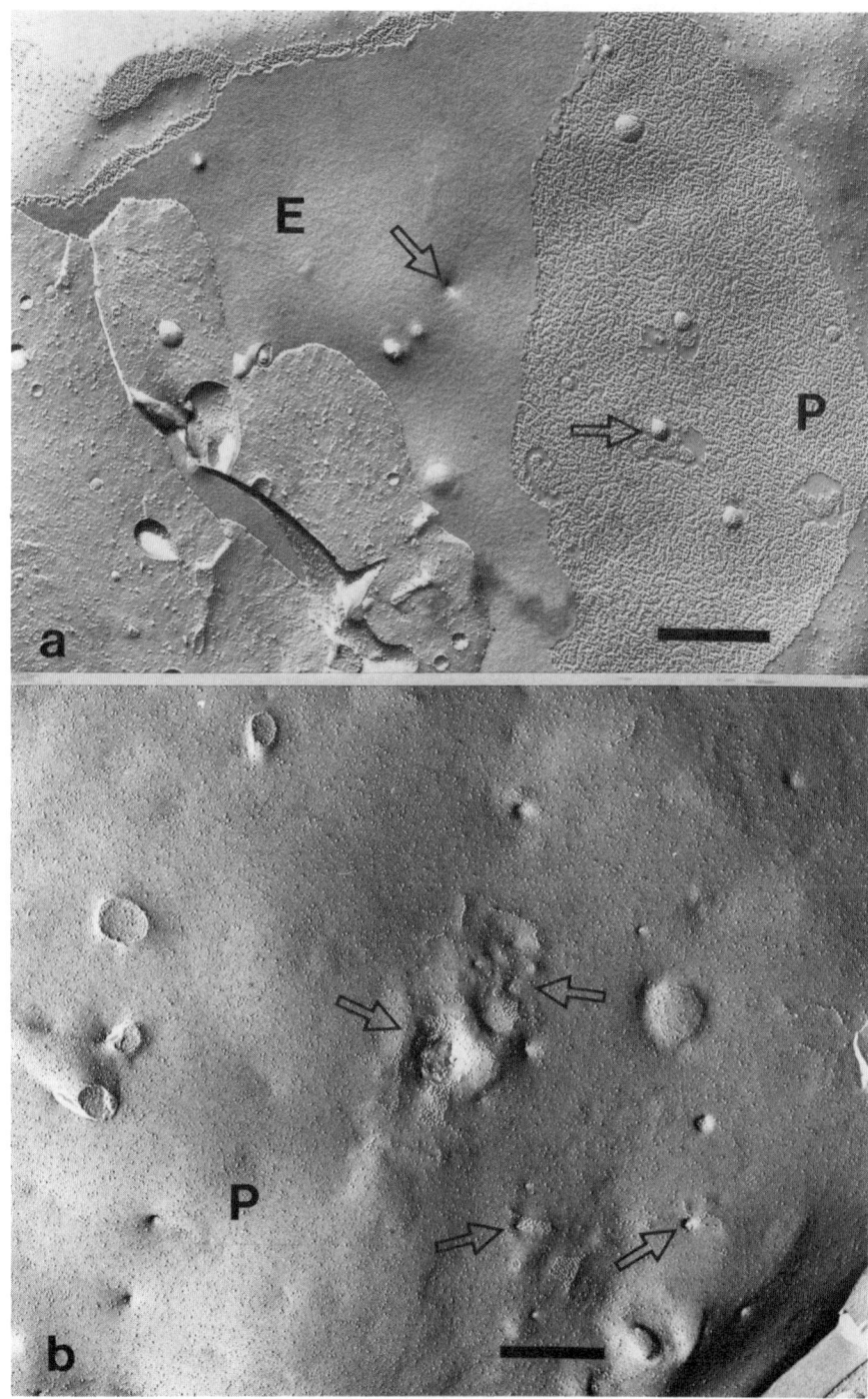

Figure 4 *(See facing page for legend.)*

Time-lapse studies of these cells indicated that typical epithelial monolayers were formed after seeding single cells in clean dishes at confluent density but that almost as soon as the monolayer was established, single cells could be observed to rise, round, and detach from the monolayer to float into the medium. During the course of the next 24–48 hours, so many of these cells had floated into the medium that large clumps of cells began to accumulate above the monolayer and the medium itself increased dramatically in turbidity. Time-lapse cinematography also demonstrated that some cells returned to the dish to maintain a continuous monolayer upon the plastic but that almost 90% of cells that rose and detached almost immediately underwent cell division. Trypan Blue dye exclusion usually indicated viability between 90 and 98% and single floating cells could be used to reestablish viable continuous monolayers with high efficiency (Murray et al. 1981).

Interestingly and not unexpectedly, quantitative analysis of gap junctions in these cells revealed large numbers of gap junctions in 24-hour monolayers ($\sim$20/cell) and very few in cells within clumps derived from floating single cells detaching from the monolayer over a 24-hour period (0.28/cell). In contrast we demonstrated that these floating cells were filled with internalized gap junction vesicles sometimes appearing to fuse with lysosomes, containing reaction product indicative of acid phosphatase activity, and occasionally appearing to be in a terminal stage of degradation. These results support the possibility that the downregulation of gap junctions within individual cells of the monolayer may be causally related to cell disaggregation and the onset of mitosis, and so experiments are now underway to study the modulation of gap junctions in cell lines responding to tumor promoters and antipromoters to analyze more directly the degree of intimacy between junctional modulation and the development of malignant cell behaviors.

Granulosa Cells in Preovulatory Follicles and Their Gap Junctions

Granulosa cells in mammalian systems possess enormous gap junctions, and these structures are readily apparent within almost all freeze-fractured granulosa cell membranes. We have estimated that they occupy between 7 and 8% of the total plasma membrane in rabbit granulosa cells from mature Graafian follicles (Larsen et al. 1981) and as much as 3–4% of the plasma membrane in both cumulus and membrana granulosa cells in mature rat follicles (Larsen et al. 1984a,b). For this reason and because the follicular vesicle in which the egg resides is nonvascular, it has often been suggested that these gap junctions may serve as an alternative intercellular circulatory system. The finding of gap junctions between cumulus cells immediately surrounding the oocyte and the oocyte surface has implied a possible function related to activities of the ovum itself. Based upon a number of interrelated observations,

Figure 4
(a) Large gap junction in rabbit membrana granulosa cell membrane. Several small blebs (arrows) are present and may indicate endocytotic activity. Scale bar, 0.5 μm. (b) Clusters of junctional particles in rat cumulus cell membrane after an ovulatory stimulus. Here too endocytotic activity is in evidence (arrows). Scale bar, 0.5 μm.

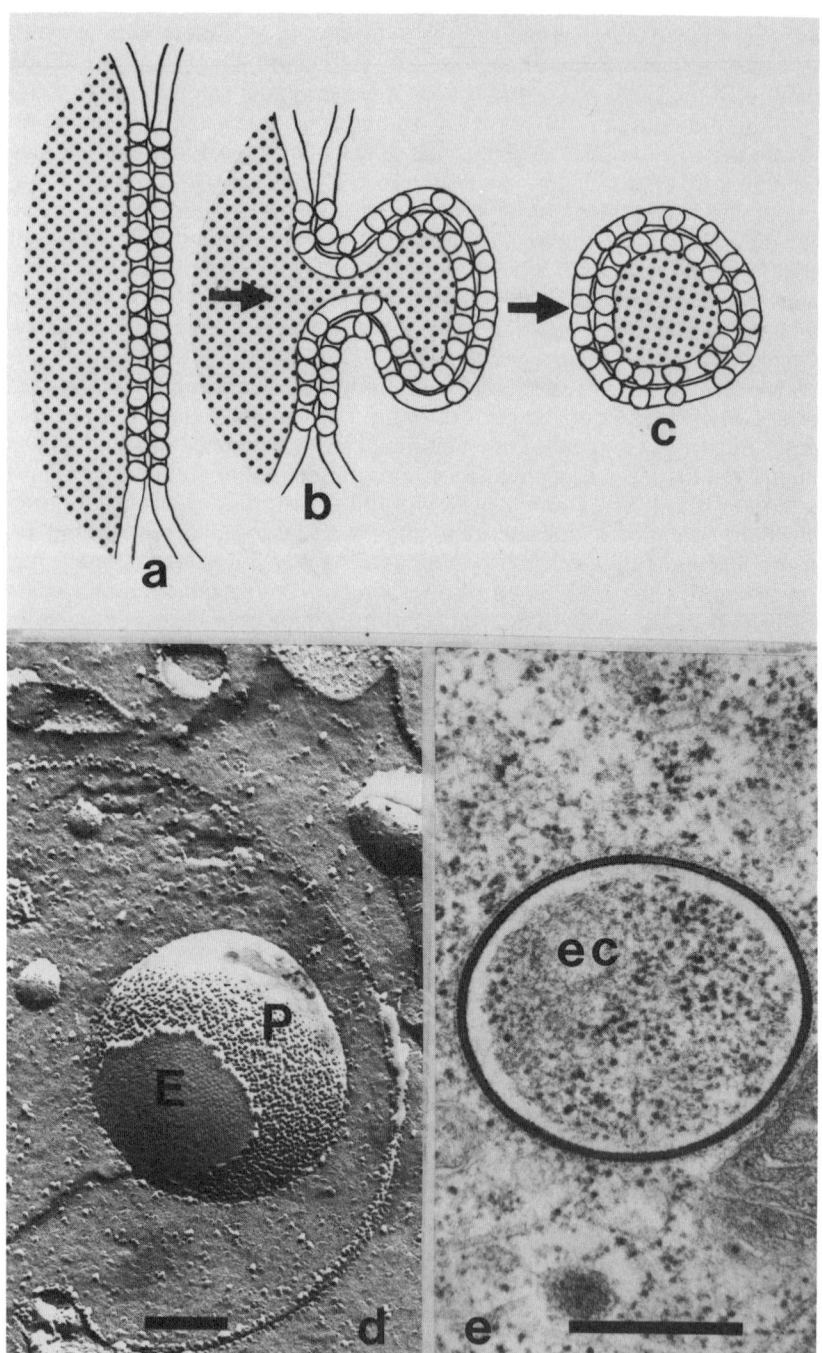

Figure 5 (*See facing page for legend.*)

several investigators have recently suggested that granulosa cells produce an inhibitor that may be transferred to the ovum through gap junctions, thereby maintaining the ovum in meiotic arrest at the dictyate stage of the first meiotic prophase (Tsafriri et al. 1976, 1982; Hillensjo et al. 1979; Eppig et al. 1983; Freter and Schultze 1984). When isolated denuded oocytes or cumulus oocyte complexes are removed from the follicle and placed in culture medium, for example, they rapidly resume meiotic maturation even in the absence of appropriate stimulating hormones (Pincus and Enzmann 1935). If, however, the isolated cumulus oocyte complex is placed upon a bed of granulosa cells within the hemisected follicle, meiotic maturation does not resume (Thibault 1972; Tsafriri and Channing 1975; Liebfried and First 1980), supporting the hypothesis that meiotic inhibition is communicated from the membrana granulosa cells to the oocyte via the oocyte-encapsulating cumulus cells. In addition, a small-molecular-weight oocyte meiosis inhibitor has been found to be effective on isolated oocytes only when they were surrounded by their typical cumulus cell envelope (Hillensjo et al. 1979). Gap junctions are frequently considered to convey the inhibitory signal on the basis largely of their unusual abundance in this system as well as the large body of work that supports a role for these structures in intercellular communication. Within this context a number of investigators have considered the possibility that the disruption of gap junctions between the mass of granulosa cells and the oocyte could play a pivotal role in signaling the ovum to resume meiotic maturation (Gilula et al. 1978; Dekel et al. 1981). Several laboratories (Moor et al. 1980; Heller and Schultz 1980; Dekel et al. 1981; Eppig 1982; Salustri and Siracusa 1983) have specifically pursued this hypothesis by studying the degree of intercellular communication over the preovulatory period, that is, the period beginning with stimulation and ending with ovulation (usually about 12 hr in the rodent models utilized in these studies). In one type of study, it has been demonstrated that mammalian eggs will incorporate radiolabeled uridine or choline only when surrounded by an intact mass of cumulus cells, but that the percentage of label incorporated by the ovum itself (during short incubation periods) compared with the amount of label incorporated by the entire cumulus oocyte complex does not begin to diminish until meiosis has resumed. These results therefore led several investigators to consider alternative hypotheses to explain the apparent granulosa-generated signal to resume meiosis. Prominent among explanations at the present time is the suggestion that the signal to

Figure 5
(a,b) Endocytosis of gap junction membrane may occur as one cell bites off a part of its neighboring cell at the junction, enclosing the adjacent cell's cytoplasm (c) (stippled area) and its membrane contribution to the junction within the resulting cytoplasmic vesicle. (Reprinted, with permission, from Larsen 1977b.) (d) Freeze-fracture replica of gap junction vesicle in a rabbit granulosa cell demonstrating typical P-face particles and E-face pits. Scale bar, 0.1 μm. (Reprinted, with permission, from Larsen and Tung 1978.) (e) Annular gap junction profile of a gap junction vesicle in a cultured SW-13 cell with enclosed cytoplasm (ec) from a neighboring cell. Scale bar, 0.5 μm. (Reprinted, with permission, from Murray et al. 1981.)

resume meiosis is more likely dependent upon a significant drop in the production of meiotic maturation inhibitor by surrounding granulosa cells (Eppig et al. 1983; Freter and Schultz 1984).

Previously, we have demonstrated that granulosa gap junctions were dramatically downregulated in the rabbit over the 12-hour preovulatory period (Larsen et al. 1981), and we decided to extend our studies to another model from which we could obtain data more economically at several critical intermediate time points. In our recent experiments with the superovulated immature rat, we have determined that massive downregulation of cumulus cell gap junctions in this system is related more intimately with the resumption of meiotic maturation than has been demonstrated in the metabolic coupling studies just alluded to (Larsen et al. 1984 and in prep.). We not only find a close correlation between the total area of gap junction membrane at the granulosa cell surface and the breakdown of the dictyate germinal vesicle (a 10-fold reduction between 2 and 3 hr after stimulation), but, in addition, an even sharper drop in the mean area of individual gap junctions (sixfold between 2 and 2.5 hr after stimulation) which falls to a near-minimum value even before much of the junctional membrane is lost from the cell surface. We have also demonstrated a correlation between the rapid loss of cumulus cell gap junctions in this system and the expansion of the cumulus oophorus which results from cumulus cell disaggregation. This suggests that in addition to a possible role in meiotic regulation, the disruption of gap junctions in this system may play a role in cell cohesion as in the cultured SW-13 cells discussed above. In light of this finding it seems possible that cumulus disaggregation would be essential for the dehiscence and expulsion of the oocyte from the follicle at ovulation and would also contribute to the breakdown of physical barriers that would otherwise interfere with sperm penetration during fertilization.

It should be mentioned, however, that there may be other populations of gap junctions in the follicle that play a role in ovulation. There are those of the membrana granulosa cells that our preliminary results suggest are downregulated also, but at a more gradual rate, and those at the oocyte surface itself. The time points we have examined in the case of the membrana granulosa gap junctions tell us that they are reduced to 10% of control values by about 7 hours after an ovulatory stimulus but have dropped to only 50% of unstimulated control values by the end of the 1-hour period when germinal vesicle breakdown takes place (from 2 to 3 hr after stimulation). Analysis of the cumulus-oocyte junctions has proven to be more difficult than those between granulosa cells. We feel, however, that these data are particularly important since if it were demonstrated that gap junctions persisted between the oocyte and the innermost cumulus layer for longer periods of time, the discrepancy between the apparently rapid and dramatic loss of cumulus–cumulus gap junctions and the metabolic coupling results published earlier by other laboratories might be explained.

SUMMARY AND CONCLUSIONS

There is growing enthusiasm within the gap junction field to explore the functional implications of gap junctions in particular cellular systems with respect to specific cellular behaviors and activities. Fortunately, the sophistication and resolution of new probes of structure and function in this field are improving rapidly. With these new morphological, physiological, molecular, and genetic approaches (many discussed in the present volume), it seems possible that we will be able to attack the problems outlined in this chapter with approaches similar to those now utilized more routinely in studies of such entities as low-density lipoprotein receptors, hormone receptors, and other well-characterized membrane specializations. It also seems very likely that the outcome of these future studies will be fundamental and will be of interest to a broad range of cell biologists.

ACKNOWLEDGMENTS

I am grateful for the skilled technical assistance of George Brunner and the collaborative contributions of E. Jacobson, S. Joffe, K. Tabata, S. Murray, H. Tung, M.A. Risinger, and S. Wert. I am also thankful for the expert typing of Delores Bolden, to Emma Lou Cardell and C. Skowron for critically reading the manuscript, and for grant support from the American Cancer Society (PDT-84) and the National Science Foundation (PCM-7816189 and PCM-8023028).

REFERENCES

Albertini, D.F. and E. Anderson. 1975. Structural modifications of lutein cell gap junctions during pregnancy in the rat and mouse. *Anat. Rec.* **181:** 171–194.

Blennerhassett, M.G. and S. Caveney. 1983. How selective is gap junctional permeability at the compartment border? *J. Cell Biol.* **97:** 125a.

Decker, R.S. 1984. Modulation of gap junctions in cultured fetal adrenal cortical cells. *J. Cell Biol.* **99:** 345a.

Dekel, N., T.S. Lawrence, N.B. Gilula, and W.H. Beers. 1981. Modulation of cell-to-cell communication in the cumulus-oocyte complex and the regulation of oocyte maturation by LH. *Dev. Biol.* **86:** 356–362.

Dermietzel, R., A. Liebstein, U. Frixen, U. Janssen-Timmen, O. Traub, and K. Willecke. 1984. Gap junctions in several tissues share antigenic determinants with liver gap junctions. *EMBO J.* **3:** 2261–2270.

Eppig, J.J. 1982. The relationship between cumulus cell-oocyte coupling, oocyte meiotic maturation, and cumulus expansion. *Dev. Biol.* **89:** 268–272.

Eppig, J.J., R.R. Freter, P.F. Ward-Bailey, and R.M. Schultz. 1983. Inhibition of oocyte maturation in the mouse: Participation of cAMP, steroids, and a putative maturation inhibitory factor. *Dev. Biol.* **100:** 39–49.

Epstein, M.L. and N.B. Gilula. 1977. A study of communication specificity between cells in culture. *J. Cell Biol.* **75:** 769–787.

Epstein, M.L., J.D. Sheridan, and R.G. Johnson. 1977. Formation of low-resistance junctions *in vivo* in the absence of protein synthesis and ATP production. *Exp. Cell Res.* **104:** 25–30.

Finbow, M.E. and S.B. Yancey. 1981. The roles of intercellular junctions. In *Biochemistry of cellular regulation* (ed. P. Knox), p. 215–248. CRC Press, Florida.

Freter, R.R. and R.M. Schultz. 1984. Microinjection of murine oocytes with a low molecular weight fraction of bovine follicular fluid inhibits meiosis. *J. Cell Biol.* **99:** 389a.

Gilula, N.B., M.L. Epstein, and W.H. Beers. 1978. Cell-to-cell communication and ovulation. A study of the cumulus-oocyte complex. *J. Cell Biol.* **78:** 58–75.

Ginzberg, R.D. and N.B. Gilula. 1979. Modulation of cell junctions during differentiation of the chicken oocyte sensory epithelium. *Dev. Biol.* **68:** 110–129.

Girsch, S.J. and C. Peracchia. 1983. Lens junction protein (MIP26) self-assembles in liposomes forming large channels regulated by calmodulin (CaM). *J. Cell Biol.* **97:** 83a.

Goodenough, D.A. 1979. Lens gap junctions: A structural hypothesis for non-regulated low resistance intercellular pathways. *Invest. Opthalmol. Visual Sci.* **18:** 1104–1122.

Gorin, M.B., S.B. Yancey, J. Cune, J.-P. Revel, and J. Horwitz. 1984. The major intrinsic protein (MIP) of the bovine lens fiber membrane: Characterization and structure based on cDNA cloning. *Cell* **39:** 49–59.

Gros, D.B., B.J. Nicholson, and J.-P. Revel. 1983. Comparative analysis of the gap junction protein from rat heart and liver: Is there a tissue specificity of gap junctions? *Cell* **35:** 539–549.

Hanna, R.B., G.D. Pappas, and M.V.L. Bennett. 1984. The fine structure of identified electrotonic synapses following increased coupling resistance. *Cell Tiss. Res.* **235:** 243–249.

Heller, D.T. and R.M. Schultz. 1980. Ribonucleoside metabolism by mouse oocytes: Metabolic cooperativity between the fully grown oocyte and cumulus cells. *J. Exp. Zool.* **214:** 355–364.

Hertzberg, E.L. 1984. A detergent-independent procedure for the isolation of gap junctions from rat liver. *J. Biol. Chem.* **259:** 9936–9943.

Hertzberg, E.L. and R.V. Skibbens. 1984. A protein homologous to the 27,000 dalton liver gap junction protein is present in a wide variety of species and tissues. *Cell* **39:** 61–69.

Hillensjo, T., A.S. Kripner, S.H. Domerants, and C.P. Channing. 1979. Action of porcine follicular fluid oocyte maturation inhibitor *in vitro:* Possible role of the cumulus cells. *Adv. Exp. Med. Biol.* **112:** 283–291.

Johnson, R.G., M. Hammer, J.D. Sheridan, and J.-P. Revel. 1974. Gap junction formation between reaggregated Novikoff hepatoma cells. *Proc. Natl. Acad. Sci.* **71:** 4536–4540.

Lane, N.J. and L.S. Swales. 1980. Dispersal of junctional particles, not internalization, during *in vitro* disappearance of gap junctions. *Cell* **19:** 579–586.

Larsen, W.J. 1977a. Structural diversity of gap junctions: A review. *Tissue Cell* **9:** 373–394.

———. 1977b. Gap junctions and hormone action. In *Transport of ions and water in epithelia* (ed. B.L. Gupta et al.), p. 333–361. Academic Press, London.

———. 1983. Biological implications of gap junction structure, distribution, and composition: A review. *Tissue Cell* **15:** 645–671.

Larsen, W.J. and M.A. Risinger. 1985. The dynamic life histories of intercellular membrane proteins. *Mod. Cell Biol.* **4:** (in press).

Larsen, W.J. and H.N. Tung. 1978. Origin and fate of cytoplasmic gap junctional vesicles in rabbit granulosa cells. *Tissue Cell* **10:** 585–598.

Larsen, W.J., H.N. Tung, and C. Polking. 1981. Response of granulosa cell gap junctions to human chorionic gonadotropin (hCG) at ovulation. *Biol. Reprod.* **25:** 1119–1134.

Larsen, W.J., S.E. Wert, and G.D. Brunner. 1984. The disruption of rat cumulus cell gap junctions could provide a signal to the egg to resume meiotic maturation. *J. Cell Biol.* **99:** 345a.

Larsen, W.J., H.N. Tung, S.A. Murray, and C.A. Swenson. 1979. Evidence for the participation of action microfilaments and bristle coats in the internalization of gap junction membrane. *J. Cell Biol.* **83:** 576–587.

Leibfried, L. and N.L. First. 1980. Follicular control of meiosis in the porcine oocyte. *Biol. Reprod.* **23:** 705–709.

Loewenstein, W.R., Y. Kanno, and S.J. Socolar. 1978. Quantum jumps of conductance during formation of membrane channels at cell–cell junction. *Nature* **274:** 133–136.

Makowski, L., D.L.D. Caspar, W.C. Phillips, T.S. Baker, and D.A. Goodenough. 1984. Gap junction structures VI. Variation and conservation in connexon conformation and packing. *Biophys. J.* **45:** 208–218.

Meyer, D.J., B. Yancey, and J.-P. Revel. 1981. Intercellular communication in normal and regenerating rat liver: A quantitative analysis. *J. Cell Biol.* **91:** 505–523.

Moor, R.M., M.W. Smith, and R.M.C. Dawson. 1980. Measurement of intercellular coupling between oocytes and cumulus cells using intracellular markers. *Exp. Cell Res.* **126:** 15–29.

Murray, S.A., W.J. Larsen, J. Trout, and S.T. Donta. 1981. Gap junction assembly and endocytosis correlated with patterns of growth in a cultured adrenal cortical tumor cell (SW-13). *Cancer Res.* **41:** 4063–4074.

Nicholson, B.J., L.J. Takemoto, M.W. Hunkapiller, L.E. Hood, and J.-P. Revel. 1983. Differences between liver gap junction and lens MIP26 from rat: Implications for tissue specificity of gap junctions. *Cell* **32:** 967–978.

Peracchia, C. 1980. Structural correlates of gap junction permeation. *Int. Rev. Cytol.* **66:** 81–146.

Peracchia, C. and G. Bernardini. 1984. Gap junction structure and cell-to-cell coupling regulation: Is there a calmodulin involvement? *Fed. Proc.* **43:** 2681–2691.

Peracchia, C. and L.L. Peracchia. 1980. Gap junction dynamics: Reversible effects of hydrogen ions. *J. Cell Biol.* **87:** 719–727.

Pincus, G., and E.V. Enzmann. 1935. The comparative behavior of mammalian eggs *in vivo* and *in vitro*. I. The activation of ovarian eggs. *J. Exp. Med.* **62:** 655–675.

Preus, D., E. Kam, J.D. Sheridan, and R.G. Johnson. 1983. Formation of functional gap junctions at 4°C. *J. Cell Biol.* **97:** 83a.

Risinger, M.A. and W.J. Larsen. 1983. Interaction of filipin with junctional membrane at different stages of the junction's life history. *Tissue Cell* **15:** 1–15.

Safranyos, R.G. and S. Caveney. 1983. Rates of diffusion of fluorescent molecules via intercellular membrane channels. *J. Cell Biol.* **97:** 82a.

Salustri, A. and G. Siracusa. 1983. Metabolic coupling, cumulus expansion and meiotic resumption in mouse cumuli oophori cultured *in vitro* in the presence of FSH or dcAMP, or stimulated *in vivo* by hCG. *J. Reprod. Fertil.* **68:** 335–341.

Sheridan, J.D., M. Hammer-Wilson, D. Preus, and R.G. Johnson. 1978. Quantitative analysis of low-resistance junctions between cultured cells and correlation with gap junctional areas. *J. Cell Biol.* **76:** 532–544.

Spray, D.C., A.L. Harris, and M.V.L. Bennett. 1981. Gap junctional conductance is a simple and sensitive function of intracellular pH. *Science* **211:** 712–715.

Spray, D.C., J.H. Stern, A.L. Harris, and M.V.L. Bennett. 1982. Gap junctional conductance: Comparison of sensitivities to H and Ca ions. *Proc. Natl. Acad. Sci.* **79:** 441–445.

Tadvalkar, G. and P. Pinto da Silva. 1983. *In vitro*, rapid assembly of gap junctions is induced by cytoskeleton disrupters. *J. Cell Biol.* **96:** 1279–1287.

Thibault, C.G. 1972. Final stages of mammalian oocyte maturation. In *Oogenesis* (ed. J.D. Bigger and A.W. Schuetz), p. 397–411. University Park Press, Baltimore, Maryland.

Tsafriri, A. and C.P. Channing. 1975. An inhibitory influence of granulosa cells and follicular fluid upon porcine oocyte meiosis *in vitro*. *Endocrinology* **96:** 922–927.

Tsafriri, A., N. Dekel, and S. Bar-Ami. 1982. The role of oocyte maturation inhibitor in follicular regulation of oocyte maturation. *J. Reprod. Fertil.* **64:** 541–551.

Tsafriri, A., S.H. Pomerantz, and C.P. Channing. 1976. Inhibition of oocyte maturation by porcine follicular fluid: Partial characterization of the inhibitor. *Biol. Reprod.* **14:** 511–516.

Turin, L. and A.E. Warner. 1980. Intracellular pH in *Xenopus* embryos: Its effect on current flow between blastomeres. *J. Physiol.* **300:** 489–504.

Unwin, P.N.T. and G. Zampighi. 1980. Structure of the junction between communicating cells. *Nature* **283:** 545–549.

Warner, A.E. and P.A. Lawrence. 1982. Permeability of gap junctions at the segmental border in insect epidermis. *Cell* **28:** 243–252.

Yancey, S.B., B.J. Nicholson, and J.-P. Revel. 1981. The dynamic state of liver gap junctions. *J. Supramol. Struct. Cell. Biochem.* **16:** 221–232.

Junctional Communication and Oocyte Maturation

William H. Beers and Paula J. Olsiewski[*]
*Department of Biology, New York University
New York, New York 10003*

Although it is now well recognized that gap junctions provide an intercellular passageway for ions and relatively low-molecular-weight molecules, the issue of the existence of intercellular communication of physiologically meaningful signals or messengers other than ions has been a difficult one to study. Even more elusive has been the identification of these communicated substances in situations where there is evidence that they appear to operate.

Several years ago, we attempted to fabricate a model system to determine whether some intercellular "communicator" of hormonal stimulation could be demonstrated (Lawrence et al. 1978). For these studies two different cell types, rat granulosa and mouse myocardial, were chosen for the following reasons. First, both form gap junctions readily in culture. Second, each one responds to hormones not recognized by the other: follicle stimulating hormone (FSH) in the case of the granulosa, catecholamines in the case of myocardial cells. Third, the hormonal responses of these cells are characteristic for each cell type: Increased synthesis of a protease, plasminogen activator, in granulosa and increased frequency of contraction in the myocardial cells. Finally, the responses in both types of cells are correlated with an increase in intracellular cAMP. Based on our knowledge at that time, it appeared that cAMP had the potential of being able to pass through junctions, and it, or something else, might be able to mediate a response in a cell that was unresponsive to a given hormone if it were communicating with one that was responsive. Of course, in order to perform these studies, a requisite condition was that the two different

---

[*]Present address: Enzo Biochem Inc., 325 Hudson Street, New York, New York 10013.

types of cells would be able to establish heterologous communication in coculture. This did occur and in fact the frequency of beating of myocardial cells that were connected to each other only by granulosa, and not myocardial–myocardial cell contacts, became synchronized. Gap junctions, ionic coupling, and metabolic cooperation could also be demonstrated between the two cell types. Having established that heterologous cell–cell communication existed we could then attempt to answer the initial question of whether exposure of the cocultures to a hormone specific to one cell type would cause both cell types to respond. In these experiments FSH caused, in addition to an increased synthesis of plasminogen activator, a dose-dependent increase in myocardial cell beat frequency. Likewise, exposure of the cocultures to norepinephrine resulted in a response by both cell types. Controls to determine the existence of an extracellular mediator of these effects were negative. In addition, our results indicated that neither cell fusion nor the migration of receptors could account for the results that were obtained. Thus, it appeared that a "communicator" of hormonal stimulation had been transferred from one cell to another. It was tempting to suggest that cAMP was the communicated molecule; however, in this system it would be extremely difficult or impossible to prove that proposal.

Nonetheless, it appeared that something that was physiologically meaningful to both cells was junctionally communicated in this artificial system. To determine whether a similar mechanism might operate under less contrived conditions, we turned our attention to a naturally occurring heterologous communicating system—the cumulus–oocyte complex. For the reasons given below, it appeared that some intracellular messenger, perhaps cAMP, might be communicated from the cumulus cells to the oocyte to control oocyte maturation.

RESULTS AND DISCUSSION

Oocyte Maturation and Communication In Vivo

In mammals, the oocyte is surrounded by several layers of modified granulosa cells collectively called the cumulus oophorus. These cells are separated from the oocyte by an acellular zone known as the zona pellucida. Thin processes emanate from the cumulus cells, penetrate the zona, and make contact with the oolemma. In the regions of contact, gap junctional structures are observed (Anderson and Albertini 1976; Gilula et al. 1978).

Prior to ovulation, mammalian oocytes are arrested in the dictyate stage of the first meiotic division. If, following the preovulatory gonadotropin surge, an oocyte is chosen for ovulation, the maturational arrest is relieved and maturation ensues. An early morphological manifestation of this is the disappearance of the germinal vesicle, or nucleus, of the oocyte. At the same time the cumulus oophorus begins to become disorganized and the cells of the cumulus start to separate from one another. This pattern is only seen in cumulus–oocyte complexes that are destined for ovulation; other complexes, even those in large follicles in the same ovary, retain their well-organized appearance and germinal vesicles.

It is possible to demonstrate communication between the cells of the cumulus and the oocyte by dye transfer and ionic coupling. However, as preovulatory follicles (i.e., those that have been chosen for ovulation) approach the time of ovulation, communication is interrupted (Gilula et al. 1978). This observation suggested the possibility that the maintenance of meiotic arrest and the integrity of cumulus–oocyte communication might somehow be related.

Oocyte Maturation In Vitro

In contrast to their behavior in vivo, oocytes, or cumulus–oocyte complexes removed from their follicles and placed under tissue culture conditions resume maturation spontaneously. This occurs in the absence of hormones. However, inclusion of membrane-permeable derivatives of cAMP or cyclic nucleotide phosphodiesterase inhibitors in the culture medium will maintain meiotic arrest (Nekola and Moor-Smith 1975). However, if gonadotropins are added at physiological concentrations to these cultures, maturational arrest will be relieved (Dekel and Beers 1978). On the other hand, so-called naked or denuded oocytes that have had their cumulus cells removed are insensitive to gonadotropins. In addition, the kinetics of spontaneous maturation in vitro of denuded oocytes are the same as those of luteinizing hormone (LH)-treated cumulus–oocyte complexes which are more rapid than control cultures of cumulus–oocyte complexes (Dekel and Beers 1980). Two additional findings are relevant to this discussion. First, gonadotropin receptors, measured as receptors for human chorionic gonadotropin (hCG), which has LH activity, are absent or present in extremely low density on the oocyte as compared with the cumulus cells (Lawrence et al. 1980). Second, in vitro LH causes an interruption in communication between the cumulus and the oocyte (Dekel et al. 1981).

Given these observations and others contained in the above references, we proposed the following as a possible model for the mechanism responsible for the control of oocyte maturation. This model proposed that cAMP synthesized in the cumulus, or perhaps the granulosa and cumulus, is communicated at some tonic level to the oocyte and this maintains meiotic arrest. As communication is terminated, which occurs in response to elevated levels of gonadotropin, the supply of cAMP to the oocyte is terminated, allowing arrest to be relieved and the resumption of meiosis.

Cumulus and Oocyte cAMP Synthesis

A direct prediction of this sort of mechanism is that the oocyte should not be able to synthesize appreciable quantities of cAMP. As seen in Figure 1, cumulus–oocyte complexes synthesize, in vitro, cAMP in response to gonadotropins, choleratoxin, and the diterpene, forskolin (Olsiewski and Beers 1983). The use of these three types of reagents allows stimulation of the cyclase system at the level of receptor (gonadotropins), regulatory subunit (choleratoxin) (Gill 1978), and catalytic subunit (forskolin) (Seamon et al. 1981), although the absolute specificity of forskolin is a subject of controversy (Darfler et al. 1982; Stengel et al. 1982). On the other hand, when denuded oocytes

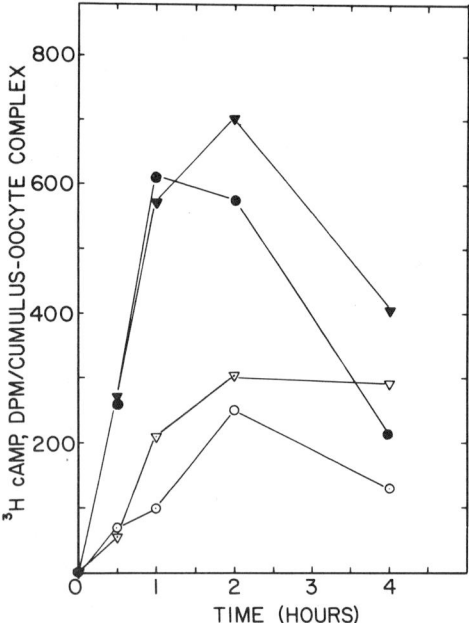

Figure 1
Time course of [^{3}H]cAMP synthesis by cumulus–oocyte complexes. Complexes were labeled for the indicated times in the presence of the indicated substances as described in Olsiewski and Beers (1983). In the case of choleratoxin, there was a 30-min preincubation with the toxin prior to the addition of the precursor label, [^{3}H]adenine. In all other cases the agent under study was added to the culture medium with the label. Forskolin, 100 μM (▼); 100 μg choleratoxin/ml (▽); 1 μg rFSH/ml (●); 1 μg rLH/ml (○). rFSH, highly purified rat follicle stimulating hormone; rLH, highly purified rat luteinizing hormone.

were examined in the same in vitro system, the results seen in Table 1 were obtained. In these studies, forskolin was the most effective agent in stimulating cAMP synthesis; however, exposure of FSH also yielded a positive response. In either case, however, the amount of synthesis was 1% or less than that seen in the complexes. LH and choleratoxin did not elevate cAMP levels above those of the control incubations. The absence of a response to choleratoxin was an unexpected result since a response to gonadotropin would suggest the presence of a functional regulatory subunit. However, the presence of a regulatory subunit, if not a typical one, is suggested by the fact that choleratoxin will cause a twofold augmentation of the response to FSH (Olsiewski and Beers 1983). A similar situation is seen in turkey erythrocytes (Field 1974). Thus, although it is poorly understood, this pattern of response is not without precedent. Nonetheless, we felt that these results were encouraging, because under optimal conditions the oocyte is only capable of synthesizing small amounts of cAMP relative to the cumulus and under physiological conditions may not produce any significant amounts of this nucleotide. This condition,

Table 1
[³H]cAMP Synthesis by Naked Oocytes

Conditions for labeling	dpm/oocyte	Total oocytes analyzed
No supplement	0.13	1467
0.2 mM MIX[a]	0.37	2283
0.2 mM MIX + 1 µg rLH[b]/ml	0.39	1688
0.2 mM MIX + 100 mg cholera toxin/ml	0.34	1545
0.2 mM MIX + 100 µM forskolin	3.5	2607
0.2 mM MIX + 1 µg rFSH[c]/ml	2.7	2863

Cumulus-free, naked oocytes were labeled in the presence of the indicated agents as described in Olsiewski and Beers (1983). The values reported are the average of at least three separate determinations for each experimental condition and fall within a range of ± 15%.
[a]MIX, 3-isobutyl-1-methylxanthine.
[b]rLH, Rat luteinizing hormone.
[c]rFSH, Rat follicle-stimulating hormone.

while in keeping with the proposed model, also provides an ideal system to examine the ability of cAMP to be transferred from cell to cell by a junctional mechanism.

Lack of Transfer of cAMP

If intact ovarian follicles are incubated under tissue culture conditions, meiotic arrest and cumulus–oocyte communication are maintained. If these cultures are exposed to gonadotropin, communication is interrupted and maturation ensues (Dekel et al. 1981). Using this system we attempted to determine whether cAMP synthesized in the cumulus oophorus could be transferred to the oocyte. If follicles are incubated in Liebowitz's L-15 medium containing 10% fetal bovine serum and 100 mCi/ml [³H]adenine (15 Ci/mmole), [³H]cAMP levels rise until approximately 4 hours of incubation and then plateau at a level of approximately 50 dpm/complex. In these experiments, [³H]cAMP was quantitated as reported previously (Olsiewski and Beers 1983). As seen in Table 2, complexes isolated from follicles incubated for 4 hours in medium supplemented with [³H]adenine contained 47 dpm cAMP/complex. Naked oocytes isolated from follicles incubated under the same conditions contained approximately 2 dpm cAMP/oocyte. In an attempt to create a situation that might favor an increased ratio between oocyte cAMP/cumulus–oocyte cAMP, if transfer were occurring, a pulse-chase protocol was adopted wherein the follicles were incubated for 2 hours in the presence of [³H]adenine and then for 2 hours more in the presence of 10^{-3} M cold adenine. Under these conditions similar results were obtained. The relative volume of the oocyte to the collective cells of the cumulus oophorus is on the order of 1:1. Thus, if rapid equilibrium were being established between the cumulus and oocyte, it would be expected that the oocyte would contain 40–60% of the radioactive cAMP of the complex. In these experiments the oocyte contained approximately 5%. This level of radioactivity is reasonably within the limits of contamination of

Table 2
Newly Synthesized [³H]cAMP in Oocytes and Cumulus–Oocyte Complexes

Preincubation (20 hr)	Additions during preincubation	Pulse or preincubation period (hr)	Chase (hr)	dpm/complex	dpm/oocyte
−	—	4	—	47	2
−	—	2	2	48	3
+	no additions	2	2	57	6
+	10 μg oLH[a]/ml	2	2	354	5

Approximately 100 complexes and 200 oocytes were used per determination. Each value reported is the average of three experiments except for the condition with exposure to LH which is the average of two. [³H]cAMP was analyzed as described in Olsiewski and Beers (1983).
[a]oLH, ovine LH.

Table 3
Analysis of cAMP in Oocytes and Cumulus—Oocyte Complexes In Vivo

Hours After hCG[a] injection	cAMP/complex (fmoles)	cAMP/oocyte (fmoles)
0	8.0	2.5
4 (preovulatory)	10.9	4.5
4 (nonpreovulatory)	5.4	4.1
8 (preovulatory)	311	5.3
8 (nonpreovulatory)	12.3	3.7

cAMP was determined by radioimmunoassay. The notations preovulatory and nonpreovulatory refer to complexes and oocytes isolated from follicles that, based on morphological criteria, are destined to ovulate or not. Determinations at 0 and 8 hr are the average of three determinations; those at 4 hr are the average of two. For each determination, except those 8 hr after hCG, 200–300 oocytes or complexes were used. The determination made 8 hr after the hCG injection utilized 30–80.

[a]hCG, Human chrionic gonadotropin.

the naked oocyte preparation. However, to check this possibility, the following experiment was conducted. Follicles were incubated for 20 hours in the presence or absence of 10 μg ovine LH (oLH)/ml. Under these conditions, LH terminates cumulus–oocyte communication (Dekel et al. 1981). Following this preincubation, the pulse-chase protocol described above was carried out. Using this procedure, no significant difference in the amount of [^{3}H]cAMP was found in the LH-treated or untreated preparations (Table 2), indicating that no significant transfer of newly synthesized cAMP was demonstrable between the cumulus and oocyte.

As a further check on the relationship between cAMP concentration in the oocyte and the resumption of maturation, the experiments in Table 3 were carried out. In immature rats, it is possible to induce ovulation at a predictable time with a regimen of injected gonadotropins. In the system used here (Gilula et al. 1978), ovulation occurs approximately 12–14 hours following an injection of hCG which is administered 48 hours after an injection of pregnant mare serum gonadotropin. Using this system, the amount of cAMP in cumulus–oocyte complexes was determined using standard radioimmunoassay techniques. As seen in Table 3, no decrease in oocyte cAMP was seen 8 hours after the hCG injection, at which time a significant proportion of oocytes had lost their germinal vesicles.

CONCLUSION

Although the results presented here and elsewhere (e.g., Freter and Schultz 1984) would argue for an involvement of junctional communication and cAMP in the control of oocyte maturation, the mechanism cannot be as simple as we originally proposed. Whether a precursor to an inhibitor is exhausted in response to elevated cumulus cAMP as proposed by Freter and Schultz (1984) or whether the communication of some other regulatory substance is interrupted remains to be seen and is the subject of active investigation.

ACKNOWLEDGMENTS

We would like to thank Kathleen McCarthy-Doino for excellent technical assistance and Jennifer Graham and David Haas for their help in preparing oocytes and complexes. Rose Belizario's contribution, that of typing this manuscript, is also greatly appreciated. This work was supported by NIH grant HD16182. P.J.O. was a recipient of an NIH Postdoctoral Fellowship HD05923.

REFERENCES

Anderson, E. and D.F. Albertini. 1976. Gap junctions between the oocyte and companion follicle cells in the mammalian ovary. *J. Cell Biol.* **71**: 680–686.

Darfler, F.J., L.C. Mahan, A.M. Koachman, and P.A. Insel. 1982. Stimulation by forskolin of intact S49 lymphoma cells involves the nucleotide regulatory protein of adenylate cyclase. *J. Biol. Chem.* **257**: 11901–11907.

Dekel, N. and W.H. Beers. 1978. Rat oocyte maturation *in vitro:* Relief of cyclic AMP inhibition by gonadotropins. *Proc. Natl. Acad. Sci.* **75**: 4369–4373.

———. 1980. Development of the rat oocyte *in vitro:* Inhibition and induction of maturation in the presence or absence of the cumulus-oophorus. *Dev. Biol.* **75**: 247–254.

Dekel, N., T.S. Lawrence, N.B. Gilula, and W.H. Beers. 1981. Modulation of cell to cell communication in the cumulus-oocyte complex and the regulation of oocyte maturation by LH. *Dev. Biol.* **86**: 356–362.

Field, M. 1974. Mode of action of cholera toxin: Stabilization of catecholamine-sensitive adenylate cyclase in turkey erythrocytes. *Proc. Natl. Acad. Sci.* **71**: 3299–3303.

Freter, R.R. and R.M. Schultz. 1984. Regulation of murine oocyte meiosis: Evidence for a gonadotropin-induced, cAMP-dependent reduction in a maturation inhibitor. *J. Cell Biol.* **98**: 1119–1128.

Gill, D.M. 1978. Mechanism of action of cholera toxin. *Adv. Cyclic Nucleotide Res.* **8**: 85–118.

Gilula, N.B., M.L. Epstein, and W.H. Beers. 1978. Cell to cell communication and ovulation. A study of the cumulus-oocyte complex. *J. Cell Biol.* **78**: 58–75.

Lawrence, T.S., W.H. Beers, and N.B. Gilula. 1978. Transmission of hormonal stimulation by cell to cell communication. *Nature* **272**: 501–506.

Lawrence, T.S., N. Dekel, and W.H. Beers. 1980. Binding of human chorionic gonadotropin by rat cumuli oophori and granulosa cells: A comparative study. *Endocrinology* **106**: 1114–1118.

Nekola, M.V. and D.M. Moor-Smith. 1975. Failure of gonadotropins to induce *in vitro* maturation of oocytes treated with dibutyryl cyclic AMP. *J. Exp. Zool.* **194**: 529–534.

Olsiewski, P.J. and W.H. Beers. 1983. cAMP synthesis in the rat oocyte. *Dev. Biol.* **100**: 287–293.

Seamon, K.B., W. Padgett, and J.W. Daly. 1981. Forskolin: Unique diterpene activator of adenylate cyclase in membranes and in intact cells. *Proc. Natl. Acad. Sci.* **78**: 3363–3367.

Stengel, D., L. Guenet, M. Desimier, P. Insel, and J. Hanoune. 1982. Forskolin requires more than the catalytic unit to active adenylate cyclase. *Mol. Cell. Endocr.* **28**: 681–690.

Importance of Electrical Cell-Cell Communication in Secretory Epithelia

Ole H. Petersen
MRC Secretory Control Research Group
The Physiological Laboratory
University of Liverpool, Liverpool L69 3BX, England

Gap junctions have been found in all secretory epithelia investigated (Petersen 1980). In the liver, electrical communication occurs over relatively long distances (500 μm) (Graf and Petersen 1978), whereas in exocrine glands with secretory end pieces, clusters of acinar cells form communicating units within which there is little electrotonic decrement from cell to cell, but these units appear to be electrically isolated from each other (Iwatsuki and Petersen 1978a). Secretory epithelial cells have generally been regarded as electrically inexcitable (Petersen 1980) and the importance of electrical communication has therefore been obscure. However in 1983 voltage-dependent, single-channel currents were observed directly in patch-clamp studies on the basolateral acinar plasma membrane from several mammalian salivary glands (Maruyama et al. 1983a), and similar voltage-gated K^+ channels have since been found in several other secretory epithelia such as the pancreatic acini from pig and humans (Maruyama et al. 1983b; Petersen et al. 1985) and the exorbital lacrimal gland from rat and mouse (Findlay 1984; Trautmann and Marty 1984; Marty et al. 1984). In view of these recent new findings it seems important to reassess the functional implications of the electrical communication in secretory epithelia.

THE COMMUNICATION NETWORK IN THE EXOCRINE ACINAR TISSUE

The electrical communication network has been investigated in some detail in the mouse pancreatic acinar tissue using simultaneous intracellular recording with two microelectrodes and direct microscopical control of the localizations of the electrode tips (Iwatsuki and Petersen 1978a). All cells within one acinus are closely electrically coupled, whereas cells in what may appear to be two different acini are sometimes coupled, but in other cases are completely electrically isolated from each other. Electrical coupling is almost an "all-or-nothing" phenomenon and this can most easily be explained by assuming the existence of electrical acinar units containing cells closely linked together by cell–cell channels, but completely electrically isolated from other acinar units. On the basis of the spatial spread of electrical current pulses in the tissue as well as measurement of the total membrane capacitance in the acinar units, it can be estimated that several hundred cells make up one unit (Iwatsuki and Petersen 1978a).

An alternative method used to map the communication network in acinar tissue is the observation of fluorescent tracer movements. Direct visualization of cell–cell communication was achieved by demonstrating transfer of fluorescein, Procion Yellow, and the intensely fluorescing dye Lucifer Yellow from the injected acinar cell to neighboring cells in the living tissue (Hammer and Sheridan 1978; Kater and Galvin 1978; Iwatsuki and Petersen 1979; Findlay and Petersen 1982). Sustained intra-acinar, microiontophoretic injection of Lucifer Yellow revealed a finite limit to the extent of dye spread. In two mouse pancreatic preparations for which complete sets of serial sections could be obtained, the dye-coupled intercommunicating acinar units consisted of 110 and 230 indvidual acinar cells; Lucifer Yellow could only be detected in the acinar cells, whereas the duct cells did not contain the dye (Findlay and Petersen 1983). These investigations of dye-coupling in the acinar tissue are in good agreement with the electrical studies and demonstrate very clearly that acinar cells are interconnected and that the functional unit is relatively large, consisting of more than 100 cells.

It is possible to uncouple acinar cells that normally communicate with each other by a number of different procedures. Large doses of pancreatic secretagogues like acetylcholine, cholecystokinin, and bombesin can evoke electrical uncoupling of cells within functional acinar units and such effects are completely and rapidly reversible (Fig. 1). More recently it has been shown that acetylcholine-evoked electrical uncoupling restricts the passage of Lucifer Yellow between the acinar cells (Findlay and Petersen 1982). Intracellular Ca^{++} injection can electrically isolate the injected cell (not reversible), and intracellular acidification, brought about either by direct intracellular H^+ injection or by exposure of the cells to solutions with high CO_2 concentrations, can evoke electrical uncoupling; this effect is fully reversible (Petersen 1980).

Electron microscopy of freeze-fractured glutaraldehyde-fixed samples of acinar tissue have revealed large membrane areas closely packed with gap junction particles, confirming the tight coupling of adjacent cells within acinar units (Kater and Galvin 1978; Meda et al. 1983). An attempt has been made to find ultrastructural changes in the gap junctions related to the uncoupling of acinar cells. Exposure of pancreatic acinar tissue to solutions equilibrated with 100% CO_2 has been shown to evoke complete electrical uncoupling of all acinar cells within a unit, but this rapid and reversible uncoupling does not evoke any marked change in the morphology of gap junctions as investigated by

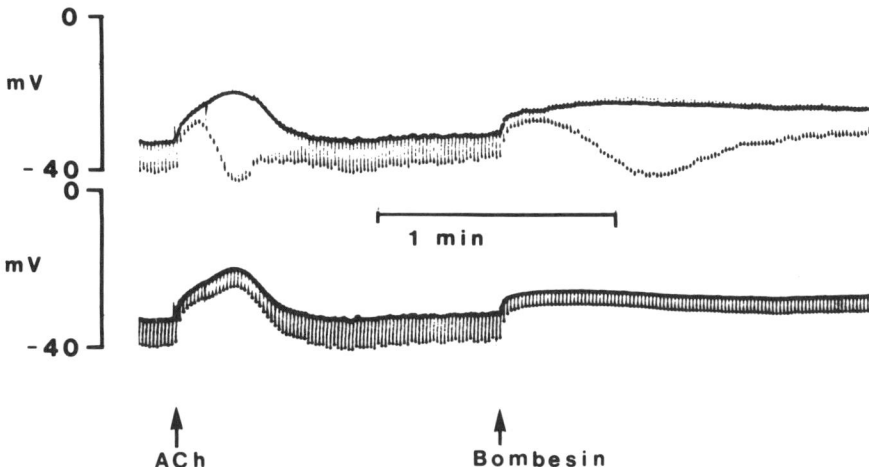

Figure 1
Mouse pancreatic acinar unit. Membrane potential measurements from two neighboring cells obtained with two separate intracellular microelectrodes. Rectangular hyperpolarizing current pulses (2.5 nA, 100 msec) were applied repetitively through the microelectrode from which the upper membrane potential recording was obtained (the electrode resistance was balanced out). At arrow marked ACh, an ejecting current pulse (80 nA, 10 sec) was applied to an extracellular acetylcholine-chloride (2 M)-containing micropipette with its tip close (<10 μM) to the surface of the acinar cells from which the recording was made. At arrow marked Bombesin, 10 pmoles of the peptide was injected into the tissue bath (20 ml, flow 10 ml/min). The two cells are tightly coupled before stimulation, but after the initial stimulus-evoked depolarization and membrane resistance reduction there is a transient large increase in current-pulse evoked hyperpolarizations in the cell of current injection (*upper trace*) with little increase in the potentials transmitted to the neighboring cell (*below*). (Reprinted, with permission, from Iwatsuki and Petersen 1978b.)

electron microscopy of freeze-fractured and glutaraldehyde-fixed acinar tissue. Careful quantification does reveal, however, small increases (2-6%) in particle diameter and spacing within gap junctions of uncoupled cells. These minor structural changes are reversible upon recoupling (Meda et al. 1983).

THE RESTING MEMBRANE POTENTIAL AND K+ CHANNELS IN ACINAR CELLS

The resting membrane potential in exocrine acinar cells varies between -30 and -70 mV, depending on the particular gland and species chosen, but is in all cases mainly dependent on the transmembrane K^+ gradient (Petersen 1980). In secretory epithelia, as in most other tissues, the negative resting potential must therefore be due to the existence of selective membrane K^+ channels and a high intracellular K^+ concentration.

Very selective K^+ channels have now been studied directly with the patch-clamp technique in acinar cells from a large number of exocrine glands (Ma-

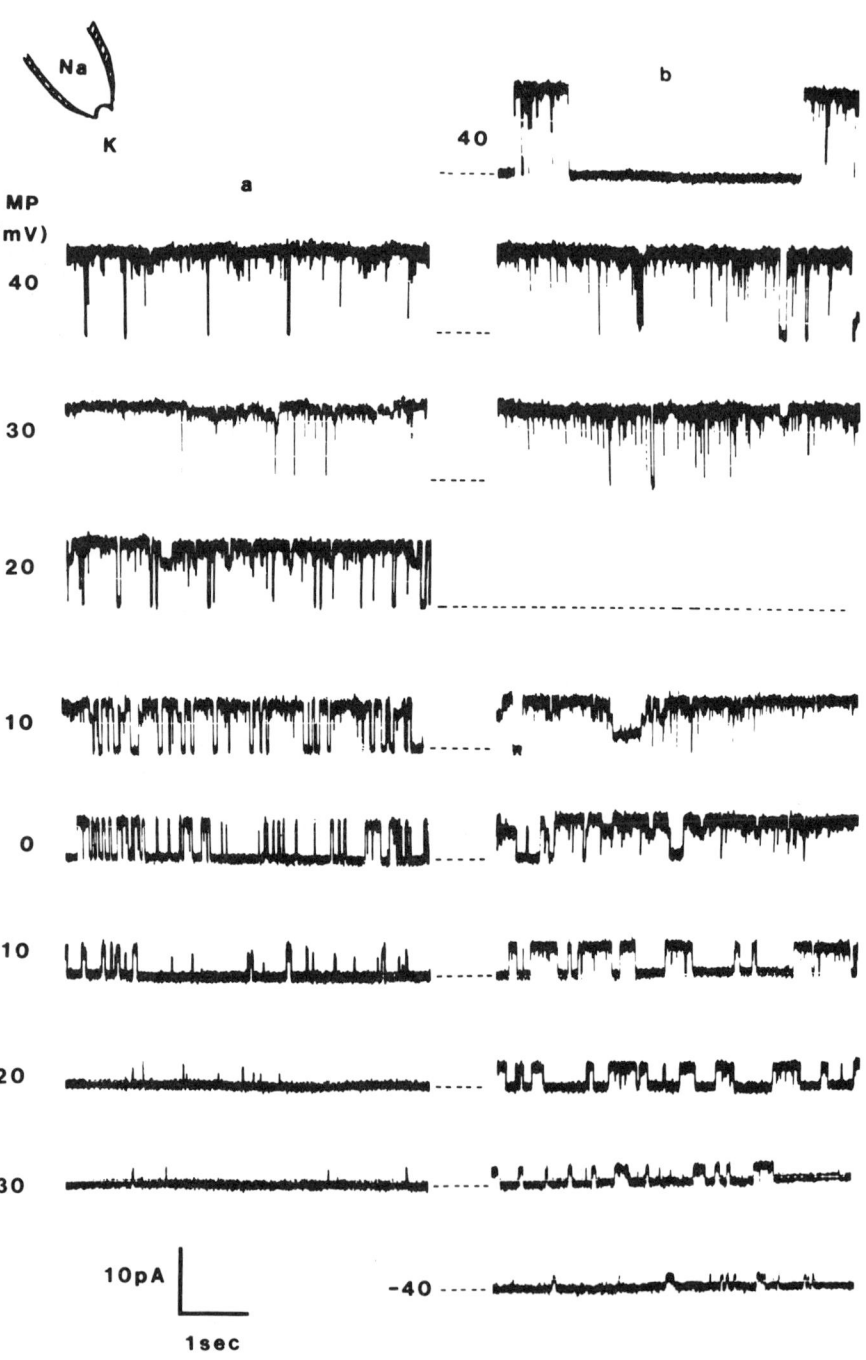

Figure 2 (See facing page for legend.)

ruyama et al. 1983a,b; Findlay 1984; Trautmann and Marty 1984). The properties of acinar cell K^+ channels can most easily be assessed by analyzing single-channel current activity in an excised patch of plasma membrane exposed to solutions of known composition. Figure 2 shows results obtained from such an excised membrane patch where the inside of the plasma membrane faces the K^+-rich (intracellular) bathing solution and the outside the Na^+-rich (extracellular) pipette solution. The left column (a) shows the current obtained when the ionized Ca^{++} concentration in the bath ($[Ca^{++}]_i$) is 10^{-8} M. At the negative membrane potentials, the current steps are small and infrequent whereas at positive potentials the channels are open much of the time but interrupted by closings revealing the large single-channel current. The null potential has not been attained, but it is clear that it is more negative than -40 mV and since K^+ is the only ion with a negative equilibrium (Nernst) potential in this experimental situation the channel must be K^+ selective. The relationship between the probability of the channel being open (p) and the membrane potential can also be obtained from Figure 2. Although channel openings are clearly occurring at a membrane potential of -30 mV, it is seen that at $[Ca^{++}]_i = 10^{-8}$ M p is very low indeed. Increasing $[Ca^{++}]_i$ to 10^{-7} M activates the channel. As seen in the right-hand column (b) of Figure 2, the channel spends much more time in the open state at the negative membrane potentials than at the lower Ca^{++} concentration, however, the amplitude of the single-channel currents is unaffected by the change in $[Ca^{++}]_i$.

It is difficult to investigate single-channel currents in excised patches at physiological membrane potentials (i.e., about -60 mV) under conditions with quasi-physiological ion gradients because the driving force for K^+ outflow (the difference between the K^+ equilibrium [Nernst] potential and the membrane potential) is so small that the single-channel currents are barely detectable. However, when the K^+ concentration is raised to intracellular levels in the pipette and recordings are made from intact cells as in the experiments represented by Figure 3, clear channel openings can be observed at the normal resting potential although the probability of channel opening is low. In this type of experiment the extracellular patch pipette is filled with an intracellular solution, i.e., $[K^+]$ in the pipette (150 mM) is very close to $[K^+]_i$, but the cells are otherwise exposed to a standard extracellular (Na^+-rich) physiological saline solution. When no potential is applied to the pipette ($V_p = 0$ mV), inward-going,

Figure 2
Single-channel current traces from an excised inside-out pig pancreatic acinar membrane patch exposed to extracellular solution on the outside (pipette) and intracellular solution on the inside (bath). The cartoon shows the recording configuration. In the part of the experiment shown in a the ionized Ca^{++} concentration in the bath ($[Ca^{++}]_i$) was 10^{-8} M. At the time when the traces shown in b were obtained from the same patch, $[Ca^{++}]_i$ was adjusted to 10^{-7} M. The membrane potentials (MP) at which the individual traces were obtained are indicated. The current level when the channel is closed is indicated by the horizontal dashed lines. Upward deflections represent outward current. Two traces represent the 10^{-7} M situation at a membrane potential of 40 mV. The upper trace shows that there were times when the channel went from the open state into a state of prolonged closure. (Reprinted, with permission, from Maruyama et al. 1983b.)

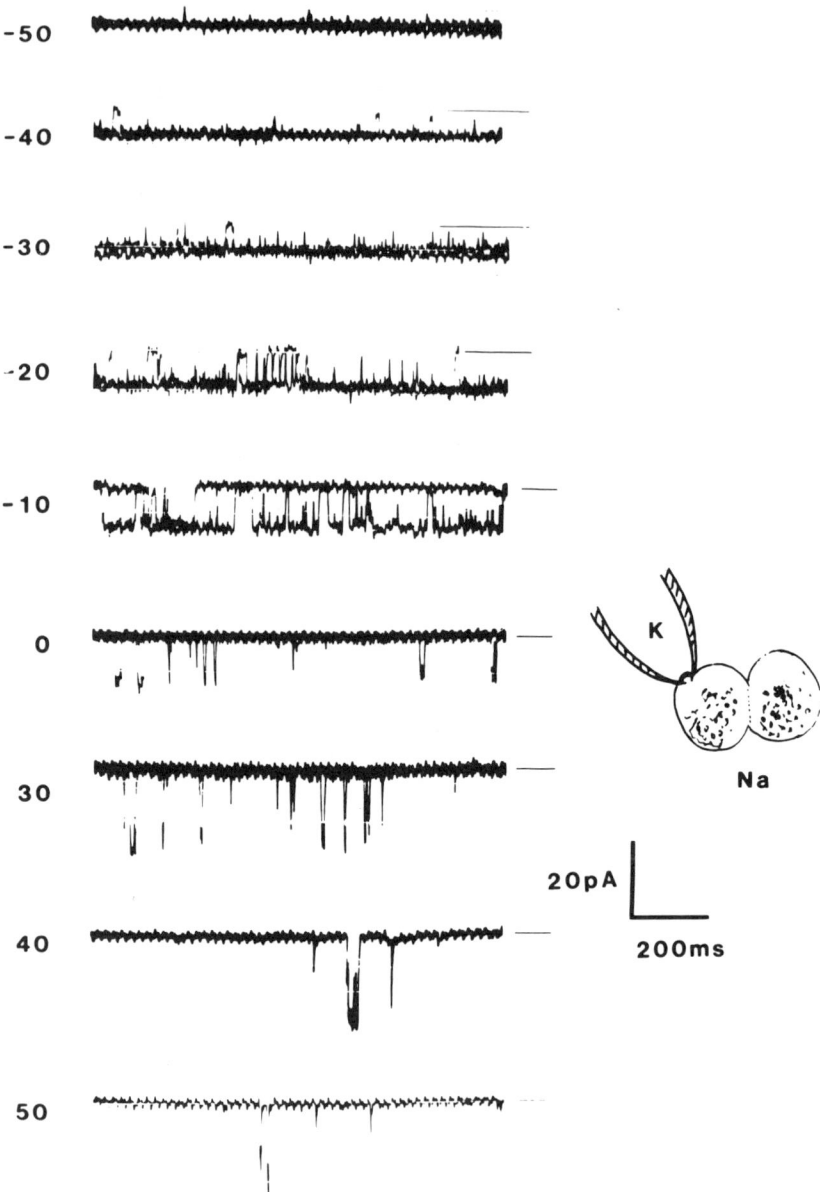

Figure 3
Single-channel current recording from isolated pig acinar cells. The cartoon shows the recording configuration. The recording pipette is fitted with intracellular solution (high K^+ concentration) whereas the bath fluid is an extracellular solution (high Na^+ concentration). The traces show single-channel currents through the K^+ channel in the isolated patch area at different pipette potentials (V_p). Two superimposed oscilloscope beam sweeps are shown at each pipette potential (except 40 and 50 mV). (From Y. Maruyama and O.H. Petersen, unpubl.)

single-channel currents are observed. The driving force for this current must be the intracellular negativity as there is no K+ gradient over the isolated patch membrane. When V_p is made positive, the inward-going, single-channel currents have larger amplitude (they also become less frequent as the probability of channel opening is dependent on the membrane potential). Making V_p negative reduces the single-channel current amplitude and the channel now spends more time in the open state.

A very useful extension of the patch-clamp technique for single-channel recording is the tight-seal, whole-cell recording (Marty and Neher 1983). Starting out from the recording situation shown in Figure 3 (cell-attached configuration), it is possible to break the isolated membrane patch and establish continuity between the pipette interior and the cell interior. The membrane potential can be clamped close to the resting level (−60 mV) and the currents associated with depolarizing and hyperpolarizing voltage jumps recorded. In exocrine acinar cells, such experiments show that the whole plasma membrane acts as a rectifier. Depolarizing voltage jumps are associated with large outward currents whereas hyperpolarizing jumps of the same magnitude generate very little inward current (Maruyama et al. 1983b; Maruyama and Petersen 1984; Trautmann and Marty 1984). This extreme rectifier property of the acinar cell membrane can be explained by the characteristics of the single voltage-gated K+ channels. Depolarization opens, whereas hyperpolarization closes, these channels. The K+ channels of the plasma membrane therefore act as a very efficient machinery for preventing depolarization, because a depolarizing influence will activate the K+ conductance and this evokes repolarization.

COUNTING THE NUMBER OF K+ CHANNELS PER CELL

The whole-cell, voltage-clamp currents can be regarded as the sum of many small elementary currents flowing through individual K+ channels of the type characterized in the single-channel studies described above. However, it is not easy to prove that this attractive hypothesis is correct. There are a number of strong arguments that support this conclusion:

1. The whole-cell K+ current is Ca^{++}-activated and the Ca^{++}-sensitivity is similar to that described for the single-channel currents (Maruyama and Petersen 1984).
2. The activation of the whole-cell currents occurs over the same voltage ranges as described for the single-channel currents (Maruyama et al. 1983b; Maruyama and Petersen 1984).
3. Ca^{++}-activated K+ currents can be relatively specifically blocked by low concentrations (about 2 mM) of tetraethylammonium (TEA) from the outside of the plasma membrane, and in the pig pancreatic acinar cells it has recently been shown that TEA applied in this way blocks both single-channel and whole-cell K+ currents (Iwatsuki and Petersen 1985).

If one accepts, on the basis of these three arguments that the whole-cell current in the acinar cells is the sum of the elementary currents flowing through individual K+ channels when they are open, it follows that the relationship between the whole cell current (I) and the single-channel current (i) is given by: $I = N p i$, where N is the total number of operational K+ channels and p the open-state probability. At positive membrane potentials where p is close to 1: $N = I/i$.

In the resting pig acinar cells this approach has indicated that the total channel number is quite small, only about 50 channels/cell (Maruyama et al. 1983b). In the rat lacrimal acinar cells, ensemble noise analysis indicates the presence of 50–150 channels/cell (Trautmann and Marty 1984). This small number of high-conductance channels would have caused problems if it had not been for the extensive cell–cell coupling in the acinar units. The open-state probability in the resting cells is very low (Fig. 3). Assuming a figure of 0.02 for the physiological situation (−60 mV) on average only one channel will be open per cell. Since, as clearly seen in Figures 2 and 3, channels open and close frequently, a cell would sometimes be in a position of having no channels open whereas in other periods one and sometimes two channels would be conducting. This situation would lead to a very unstable membrane potential in an electrically isolated cell. However, in the acinar tissue about 100 cells form a functional unit and this means that on average about 100 channels will be open in the resting situation in any one unit. With this number of channels the effect of openings and closings of single channels will be modest, as the sudden closure of one channel, for example, would only reduce the overall conductance by 1% whereas in a single cell this event would have caused a 100% loss of conductance. It would appear therefore that cell–cell junctional channels are essential for maintaining a stable resting potential in cells with small numbers of K+ channels such as the exocrine acinar cells. Many cells sharing their small number of K+ channels can therefore overcome the threat of an unstable and unpredictable membrane potential arising from the stochastic nature of channel behavior.

THE IMPORTANCE OF THE MEMBRANE POTENTIAL FOR THE FUNCTION OF TRANSPORTING EPITHELIAL CELLS

In the previous section it has been argued that the gap junctions allow many cells to share membrane K+ channels thereby allowing a stable membrane potential. A stable membrane potential is important for many transport processes. One example is illustrated in Figure 4. All cells take up amino acids, and one mechanism that has been identified for many neutral amino acids in very many different cell types is cotransport with Na+ (Petersen 1980). "Downhill" movement of Na+ is coupled to "uphill" movement of amino acid and the driving force is provided by the transmembrane electrochemical gradient for Na+, of which the membrane potential is an important part. A low intracellular Na+ activity is, of course, also essential and this is maintained by the Na+-K+ pump. In the model shown in Figure 3 Na+ recirculates via the Na+-amino acid cotransporter and Na+-K+ pump. Therefore it is necessary also to provide an exit pathway for K+ so that K+ can recirculate across the membrane. The K+ channels described in the previous sections seem ideally suited to the model represented by Figure 4, since they would be activated by the depolarization brought about by the introduction of a neutral amino acid to the extra-

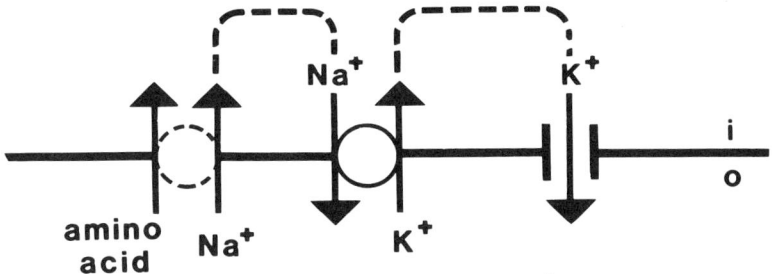

Figure 4
A simplified scheme to account for the transport of amino acids and cations across the basolateral plasma membrane of acinar cells. The Na$^+$ amino acid cotransport system is driven by the electrochemical gradient for Na$^+$ which is established by the Na$^+$-K$^+$ pump and the K$^+$ channels. The inner and outer surfaces of the plasma membrane are denoted by i and o. (Reprinted, with permission, from Singh and Petersen 1984.)

cellular solution, thereby partially repolarizing the membrane, and also provide the necessary exit pathway for K$^+$. Many secretory epithelia transport NaCl and in several thoroughly investigated cases Cl$^-$ moves through the cells whereas Na$^+$ mostly passes paracellularly, i.e., between the cells through the tight junctions (Frizzell et al. 1979; Pearson et al. 1984). There is now strong evidence in many different tissues for Na$^+$-coupled Cl$^-$ transport. Also in these cases recirculation of K$^+$ is essential in order for the transport to occur, and it is indeed possible to envisage regulation of the overall Cl$^-$ uptake by control of the opening of the K$^+$ channels (Petersen and Maruyama 1984).

In conclusion, it is suggested that it may be essential for transporting epithelia to share K$^+$ channels in a whole unit of communicating cells in order to provide the stable K$^+$ conductance essential for several key transport functions.

REFERENCES

Findlay, I. 1984. A patch-clamp study of potassium channels and whole-cell currents in acinar cells of the mouse lacrimal gland. *J. Physiol.* **350:** 179–195.

Findlay, I. and O.H. Petersen. 1982. Acetylcholine-evoked uncoupling restricts the passage of Lucifer Yellow between pancreatic acinar cells. *Cell Tissue Res.* **225:** 633–638.

———. 1983. The extent of dye coupling between exocrine acinar cells of the mouse pancreas. *Cell Tissue Res.* **232:** 121–127.

Frizzell, R.A., M. Field, and S.G. Schultz. 1979. Sodium-coupled chloride transport by epithelial tissues. *Am. J. Physiol.* **236:** F1–F8.

Graf, J. and O.H. Petersen. 1978. Cell membrane potential and resistance in liver. *J. Physiol.* **284:** 105–126.

Hammer, M.G. and J.D. Sheridan. 1978. Electrical coupling and dye transfer between acinar cells in rat salivary glands. *J. Physiol.* **275:** 495–505.

Iwatsuki, N. and O.H. Petersen. 1978a. Electrical coupling and uncoupling of exocrine acinar cells. *J. Cell Biol.* **79:** 533–545.

———. 1978b. in vitro action of bombesin on amylase secretion, membrane potential and membrane resistance in rat and mouse pancreatic acinar cells. *J. Clin. Invest.* **61:** 41–46.

———. 1979. Direct visualization of cell to cell coupling: Transfer of fluorescent probes in living pancreatic acini. *Pflueger's Arch.* **380:** 277–281.

———. 1985. Action of tetraethylammonium on calcium-activated potassium channels in pig pancreatic acinar cells studied by patch-clamp single-channel and whole-cell current recording. *J. Membr. Biol.* **86:** (in press).

Kater, S.B. and N.J. Galvin. 1978. Physiological and morphological evidence for coupling in mouse salivary gland acinar cells. *J. Cell Biol.* **79:** 20–26.

Marty, A. and E. Neher. 1983. Tight-seal whole-cell recording. In *Single-channel recording* (ed. B. Sakmann and E. Neher), p. 107–122. Plenum Press, New York.

Marty, A., Y.P. Tan, and A. Trautmann. 1984. Three types of calcium-dependent channel in rat lacrimal glands. *J. Physiol.* **357:** 293–325.

Maruyama, Y. and O.H. Petersen. 1984. Control of K^+ conductance by cholecystokinin and Ca^{2+} in single pancreatic acinar cells studied by the patch-clamp technique. *J. Membr. Biol.* **79:** 293–300.

Maruyama, Y., D.V. Gallacher, and O.H. Petersen. 1983a. Voltage and Ca^{2+}-activated K^+ channel in baso-lateral acinar cell membranes of mammalian salivary glands. *Nature* **302:** 827–829.

Maruyama, Y., O.H. Petersen, P. Flanagan, and G.T. Pearson. 1983b. Quantification of Ca^{2+}-activated K^+ channels under hormonal control in pig pancreatic acinar cells. *Nature* **305:** 228–232.

Meda, P., I. Findlay, E. Kolod, L. Orci, and O.H. Petersen. 1983. Short and reversible uncoupling evokes little change in the gap junctions of pancreatic acinar cells. *J. Ultrastruct. Res.* **83:** 69–84.

Pearson, G.T., P.M. Flanagan, and O.H. Petersen. 1984. Neural and hormonal control of membrane conductance in the pig pancreatic acinar cell. *Am. J. Physiol.* **247:** G520–G526.

Petersen, O.H. 1980. *The electrophysiology of gland cells*. Academic Press, New York.

Petersen, O.H. and Y. Maruyama. 1984. Calcium-activated potassium channels and their role in secretion. *Nature* **307:** 693–696.

Petersen, O.H., I. Findlay, N. Iwatsuki, J. Singh, D.V. Gallacher, C.M. Fuller, G.T. Pearson, M.J. Dunne, and A.P. Morris. 1985. Human pancreatic acinar cells: Studies of stimulus-secretion coupling. *Gastroenterology* **89:** 109–117.

Singh, J. and O.H. Petersen. 1984. The effects of L-alanine and acetylcholine on membrane potential, ^{45}Ca and $^{86}Rb^+$ efflux and amylase secretion in the isolated mouse pancreas. *Q. J. Exp. Physiol. Cogn. Med. Sci.* **69:** 531–540.

Trautmann, A. and A. Marty. 1984. Activation of Ca-dependent K channels by carbamoylcholine in rat lacrimal glands. *Proc. Natl. Acad. Sci.* **81:** 611-615.

Electrical Interactions and Synchronization of Cortical Neurons: Electrotonic Coupling and Field Effects

F. Edward Dudek and Robert W. Snow
Department of Physiology
Tulane University School of Medicine
New Orleans, Louisiana 70112

Synaptic interactions are critical for local neuronal integration. Local circuit interactions are thought to be increased in phylogenetically advanced areas of the mammalian brain and may be the basis of higher brain functions (Schmitt et al. 1976). For both projected and local transfer of signals between neurons, chemical synapses are generally considered to be the prominent form of communication and information processing. Nonetheless, electrotonic synapses can impart subtle and complicated integrative capabilities to a neuronal network (for review, see Bennett 1977). Several lines of evidence developed over the last two decades have indicated that electrotonic junctions are more widespread than previously thought; recent data described here support the hypothesis that electrotonic junctions are present between some neurons in higher telencephalic structures such as the hippocampus and neocortex (for review, see Dudek et al. 1983).

Synchrony of neuronal activity occurs under both normal (e.g., slow brain waves) and pathological conditions (e.g., epilepsy). The mammalian hippocampus, the simplest and most primitive region of cerebral cortex, has long been used to study basic neurophysiological mechanisms and as a model for understanding the basis of epileptiform discharges. Until recently, the possibility that electrical interactions could be an important synchronizing mechanism in the hippocampus had received little or no experimental support. Few studies had been performed in the hippocampus or neocortex concerning the presence of gap junctions and electrotonic synapses, or on the occurrence of electrical field (ephaptic) interactions. However, recent data from several laboratories now support the following hypotheses: (1) hippocampal neurons can fire synchronously without active chemical synapses; (2) dye coupling, gap junctions, and electrotonic synapses are present in the hippocampus and neocortex; and (3) electrical field effects are an important synchronizing mechanism of epileptiform bursts in the hippocampus. Detailed reviews have recently been provided on this subject (Dudek et al. 1983, 1985).

RESULTS AND DISCUSSION

Synchronization of Hippocampal Neurons after Blockade of Chemical Synapses

Although several studies had indicated the importance of recurrent excitatory chemical synapses in synchronization of epileptiform bursts (e.g., Johnston and Brown 1981; Traub and Wong 1983), it was unclear whether chemical transmission was actually necessary for synchronization of hippocampal neurons. When chemical synapses were blocked by incubating hippocampal slices in high $[Mg^{++}]$ or by addition of Mn^{++}, and excitability was increased by lowering extracellular $[Ca^{++}]$ (which also greatly contributed to blockade of chemical synapses), pyramidal and granule cells could generate synchronous activity (Jefferys and Haas 1982; Taylor and Dudek 1982b; Snow and Dudek 1984a). Local electrical stimulation could evoke prolonged afterdischarges of synchronous spikes (negative extracellular "population spikes" with simultaneous intracellular prepotentials and action potentials). Spontaneous bursts of population spikes also occurred (Fig. 1). These data indicated that one or more mechanisms, independent of chemical synapses, were capable of synchronizing hippocampal neurons. Changes in the concentration of extracellular ions were probably involved in this phenomenon, but would be expected to act on a relatively slow time scale (Konnerth et al. 1984; Taylor and Dudek 1984b). The presence of population spikes in extracellular recordings indicated synchronization of hippocampal neurons on a millisecond time scale; electrotonic coupling through gap junctions and electrical field effects were two prime candidate mechanisms for such rapid synchronization.

Dye coupling

The small fluorescent dye, Lucifer Yellow, has been shown to pass between a wide variety of coupled cells (Stewart 1981), presumably through gap junctions (see other chapters). Dye coupling occurred between hippocampal neurons (Fig. 2A) after most intracellular injections of Lucifer Yellow in vitro (MacVicar and Dudek 1980, 1982; Andrew et al. 1982) and in vivo (MacVicar et al. 1982). Generally two or three neurons were stained and the site of coupling appeared to be between dendrites and/or somata. In neocortical slices, similar results were obtained (Gutnick and Prince 1981), and the amount of dye coupling appeared higher at early stages of development (Connors et al. 1983). Injection of Lucifer Yellow into glia yielded widespread dye coupling (Gutnick et al. 1981). In these studies on hippocampus and neocortex, extracellular ejections of Lucifer Yellow rarely stained neurons.

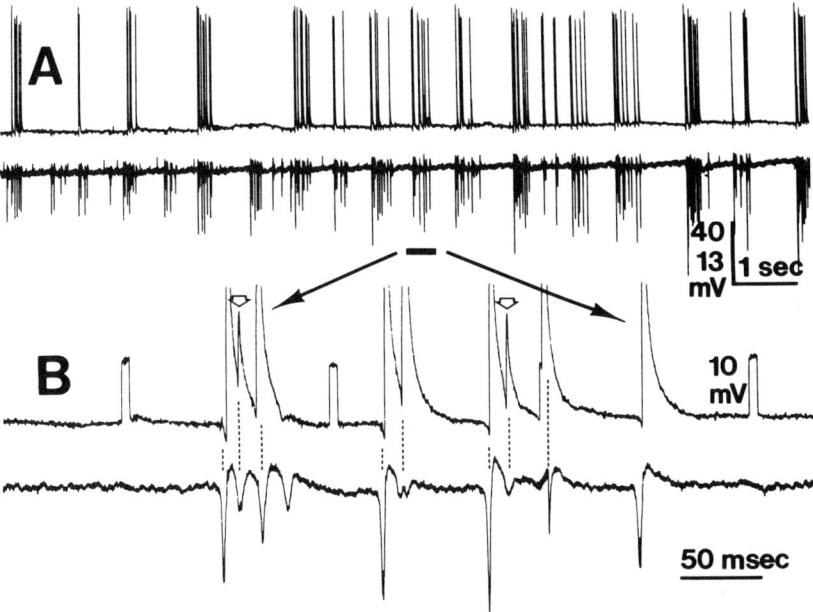

Figure 1
Spontaneous bursts in hippocampal pyramidal cells during blockade of chemical synaptic transmission. (*A*) Slow time scale. Bursts of action potentials in a neuron (*upper trace*) were synchronous with extracellular population spikes (*lower trace*). (*B*) Fast time scale. Recording is of period underlined in *A*. Individual action potentials and subthreshold depolarizations (open arrows) were always aligned (dashed lines) with field potential spikes. (Reprinted, with permission, from Taylor and Dudek 1982b.)

It has been proposed that dye coupling is an artifact from dye passage along the electrode track (Knowles et al. 1982; Alger et al. 1983); however, exposure of neocortical slices to CO_2 significantly reduced neuronal and glial dye coupling (Gutnick and Lobel-Yaakov 1983; Connors et al. 1984), and incubation of hippocampal slices in propionate led to a significant reduction in dye coupling between pyramidal cells (MacVicar and Jahnsen 1985). These treatments would be expected to lower intracellular pH and reduce gap junctional communication (see Spray et al., this volume) and would not be expected to change the frequency of an artifact. Although some fraction of the observed dye coupling may be due to artifact, the observation of significant reductions in dye coupling with treatments expected to lower intracellular pH argues that at least some dye transfer occurs through gap junctions. However, the preparation of brain slices and the associated damage to axons and dendrites may cause formation of new junctions or strengthen existing ones; this is known to occur among specific neurons in the snail, *Helisoma* (Bulloch et al. 1980; Bulloch and Kater 1981), and recent studies measuring dye coupling in neocortical slices support this hypothesis (M.J. Gutnick, pers. comm.).

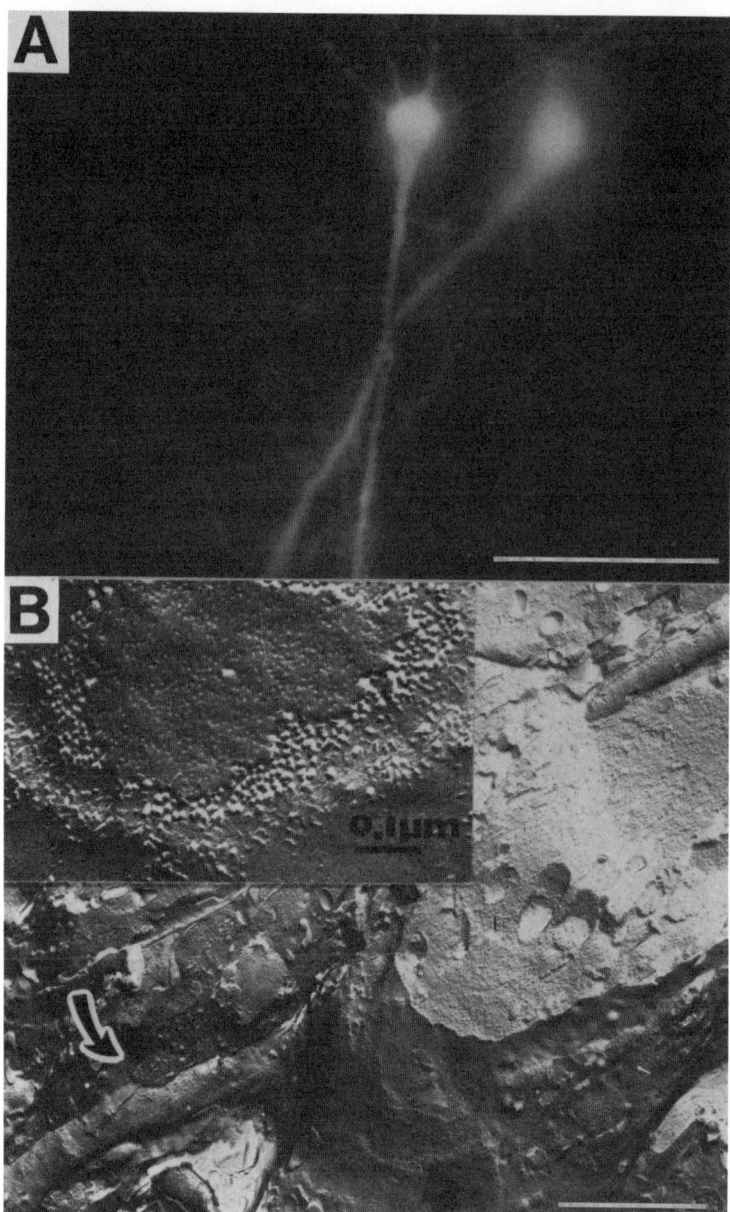

Figure 2
Evidence for dye transfer through gap junctions between hippocampal pyramidal cells. (A) Dye coupling after intracellular injection with Lucifer Yellow. A pair of CA1 pyramidal cells was stained after a single injection; the site of contact was apical dendrites, and the somata were clearly separated. Bar, 50 μm. (Reprinted, with permission, from Dudek et al. 1983.) (B) Gap junction in freeze-fracture replica of pyramidal cell. Gap junction is indicated by straight arrow, and a dendrite with chemical synapse is shown by curved arrow. Inset is enlargement of gap junction and shows particles on P face and pits on E face. White bar, 1 μm. (Reprinted, with permission, from Schmalbruch and Jahnsen 1981.)

Gap Junctions

Anatomical observations of close neuronal appositions, particularly in cell body layers, had suggested the occurrence of electrical interactions among neurons in the hippocampus (Green and Maxwell 1961) and neocortex. Gap junctions were observed in freeze-fracture replicas (Fig. 2B) of CA3 pyramidal cells (Schmalbruch and Jahnsen 1981) and dentate granule cells (MacVicar and Dudek 1982); some of these junctions were relatively large ($\sim$600 particles) and they were either on promixal dendrites or cell bodies. A single large gap junction between two CA3 neurons has been predicted to produce a coupling ratio of 0.7 (Taylor and Dudek 1982a; Dudek et al. 1983). In thin sections, however, gap junctions were only seen on glia and putative interneurons, but not on pyramidal or granule cells (Schmalbruch and Jahnsen 1981; Kosaka 1983a,b; but also see Schwartzkroin 1983). Gap junctions have also been reported among neocortical neurons (see Dudek et al. 1983, 1985 for refs.). Thus, the available data suggest that at least a few gap junctions are present on mammalian cortical neurons; some of these junctions may be large and might provide strong electrotonic coupling among small groups of neurons.

Electrotonic Coupling

Two types of electrophysiological evidence have been obtained to support the hypothesis that some hippocampal and neocortical neurons are electrotonically coupled through gap junctions. First, antidromic stimulation evoked short-latency depolarizations in hippocampal pyramidal cells (Knowles et al. 1982; Taylor and Dudek 1982a), dentate granule cells (MacVicar and Dudek 1982), and neocortical neurons (Gutnick and Prince 1981), but these events were only seen in a relatively small fraction of neurons. A previous action potential did not occlude some of these depolarizations (Fig. 3A), thus arguing that they were electrotonic synaptic potentials from coupled neurons. Second, dual intracellular recordings have shown low coupling ratios (0.1–0.4, Fig. 3B) between a small percentage of pyramidal (MacVicar and Dudek 1981; Wong et al. 1984) and granule cells (MacVicar and Dudek 1982). High coupling ratios have also been observed (MacVicar and Dudek 1981; Knowles et al. 1982), which could represent either simultaneous impalements of the same cell or dual recordings from coupled cells. Thus, in both hippocampus and neocortex, dye coupling has been observed frequently, but convincing evidence for electrotonic coupling has been rare. Knowles et al. (1982) and Alger et al. (1983) have argued that these and other observations mean dye coupling is artifactual; however, neither group has yet suggested how dye, but not ions, could be transferred between neurons. Only such a result could give rise to dye coupling without electrotonic coupling. Furthermore, the antidromic stimulation technique will not detect coupling between neurons if it is too weak or too strong, and dual intracellular recording suffers from a severe sampling problem and from the difficulty of proving conclusively whether one is in the same or different cells. Caveats aside, these results support the hypothesis that small groups of hippocampal and neocortical neurons are coupled; some weak coupling appears to be present, and strong coupling is also possible.

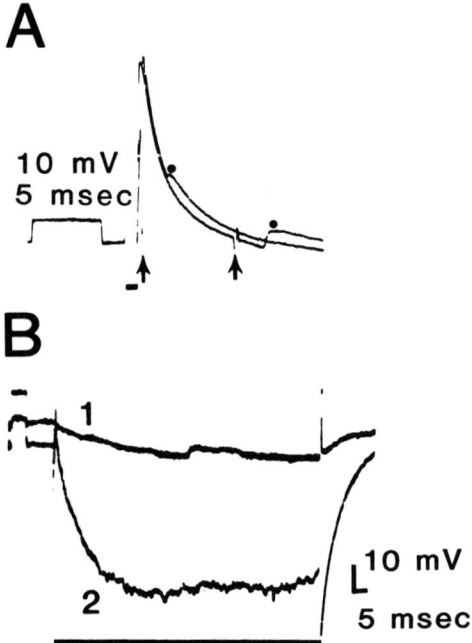

Figure 3
Electrophysiological evidence for coupling between pyramidal cells. (*A*) Antidromic test in CA1 pyramidal cell. Synaptic transmission was blocked by adding Mn^{++} (2.3 mM) and lowering $[Ca^{++}]$ (0.5 mM). The neuron was directly stimulated by a brief depolarizing pulse (bar). Antidromic stimuli were delivered at arrows. Short latency depolarizations (●) evoked by antidromic stimuli were not blocked by preceding action potential, arguing that they arose in a coupled neuron. (Reprinted, with permission, from Taylor and Dudek 1982a.) (*B*) Dual intracellular recording. Two CA3 pyramidal neurons were penetrated independently. Hyperpolarizing current (3 nA) injected into cell 2 caused a hyperpolarization in cell 1. Both electrodes contained horseradish peroxidase, and two stained neurons were recovered histologically. (Reprinted, with permission, from MacVicar and Dudek 1981.)

Hypothetical Relation between Electrotonic and Dye Coupling among Neurons

Recently, Murphy et al. (1983) showed that the probability of occurrence of dye coupling among identified *Helisoma* neurons was directly related to the strength of electrotonic coupling; that is, dye coupling consistently occurred between neurons with strong electrotonic coupling, but was not present among weakly coupled neurons. The dye coupling among small clusters of neurons in hippocampus and neocortex may represent transfer through gap junctions of high conductance that are close to the somatic impalement site; this is consistent with the ultrastructural data on gap junctions (see above). On the other hand, the lack of dye coupling in the inferior olive (R. Llinás, pers. comm.) may reflect more remote and weakly transmitting junctions; the resulting weak (but frequent) coupling might then consistently yield positive results with the antidromic test (Llinás et al. 1974). This interpretation of the data suggests the hypothesis that the antidromic test and dye coupling reveal weak and strong junctions, respectively, among mammalian neurons. Future studies (see below) may be able to test this hypothesis.

Possible Contribution of Electrotonic Coupling to Synchronization

Because the number and strength of electrotonic junctions among neurons in hippocampus and neocortex is unknown, the contribution of these interactions to synchronization is hard to evaluate. Depolarizations that could be electrotonic coupling potentials were seen during synchronous bursting in hippocampus (Figs. 1 and 4), but these depolarizations could also have been generated by the impaled neuron (i.e., in axons or dendrites) (Taylor and Dudek 1982b, 1984b). Computer simulations have suggested that coupling in small clusters of hippocampal neurons could enhance synchronization when electrical field effects were of intermediate strength (Traub et al. 1985). Addi-

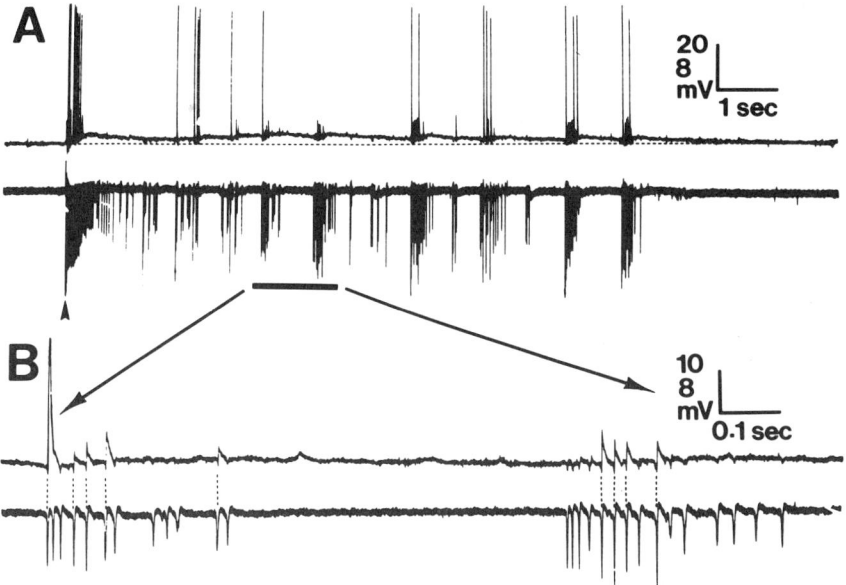

Figure 4
Possible electrotonic coupling potentials during synchronous activity of hippocampal pyramidal cells. (A) Stimulated afterdischarge in hyperexcitable preparation after blockade of chemical synaptic transmission. (B) Possible coupling potentials. Subthreshold depolarizations were synchronous with population spikes (dashed lines) and could represent electrotonic synaptic potentials from a coupled neuron or partial spikes from an electrically remote part of the impaled cell. Period under bar in A is expanded in B. (Reprinted, with permission, from Dudek et al. 1983.)

tional studies are needed to define quantitatively the extent and strength of electrotonic coupling among neurons in hippocampus and neocortex before their relative importance to synchronization can be assessed under normal and abnormal conditions.

Electrical Field Effects and Their Role in Synchronization

The close neuronal apposition among pyramidal cells in the hippocampus (Green and Maxwell 1961), combined with their parallel arrangement, provides an ideal anatomical condition for electrical field effects. Differential recording during antidromic or orthodromic activation (Fig. 5) showed that synchronous spike activity ("population spikes") tends to depolarize and synchronize nearby inactive neurons in vitro (Richardson et al. 1984; Taylor and Dudek 1982b, 1984a; R.W. Snow et al., in prep.) and in vivo (Taylor et al. 1984). Furthermore, similar rapid depolarizations, which have been attributed to electrical field effects, were readily recorded in hippocampal pyramidal cells during synchronous firing when chemical synapses were blocked (Taylor and Dudek 1982b, 1984b; Haas and Jefferys 1984) or when excitatory chemical synapses were functional during picrotoxin-induced epileptiform bursts (Fig. 6) (Snow and Dudek 1984b). These data support the hypothesis that electrical

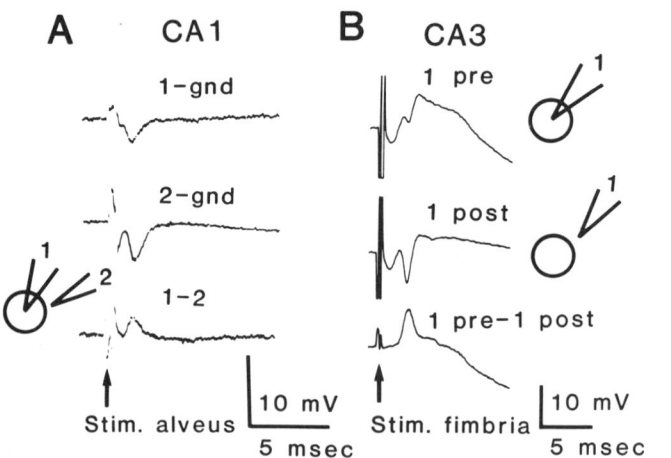

Figure 5
Electrical field effects in hippocampal pyramidal cells. (*A*) Antidromic stimulation with chemical synapses blocked. Stimulation of alveus gave rise to extracellular population spikes (electrode 2-gnd). Subtraction of the field potential from the intracellular recording (electrode 1-gnd) revealed a field effect depolarization in the impaled cell (recording 1-2, *bottom trace*). (*B*) Orthodromic activation. Averaged extracellular recording obtained outside the cell (1 post) after withdrawal of the electrode was digitally subtracted from averaged intracellular recording (1 pre). The resulting transmembrane recording showed a large field effect depolarization during the extracellular population spike. (Reprinted, with permission, from Dudek et al. 1985.)

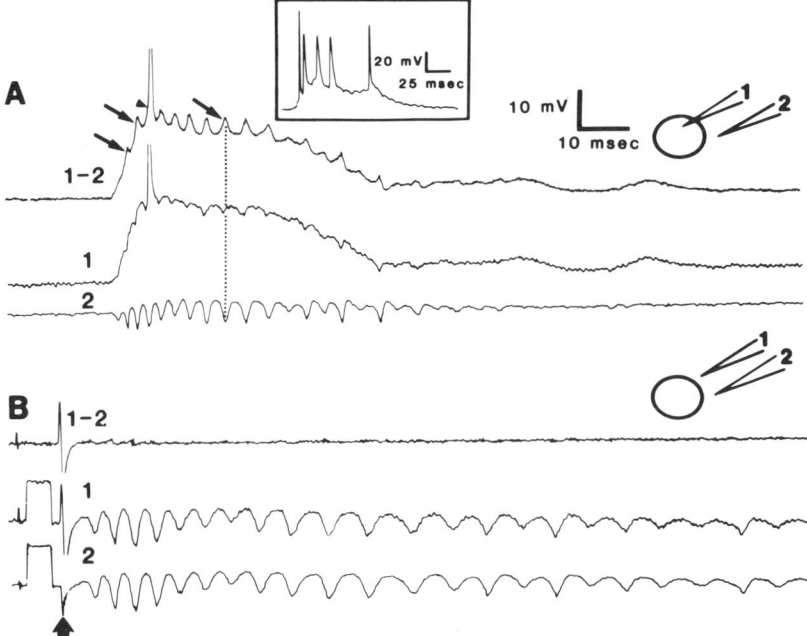

Figure 6
Field effect depolarizations during synchronous firing of hippocampal neurons. Paroxysmal depolarization shifts in CA1 pyramidal cells were induced with picrotoxin. (A) Field effect depolarizations during depolarization shift. Inset shows intracellular recording (with reference to ground) of a spike burst at resting potential. Transmembrane recording (1-2) shows field effect depolarizations (marked with arrows) synchronous with extracellular population spikes (electrode 2). The cell was hyperpolarized to block spikes. One spike appeared to arise from a field effect depolarization (arrowhead). Intracellular recording (with reference to ground, trace 1) shows small negativities during population spikes. (B) Control recording. The lack of difference between electrodes 1 and 2 after withdrawal from the cell is shown; thus, electrode 2 had previously recorded (i.e., during A) the field potential just outside the impaled neuron (Reprinted, with permission, from Dudek et al. 1985.)

field effects can have a powerful synchronizing influence among hippocampal neurons, particularly when enough neurons fire synchronously and generate population spikes. Intracellular electrophysiological data concerning electrical field effects among neocortical neurons are not yet available, but the population spikes tend to be smaller in the neocortex than in the hippocampus so that field interactions are likely to be weaker.

CONCLUSIONS AND FUTURE STUDIES

Both anatomical and physiological studies support the hypothesis that dye coupling, gap junctions, and electrotonic synapses are present among some hippocampal and neocortical neurons. A quantitative description of the characteristics of the coupling is unavailable, primarily because it is unknown whether some of the reported coupling in these cortical areas is artifactual. Several new techniques described here in other chapters should elucidate the issue of strength and number of electrotonic junctions in hippocampus and elsewhere in the brain. For example, studies using direct visualization of dye transfer between neurons may now be feasible, and both large and small molecules might be injected simultaneously. Further work with treatments that block gap junctions, such as intracellular injection of gap junction antibodies (see Hertzberg and Spray, both this volume), may provide stronger evidence concerning the presence, number, and strength of electrotonic junctions. These types of studies, particularly comparing neurons and glia (e.g., Connors et al. 1984), should indicate the degree to which the regulatory mechanisms for gap junctions in the brain are similar to those in other mammalian tissues, such as heart and liver. The relative paucity of these junctions among neurons, especially when compared with glia, will make this a difficult task.

Electrotonic junctions are generally considered to be "synchronizing synapses" (Bennett 1977). Although electrotonic synapses are present in hippocampus and neocortex, the relatively small number of them suggested by the available data raises the issue of their actual importance to neuronal integration and synchronization. More accurate measures of the number and strength of electrotonic junctions, a better ability to regulate specifically junctional characteristics, and computer simulations with quantitative models of neuronal networks should together indicate whether electrotonic junctions contribute importantly to synchronization. Further assessment of the importance of electrotonic junctions in normal brain function awaits our understanding of how local neuronal interactions are involved in higher brain processing.

ACKNOWLEDGMENTS

We are grateful to Drs. R.D. Andrew, B.A. MacVicar, C.P. Taylor, and R.D. Traub for their important contributions to the collaborative research summarized here and to Drs. H. Schmalbruch and H. Jahnsen for providing Figure 2B. We are also grateful to B. Farmer and J. Shippey for secretarial and photographic assistance. This manuscript was prepared under the support of grant NS 16683 from the National Institutes of Health.

REFERENCES

Alger, B.E., M. McCarren, and R.S. Fisher. 1983. On the possibility of simultaneously recording from two cells with a single microelectrode in the hippocampal slice. *Brain Res.* **270:** 137–141.

Andrew, R.D., C.P. Taylor, R.W. Snow, and F.E. Dudek. 1982. Coupling in rat hippocampal slices: Dye transfer between CA1 pyramidal cells. *Brain Res. Bull.* **8:** 211–222.

Bennett, M.V.L. 1977. Electrical transmission: A functional analysis and comparison to chemical transmission. In *Handbook of physiology, section 1: The nervous system* (ed. E.R. Kandel), vol. 1, p. 357–416. American Physiological Society, Bethesda, Maryland.

Bulloch, A.G.M. and S.B. Kater. 1981. Selection of a novel connection by adult molluscan neurons. *Science* **212:** 79–81.

Bulloch, A.G.M., S.B. Kater, and A.D. Murphy. 1980. Connectivity changes in an isolated molluscan ganglion during in vivo culture. *J. Neurobiol.* **11:** 531–546.

Connors, B.W., L.S. Benardo, and D.A. Prince. 1983. Coupling between neurons of the developing rat neocortex. *J. Neurosci.* **3:** 773–782.

―――. 1984. Carbon dioxide sensitivity of dye-coupling among glia and neurons of the neocortex. *J. Neurosci.* **4:** 1324–1330.

Dudek, F.E., R.W. Snow, and C.P. Taylor. 1985. Role of electrical interactions in synchronization of epileptiform bursts. In *Basic mechanisms of the epilepsies* (ed. A.V. Delgado-Escueta et al.). Raven Press, New York. (In press.)

Dudek, F.E., R.D. Andrew, B.A. MacVicar, R.W. Snow, and C.P. Taylor. 1983. Recent evidence for and possible significance of gap junctions and electrotonic synapses in the mammalian brain. In *Basic mechanisms of neuronal hyperexcitability* (ed. H.H. Jasper and N.M. van Gelder), p. 31–73. A.R. Liss, New York.

Green, J.D. and D.S. Maxwell. 1961. Hippocampal electrical activity I. Morphological aspects. *Electroencephalogr. Clin. Neurophysiol.* **13:** 837–846.

Gutnick, M.J. and R. Lobel-Yaakov. 1983. Carbon dioxide uncouples dye-coupled neuronal aggregates in neocortical slices. *Neurosci. Lett.* **42:** 197–200.

Gutnick, M.J. and D.A. Prince. 1981. Dye coupling and possible electrotonic coupling in the guinea pig neocortical slice. *Science* **211:** 67–70.

Gutnick, M.J., B.W. Connors, and B.R. Ransom. 1981. Dye-coupling between glial cells in the guinea pig neocortical slice. *Brain Res.* **213:** 486–492.

Haas, H.L. and J.G.R. Jefferys. 1984. Low-calcium field burst discharges of CA1 pyramidal neurones in rat hippocampal slices. *J. Physiol.* **354:** 185–201.

Jefferys, J.G.R. and H.L. Haas. 1982. Synchronized bursting of CA1 hippocampal pyramidal cells in the absence of synaptic transmission. *Nature* **300:** 448–450.

Johnston, D. and T.H. Brown. 1981. Giant synaptic potential hypothesis for epileptiform activity. *Science* **211:** 294–297.

Knowles, W.D., P.G. Funch, and P.A. Schwartzkroin. 1982. Electrotonic and dye coupling in the hippocampal slice. *Neuroscience* **7:** 1713–1722.

Konnerth, A., U. Heinemann, and Y. Yaari. 1984. Slow transmission of neural activity in hippocampal area CA1 in absence of active chemical synapses. *Nature* **307:** 69–71.

Kosaka, T. 1983a. Gap junctions between non-pyramidal cell dendrites in the rat hippocampus (CA1 and CA3 regions). *Brain Res.* **271:** 157–161.

―――. 1983b. Neuronal gap junctions in the polymorph layer of the rat dentate gyrus. *Brain Res.* **277:** 347–351.

Llinás, R., R. Baker, and C. Sotelo. 1974. Electrotonic coupling between neurons in cat inferior olive. *J. Neurophysiol.* **37:** 560–571.

MacVicar, B.A. and F.E. Dudek. 1980. Dye-coupling between CA3 pyramidal cells in slices of rat hippocampus. *Brain Res.* **196:** 494–497.

―――. 1981. Electrotonic coupling between pyramidal cells: A direct demonstration in rat hippocampal slices. *Science* **213:** 782–785.

―――. 1982. Electrotonic coupling between granule cells of rat dentate gyrus: Physiological and anatomical evidence. *J. Neurophysiol.* **47:** 579–592.

MacVicar, B.A. and H. Jahnsen. 1985. Uncoupling of CA3 pyramidal neurons by propionate. *Brain Res.* **330:** 141–145.

MacVicar, B.A., N. Ropert, and K. Krnjević. 1982. Dye-coupling between pyramidal cells of rat hippocampus in vivo. *Brain Res.* **238:** 239–244.

Miles, R. and R.K.S. Wong. 1983. Single neurones can initiate synchronized population discharge in the hippocampus. *Nature* **306:** 371–373.

Murphy, A.D., R.D. Hadley, and S.B. Kater. 1983. Axotomy-induced parallel increases in electrical and dye coupling between identified neurons of *Helisoma*. *J. Neurosci.* **3:** 1422-1429.

Richardson, T.L., R.W. Turner, and J.J. Miller. 1984. Extracellular fields influence transmembrane potentials and synchronization of hippocampal neuronal activity. *Brain Res.* **294:** 255-262.

Schmalbruch, H. and H. Jahnsen. 1981. Gap junctions on CA3 pyramidal cells of guinea pig hippocampus shown by freeze-fracture. *Brain Res.* **217:** 175-178.

Schmitt, F.O., P. Dev, and B.H. Smith. 1976. Electrotonic processing of information by brain cells. *Science* **193:** 114-120.

Schwartzkroin, P.A. 1983. Local circuit considerations and intrinsic neuronal properties involved in hyperexcitability and cell synchronization. In *Basic mechanisms of neuronal hyperexcitability* (ed. H.H. Jasper and N.M. van Gelder), p. 75-108. A.R. Liss, New York.

Snow, R.W. and F.E. Dudek. 1984a. Synchronous epileptiform bursts without chemical transmission in CA2, CA3 and dentate areas of the hippocampus. *Brain Res.* **298:** 382-385.

―――. 1984b. Electrical fields directly contribute to action potential synchronization during convulsant-induced epileptiform bursts. *Brain Res.* **323:** 114-118.

Stewart, W.W. 1981. Lucifer dyes—highly fluorescent dyes for biological tracing. *Nature* **292:** 17-21.

Taylor, C.P. and F.E. Dudek. 1982a. A physiological test for electrotonic coupling between CA1 pyramidal cells in rat hippocampal slices. *Brain Res.* **235:** 351-357.

―――. 1982b. Synchronous neural afterdischarges in rat hippocampal slices without active chemical synapses. *Science* **218:** 810-812.

―――. 1984a. Excitation of hippocampal pyramidal cells by an electrical field effect. *J. Neurophysiol.* **52:** 126-142

―――. 1984b. Synchronization without active chemical synapses during hippocampal afterdischarges. *J. Neurophysiol.* **52:** 143-155.

Taylor, C.P., K. Krnjević, and N. Ropert. 1984. Facilitation of CA3 pyramidal cell firing by electrical fields generated antidromically. *Neuroscience* **11:** 101-109.

Traub, R.D., and R.K.S. Wong. 1983. Synaptic mechanisms underlying interictal spike initiation in a hippocampal network. *Neurology* **33:** 257-266.

Traub, R.D., F.E. Dudek, C.P. Taylor, and W.D. Knowles. 1985. Simulation of hippocampal afterdischarges synchronized by electrical interactions. *Neuroscience* **14:** 1033-1038.

Wong, R.K.S., R.D. Traub, and R. Miles. 1984. Epileptic mechanisms as revealed by studies of the hippocampal slice. In *Electrophysiology of epilepsy* (ed. P.A. Schwartzkroin and H. Wheal), pp. 253-275. Academic Press, New York.

Electrotonic Transmission in the Mammalian Central Nervous System

Rodolfo R. Llinás
Department of Physiology and Biophysics
New York University Medical Center
New York, New York 10016

Electrotonic coupling between neurons in mammalian brain was described long after its initial discovery (Furshpan and Potter 1959) and thorough study in the nervous system of invertebrates and lower vertebrates (cf. Bennett 1966, 1977). There were two major reasons for this delay. First, following the demonstration of chemical transmission in mammalian central nervous system (CNS) (cf. Eccles 1964) it was considered, a priori, that electrical junctions would not be present in mammalian forms as they probably represented a more primitive form of cell–cell interaction. Second, for technical reasons having to do with the size and fragility of neuronal elements in mammals, an unambiguous experimental determination of electrotonic coupling was difficult to implement. Ultimately, the presence of electrotonic transmission was inferred from experimental results in in vivo studies of the mammalian CNS at three sites. These are the mesencephalic root of the Vth cranial nerve (MSN) in the rat (Baker and Llinás 1971), the vestibular nucleus of the rat (Korn et al. 1973; Wylie 1973), and the inferior olive of the cat (Llinás et al. 1974). In these three examples, the physiological signs of electrical coupling were supported by parallel ultrastructural studies. The ultrastructural correlate for electrotonic coupling is considered today to be the presence of "gap junctions" (Robertson 1963; Revel and Karnovsky 1967; Payton et al. 1969; Pappas et al. 1971; Sotelo et al. 1974; cf. Bennett and Goodenough 1978; Sotelo and Korn 1978). Such gap junctions have been found in at least nine different regions of the CNS, forming axo-somatic, axo-axonic, soma-somatic, and dendro-dendritic membrane appositions (cf. Sotelo 1977).

DISCUSSION

In Vivo Studies of Electrotonic Coupling in the Mammalian Brain

General Comments

In principle, the only definitive criterion for the identification of electrotonic coupling between cells is provided by the direct demonstration of a low-resistance pathway between cellular elements (Furshpan and Potter 1959; Watanabe and Bullock 1960; Watanabe and Grundfest 1961; Hagiwara and Morita 1962; Eckert 1963; Martin and Pilar 1963; Bennett et al. 1964; Auerbach and Bennet 1969; Levitan et al. 1970; Nicholls and Purves 1970; Spira and Bennett 1972) using simultaneous intracellular recordings (cf. Bennett 1966). This criterion is most commonly met by the injection of a current pulse across the membrane of one of the impaled cells while recording a close-to-simultaneous potential, of the same polarity, in the second neuron (Bennett et al. 1964; Bennett 1966). This direct demonstration has not been possible in in vivo studies of the mammalian CNS, among other reasons because the neurons must be impaled blindly.

Given these limitations, indirect criteria were used in the initial mammalian studies. The studies of the three CNS nuclei mentioned above took advantage of the properties of action currents that generate antidromically or orthodromically activated action potentials. Thus, if a set of neurons is electrotonically coupled, an action potential in any one of them will inject a certain amount of its action current into the other members of the set. This transfer of charge, occurring via the low-resistance junctions, produces a potential in the coupled cell group. This potential with antidromic stimulation has a latency only slightly longer than that of antidromic invasion and has been referred to as the short latency depolarization (SLD). The question of latency is very important since many cells have axon collaterals that may return to neighboring neurons and that may mimic, by chemical synaptic transmission, true SLDs. Thus, the latency of any electrotonically mediated SLD should, in general, be shorter than that of chemical transmission since no synaptic delay is present (Katz 1969). However, this postulate does not always hold true since the cable properties of the coupling can cause delays comparable to those observed in chemical transmission (cf. Bennett 1977).

An additional criterion for identifying an SLD as due to electrotonic coupling is its lack of a reversal potential. Indeed, the electromotive force (emf) generating the SLD is dependent on the properties of the actively invaded cell, rather than that of the passive neurons. For example, if SLDs are generated by the all-or-nothing action potentials in neighboring neurons, a reversal potential (Eccles 1964; Eccles et al. 1966; Smith et al. 1967; Kuno and Llinás 1970; Llinás and Baker 1972) should not be demonstrable in these cases. Unfortunately, the proper testing of the absence of an E_{SLD} (being a negative result) is difficult to determine unambiguously, particularly since membrane rectification can prevent the generation of a sufficiently large membrane potential change at remotely placed junctions. Also, electrical junctions are often electrically modulated (Harris et al. 1983); thus, an SLD appearing about to reverse as the membrane is depolarized may, in fact, just be turning off gap junctions or activating delayed rectification which, by shunting the cell, will reduce the size of the SLD. For these reasons, extrapolating reversal potentials for SLD is clearly not permissible. Finally, if SLDs are generated by synaptic potentials in neighboring cells, a reversal potential may be observed.

The Mesencephalic Root of the Vth Nucleus

The initial evidence for electrotonic coupling at the mammalian CNS site was the ultrastructural demonstration by Hinrichsen and Larramendi (1968) of gap junctions between the neurons in the mesencephalic root of the Vth nucleus (MSN) in rats (see also Brightman and Reese 1969). A subsequent electrophysical study (Baker and Llinás 1971) demonstrated the presence of SLDs following the invasion of MSN cells from masseteric nerve stimulation (Fig. 1A and B). This depolarization had a short latency and a gradable amplitude (not illustrated here) indicating that cells were coupled to more than one other element. On some occasions full spikes were generated from the SLDs (Fig. 1B).

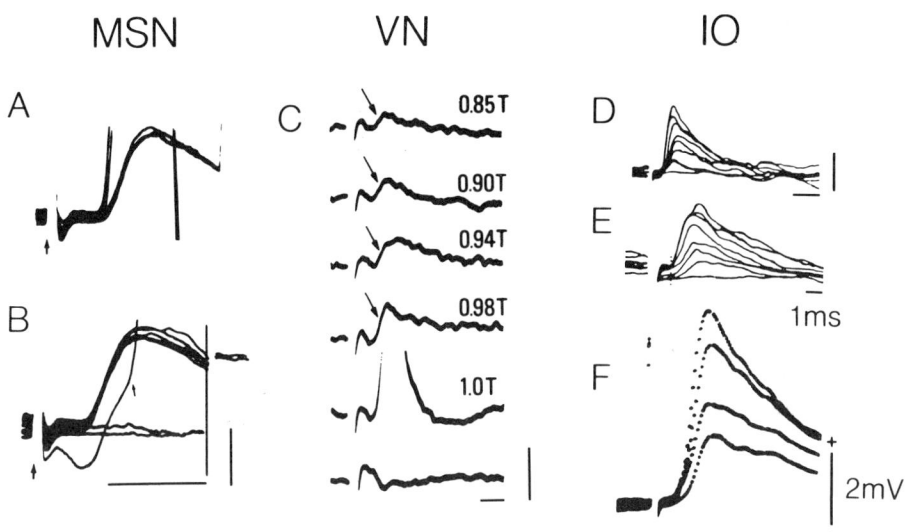

Figure 1
Short latency depolarization (SLD) in mammalian CNS. (A, B) Coupling between neurons in the rat mesencephalic nucleus (MSN). (A) Demonstration of SLD by threshold trigeminal nerve stimulation. When invasion fails, the underlying SLD is seen. The SLD may activate an action potential (arrow, B). (C) Electrotonic coupling between vestibular nucleus neurons (VN) in the rat. SLDs are seen in vestibular nucleus cells following antidromic stimulation of the vestibulospinal tract at spinal cord level. Threshold activation (1.0T) generates a full spike. Subthreshold stimulation (0.98-0.85T) produces a graded reduction of the SLD (arrows). The bottom trace shows the extracellular field potential. (D-E) Electrotonic coupling between inferior olivary cells of the cat. Graded SLD is seen following antidromic activation of neighboring neurons. (F) Computer average for four stimulation levels. Calibrations are identical throughout the figure: voltage, 2 mV; time, 1 msec. (Modified from Baker and Llinás 1971 [A-B]; Korn et al. 1973 [C]; Llinás et al. 1974 [D-F].)

To exclude the possibility of mistakenly identifying an M spike (Eccles 1955) as an SLD, action potentials activated directly by current injection at the soma were collided with antidromic invasions of the neuronal ensemble (Fig. 2A–C). In this design the somatofugal action potential (direct stimulation) would collide with the antidromic somatopetal spike (Fig. 2A–C) obviating the problem of separating M spikes from SLDs; this method also revealed SLDs in neurons with low thresholds for antidromic stimulation.

The Vestibular Nucleus

A second example of electrotonic coupling in the mammalian CNS was described by Korn et al. (1973) in the vestibular (Deiters') nucleus of the rat. The demonstration was also indirect, utilizing techniques similar to those used for the MSN. Gap junctions appear to be more numerous than in the MSN. As

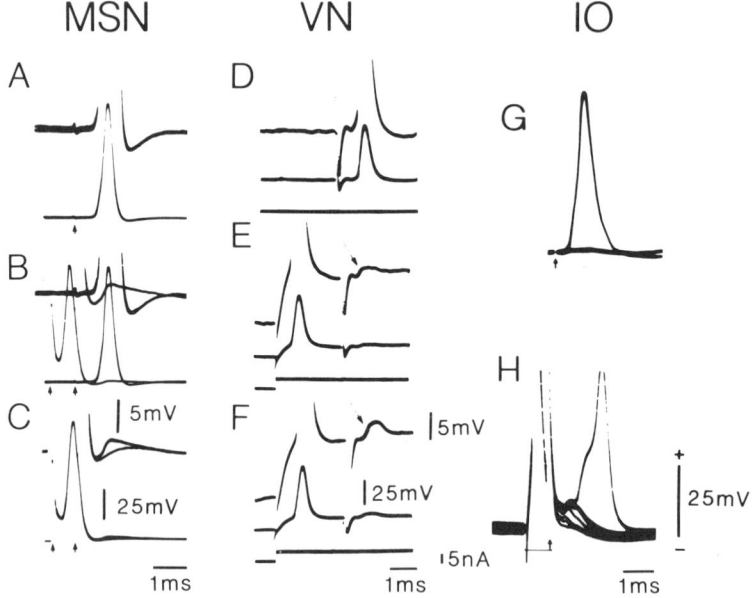

Figure 2
Collision experiments in mammalian CNS. (A–C) Rat mesencephalic nucleus neuron (MSN) (higher gain on upper trace). (A) Antidromic activation. (B) Direct activation (first arrow) precedes antidromic invasion unmasking a short latency depolarization (SLD) (superimposed sweeps with and without direct stimulation). (C) Peripheral stimulation is absent in one sweep to illustrate the SLD amplitude. (D–F) Collision experiments as in A–C for a rat vestibular neuron (VN) (higher gain on upper trace). Direct stimulation is provided by a prolonged current injection (lower trace). (G–H) Inferior olivary neuron of cat. Arrow indicates antidromic stimulus, horizontal line in H the duration of the direct current injection. In G with antidromic stimulation threshold for the cell's axon, a small SLD is seen. When the direct stimulus blocked the cell's antidromic response, the SLD became larger until an "orthodromic" spike was generated. (Modified from Baker and Llinás 1971 [A–C]; Korn et al. 1973 [D–F]; Llinás et al. 1974 [G–H].)

could be expected from the anatomy, SLDs were commonly observed and could be recorded synchronously with the antidromic invasion of the vestibular cells following spinal cord stimulation.

SLDs for the vestibular neuron are shown in Figure 1C. Its graded nature was shown by increasing the amplitude of the antidromic volley that generated progressively large SLDs. The strength of vestibulo-spinal tract stimulation is given to the right of each record as the percentage of the threshold (T) for antidromic activation of the impaled cell. Collision experiments ruled out the possibility of M spike activation. The latencies of these SLDs are within the range expected for electrotonic coupling potentials. In this case, as in that of the MSN, direct somatofugal action potentials did not block the SLD underlying the direct antidromic invasion (Fig. 2D–E). In comparing the ultrastructural characteristics of Deiters' nucleus gap junctions with those in other parts of the CNS, one feature is quite striking—their location. In Deiters' nucleus, gap junctions occur between the presynaptic fibers and their postsynaptic vestibular nuclear cells. The actual coupling does not occur, therefore, directly between vestibular cells, but rather indirectly through their common presynaptic fiber. This organization had been previously reported in various nuclei of other species (Bennett 1972).

The Inferior Olive

Gap junctions in the inferior olive have been reported in a number of species, including cat, rat, opossum, and monkey (cf. Sotelo and Korn 1978). The original electrophysiological demonstration of electrotonic coupling in this nucleus was made in the cat (Llinás et al. 1974). As in the two previous examples, the presence of coupling was identified electrophysiologically by the presence of an SLD (Fig. 1F, G) and the failure to block the SLD when antidromic and direct stimuli were paired (Fig. 2G, H). Coupling in this nucleus was encountered more often than in the MSN or in Deiters' nucleus and showed a finely graded SLD (Fig. 1F, G), which was often capable of activating the inferior olivary cells.

In Vitro Studies of Electrotonic Coupling in the Mammalian Brain

The development of the brain slice technique greatly facilitated the study of electrotonic coupling in the mammalian CNS; as with in vivo studies, both anatomical and electrophysiological approaches have been utilized. Morphological evidence has been obtained using the fluorescent dye, Lucifer Yellow (Stewart 1978), which crosses gap junctions. Although Lucifer Yellow has become an important and readily available tool, both false-positive and false-negative findings can be generated. For this reason dye-coupling cannot be used as the sole criterion to demonstrate electrotonic coupling. Dye-coupling has been observed between cells in guinea pig neocortical slices (Gutnick et al. 1981; Gutnick and Prince 1981) and between pyramidal cells in hippocampal slices of rat brain (Stewart 1978; MacVicar and Dudek 1980b).

The ambiguity of the dye-coupling technique is obviated, of course, by the direct electrophysiological demonstration of electrotonic coupling. Such a demonstration has been made between inferior olive cells in slices of guinea pig brainstem (Llinás and Yarom 1979, 1981) and between pyramidal cells in rat hippocampal slices (MacVicar and Dudek 1981). In both cases simultaneous, intracellular recordings were made from neuron pairs. Indirect electrophysiological evidence for coupling has been reported for cells in guinea pig neocortical slices (Gutnick and Prince 1981). In these experiments extracellular stimulation elicited subthreshold depolarizations that were resistant to Ca^{++} channel blockade by extracellular Mn^{++} and that did not collide with antidromic spikes.

Hippocampal CA3 Pyramidal Cells

The first direct electrophysiological evidence for electrotonic coupling in the hippocampus was provided by MacVicar and Dudek (1981). They impaled two pyramidal cells simultaneously, as shown in Figure 3. Depolarization of one cell (A1) produced two action potentials in the directly activated cell, as well as a postanodal spike, while SLDs were recorded in the coupled cell (A2).

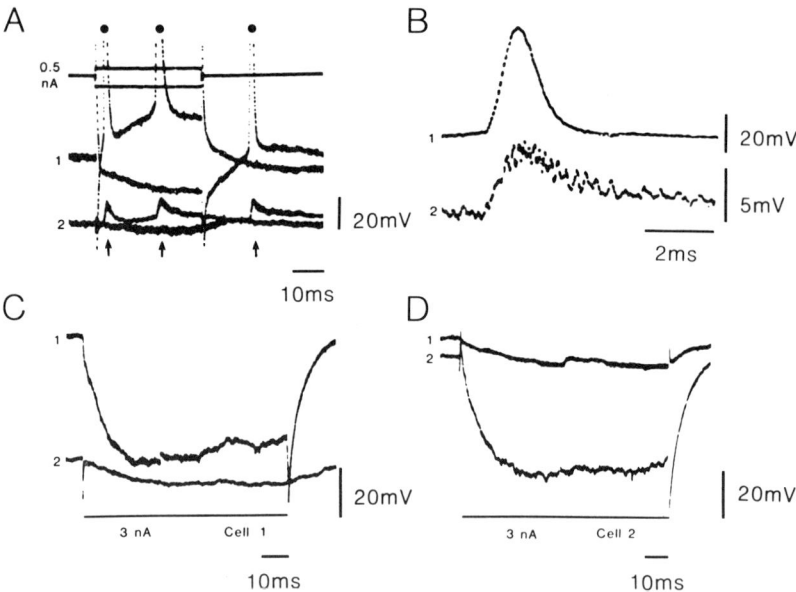

Figure 3
Electrotonic coupling between pyramidal cells in slice from rat hippocampus. Simultaneous intracellular recording from two pyramidal cells. (A) Depolarizing or hyperpolarizing current pulses injected into cell 1 (A1) caused similar but smaller voltage changes in the coupled cell (A2). Electrotonic interactions were bidirectional; the coupling ratio for these cells was near 0.2. Action potentials in the injected cell (dots in A1) were seen as SLD in the coupled cell (A2). (B) Action potential (B1) and the SLD in the coupled cell (B2 at higher gain) are illustrated at higher sweep speed. (C–D) Bidirectional electrotonic coupling between two pyramidal cells subsequently stained with horseradish peroxidase. Hyperpolarizing current (bars) injected into either cell caused a smaller voltage change in the other coupled cell. (Modified from MacVicar and Dudek 1981.)

Hyperpolarization of cell 1 was also seen across the membrane of cell 2. The activated spike and the SLD are shown at a higher gain in B.

This coupling was shown to be bidirectional (Fig. 3C, D). Hyperpolarizing current injected across cell 1 spreads to cell 2 (Fig. 3C) and hyperpolarization in cell 2 spreads to cell 1 (Fig. 4D). That this voltage was truly recorded across the membrane of the coupled cell was shown since, after the electrode was removed from the cell, current passed into the other cell was not seen by the extracellular electrode. One potential problem in this type of study is the possibility of double penetration of the same cell. This problem is obviated by intracellular staining of the impaled neurons with horseradish peroxidase. MacVicar and Dudek (1981) stained four pairs of coupled neurons. In over 400 neuron pairs they identified 47 that were electrotonically coupled (i.e., less than 12%).

Inferior Olivary Cells

The inferior olive nucleus provided the first direct demonstration of electrical coupling in mammalian CNS (Llinás and Yarom 1979, 1981). Two paradigms were used in the in vitro studies of coupling in this nucleus. Antidromic stimulation of an inferior olive cell at rest generated the action potential (Fig. 4A); if the cell was hyperpolarized, graded SLDs were recorded (Fig. 4B). Following the addition of Mn^{++} to the bath, the two types of Ca^{++} action potentials present in these cells (Fig. 4C), as well as chemical synaptic transmission,

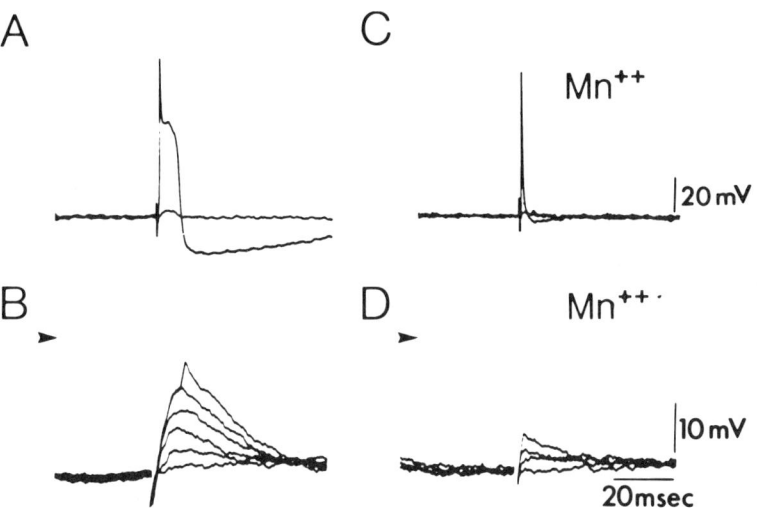

Figure 4
Electrotonic coupling between inferior olivary cells in brainstem slice from guinea pig. (A) Inferior olivary antidromic spike and subthreshold potential elicited at rest by periolivary white matter stimulation. (B) Membrane hyperpolarization reveals subthreshold SLD responses (resting potential is given by arrowhead). (C–D) Records obtained as in A and B after addition of 5 mM $MnCl_2$ to the bath. Persistence of graded SLD indicates electrotonic coupling between inferior olivary neurons. (Modified from Llinás and Yarom 1981.)

were blocked; the SLDs remained (Fig. 4D), although reduced in amplitude. Note that the afterdepolarization following the action potential in Figure 4A is greatly shortened and that the magnitude and duration of the afterhyperpolarization is markedly reduced in Figure 4C. This change is due to the blockage of Ca^{++} conductance (g_{Ca}) by Mn^{++}.

A direct demonstration of electrotonic coupling between olivary cells is shown in Figure 5, where simultaneous recordings from a pair of neighboring neurons show nonrectifying electrotonic coupling. The spikes in each cell could be independently activated by white matter stimulation (Fig. 5A, 1 and 2). In Figure 5B a directly evoked action potential in cell 1 produced a clear depolarization in cell 2. The depolarization was generated across the membrane in cell 2 since after removing the microelectrode from this cell, no potential could be observed when cell 1 was activated. In another cell pair (Fig. 5C, D), direct hyperpolarization of one cell (1) generated a hyperpolarization in a second cell (2). As expected from the electrotonic spread of this potential, the amplitude in cell 2 was smaller than that in cell 1 and the time course was slower. However, there was no obvious delay between the two potentials. A reversal of the direction of current injection produced a depolarization of both cells, indicating nonrectifying coupling between these neurons (Fig. 5D). The absence of a response following removal of the electrode from cell 2 (bottom trace, Fig. 5D) showed that the previously recorded potentials were generated across the membrane of cell 2. Coupling could be demonstrated by only 10%

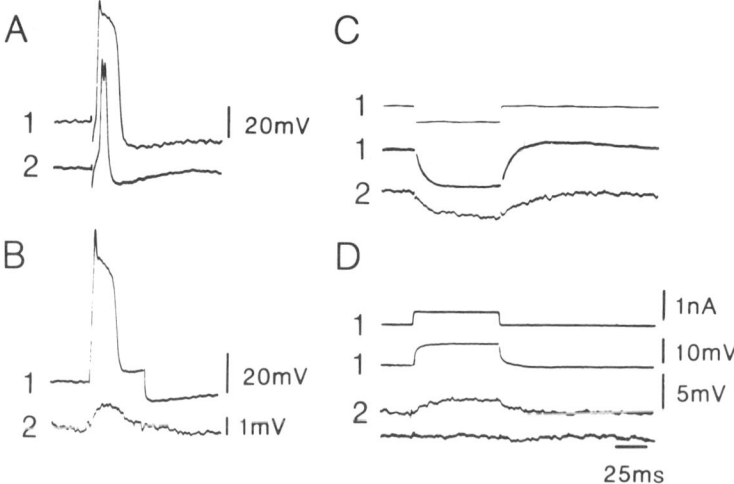

Figure 5
Electrotonic coupling between inferior olivary cells in brainstem slice from guinea pig. (A) Action potentials obtained from two neurons following antidromic stimulation in periolivary white matter. (B) Direct stimulation of cell 1 elicits typical olivary spike in cell 1 and short latency depolarization in cell 2. (C, D) Subthreshold stimulation in another pair of coupled neurons. Negative (C) or positive (D) current pulse in cell 1 (*upper trace*) elicited hyperpolarization (C) or depolarization (D) of both cells. The response in cell 2 is decreased in amplitude and slowed in time course. The last trace in (D) was recorded after removing the electrode from cell 2 only and injecting current into cell 1. (Modified from Llinás and Yarom 1981.)

of the cell pairs studied (6 of 60 pairs). This is probably due to the electrotonic decrement produced by the widespread distribution of the junctions and the fact that the coupling ratio is rather low.

Functional Role of Electrotonic Synaptic Transmission in the Mammalian Brain

As outlined above, in vitro studies have provided direct evidence for electrotonic coupling in the mammalian CNS studies that corroborated the earlier indirect evidence obtained in vivo. The remaining question concerns, therefore, not the existence of electrotonic coupling, but rather its significance in CNS function.

The Hippocampus

Relating to this structure, it has been suggested that electrotonic coupling is of demonstratable physiological significance and that, in addition, this coupling may be related to the ease with which epileptic discharges may be initiated in this cortex (MacVicar and Dudek 1981). According to these authors, electrotonic coupling would function as low-pass filters for the transmission of signals between neuronal elements (Bennett 1977); this would favor the transmission of slow potential changes associated with synaptic inputs and burst discharge (Wong and Prince 1978; Johnston et al. 1980; Johnston and Brown 1981). Also, because of their electrotonic interactions, pyramidal neurons would tend to respond more readily to synchronous synaptic inputs (which would lead to spike bursts), while asynchronous input would be shunted throughout the network (Bennett 1972). At the same time transmission would be shunted by the increased membrane conductance due to recurrent inhibition (Spira and Bennett 1972) tending to prevent epilepsy. The tendency to synchronicity does facilitate the genesis of epileptic discharge in this system. Indeed, convulsive agents (such as penicillin) that reduce recurrent inhibition (Wong and Prince 1979; Dingledine and Gjerstad 1979; Schwartzkroin and Prince 1978, 1980) should enhance coupling by eliminating the powerful current shunt provided by the inhibitory postsynaptic potentials (IPSPs). Also, in networks such as the hippocampus, which tend to reverberate, the positive feedback from the spread of action potentials between coupled neurons could lead to prolonged bursting. In short, then, electrotonic coupling in the hippocampus, particularly when combined with recurrent excitation (MacVicar and Dudek 1980a), could contribute to synchrony and spread during both normal brain rhythms and epileptogenic events.

The Inferior Olive

The functional significance of the extensive electrotonic coupling in the inferior olive had been difficult to approach even at the level of an "explanation sketch". In fact, the electrical coupling in this system remained for a number of years an electrophysiological curiosity rather than a meaningful finding from a physiological point of view. The difficulty in understanding the functional significance of this particular coupling organization was probably related to the one-to-one organization of the climbing fiber–Purkinje cell system. In-

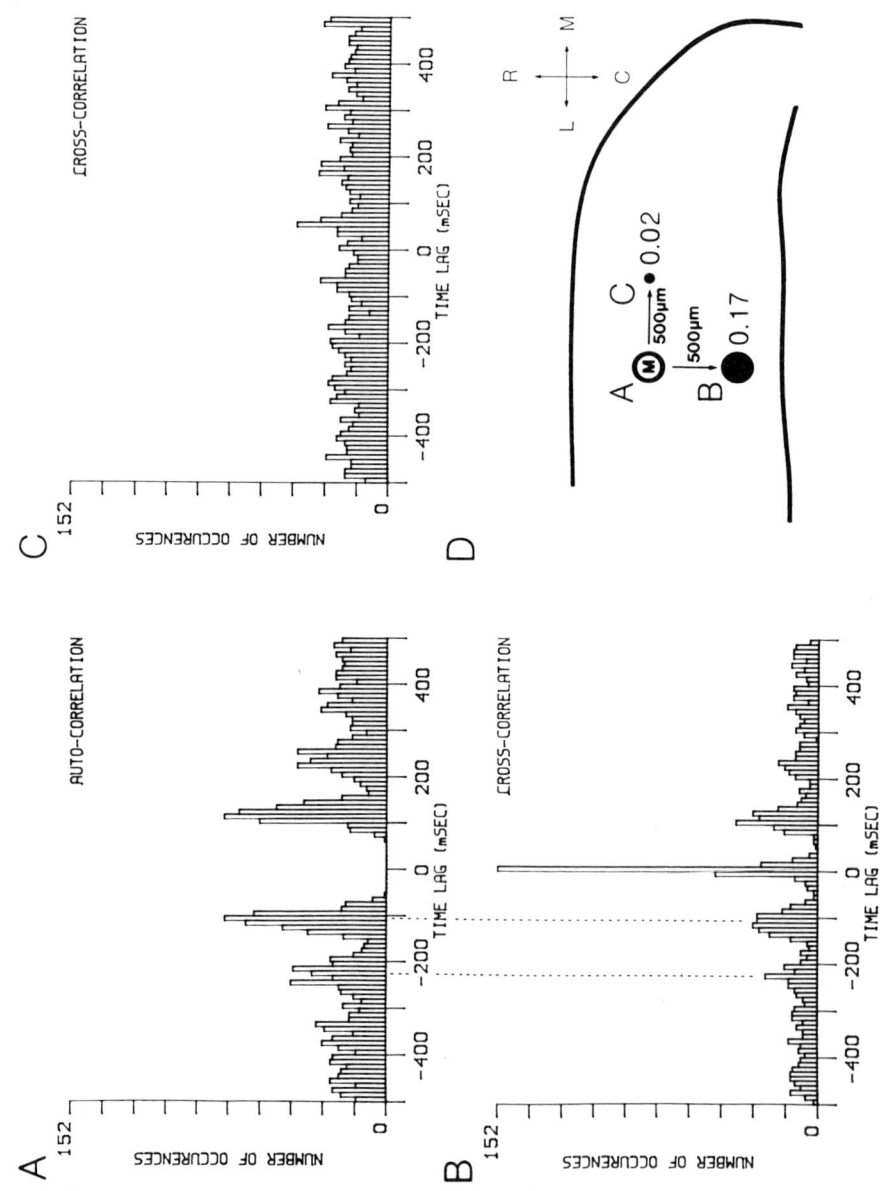

Figure 6 (See facing page for legend.)

deed, all measurements attempting to correlate climbing fiber activity with any aspect of motor (or sensory) function found only very loose correlations (cf. Llinás and Simpson 1981). Given the above, it was thought that the olivocerebellar system could only be understood when sufficiently large numbers of inferior olivary units could be recorded simultaneously (Llinás 1974). Recently this approach was finally realized by the development of a multielectrode recording technique (Bower and Llinás 1982, 1983; Saski and Llinás 1985). Some suggestive findings relating to the functional significance of the coupling are reviewed below.

Multiple Unit Recordings

One of the central problems in attempting to study the significance of the electrical coupling in a central nucleus is that of accessibility of neurons to individual recording paradigms. Fortunately, in the case of the inferior olive, electrical coupling can be studied indirectly at the cerebellar cortex by determining the correlation between the climbing fiber activation of individual Purkinje cells. This is because inferior olivary axons reach individual Purkinje cells in a one-to-one manner and generate easy-to-recognize "complex spikes" in these large neurons (Eccles et al. 1966). Indeed, if climbing fiber activation of Purkinje cells occurs continuously in a close to synchronous manner, it can be assumed that such coincidences indicate direct functional interactions between the inferior olivary neurons.

Two types of inferior olivary interactions should, in principle, be capable of generating complex spike synchrony: (1) common excitatory synaptic inputs which activate the cells more or less simultaneously, and (2) electrotonic interactions as demonstrated in the in vivo and in vitro studies discussed above.

The precise experimental paradigm developed to investigate climbing fiber activation of Purkinje cells consisted initially of 16 individually inserted extracellular micropipettes which could record simultaneously the activity of 16 Purkinje cells located at 250-μm intervals (Bower and Llinás 1982, 1983). More recently this technique has been extended to 32 Purkinje cells (Sasaki and Llinás 1985). Recording their climbing fiber responses demonstrated that: (1) A clear auto- and cross-correlation can be obtained from each of the recorded cells. The spontaneous activity of individual inferior olivary cells (recorded under ketamine or barbiturate analgesia) has a frequency of approximately 10/second (Fig. 6A). (2) A cross-correlation between Purkinje cell pairs

Figure 6
Correlation of spontaneously occurring climbing fiber activation of Purkinje cells simultaneously recorded in rat cerebellar cortex. (*A*) Autocorrelogram of Purkinje cell firing via climbing fiber activation (CFA) demonstrating rhythmicity of about 9/sec. (*B*) Cross-correlogram generated by comparing spontaneous CFA of cell in *A* with that of another Purkinje cell located 500 μm caudally in the cerebellar cortex. Here cross-correlation has a coefficient of 0.17. (*C*) Cross-correlation of CFA with a Purkinje cell located 500 μm in the medial direction showing a correlation coefficient of 0.02. (*D*) Diagram of Crus II in rat cerebellum showing location of Purkinje cells in *A, B,* and *C*. M signifies master Purkinje cell. The rostrocaudal and mediolateral directions are shown to the right of the diagram (K. Sasaki and R. Llinás, unpubl.).

located at different points over the surface of the cortex reveals a very clear synchrony of firing between cells in the rostrocaudal direction (Fig. 6B). This cross-correlation may extend for more than 1 mm rostrocaudally but is almost absent in the mediolateral direction (Fig. 6C). The results suggest, therefore, that the spontaneous activity of the set of inferior olivary neurons with axons (climbing fibers) terminating in a particular area of the cerebellar cortex is such that these inferior olivary neurons fire close to synchronously. This synchrony can be demonstrated for cells lying in the rostrocaudal direction at any level in the mediolateral axis, implying that inferior olivary cells are coupled so as to generate close to isochronous activation of rostrocaudal Purkinje cells bands. The bands of active Purkinje cells can be seen as generating an organized and close-to-simultaneous inhibition of cerebellar nuclear cells. That these bands of activity are not produced by common synaptic input can be easily demonstrated by determining the spatial distribution, over the cerebellar cortex, of the climbing fiber reflex which follows antidromic activiation of the inferior olive from the cerebellar cortex (Eccles et al. 1966). Indeed, present results indicate that this inferior olivary reflex is organized in similar bands as above. Given that interactions between inferior olivary neurons are produced via their electrical coupling (Llinás and Yarom 1981), we take this as strong evidence that such banding reflects the organization of this coupling (Sasaki and Llinás 1985).

The spatial organization of this Purkinje cell synchrony becomes more pronounced after injection of harmaline, without demonstrating any loss of the rostrocaudal order described above (Fig. 7). Accompanying this increased synchrony in the inferior olive, there is a 9- to 12-Hz well-defined tremor evident throughout the musculature of the animal which is characterized by the alternate activation of the flexor and extensor musculature (Villablanca and Riobo 1970; Lamarre et al. 1971; Llinás and Volkind 1973).

At this juncture two questions are of great interest: (1) What is the mechanism for the generation of rostrocaudal bands of synchronous Purkinje cell activity? and (2) What is the functional significance of such synchrony?

How Are the Rostrocaual Bands of Purkinje Cell Activity Generated?

Because Purkinje cell firing via the climbing fiber can be deeply modified by drugs such as picrotoxin that block γ-aminobutyric acid (GABA) inhibition (Takeuchi and Takeuchi 1969), we have proposed that banding represents not just a morphological characteristic of the connectivity at the inferior olivary level but a well-defined functional state regulated by inhibitory inputs (Sasaki and Llinás 1985). Indeed, since the banding of Purkinje cell activity disappears when inhibitory potentials are blocked, one must conclude that the banding is due to functional decoupling of select electrotonic junctions by inhibition, as suggested for *Navanax* by Spira and Bennett (1972) and for the inferior olive by Llinás et al. (1974).

Two important points arise from these findings: (1) electrical coupling seems to be important for the generation of the synchronous Purkinje cell activity initiated by climbing fiber activation and, (2) the lack of isochronicity in the

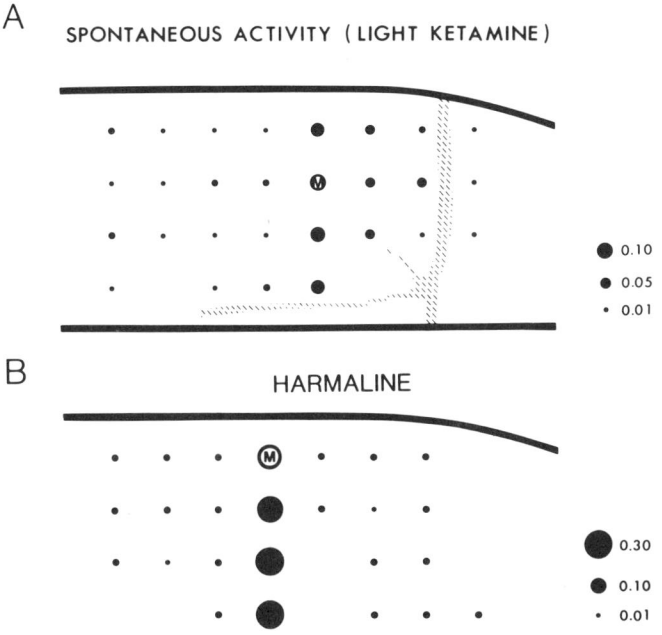

Figure 7
(A) Cross-correlogram for spontaneously occurring climbing fiber activity (CFA) over a folium, shown as the value of correlation coefficient for CFAs between the master Purkinje cell (M) and 25 other Purkinje cells in the lobule recorded at 250-μm distances from each other. The coefficient is indicated by the area of the circle at each one of the impaling sites. See calibration to the right. Hatched area shows the location of a vessel which prevented obtaining recordings for all 32 microelectrodes. (B) (In another experiment): Correlation coefficient between CFAs of Purkinje cells after addition of harmaline. Note the increased coefficient (0.3) and the preservation of the spatial organization of the cross-correlation which is seen in both A and B as a rostrocaudal band of large dots indicating high cross-correlation coefficients (K. Sasaki and R. Llinás, unpubl.).

mediolateral direction seems to be generated by inhibitory decoupling of inferior olivary neurons.

How Is Electrotonic Coupling in the Inferior Olive Related to the Function of This Nucleus?

In a recent paper, it was proposed that physiological tremor is generated, to a great extent, by the synchronous activation of the inferior olivary cells (Llinás 1984). These neurons, by activating the cerebellar and vestibular nuclei, produce synchronized synaptic inputs onto motor neurons, which ultimately produce movement. Indeed, if this were the case, it would be expected that a tremor would occur characterized by the 9- to 10-Hz frequency of inferior olivary firing. In fact, the frequency of physiological tremor (9–10/sec) matches very well the spontaneous rhythmicity in the inferior olivary nucleus (Marsden 1984). Also, when the inferior olive is lesioned (such as in myoclonus), well-defined, isochronous myoclonic movements occur which include the midline and less frequently, the lateral musculature (Marsden 1984). These activations are jerk-like and repetitious, and have a clear rhythmicity. Finally, drugs that increase the oscillation of the inferior olive, such as harmaline or serotonin, generate tremor only if the inferior olive is intact. No tremor is seen when the inferior olivary nucleus is damaged. From the above then, it seems reasonable to suggest that the inferior olive serves as a tremor-generating system, one which serves to coordinate movement by allowing motor neurons to be activated with a certain isochronicity (Llinás 1984).

SUMMARY AND CONCLUSIONS

Clearly, much remains to be understood about the significance of electrical coupling in mammalian CNS. At this time, two conclusions are unavoidable, however. (1) Electrical coupling is unquestionably present in the mammalian CNS. (2) Such coupling seems to be related to two main unit properties: the synchronous activity of neuronal ensembles and the rejection of asynchronous inputs. More specifically, the two examples given indicate that (1) electrical coupling may serve as a low-pass filter in the hippocampus and (2) it may be central in generating such fundamental properties as tremor. It is this tremor that provides the matrix upon which all coordinated movement is generated, at least in the sense that the onset of our movements is dictated by a carrier frequency upon which "successful" motor commands can then ride.

ACKNOWLEDGMENTS

This research was supported by USPHS grant NS13742 from NINCDS.

REFERENCES

Auerbach, A.A. and M.V.L. Bennett. 1969. A rectifying electrotonic synapse in central nervous system of a vertebrate. *J. Gen. Physiol.* **53:** 211–237.

Baker, R. and R. Llinás. 1971. Electrotonic coupling between neurones in the rat mesencephalic nucleus. *J. Physiol.* **212:** 45–63.

Bennett, M.V.L. 1966. Physiology of electrotonic junctions. *Ann. N.Y. Acad. Sci.* **137(2):** 509–539.

———. 1972. A comparison of electrically and chemically mediated transmission. In *Structure and function of synapses* (ed. G.D. Pappas and D.P. Purpura), pp. 221–226. Raven Press, New York.

———. 1977. Electrical transmission: A functional analysis and comparison to chemical transmission. In *Handbook of physiology,* section 1: *Cellular biology of neurons* (ed. E.R. Kandel) vol. 1, pp. 357–416. Williams and Wilkins, Baltimore.

Bennett, M.V.L. and D.A. Goodenough. 1978. Gap junctions, electrotonic coupling and intercellular communication. *Neurosci. Res. Program Bull.* **16(1):** 377–463.

Bennett, M.V.L., M. Gimenez, Y. Nakajima, and G.D. Pappas. 1964. Spinal and medullary nuclei controlling electric organ in the eel. *Electrophorus. Biol. Bull.* **127:** 362.

Bower, J. and R. Llinás. 1982. Simultaneous sampling and analysis of the activity of multiple, closely adjacent, cerebellar Purkinje cells. *Soc. Neurosci. Abstr.* **8:** 830.

———. 1983. Simultaneous sampling of the responses of multiple, closely adjacent, Purkinje cells responding to climbing fiber activation. *Soc. Neurosci. Abstr.* **9:** 607.

Brightman, M.W. and T.S. Reese. 1969. Junctions between intimately apposed cell membranes in the vertebrate brain. *J. Cell Biol.* **40:** 648–677.

Dingledine, R. and L. Gjerstad. 1979. Penicillin blocks hippocampal IPSPs, unmasking prolonged EPSPs. *Brain Res.* **168:** 205–209.

Eccles, J.C. 1955. The central action of antidromic inpulses in motor nerve fibres. *Pfluegers Arch. Gesamte Physiol.* **260:** 385–415.

———. 1964. *The physiology of synapses.* Springer-Verlag, Berlin.

Eccles, J.C., R. Llinás, and K. Sasaki. 1966. the excitatory synaptic action of climbing fibres on the Purkinje cells of the cerebellum. *J. Physiol.* **182:** 268–296.

Eckert. R. 1963. Electrical interaction of paired ganglion cells in the leech. *J. Gen. Physiol.* **46:** 573–588.

Furshpan, E.J. and D.D. Potter. 1959. Transmission at the giant motor synapses of the crayfish. *J. Physiol.* **145:** 289–325.

Gutnick, M.J. and D.A. Prince. 1981. Dye coupling and possible electrotonic coupling in the guinea pig neocortical slice. *Science* **211:** 67–70.

Gutnick, M.J., B.W. Connors, and B.R. Ransom. 1981. Dye-coupling between glial cells in the guinea pig neocortical slice. *Brain Res.* **213:** 486–492.

Hagiwara, S. and H. Morita. 1962. Electrotonic transmission between two nerve cells in leech ganglion. *J. Neurophysiol.* **25:** 721–731.

Harris, A.L., D.C. Spray, and M.V.L. Bennett. 1983. Control of intercellular communication by voltage dependence of gap junctional conductance. *J. Neurosci.* **3:** 79–100.

Hinrichsen, C.F.L. and L.M.H. Larramendi. 1968. Synapses and cluster formation of the mouse encephalic fifth nucleus. *Brain Res.* **7:** 296–299.

Johnston, D. and T.H. Brown. 1981. Giant synaptic potential hypothesis for epileptiform activity. *Science* **211:** 294–297.

Johnston, D., J.J. Hablitz, and W.A. Wilson. 1980. Voltage clamp discloses slow inward current in hippocampal burst-firing neurones. *Nature* **286:** 391–393.

Katz, B. 1969. *The release of neural transmitter substances.* C.C. Thomas, Springfield, Illinois.

Korn, H., C. Sotelo, and F. Crepel. 1973. Electrotonic coupling between neurons in the rat lateral vestibular nucleus. *Exp. Brain Res.* **16:** 255–275.

Kuno, M. and R. Llinás. 1970. Alterations of synaptic action in chromatolysed motoneurones of the cat. *J. Physiol.* **210:** 823–838.

Lamarre, Y., C. de Montigny, M. Dumont, and M. Weiss. 1971. Harmaline-induced rhythmic activity of cerebellar and lower brainstem neurons. *Brain Res.* **32:** 246–250.

Levitan, H., L. Tauc, and J.P. Segundo. 1970. Electrical transmission among neurons in the buccal ganglion of a mollusc, *Navanax inermis. J. Gen. Physiol.* **55:** 484–496.

Llinás, R. 1974. Motor aspects of cerebellar control (18th Bowditch Lecture). *Physiologist* **17:** 19–46.

———. 1984. Rebound excitation as the physiological basis for tremor: A biophysical study of the oscillatory properties of mammalian central neurons *in vitro*. In *Movement disorders: Tremor* (ed. L.J. Findley and R. Capildeo), pp. 165–182. Macmillan Press, London.

Llinás, R. and R. Baker. 1972. A chloride dependent inhibitory postsynaptic potential in cat trochlear motoneurons. *J. Neurophysiol.* **35:** 484–492.

Llinás, R. and J.I. Simpson. 1981. Cerebellar control of movement. In *Handbook of behavioral neurobiology: Motor coordination* (ed. A.L. Towe and E.S. Luschei), vol. 5, pp. 231–302. Plenum Press, New York.

Llinás, R. and R. Volkind. 1973. The olivo-cerebellar system: Functional properties as revealed by harmaline-induced tremor. *Exp. Brain Res.* **18:** 69–87.

Llinás, R. and Y. Yarom. 1979. Long-term excitability changes in mammalian inferior olive neurons in vitro. *Soc. Neurosci. Abstr.* **5:** 105.

———. 1981. Properties and distribution of ionic conductances generating electroresponsiveness of inferior olivary neurons in vitro. *J. Physiol.* **315:** 569–584.

Llinás, R., R. Baker, and C. Sotelo. 1974. Electrotonic coupling between neurons in the cat inferior olive. *J. Neurophysiol.* **37:** 560–571.

MacVicar, B. and F.E. Dudek. 1980a. Local synaptic circuits in rat hippocampus: Interactions between pyramidal cells. *Brain Res.* **184:** 220–223.

———. 1980b. Dye-coupling between CA3 pyramidal cells in slices of rat hippocampus. *Brain Res.* **196:** 494–497.

———. 1981. Electrotonic coupling between pyramidal cells: A direct demonstration in rat hippocampal slices. *Science* **213:** 782–784.

Marsden, C.D. 1984. Origins of normal and pathological tremor. In *Movement disorders: Tremor* (ed. L.J. Findley and R. Capildeo), pp. 37–84. MacMillan Press, London.

Martin, A.R. and G. Pilar. 1963. Dual mode of synaptic transmission in the avian ciliary ganglion. *J. Physiol.* **168:** 443–463.

Nicholls, J.G. and D. Purves. 1970. Monosynaptic chemical and electrical connections between sensory and motor cells in the central nervous system of the leech. *J. Physiol.* **209:** 647–667.

Pappas, G.D., Y. Asada, and M.V.L. Bennett. 1971. Morphological correlates of increased coupling resistance at an electrotonic synapse. *J. Cell Biol.* **49:** 173–188.

Payton, B.W., M.V.L. Bennett, and G.D. Pappas. 1969. Permeability and structure of junctional membranes at an electrotonic synapse. *Science* **166:** 1641–1643.

Revel, J.P. and M.J. Karnovsky. 1967. Hexagonal array of subunits in intercellular junctions of the mouse heart and liver. *J. Cell Biol.* **33:** C7–12.

Robertson, J.D. 1963. The occurrence of a subunit pattern in the unit membranes of club endings in Mauthner cell synapses in goldfish brains. *J. Cell Biol.* **19:** 201–221.

Sasaki, K. and R. Llinás. 1985. Dynamic electrotonic coupling in mammalian inferior olive as determined by simultaneous multiple Purkinje cell recording. *Biophys. J.* **47:** 53a.

Schwartzkroin, P.A. and D.A. Prince. 1978. Cellular and field potential properties of epileptogenic hippocampal slice. *Brain Res.* **147:** 117–130.

———. 1980. Changes in excitatory and inhibitory synaptic potentials leading to epileptogenic activity. *Brain Res.* **183:** 61–67.

Smith, T.G., R.B. Wuerker, and K. Frank. 1967. Membrane impedance changes during synaptic transmission in cat spinal motoneurones. *J. Neurophysiol.* **30:** 1072–1096.

Sotelo, C. 1977. Electrical and chemical communication in the central nervous system. In *International cell biology* (ed. B.R. Brinkely and K.R. Porter), pp. 83–92. Rockefeller University Press, New York.

Sotelo, C. and H. Korn. 1978. Morphological correlates of electrical and other interactions through low-resistance pathways between neurons of the vertebrate central nervous system. *Int. Rev. Cytol.* **55:** 67–107.

Sotelo, C., R. Llinás, and R. Baker. 1974. Structural study of the inferior olivary nucleus of the cat. Morphological correlates of electrotonic coupling between neurons. *J. Neurophysiol.* **37:** 541–559.

Spira, M.E. and M.V.L. Bennett. 1972. Synaptic control of electrotonic coupling between neurons. *Brain Res.* **37:** 294–300.

Spray, D. and M.V.L. Bennett. 1985. Physiology and pharmacology of gap junctions. *Annu. Rev. Physiol.* **47:** 281–303.

Stewart, W.W. 1978. Functional connections between cells as revealed by dye-coupling with a highly fluorescent napthalimide tracer. *Cell* **14:** 741–759.

Takeuchi, A. and N. Takeuchi. 1969. A study of the action of picrotoxin on the inhibitory neuromuscular junction of the crayfish. *J. Physiol.* **205:** 377–391.

Villablanca, J. and F. Riobo. 1970. Electroencephalographic and behavioral effects of harmaline in intact cats and in cats with chronic mesencephalic transection. *Psychopharmacologia* **17:** 302–313.

Watanabe, A. and T.H. Bullock. 1960. Modulation of activity of one neuron by subthreshold slow potentials of another in lobster cardiac ganglion. *J. Gen. Physiol.* **43:** 1031–1046.

Watanabe, A. and H. Grundfest. 1961. Impulse propagation at the septal and commissural junctions of crayfish lateral giant axons. *J. Gen. Physiol.* **45:** 267–308.

Wong, R.K.S. and D.A. Prince. 1978. Participation of calcium spikes during intrinsic burst firing in hippocampal neurons. *Brain Res.* **159:** 385–390.

―――. 1979. Dendritic mechanisms underlying penicillin-induced epileptiform activity. *Science* **204:** 1228–1231.

Wylie, R.M. 1973. Evidence of electrotonic transmission in the vestibular nuclei of the rat. *Brain Res.* **50:** 179–183.

Interaction of Electrical and Chemical Synapses

Michael V.L. Bennett, M.B. Zimering,
M.E. Spira,* and David C. Spray
Division of Cellular Neurobiology
Department of Neuroscience
Albert Einstein College of Medicine
Bronx, New York 10461

*Neurobiology Unit, Hebrew University
Jerusalem, Israel

Numerous electrical synapses have been described between neurons. Gap junctions are the usual morphological substrate, although a few instances where cells interact electrically across extracellular space are known or hypothesized (Bennett 1977; Korn and Faber 1979; Traub et al. 1985). In most instances the gap junctional membrane is electrically linear, i.e., it behaves like a fixed resistance connecting the coupled cells. Transmission is comparable to electrotonic spread, and the term "electrotonic synapse" is often used. A few electrical synapses are known in which the junctional membrane rectifies unidirectionally (cf. Giaume and Korn, this volume). Symmetrical voltage dependence like that between amphibian blastomeres has been seen in the short-lived Rohon-Beard neurons of the *Xenopus* embryo (Spitzer 1982) and in retinula cells of the *Limulus* lateral eye (Smith and Baumann 1969). Although electrotonic synapses can exhibit many of the properties of excitatory chemical synapses, initially it appeared that "plasticity" was more a property of chemical than electrical transmission (Bennett 1977). To be sure, temporal or spatial summation could lead to facilitation of impulse transmission. But changes in electrical postsynaptic potentials (PSPs) themselves were unusual except for a few cases of modulation due to broadening of the presynaptic action potential or propagation of action potentials closer to the synapses themselves (Bennett and Pappas 1983).

One reason for the scarcity of plasticity of electrotonic synapses in the systems studied may have been that these systems display highly synchronous activity: Reciprocal activation, simply achieved by ohmic behavior, is what is required. This paper concerns systems in which there are functional requirements for both synchronous and asynchronous behavior of the same neuronal ensemble. The degree of coupling is under cellular control, apparently not by modulation of junctional conductance but by changes in nonjunctional membrane resulting from activity of inhibitory chemical synapses.

NAVANAX EXPANSION MOTOR NEURONS

The expansion motor neurons (EMNs) controlling the radial muscles of *Navanax* are responsible for expansion of the pharynx. The radial muscles shorten, thinning the pharyngeal wall and pushing on each other sideways. This movement can be very rapid and underlies the unusual (for a gastropod) prey capture method of *Navanax*: It sucks in its prey whole. Because rapid pharyngeal expansion could well require synchronous motor neuron activity, it was not surprising to us that the motor neurons were coupled (Spira and Bennett 1972), although others had questioned whether the coupling was strong enough to affect firing (Levitan et al. 1970). If cells are depolarized simultaneously by distributed synaptic inputs, as would be expected during expansion, even weak coupling can provide considerable synchronization (Fig. 1).

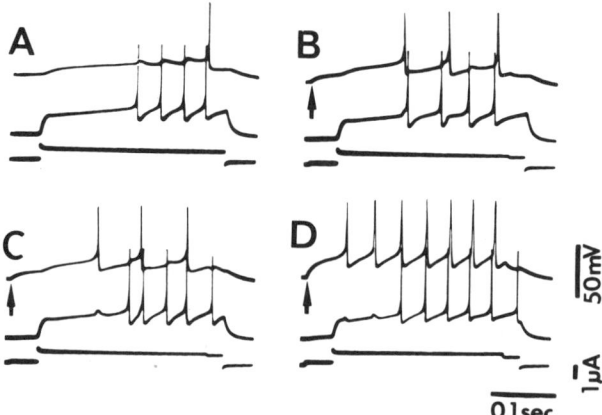

Figure 1
Synchronization of firing of EMNs by electrotonic coupling. Recording from ipsilateral M and G cells on first and second traces, current on third trace. (*A*) Depolarization of the G cell spread to the M cell. Brief depolarizations of the M cell, electrical PSPs, followed the G-cell impulses; the last PSP fired the M cell. (*B*) Depolarization was also applied in the M cell (starting at arrow). The M cell fired first and the resulting PSP excited the G cell. (*C,D*) Increased depolarizing current in the M cell caused it to fire before the G cell was sufficiently depolarized to be excited by the M-cell PSPs. Later in the course of the pulses, the firing could be quite synchronous or either cell could excite the other. (Reprinted, with permission, from Spira et al. 1980.)

SYNAPTIC CONTROL OF ELECTROTONIC COUPLING

Electrical coupling between EMNs can be blocked completely by synaptic inputs, elicited, for example, by a brief train of stimuli to the pharyngeal nerve (Fig. 2) or by tactile stimulation of the pharyngeal wall (Spira and Bennett 1972; Spira et al. 1980). The disappearance of coupling is associated with a small depolarization and a moderate increase in input conductance. The cells are receiving inhibitory inputs whose reversal potential is close to the resting potential (actually the inhibitory postsynaptic potentials [IPSPs] are slightly depolarizing; Spray et al. 1980). How inhibition reduces coupling may be understood with respect to the simple equivalent circuit for two isopotential cells joined by a coupling conductance. For this circuit the coupling coefficient k_{12} for electrotonic spread to cell 2 when current is applied in cell 1 is given by $V_2/V_1 = g_j/(g_j + g_{nj2})$, where g_j and g_{nj} are junctional and nonjunctional conductances, respectively. For moderate to small coupling coefficients ($g_j < g_{nj}$), the proportional decrease in coupling coefficient approaches the proportional increase in g_{nj}. In Figure 2 the input conductance increases by a factor of about 2, but the decrease in coupling coefficient is much greater.

A simple modification of the equivalent circuit can explain the large decrease in coupling coefficient observed. Suppose the cells are coupled through an axonal or dendritic pathway with an axial conductance that is not very large (Fig. 2, diagram). Then inhibitory synapses that greatly increase the conductance at, say, the middle of the coupling pathway can short-circuit electrotonic

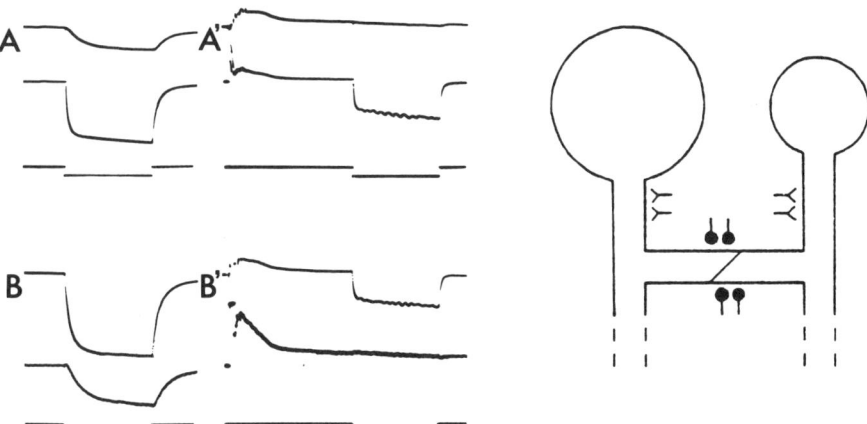

Figure 2
Control of coupling by inhibitory synapses. (*A,B*) Direct coupling of M and G EMNs (current on lower trace). (*A',B'*) Following a brief train of stimuli to the pharyngeal nerve, the neurons were no longer coupled. Their input conductances were increased, and there was a small depolarization owing to activation of inhibitory synapses with a reversal potential just positive to the resting potential. The loss of coupling resulted from short circuiting by inhibitory synapses along the coupling pathway (*right*). Higher gain recording in G cell in *B*, *B'*. (Reprinted, with permission, from Spira and Bennett 1972.)

spread without increasing the input conductance of the cell beyond that of the axial conductance to the site of inhibition.

The decrease in coupling could, of course, be produced by decrease in junctional conductance as well. The changes in coupling observed are sufficiently rapid in onset and recovery as to make unlikely a mechanism like that described for retinal horizontal cells (Teranishi et al. 1984; Lasater and Dowling; Neyton et al.; both this volume).

There is another test of the hypothesis that reduced coupling is caused by inhibitory shunting. For close coupling, the axial conductance must be as large as possible. To prevent input conductance g_{in} from becoming too large during the uncoupling produced by inhibition, the axial conductance must not be too large. (A very high input conductance would prevent the cell from being excited by other inputs, which it can be, as noted below.) Assuming symmetry between the coupled cells and very close coupling at the junctions themselves (large g_j), one can show that the ratio of input conductances in coupled and uncoupled states is given by $g_{in}/g_{in}^* < 1 - k$, where g_{in} and g_{in}^* are the input conductances before and during the uncoupling and k is the coupling coefficient. If the ratio of input conductances were larger than $1 - k$, one would have to appeal to another mechanism, such as decrease in junctional conductance. The ratio of input conductances could be smaller than $1 - k$ if inhibitory synapses were closer to the cell bodies or if g_j were significant. Experimental data satisfy the equation (0.5 < 0.7 in Fig. 2), and thus are consistent with the suggestion that reduction in coupling is due to inhibitory conductance increase along the coupling pathway. Of course they do not exclude that there are also decreases in junctional conductance.

It was easy to hypothesize a functional role for the synaptic uncoupling. When coupled, the cells' firing would tend to be synchronized, which would be appropriate for the rapid pharyngeal expansion of prey capture. When uncoupled, the cells would be more capable of firing asynchronously, which could be appropriate for peristalsis, swallowing, or regurgitation. A number of the smaller EMNs do fire during uncoupling, and these responses can be evoked by tactile stimulation of the pharyngeal wall, a physiological stimulus that evokes peristalsis. The giant neurons, each of which innervates the entire pharynx, do not fire during uncoupling and they would not be capable of localized expansion.

To demonstrate the validity of these hypotheses (or suggestions) requires the monitoring of unit activity during actual behavior. The ganglia do exhibit the requisite behavior when severely dissected and a promising optical technique for multiunit recording (100 or more) is applicable to *Navanax* (London et al. 1984). Until these experiments are performed (and perhaps after), the physiological significance of the uncoupling will be undecided.

We do want to distinguish synaptic uncoupling from uncoupling, that is a simple by-product of inhibition. With the exception of through-conducting axoaxonic synapses, most coupling in invertebrates is mediated by neuritic contacts in the neuropil. Thus, inhibitory synapses, which are also in the neuropil, will, in general, cause uncoupling, and the basic connectivity for uncoupling illustrated in Figure 2 is widespread. We would like synaptic uncoupling to mean something more, that is, uncoupling with the purpose or adaptive value of allowing asynchronous firing evoked by other inputs. Inhibition that is not hyperpolarizing is more likely to permit independent firing during the uncoupling.

Synaptically mediated increase in coupling is also possible where chemical synapses act by decreasing nonjunctional conductance (Carew and Kandel 1976).

A prominent example of dendrodendritic electrical synapses in mammals is in the inferior olive. Pharmacological block of inhibition causes more widely spread synchronous firing, suggesting that inhibition divides the olive into groups of synchronously firing cells (Llinás, this volume). Whether physiological inputs also change spatial patterns of firing remains to be determined. Electrical synapses are also found in the olfactory bulb between mitral and granule cell dendrites (Pinching and Powell 1971). Since the granule cells are inhibitory to the mitral cells, there is considerable possibility for modulation of interactions, perhaps more than simple recurrent inhibition which is presumed to operate in sensory processing.

A number of motor neuron pools in lower vertebrates are electrically coupled, and analysis of excitatory and inhibitory inputs during different patterns of movement may reveal inhibitory modulation of coupling (cf. Bennett 1977). Several additional sites of dendrodendritic coupling in mammals—somatosensory cortex, cerebellar cortex, and hippocampus—are also possible sites of this form of neural interaction (cf. Bennett 1977; Korn and Faber 1979; Dudek et al. 1983).

REVERSED COUPLING

Another quite dramatic form of interaction of chemical and electrical synapses in the *Navanax* buccal ganglia involves the circumferential motor neurons (CMNs), which control the circumferential muscles. These neurons are involved in pharyngeal emptying, whether for swallowing or regurgitation, and also in closing of the lips and the pharyngeal esophageal sphincter.

Two CMNs sitting quietly, i.e., not active and without synaptic inputs, generally are electrotonically coupled if they innervate neighboring muscles (and an extensive study of the interconnections has been made; Zimering 1983). Passing current pulses, either hyperpolarizing or depolarizing, in one of the cells may then initiate repetitive IPSPs (Fig. 3A; Spira et al. 1976). Also the IPSPs may start up spontaneously, or in response to pharyngeal nerve stimulation. These IPSPs may occur in one or both cells. Although they are sometimes so regular as to suggest unitary origin (see Fig. 5F), they generally clearly consist of multiple components (Fig. 3A). We termed the inhibitory neurons the C group when we first found IPSPs, since at this point we called the coupled CMNs A and B. When the cells are generating C-group IPSPs, there is a remarkable change in the relation between the two cells. Now hyperpolarization of one cell causes depolarization of the other cell, and depolarization causes hyperpolarization (Fig. 3). The sign of coupling between the cells is reversed.

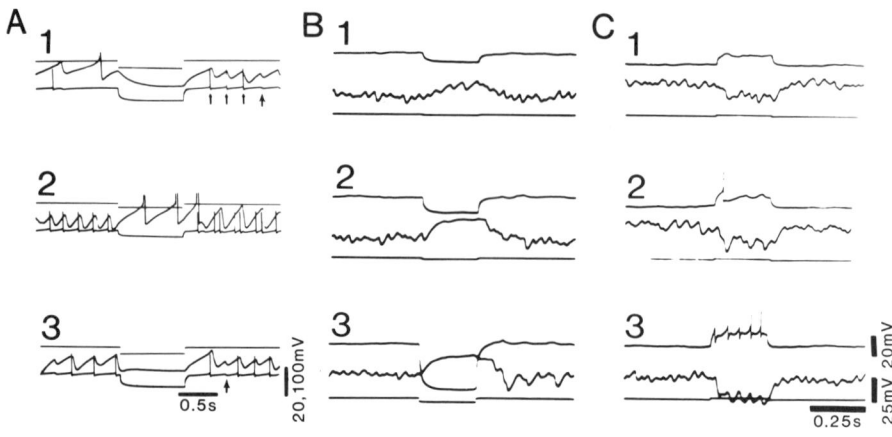

Figure 3
Reversed coupling between pairs of unidentified CMNs. (A) In this pair of cells, there was initially little C-group activity. Penetration caused firing of the lower trace cell and IPSPs in both cells followed many of the impulses (small arrows). Occasional IPSPs occurred independently of the impulses (large arrows). Initially the cells were directly coupled (A1), but during higher-frequency IPSPs, hyperpolarization of the lower trace cell blocked the IPSPs leading to depolarization and firing of the upper trace cell; the sign of coupling was reversed (A2). Stronger hyperpolarization in the lower-trace cell overcame the effect of disinhibition and hyperpolarized the upper-trace cell (A3). (Reprinted, with permission, from Spira et al. 1976.) (B,C) There was considerable C-group activity in this different pair of cells. Progressive hyperpolarization of the upper trace cells (B) decreased IPSP activity in both cells and depolarized the lower trace cell. Progressive depolarization of the upper-trace cell (C), even when subthreshold for impulses, increased hyperpolarization of the lower-trace cell. Impulses in the upper-trace cell caused larger IPSPs in the other cell.

A simple neural circuit can explain this bizarre result (Fig. 4A). The first cell is coupled electrotonically to the second cell and also (presumably electrotonically) to a pool of inhibitory interneurons that inhibit the second cell (and in some pairs the first cell as well; dotted line in Fig. 4A). These interneurons are coupled to each other (inferred from the tendency toward synchrony of the IPSPs). When the C group is inactive, the cells are simply electrotonically coupled. When the C group is active, hyperpolarization of the first cell suppresses activity of the inhibitory neurons. The resulting disinhibition depolarizes the second cell and may cause it to fire (Figs. 3A2, 5A,B). If the first cell is depolarized, the inhibitory interneurons fire more causing increased hyperpolarization of the second cell (Fig. 3C).

Reversed coupling in this situation differs from ordinary driving of an inhibitory interneuron in several ways: There is a pool of more or less synchronously active interneurons that are electrotonically coupled to the first cell, thus hyperpolarization and subthreshold depolarization (as well as impulses) in the first cell modulate their firing frequency and produce opposite changes of either sign in the second cell. When there is moderate level of C-group activity,

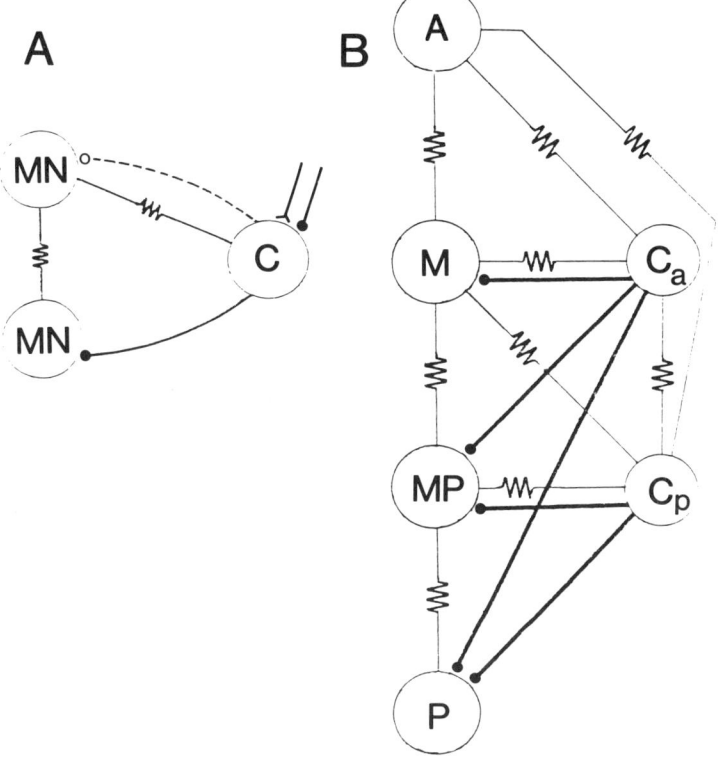

Figure 4
Neural circuits accounting for reversed coupling and its spatial organization. (A) Two motor neurons are directly coupled to each other; one is coupled to the C group of inhibitory neurons which inhibits one or both of the cells. Synaptic inputs to the C group regulate its excitability. If the C group is inexcitable, the motor neurons are directly coupled. If the C group is tonically active, driving of the C group by polarization in the upper motor neuron modulates the inhibitory activity so that the sign of coupling between the two motor neurons is reversed. (B) A circuit explaining the anterior posterior organization of reversed coupling (see text). (Reprinted, with permission, from Zimering 1983.)

both electrotonic spread of hyperpolarization and disinhibition can be seen (Fig. 3A2 and 3). When there is a great deal of inhibitory activity, direct coupling is not seen; evidently the CMNs are synaptically uncoupled in the same way as EMNs are be uncoupled (Fig. 3B,C). Quieting down of the IPSPs or high Mg^{++} treatment to block them will then reveal the direct coupling.

To analyze the circuitry underlying reversed coupling, it was necessary to identify which motor neurons were being studied. There are 12 CMNs that innervate four distinct regions of muscle termed anterior, middle, midposterior, and posterior. It is convenient to denote the CMNs by R and L for right and left, B for bilateral if they innervate muscle bilaterally, and A, M, MP, and P for the muscle region innervated. Eight CMNs are bilaterally symmetrical and in-

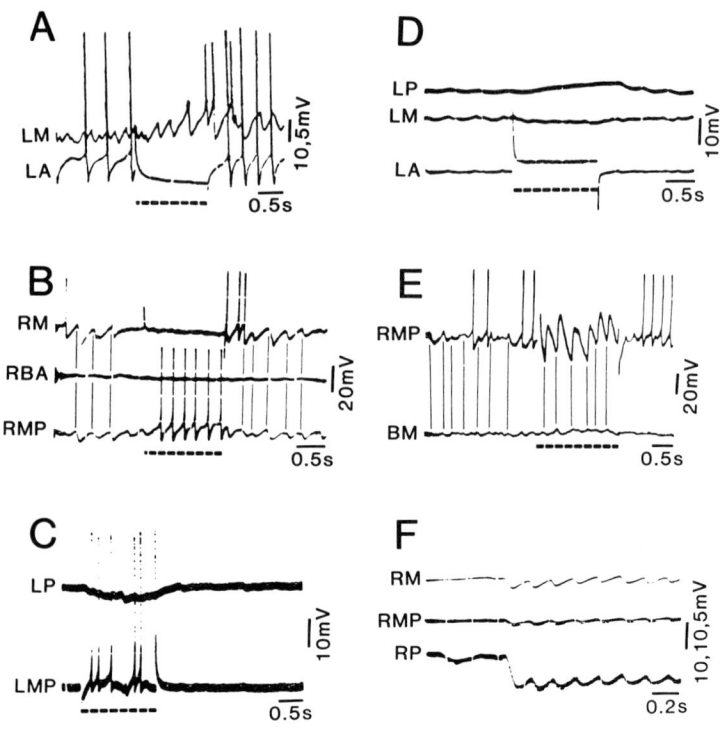

Figure 5
Reversed coupling among identified CMNs. Each trace is labeled as to the CMN from which the recording was made. Time of current application is indicated by the dotted line below each record (hyperpolarizing in A,B,D,E; depolarizing in C). The vertical lines in B and E indicate correlated IPSPs in the CMNs from which recordings were made (explanation in text). (Reprinted, with permission, from Zimering 1983.)

nervate muscles ipsilaterally. Four innervate muscles bilaterally, a bilateral pair innervates A muscle (RBA and LBA), one on the right innervates M muscle (RBM), and one on the left (LBP) innervates P muscle.

A striking finding was that reversed coupling is polarized from anterior to posterior. It is found from A cells to M, MP, and P, from M cells to MP and P, and from MP cells to P (except for LBP which does not exhibit coupling with any of the other neurons either direct or reversed). Examples are shown in Figure 5, A–C. Reversed coupling does not go from a given CMN to a CMN innervating a more anterior muscle (Fig. 5B,E). Examination of the IPSP activity leads to a possible circuit (Fig. 4B). Repetitive IPSPs do not occur in A or BA neurons along with those in more posterior neurons (Fig. 5B). Thus, these neurons cannot receive reversed coupling. M, MP, and P neurons often receive synchronous IPSPs (Fig. 5B,F) that are decreased by hyperpolarization of M. Since MP and P do not cause reversed inhibition in M, this C-group (C_a for anterior) must not be coupled to MP or P. To account for reversed coupling from MP to P, it is necessary to postulate a second C group (C_p for posterior). These neurons end largely on the P cells, but are not directly coupled to them. The C_p cells also are coupled to A and perhaps M neurons as well, because A can exhibit simultaneously reversed coupling to P and direct coupling to M (Fig. 5D). This circuit also accounts for the observation that hyperpolarization of M decreases the IPSPs in it as well as in MP and P (Fig. 5B). In contrast, the frequency of IPSPs in MP cells is not markedly affected by hyperpolarization (Fig. 5E); this insensitivity may be contributed to by lack of IPSPs from C_p to MP as well as by lack of coupling between MP and C_a. Since IPSPs are often synchronous in M, MP, and P, there is likely to be enough coupling between C_a and C_p to mediate synchronization of the two groups. The coupling however cannot be close since polarization of MP does not affect the C_a-group IPSPs in M.

A few (5) cells that may have been C-group interneurons were recorded in a specific region of the ganglion. Stimulation of these cells did not cause pharyngeal movement, but did cause IPSPs in more posterior CMNs (Fig. 6A,B). These evoked IPSPs were irregular in amplitude, consistent with coupling of the C group but only partially synchronous firing among them. During repetitive activity, IPSPs could precede firing of the presumptive C-group neuron and could even occur when it failed to fire (Fig. 6D). However in the latter case a small depolarization was present in the interneuron, again consistent with a coupled group of inhibitory neurons. Furthermore, hyperpolarization of the

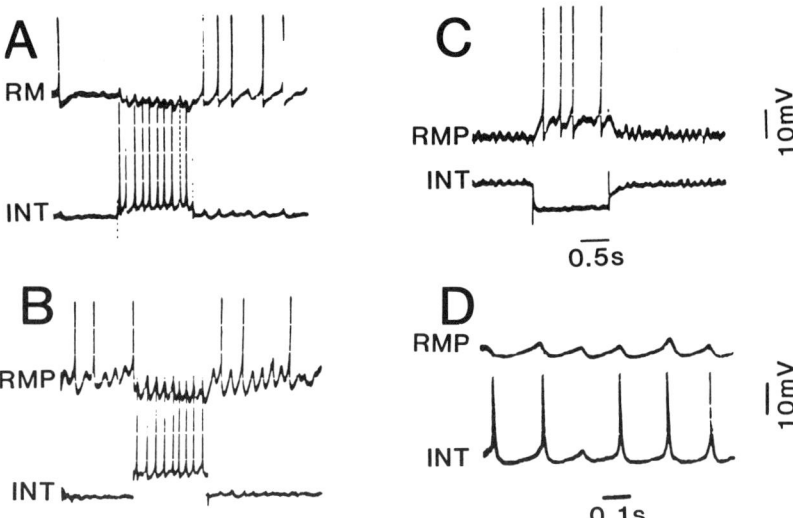

Figure 6
Recordings from possible C-group interneurons (INT). (A,B) Stimulation of this INT caused IPSPs in RM and RMP. The IPSPs were irregular in amplitude, although they occurred one for one with the INT spikes. (C) Hyperpolarization of the same INT depolarized RMP. (D) Repetitive spikes in this INT were associated with IPSPs in RMP, but a missed spike did not prevent the IPSP. The subthreshold depolarization in INT at the time of the IPSP suggests a coupled group of inhibitory neurons. (Reprinted, with permission, from Zimering 1983.)

interneurons could reduce low-voltage oscillatory activity in them and also reduce IPSPs in CMNs, depolarizing them (Fig. 6C). Although the interneuron in Figure 6 inhibited the M neurons and was coupled to several CMNs, insufficient data were obtained on the strength of direct coupling to different CMNs to test further the circuit of Figure 4B.

The anterior–posterior organization of reversed coupling strongly implies a role in patterned movements. During C-group activity, firing of anterior motor neurons inhibits more posterior motor neurons. This situation could facilitate swallowing, starting with contraction of A CMNs controlling the anterior sphincter. It is unclear how M neurons become active, but other interneurons could excite these cells and inhibit a C_a subgroup inhibiting the M cells. Successive stages of progressively more posterior excitation of CMNs and inhibition of C-group neurons would be required. There would still have to be a higher-level pattern generator to orchestrate the movement. The reversed coupling would act at a more restricted level as nonreciprocal inhibition of temporary antagonists.

Reversed coupling in which hyperpolarization of one cell depolarizes more posterior cells would not operate in this scheme; responses like those illustrated in Figure 3B would require inhibitory inputs to the CMNs in addition to the C group. Although non-C-group IPSPs are sometimes seen, there are insufficient data to attempt to include them at this time.

Direct coupling as observed in the absence of C-group activity presumably simply increases synchrony of CMN activity. As peristalsis proceeded posteriorly, anterior neurons might become coupled; also, mutual excitation could increase the efficacy of pharyngeal emptying prior to an expansion.

As with expansion motor neurons, multiunit recording is required to analyze more fully the significance of the highly ordered circuitry.

OTHER MODES OF INTERACTION

Given that cells can be connected by chemical excitatory, chemical inhibitory, and linear electrical synapses, the number of possible circuits increases very rapidly with the number of cells involved. With two cells there are 19 combinations (because a single cell can transmit as well as receive one, two, or three ways), and with three a much larger number. Other possibilities arise where cells are not isopotential. For example, teleost occulomotor neurons appear to be coupled at the level of their cell bodies, tending to synchronize firing during saccades, while impulses mediating slow, graded movements arise in the dendrites where the cells are not coupled (Korn and Bennett 1975). Consideration of all possible circuits is unlikely to have much predictive value, and a case-by-case evaluation appears at this time a more rewarding approach. Of the examples considered here, synaptic control of coupling may be a fairly common arrangement. It is a natural result where cells are electrically coupled and also receive inhibition. Reversed coupling requires an added connection of one of the coupled cells to the inhibitory neurons. Reciprocal connections involving electrical coupling and inhibition occur in the lobster stomatogastric ganglion (Eisen and Marder 1982) and mammalian olfactory bulb (Pinching and Powell 1971). Modulation of the inhibitory elements' excitability to change between direct coupling and inhibition or reversed coupling is possible. However one might well not use the same terminology. For us, the choice followed naturally from the dramatic reversal from direct coupling to coupling of the opposite sign, which is, of course, simply an inhibitory relation with modulation of tonic activity allowing depolarization as well as hyperpolarization of the inhibited cell.

The findings reported here may not be widely generalizable in their particular details. They do indicate further that electrical synapses can be involved in complex neuronal interactions, even where the junctional membranes remain, as far as is known, fixed in their properties.

ACKNOWLEDGMENTS

This work was supported in part by National Institutes of Health grants NS 07512 and NS 12627. M.B.Z. was supported by NIH Medical Scientist Training Grant GM 1674.

REFERENCES

Bennett, M.V.L. 1977. Electrical transmission: A functional analysis and comparison with chemical transmission. In *Handbook of physiology*, section 1: *The nervous system* (ed. E.R. Kandel), vol. 1, p. 357–416. Williams and Wilkins, Baltimore, Maryland.

Bennett, M.V.L. and G.D. Pappas. 1983. The electromotor system of the stargazer: A model for integrative actions at electrotonic synapses. *J. Neurosci.* **3:** 748–761.

Carew, T.J. and E.R. Kandel. 1976. Two functional effects of decreased conductance EPSPs: Synaptic augmentation and increased electrotonic coupling. *Science* **192:** 150–153.

Dudek, F.E., R.D. Andrew, B.A. MacVicar, R.W. Snow, and C.P. Taylor. 1983. Recent evidence for and possible significance of gap junctions and electrotonic synapses in the mammalian brain. In *Basic mechanisms of neuronal hyperexcitability* (ed. H.H. Jaspar and N.M. van Gelder), p. 31–73. A.R. Liss, New York.

Eisen, J.S. and E. Marder. 1982. Mechanisms underlying pattern generation in lobster stomatogastric ganglion as determined by selective inactivation of identified neurons. III. Synaptic connections of electrically coupled pyloric neurons. *J. Neurophysiol.* **48:** 1392–1415.

Korn, H. and M.V.L. Bennett. 1975. Vestibular nystagmus and teleost oculomotor neurons: Functions of electrotonic coupling and dendritic impulse initiation. *J. Neurophysiol.* **38:** 430–451.

Korn, H. and D.S. Faber. 1979. Electrical interactions between vertebrate neurons: Field effects and electrotonic coupling. In *The neurosciences: Fourth study program* (ed. F.O. Schmitt and F.G. Worden), p. 333–358. MIT Press, Cambridge, Massachusetts.

Levitan, H., L. Tauc, and J.P. Segundo. 1970. Electrical transmission among neurons in the buccal ganglion of the mollusc, *Navanax inermis. J. Gen. Physiol.* **55:** 484–496.

London, J.A., D. Zecevic, and L.B. Cohen. 1984. Optical monitoring of activity from buccal ganglia during pharyngeal expansion (feeding) in a minimally dissected *Navanax. Soc. Neurosci. Abstr.* **10:** 508.

Pinching, A.J. and T.P.S. Powell. 1971. The neuropil of the glomeruli of the olfactory bulb. *J. Cell Sci.* **9:** 347–377.

Smith, T.G. and F. Baumann. 1969. The functional organization within the ommatidium of the lateral eye of the *Limulus. Prog. Brain Res.* **31:** 313–349.

Spira, M.E. and M.V.L. Bennett. 1972. Synaptic control of electrotonic coupling between neurons. *Brain Res.* **37:** 294–300.

Spira, M.E., D.C. Spray, and M.V.L. Bennett. 1976. Electrotonic coupling: Effective sign reversal by inhibitory neurons. *Science* **194:** 1065–1067.

———. 1980. Synaptic organization of expansion motoneurons of *Navanax inermis*. *Brain Res.* **195:** 241–269.

Spitzer, N. 1982. Voltage and stage dependent uncoupling of Rohon-Beard neurones during embryonic development of *Xenopus* tadpoles. *J. Physiol.* **330:** 145–162.

Spray, D.C., M.E. Spira, and M.V.L. Bennett. 1980. Synaptic connections of buccal mechanosensory neurons in the opisthobranch mollusc, *Navanax inermis*. *Brain Res.* **182:** 271–286.

Teranishi, T., K. Negishi, and S. Kato. 1984. Regulatory effects of dopamine on spatial properties of horizontal cells in carp retina. *J. Neurosci.* **4:** 1271–1280.

Traub, R.D., F.E. Dudek, C.P. Taylor, and W.D. Knowles. 1985. Simulation of hippocampal afterdischarges by electrical interactions. *Neuroscience* **14:** 1033–1038.

Zimering, M.B. 1983. "Synaptic connectivity of the circumferential motoneurons in the buccal ganglia of *Navanax* and its role in the feeding behavior." Ph.D. thesis, Albert Einstein College of Medicine. University Microfilms, Ann Arbor, Michigan.

Junctional Voltage-dependence at the Crayfish Rectifying Synapse

Christian Giaume and Henri Korn
Laboratoire de Neurobiologie Cellulaire
Département de Biologie Moléculaire
Institut Pasteur
75724 Paris Cedex 15, France

Although electrotonic transmission at gap junctions is most often symmetric (Bennett 1977), "synaptic rectification" has been demonstrated in some coupled systems. Evidence of this property was first obtained at the crayfish giant motor synapse (GMS), for which it was thus proposed that the junction behaves like a diode rather than like a simple resistor (Furshpan and Potter 1959). This work represents the first report of a voltage-dependent electrotonic synapse, although at that time this mechanism was not defined with such terms. It is important to point out that the term "rectification" and the diode hypothesis were then attributed to the synaptic membrane, the junction being still considered as a single unit; the individual junctional channel properties were not referred to because the ultrastructure of gap junctions was not known.

Following the initial observation of a voltage difference between the resting membrane potentials of coupled cells at the GMS we investigated the consequences of modifying the junctional polarization on the synaptic rectification. We first confirmed the diode hypothesis by showing that increased junctional conductance and bidirectional transmission result when the sign of the resting polarization difference is inverted (Giaume and Korn 1983a). This plasticity of the GMS allowed us to study several properties of the voltage-dependency of electrotonic coupling between neurons in the adult. These results will be compared briefly with those obtained at other voltage-sensitive gap junctions studied in mostly nonexcitable embryonic tissues. We will also discuss an unexpected loss of rectification that occasionally follows internal acidification (Giaume and Korn 1983b) and that raises the interesting question of a possible evolution of junctional voltage dependence under physiological conditions, during development, or even in the adult.

EXPERIMENTAL DATA

Voltage Dependence and Electrotonic Coupling at the Giant Motor Synapse.

Transmission was studied between the lateral giant axon (presynaptic cell) and the giant motor fiber (postsynaptic), which comprise the GMS, by overlapping at the level of the third ganglionic root of the abdominal nerve cord (Fig. 1A). Action potentials are transmitted from the giant axon to the motor fiber and blocked in the reverse direction (Furshpan and Potter 1959). This rectification can be demonstrated directly by intracellular injections of current: Depolarizations are transmitted solely in the orthodromic direction and correspondingly, hyperpolarizations cross the junction when applied to the motor fiber (Fig. 1B_1). As stated above, the cells do not have the same resting membrane potentials: the lateral giant axon is about 15 mV more inside negative than the giant motor fiber. The reality of this voltage difference is now well established and does not seem due to injury of the postsynaptic cell since (1) it was observed in all trials (more than 70 preparations), (2) normal neuronal functions were preserved (i.e., postsynaptic action potential, spontaneous activity, response to antidromic stimulation), and (3) a similar difference is present, and of the same value ($\sim$15 mV) at rectifying synapses between giant fibers and motor neurons of the hatchetfish (Auerbach and Bennett 1969) and in the spinal cord of the lamprey (Ringham 1975). Inversion of the transjuctional difference, ΔV, induces marked changes in the electrotonic coupling coefficient (Giaume and Korn 1983a). For example, as ΔV was reversed by maintained postsynaptic hyperpolarization, the electrotonic response for presynaptic depolarization was enhanced (Fig. 1, B_2). Furthermore, as ΔV was made even more positive, hyperpolarizing current injections were also transmitted across the GMS (Fig. 1, B_3) until a bidirectional state was reached at which currents of both signs were transmitted equally (not illustrated in Fig. 1B). These results are summarized by plots of the coupling coefficients versus ΔV which indicate that synaptic rectification and electrotonic coupling are strongly dependent upon the potential difference across the junction (Fig. 1C).

The same conclusion could be obtained by observing more physiological events such as spontaneously occurring postsynaptic potentials (PSP) which can be recorded in the motor fiber but are not transmitted to the lateral axon under normal resting conditions. In contrast, when ΔV was reversed by a postsynaptic hyperpolarization, the PSPs were transmitted in the antidromic direction and recorded in the lateral axon (Fig. 1D).

These results corroborate and complete the diode hypothesis. When biased toward higher conductance by relative presynaptic positivity, the coupling conductance becomes larger and transmits small voltage changes of either sign because these changes have little effect on the conductance.

Characteristics of the Junctional Voltage-dependence

The insufficient time resolution obtained with current clamp techniques did not allow us to observe the kinetics of coupling modifications produced by inversion of the junctional polarization ΔV: The electrotonic response had already reached its maximum as soon as the first test pulse occurred during a directly applied postsynaptic hyperpolarizing pulse (Fig. 2A). In contrast, due to its shorter time constant, the double voltage-clamp technique (Fig. 2B) has permitted us to monitor directly the junctional current, I_J, with a faster time response. Because of the short synaptic delay, the onset of the increase in junctional conductance was expected to be very rapid (Fig. 2C). Indeed, voltage-clamp recordings (Fig. 2D) confirmed that the junctional channels opened within 500 μsec of a postsynaptic hyperpolarization after which time a steady state of junctional conductance was achieved. In addition, the steady-state current flowing through the junction increased as the voltage difference was enhanced by postsynaptic hyperpolarizations (Fig. $2E_1$, E_2), and this relationship is plotted in Figure 3.

Since each cell's potential could be controlled independently, it was possible to monitor I_J either from the presynaptic axon or from the motor fiber as a voltage step was delivered in the opposite cell. An example of the derived relationship between I_J and the potential difference across the junction is illustrated in Figure 3 (upper diagram): I_J was only increased when ΔV was made positive. In order to distinguish the dependency of I_J upon the transjunctional and the inside-outside potentials, the same relative value of ΔV could be obtained for different levels of both membrane potentials. For each set of experimental measurements, the junctional slope conductance curve was calculated (Fig. 3, inset) and G_J was derived from low to high conductance states of the junction, respectively. Thus, the plot for each series of measurements of the minimal and maximal junctional conductances (calculated at $\Delta V = 0$ and +40 mV, respectively) versus the two transmembrane potentials (lower diagram) demonstrates that the two derived values of G_J were not significantly modified within a range of 60 mV. This result directly confirms and extends the notion, which is implicit in the results of indirect estimates of the synaptic current (Furshpan and Potter 1959; Margiotta and Walcott 1983), that the junctional conductance is mainly controlled by the voltage difference across the junction.

Effects of Uncoupling Agents

Electrotonic coupling can be reduced by applying various drugs whose blocking effects on gap junctions are reported to be different (Spray et al. 1984). These drugs alter junctional conductance either indirectly, possibly by an increase in cytoplasmic ionic concentration (e.g., H^+ and/or Ca^{++}), or directly by an external interaction with the junctional macromolecule. The effects of such agents were tested on electrotonic coupling at the crayfish rectifying synapse in order to determine if its pharmacological responses were different from those of other gap junctions.

First, sulfhydryl reagents that uncouple the voltage-insensitive crayfish septate junctions without changing the internal pH (pH_i) (Giaume et al. 1981; Campos de Carvalho and Ramon 1982) were also effective at the GMS. For example, diamide and N-ethylmaleide (NEM) progressively blocked the electrotonic coupling of the GMS with the same time course (30–60 min) and at identical concentration (1 mM). Second, the metabolic inhibitor dinitrophenol

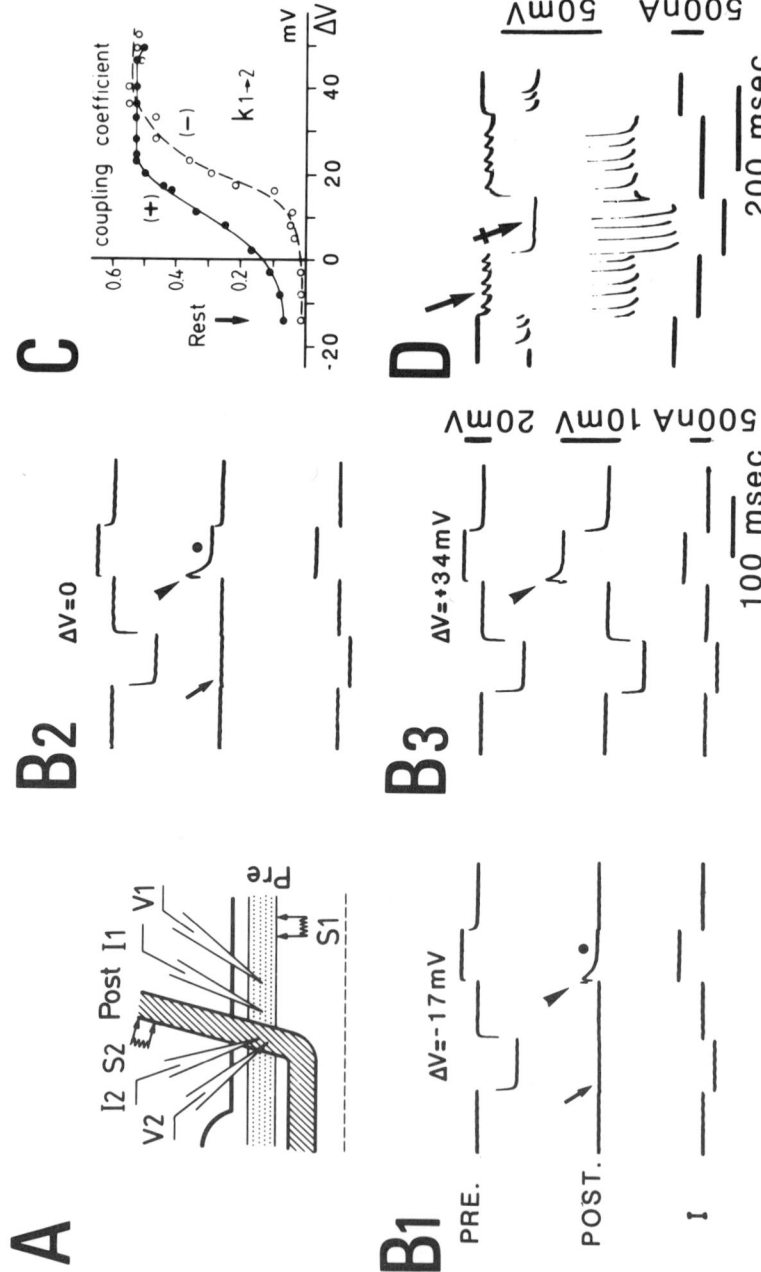

Figure 1 (See facing page for legend.)

(DNP, 1 mM), which uncouples gap junctions with a concomitant elevation of cytoplasmic calcium and an internal pH change (Rose and Rick 1978) similarly increased the junctional resistance of the giant motor synapse.

Finally, internal and external perfusions of octanol have shown that at the crayfish septate synapse, the junctional resistance was reversibly increased when the alcohol was applied extracellularly (Johnston et al. 1980). Applications of a 1 mM octanol solution (prepared by adding a 0.5 (vol/vol) of dimethylsulfoxide) also completely abolished electrotonic coupling at the rectifying synapse and coupling was restored following removal of the drug without any loss of the junction's rectifying properties (Fig. 4A).

Uncoupling Induced by Internal Acidification

Cytoplasmic acidifications, which decrease the conductance of gap junctions (Turin and Warner 1977; Giaume and Korn 1982; Campos de Carvalho et al. 1984), were performed by perfusing the nerve cord with either 100% CO_2 saturated Ringer's solution or with a solution in which sodium was replaced by acetate and the pH was fixed at 6.5. Following these treatments we observed two different kinds of results. Most of the time, the junction was uncoupled totally and, after wash, fully recovered its voltage-dependent properties (Fig. 4B). However, in several experiments (7 to 33), the cells first became uncoupled (Fig. $5A_1$, A_2), but, as coupling recovered, the junction then transmitted symmetrically (Fig. $5A_3$). In such cases, the antidromic voltage transfer function (the postsynaptic voltage versus its electrotonic response in the lateral

Figure 1
Induction of bidirectional transmission at the GMS. (A) Diagram illustrating the experimental set up for intracellular recording and for extracellular stimulation at both sides of the rectifying synapse. (B_1–B_3) Symmetrical electrotonic transmission after inversion of the resting junctional polarization ΔV. (*Upper and middle sweeps*) Voltage traces recorded from the presynaptic (PRE.) and postsynaptic (POST.) cells, respectively. (*Lower trace*) Current applied through the presynaptic electrode. (B_1) At rest, ΔV was −17 mV. Transmission of negative currents directly applied in the presynaptic cell was transmitted (dot). The depolarization generated pre- and postsynaptic spikes (not photographed here) followed, in the motor fiber, by a slow postsynaptic potential (→). (B_2) As ΔV was fixed at zero by a maintained postsynaptic hyperpolarization, transmission of a positive pulse was facilitated but hyperpolarization was still blocked. (B_3) As ΔV was inverted and held at +34 mV, test currents of both polarities were transmitted. (C) Plot of coupling coefficient (*ordinate*) versus ΔV (*abscissa*), as the postsynaptic fiber was made progressively more negative. The coupling coefficients for positive (———) and negative (--------) pulses were gradually increased and finally reached the same plateau (different experiment than that illustrated in B). (D) Evidence that PSPs in the motor fiber (dots, *middle trace*) were not transmitted in the antidromic direction and that they only spread into the presynaptic axon (arrows, *upper trace*) during hyperpolarization of the postsynaptic fiber that inverted the junctional polarization ΔV from −17 to +53 mV. A reduction of this potential difference by a short polarization of the lateral axon (crossed arrow) reversibly blocked the transmission of the PSPs despite a concomitant increase of their amplitude (due to increased postsynaptic hyperpolarization and to decreased input conductance because of decreased junctional conductance). (Modified from Giaume and Korn 1983a and 1984.)

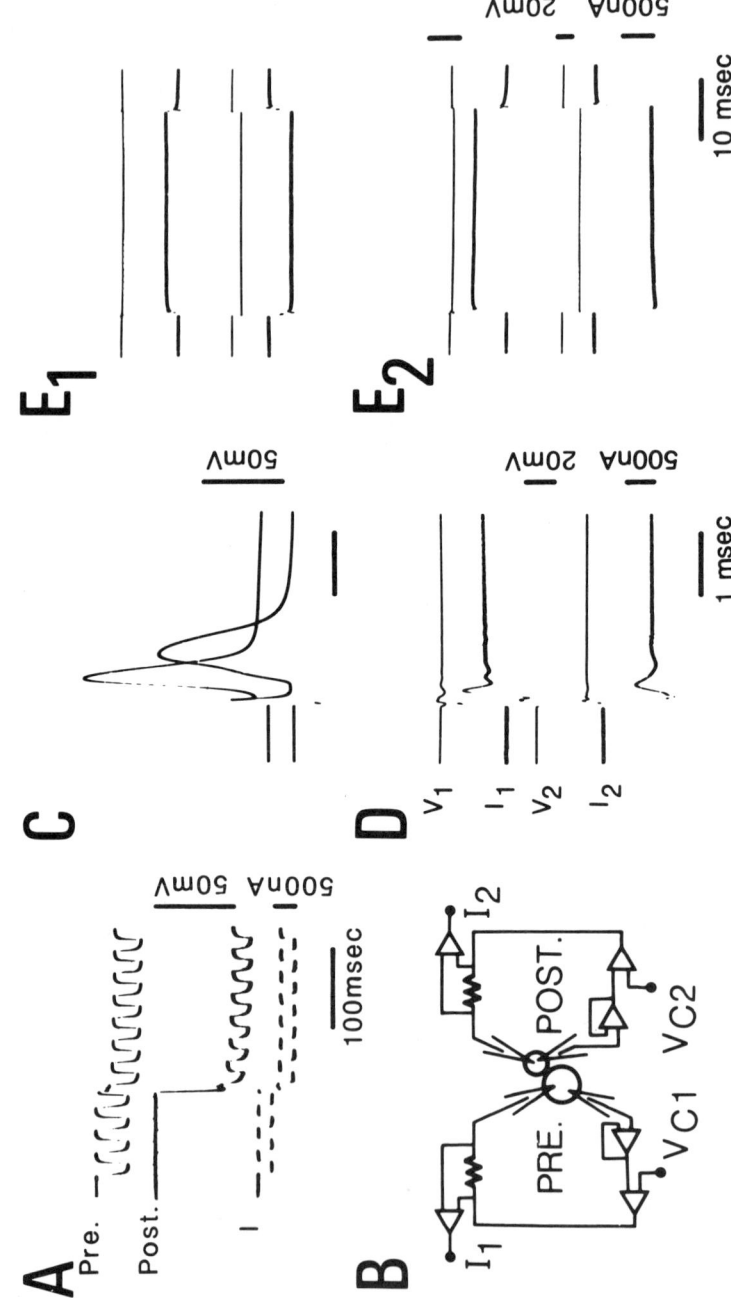

Figure 2 (See facing page for legend.)

axon), initially nonlinear (Furshpan and Potter 1959), became linear, thus indicating an absence of voltage dependency (Fig. 5C). This loss of rectification was not reversible up to the end of the experiment (about 90 min). Yet, during this state of induced symmetrical transmission the junction was still sensitive to uncoupling agents such as CO_2 or diamide (although the time course for uncoupling was longer). The slopes of the two V_{post}/V_{pre} relations obtained in the symmetrical state and the conducting portion of the rectifying state are different, indicating that the overall conductance of the junction was reduced (Fig. 5B, C). This reduction could be due either to a decrease in unitary channel conductance and/or to a reduction in the number of conducting channels, or partially to a change in nonjunctional conductances. In contrast, as shown in Figure 4, when the junction recovered its full rectifying properties after exposure to octanol or CO_2, the coupling coefficients were not changed.

DISCUSSION

The characteristic of voltage-dependent but symmetrical electrotonic junctions has been described largely in embryonic cells of vertebrates (Spray et al. 1979; White et al. 1982) and invertebrates (Obaid et al. 1983; Bennett and Spray 1984). By contrast, electrotonic transmission through adult gap junctions is not generally controlled by potential, although rectification has been observed in few instances (Bennett 1977). Because of its relative accessibility, the crayfish GMS is a convenient preparation for studying a well-developed adult form of voltage-dependent electrotonic transmission.

The application of blockers has shown that the pharmacology of the embryonic and adult junctions are similar; e.g., channels of the GMS are sensitive to various agents in the same concentration ranges and with identical time courses as those from amphibian embryonic cells (Spray et al. 1984). Moreover, their dependency upon transjunctional potential rather than cellular transmembrane potential is identical. Finally, the voltage sensitivity reported

Figure 2
Kinetics and steady state of junctional conductance at the GMS. (A) In current clamp conditions, short negative pulses (25 msec in duration) were applied in the lateral axon (*upper trace*, pre), and spread of current to giant motor fiber (*middle trace*, post) began rapidly after the onset of postsynaptic hyperpolarization (*lower sweep*: simultaneous recording of currents injected in both neurons). (B) Schematic drawing of the double voltage-clamp circuit used for measurements of junctional currents. (Modified from Spray et al. 1981.) (C) Short latency orthodromic transmission of action potentials. (D) Double voltage-clamp recording at the motor synapse. V_1, I_1 and V_2, and I_2 are the voltage and current traces recorded with electrodes in the pre- and postsynaptic neurons, respectively. A hyperpolarizing test pulse was applied in cell 2 while the membrane potential was maintained in neuron 1 by a positive current (I_1) (so that the junctional current I_J, which had crossed the synapse, was equal to I_1 but of opposite polarity). Note that the steady state was reached extremely rapidly, i.e., by no more than 500 μsec. (E_1–E_2) Slower sweeps than in D showing that the junctional current was increased as the negative pulses in the postsynaptic cell were made larger without modifications of the steady state (same disposition of traces as in D).

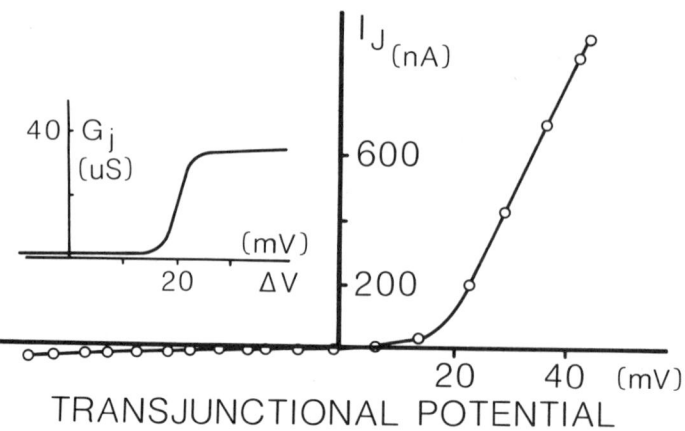

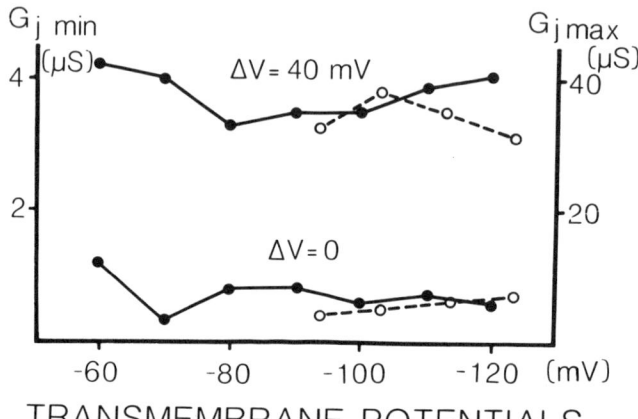

Figure 3
Conductance-voltage relation during the steady state of junctional conductance. (*Upper diagram*) Plot of the junctional current (*ordinate*) versus the potential difference across the synapse (*abscissa*). Voltage steps were produced in the postsynaptic fiber and corresponding junctional currents were recorded in the lateral axon, membrane potentials were fixed at −90 and −65 mV, respectively; the *I/V* curve (———) was plotted from experimental data (○) using a computer program that selected the best fitting hyperbola obtained with asymptotes defined by linear regression. I_J is dependent on the transjunctional voltage. (*Inset*) The junctional slope conductance was calculated as the derivative of the *I/V* relationship and plotted against ΔV. (*Lower diagram*) Plots of the two states of junctional conductance derived from the above (*ordinate*) versus both cells' membrane potentials (*abscissa*). Minimum and maximum G_J were calculated from the *I/V* relations obtained at a same ΔV but at different membrane potentials of the presynaptic (———) and postsynaptic (--------) neurons (the left-side scale refers to the minimal conductances calculated at $\Delta V = 0$, and the right one is related to $G_{J\,max}$ at $\Delta V = +40$ mV). Note that this representation does not take into account the voltage changes introduced by the pulse used to measure I_J.

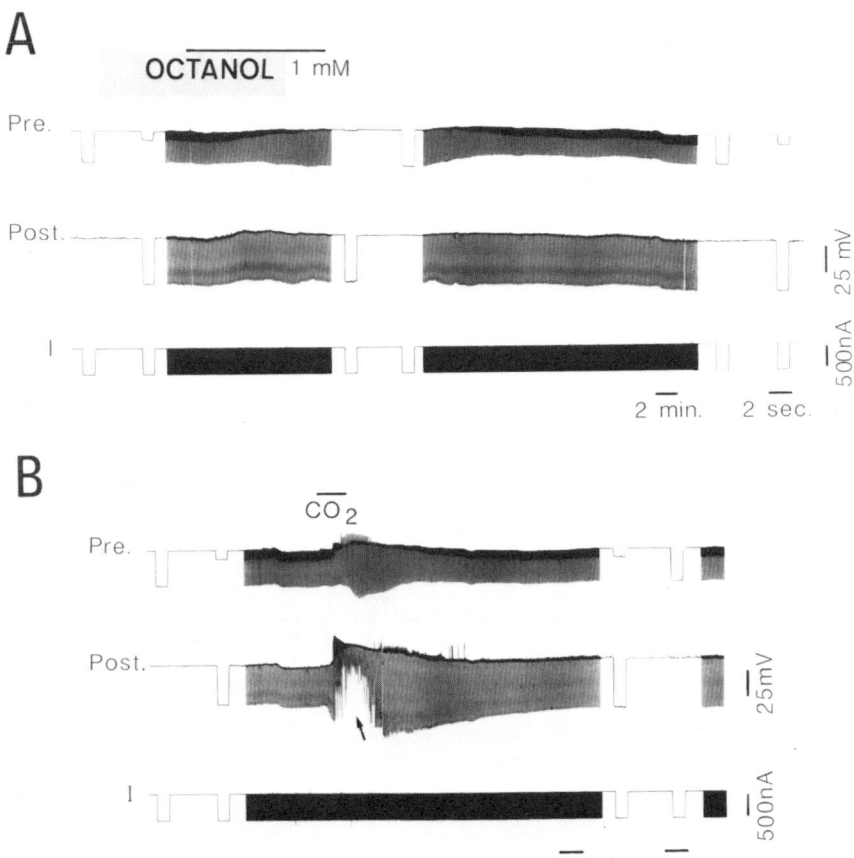

Figure 4
Effects of uncoupling agents. Electrotonic coupling was estimated by injecting hyperpolarizing current pulses (downward, negative) in the lateral axon (Pre., *upper trace*) and the motor fiber (Post., *middle trace*), alternately. The large and small vertical deflections are indicative of the input resistances of the injected cell, and of the transfer resistances of the noninjected cell, respectively. Currents injected in both elements are shown on the same sweep (*lower trace*). (*A*) Uncoupling of the GMS by a 1 mM octanol solution. (*B*) Blocking effect of a 100% CO_2-saturated Ringer's solution, with a concomitant reduction of the postsynaptic input resistance due primarily to a transient, increased, chemically mediated activity in the post fiber (arrow). Note that in both cases, the uncoupling was reversible, the rectifying properties being the same after wash and in the control, as shown by the recordings obtained with a fast sweep spread, at the beginning and at the end of each trace.

here for the main current carrying ions holds also for larger molecules such as Lucifer Yellow (Giaume and Korn 1984), just as was found for embryonic gap junctions (Spray et al. 1979; Zimmerman and Rose 1983).

However, several findings suggest that junctional channels of the GMS may

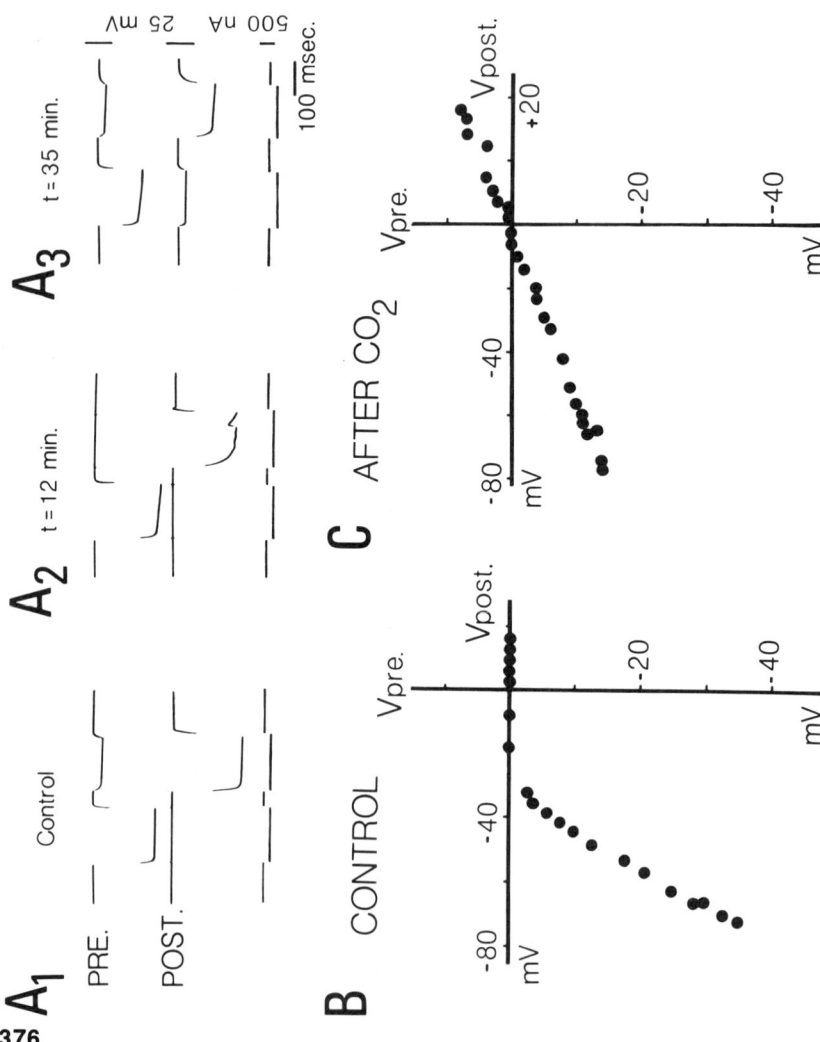

Figure 5 (See facing page for legend.)

have specific characteristics of their own. First, the junction at rest is in a low-conductance state with most of its channels presumably closed (see Margiotta and Walcott 1983; Giaume and Korn 1984), whereas other junctions studied have a high resting G_J. Second, their opening and closing times are extremely brief (less than 500 μsec), whereas those of amphibian embryos are about 1000 times longer (Harris et al. 1981). Third, GMS junctional conductance is not the same for identical transjunctional voltages of either sign, but rather it is increased only when the presynaptic cell is made relatively more positive than the postsynaptic cell. This dependency contrasts with the one present at embryonic junctions where G_J is decreased equally by transjunctional polarizations of either sign (Spray et al. 1979). Finally, the observation in some cells of prolonged loss of rectification after brief periods of acidification suggests that voltage and protons may share a common site of action, as in *Chironomus* salivary glands where voltage and pH interact (Obaid et al. 1983), which is not the case for amphibian embryonic tissues where both gating mechanisms are reported to be independent (Spray et al. 1984).

Synaptic transmission at the GMS is assumed to combine advantages of both chemical and electrical transmission, namely the functional polarity of the former and the short synaptic delay and the high frequency response of the latter. However, our observations indicate that this synapse can, in fact, mediate ionic currents in a symmetrical mode since its "one-way selectivity" is, in reality, modulated by the level and the sign of the transjunctional polarization. From this point of view, the 15-mV difference between the two resting membrane potentials can be interpreted as a "safety factor," the function of which is to prevent antidromic transmission of PSPs or action potentials. In this manner, the integrity of the tailflip, which is mediated by the giant escape circuitry (Krasne et al. 1977), would be preserved.

In addition, the experiments in which we induced cytoplasmic acidification indicate that protons can, under certain conditions not presently understood, modify the properties of the channels and bring the junction to another state, which is characterized by an ohmic relationship. Although this observation was not consistently found in all the preparations and the mechanism is un-

Figure 5
Bidirectional transmission induced in the GMS following uncoupling by low internal pH. (A_1–A_3, *lower and upper traces*) Intracellular recordings from the pre- and postsynaptic cells, respectively. Hyperpolarizing currents were transmitted solely to the adjacent neuron when applied from the motor fiber (control records, A_1). After a 3-min exposure to a CO_2 100% saturated solution, the coupling was totally abolished (A_2). After wash, the junction recoupled but with transmission of the negative currents in both directions (A_3) (current traces have been retouched by hand). (*B–C*) Voltage-transfer functions before and after exposure to CO_2. Plots of the transmembrane (V_{post}, *abscissae*) versus the transjunctional (V_{pre}, *ordinates*) potentials generated by polarizing current injected in the motor fiber, defining the voltage transfer function across the junction. (*B*) The relationship is nonohmic in the control, as expected from a rectifying synapse. (*C*) Linear relation obtained after recovering from uncoupling by application of CO_2 (same conditions as for Fig. 4B), indicating a total loss of the voltage dependence. (A_1–A_3 and *B–C*, are from two different experiments, respectively).

known, it may be similar to that underlying changes observed during the development of *Xenopus* Rohon-Beard neurons where internal acidification also eliminates the voltage-dependence of gap junctions (Spitzer 1982). Our results thus raise a more general question about a possible alteration of junctional voltage-sensitivity during development. It may be that gap junctions are commonly voltage-dependent at embryonic stages and become less sensitive to transjunctional polarization as differentiation takes place. Rectifying synapses present in adult tissues might then represent a stabilization of these junctions which have been preserved from such drastic modification. If so, this persistence would be however only partial, as indicated by the different properties of embryonic and adult rectifying junctions that have been demonstrated in this report.

ACKNOWLEDGMENTS

We wish to thank Dr. Paul Funch for critical comments of the manuscript and Josseline Nicolet for technical help and secretarial assistance.

REFERENCES

Auerbach, A.A. and M.V.L. Bennett. 1969. A rectifying electrotonic synapse in the central nervous system of a vertebrate. *J. Gen. Physiol.* **53:** 211–237.

Bennett, M.V.L. 1977. Electrical transmission: A functional analysis and comparison with chemical transmission. In *Cellular biology of neurons,* Section 1 (ed. E. Kandel), vol. 1, p. 357–416. Williams and Wilkins, Baltimore.

Bennett, M.V.L. and D.C. Spray. 1984. Gap junctions: Two voltage-dependent gates in series allow voltage induced steady state cycling around a circular reaction theme. *Biophys. J.* **45:** 60a.

Campos de Carvalho, A. and F. Ramon. 1982. The effect of protein reagents on the gap junctions of crayfish septate axons. *Biophys. J.* **37:** 287a.

Campos de Carvalho, A., D.C. Spray, and M.V.L. Bennett. 1984. pH-dependence of transmission at electrotonic synapse of the crayfish septate axon. *Brain Res.* **321:** 279–286.

Furshpan, E.J. and D.D. Potter. 1959. Transmission at the giant motor synapses of the crayfish. *J. Physiol.* **145:** 289–325.

Giaume, C. and H. Korn. 1982. Ammonium sulfate induced uncouplings of crayfish septate axons with and without increased junctional resistance. *Neuroscience.* **7:** 1723–1730.

———. 1983a. Bidirectional transmission at the rectifying electrotonic synapse: A voltage-dependent process. *Science* **220:** 84–87.

———. 1983b. CO_2 induced irreversible loss of voltage-dependence at a rectifying electrotonic synapse. *Soc. Neurosci.* **9:** 513. (Abstr. 149.10.)

———. 1984. Voltage-dependent dye-coupling at a rectifying electrotonic synapse of the crayfish. *J. Physiol.* **356:** 151–167.

Giaume, C., M.E. Spira, and H. Korn. 1981. Diamide, a thiol oxidizing agent, uncouples electrotonic junctions of crayfish septate axons, but not those of *Navanax* motoneurons. *Neuroscience* **6:** 2239–2247.

Harris, A.L., D.C. Spray, and M.V.L. Bennett. 1981. Kinetic properties of a voltage-dependent junctional conductance. *J. Gen. Physiol.* **77:** 95–117.

Johnston, M.F., S.A. Simon, and R. Ramon. 1980. Interaction of anesthetics with electrical synapses. *Nature* **286:** 498–500.

Krasne, F.B., J.J. Wine, and A. Kramer. 1977. The control of the crayfish escape behavior. In *Identified neurons and behavior of arthropods* (ed. G. Hoyle), p. 275–292. Plenum Press, New York.

Margiotta, J.F. and B. Walcott. 1983. Conductance and dye-permeability of a rectifying electrical synapse. *Nature* **305:** 52–56.

Obaid, A.L., S.J. Socolar, and B. Rose. 1983. Cell-to-cell channels with two independent regulated gates in series: Analysis of junctional modulation by membrane potential calcium and pH. *J. Membr. Biol.* **73:** 69–89.

Ringham, G.L. 1975. Localization and electrical characteristics of a giant synapse in the spinal cord of the lamprey. *J. Physiol.* **251:** 395–407.

Rose, B. and R. Rick. 1978. Intracellular pH, intracellular free Ca and junctional cell-cell coupling. *J. Membr. Biol.* **44:** 377–415.

Spitzer, N.C. 1982. Voltage- and stage-dependent uncoupling of Rohon-Beard neurons during embryonic development of *Xenopus* tadpoles. *J. Physiol.* **330:** 145–162.

Spray, D.C., A.L. Harris, and M.V.L. Bennett. 1979. Voltage-dependence of junctional conductance in early amphibian embryos. *Science* **204:** 432–434.

———. 1981. Equilibrium properties of a voltage-dependent junctional conductance. *J. Gen. Physiol.* **77:** 77–93.

Spray, D.C., R.L. White, A. Campos de Carvalho, A.L. Harris, and M.V.L. Bennett. 1984. Gating of gap junction channels. *Biophys. J.* **45:** 219–230.

Turin, L. and A. Warner. 1977. Carbon dioxide reversibly abolishes ionic communication between cells of early amphibian embryo. *Nature* **270:** 56–57.

White, R.L., D.C. Spray, A.Campos de Carvalho, and M.V.L. Bennett. 1982. Voltage-dependent gap junctional conductance between fish embryonic cells. *Soc. Neurosci.* (Abstr.). **8:** 944.

Zimmerman, A.L. and B. Rose.1983. Analysis of cell-to-cell diffusion kinetics: Changes in junctional permeability without accompanying changes in selectivity. *Biophys. J.* **41:** 216a.

Neurotransmitter-induced Modulation of Gap Junction Permeability in Retinal Horizontal Cells

J. Neyton,† M. Piccolino,* and H.M. Gerschenfeld†

†Laboratoire de Neurobiologie
Ecole Normale Supérieure
75230 Paris 05, France

*Istituto di Neurofisiologia del C.N.R.
56100 Pisa, Italy

In contrast with chemical synapses, signals are transmitted in electrical synapses from one neuron to the next by a flow of intracellular ions across the gap junction channels, without intervention of a synaptic transmitter (see Bennett 1977; Loewenstein 1981). Thus, electrical synapses may appear as structurally and functionally fixed pathways of interneuronal communication when compared with the much more flexible chemical synapses. However, the permeability of the gap junctions has been shown to be modulated by different physiological manipulations. For instance, a transjunctional potential difference between the electrically coupled cells may decrease the permeability of the gap junction (Spray et al. 1979, 1981a, 1984; Harris et al. 1983). A similar junctional conductance decrease was observed in a different preparation when changing the potential between the cytoplasm and exterior of the coupled cells in the absence of a transjunctional potential (Obaid et al. 1983; see also Spray et al. 1984). Moreover, decreases in the gap junction conductance were also obtained by increasing the intracellular concentration of either Ca^{++} (Rose and Loewenstein 1976; Spray et al. 1982) or H^+ (Turin and Warner 1977, 1980; Giaume et al. 1980; Spray et al. 1981b).

It has also been reported that the electrical coupling between neurons could be altered by synaptic transmitters. In most of these cases the transmitter was found to modify the electrical coupling by changing the conductance of the nonjunctional membranes of the neurons (Spira and Bennett 1972; Carew and Kandel 1976; Spira et al. 1980). That a transmitter could induce a modification of the gap junction permeability itself has been recently suggested in two cases: the acinar cells of the pancreas (Iwatsuki and Petersen 1978; Findlay and Petersen 1982) and the retinal horizontal cells of either the turtle (Gerschenfeld et al. 1982; Piccolino et al. 1982) or the carp (Teranishi et al. 1983, 1984).

The present paper is a short review of our experimental results on the modulation of the permeability of the gap junctions between turtle horizontal cells by dopamine (DA) (see also Piccolino et al. 1984). We have found that DA markedly decreases the summation area of the receptive field of the H1 horizontal cell axon terminals, alters the electrical coupling between the axon terminals in a way that is consistent with a decrease in junctional conductance, and markedly restricts the diffusion of the dye Lucifer Yellow in the axon terminal network. It was also found that all these effects of DA were due to the activation of D1 receptors to DA and that pharmacological agents that increase the intracellular cAMP mimic the effects of DA on the H1 horizontal cells. We then concluded that DA decreases the permeability of the gap junctions between turtle H1 horizontal cells and that this action is mediated by an intracellular second messenger, probably cAMP.

THE H1 HORIZONTAL CELL AXON TERMINAL NETWORK OF THE TURTLE RETINA

The H1 horizontal cells of the *Pseudemys* turtle retina are highly differentiated, spikeless neurons that play the role of associative interneurons in the distal retina. Morphologically they are axon-bearing neurons showing in Golgi preparations a small soma with radiating dendrites, and a thin axon of 0.2–0.5 μm diameter and about 100–400 μm length ending in a stubby thick terminal expansion (Leeper 1978). The cell bodies of the H1 horizontal cells (H1CB) are electrically coupled exclusively to other H1CB (Byzov 1975; Stewart 1978; Piccolino et al. 1982) in a loose manner, through small gap junctions (Witkovsky et al. 1983). In contrast, the H1 horizontal cell axon terminals (H1AT) are tightly, electrically coupled only to other H1AT (Byzov 1975; Piccolino et al. 1982) through extensive gap junctions (Witkovsky et al. 1983).

In spite of their anatomical continuity, H1CB and H1AT are electrically "disconnected" from each other since the fine axon fiber connecting them is unable to support electrical conduction because of the impedance mismatch resulting from the length and slenderness of the axon fiber in comparison with the volume of both the soma and the axon terminal. However, both parts of the H1 horizontal cells receive input from the photoreceptors. Therefore, the H1 horizontal cells of the turtle function as two different and electrically independent networks, the cell body network and the H1AT network (see Gerschenfeld et al. 1982). Both networks give hyperpolarizing responses to light spot stimuli of every wavelength. An illustration of the characteristics of the receptive field of the H1AT network is given in the top recording of Figure 1, which shows the responses of an intracellularly recorded H1AT to white light stimuli of different configurations. The resting dark potential of this H1AT was approximately −25 mV, and when spots of light of varying diameters centered on the impaled element were flashed over the retina, they evoked hyperpolarizing responses that decreased in amplitude in parallel with the gradual decrease of the diameter of the spot (Fig. 1, top). When annuli of light of 5 mm

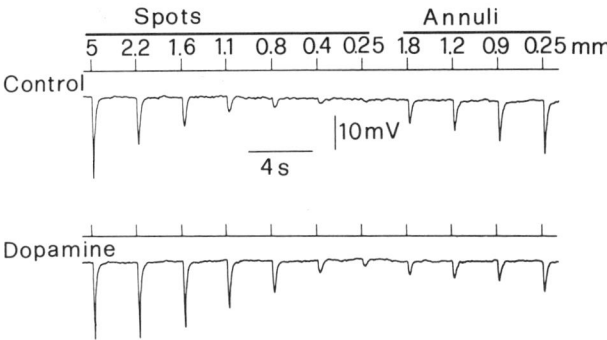

Figure 1
Effects of DA on the H1AT light responses. Intracellular recordings of the responses evoked in a H1AT by white light stimuli of fixed intensity. (*Top recording*) Control light responses; (*bottom recording*) responses to the same light stimuli in the presence of 5 μM DA in the bathing medium. In this and in the following illustrations, the trace above the recordings indicates the timing of the stimuli and the numbers above it indicate the diameter of the light spots and of the inner diameter of the light annuli used as stimuli. The outer diameter of the light annuli was 5 mm in all experiments.

outer diameter and varying inner diameter were flashed, the responses increased in amplitude with the increase of the illuminated peripheral area (Fig. 1, top).

DOPAMINE DECREASES THE PERMEABILITY OF THE GAP JUNCTIONS BETWEEN H1 CELL AXON TERMINALS BY ACTIVATING RECEPTORS LOCATED IN THEIR MEMBRANE PROPER

Effects of DA on the Receptive Field Properties of H1AT

The bottom recording of Figure 1 illustrates the changes induced by DA on the H1AT light responses. The application to the retina of a 5 μM concentration of DA for 20 minutes did not alter significantly the dark membrane potential of the H1AT. However, within a few minutes after the DA application the amplitude of the responses to spots of diameters between 2.2 and 0.25 mm became markedly increased whereas the responses to the light annuli all decreased. Thus, DA elicited a narrowing of the receptive field profile of the H1AT.

In experiments using light slit stimulation as described by Lamb (1976), we have also shown that such narrowing of the H1AT receptive field profile induced by DA corresponds to a decrease of λ, the space constant of the H1AT network, to approximately 50% of its control value (Piccolino et al. 1984).

The DA effect on the H1AT responses always had a rapid onset and was reversible, but only after prolonged (60–90 min) washing with normal saline. The minimal effective DA concentration was 15 nM and the maximal DA effect was obtained with 2–10 μM concentrations. Even at concentrations of 50 μM, DA never altered the dark membrane potential.

DA agonists such as epinine, (±)-6,7-dihydroxy-2-aminotetralin (ADTN), apomorphine, nomifensine, and dihydronomifensine had similar effects as DA on the H1AT receptive field profile.

Effects of DA on the Spread of Intracellularly Injected Current in the H1AT

The changes evoked by DA in the H1AT receptive field profile reflect a decrease in the area of effective electrical interaction in the H1AT network. Such a decrease of electrical interaction could result either from an increase in the coupling resistance of the electrical junctions or a decrease in the nonjunctional membrane resistance of the coupled cells. To investigate these points we studied the effect of DA on the spread of current in the H1AT network.

Figure 2 illustrates recordings obtained from two different H1AT simultaneously impaled by microelectrodes whose tips were separated by less than 100 μm. The recordings shown in Figure 2, left, were obtained in a normal saline and show the responses of both H1AT to different light stimuli: a large spot, a small spot, and an annulus. Thereafter, inward current pulses of 8 nA were injected into one of the axon terminals (Fig. 2a) while recording electrotonic potentials in the other (Fig. 2b). The same light and electrical stimuli were applied during the application of 5 μM DA (Fig. 2, right); the amplitude of the electrotonic potentials then became increased (Fig. 2d).

According to Lamb (1976), for many purposes, the electrical behavior of the horizontal cell network can be conveniently described with a continuous sheet model. The space constant λ of such a sheet is given by $\lambda = (R_m/R_s)^{1/2}$, in which R_m, the resistance of the lamina in the transverse direction is the equivalent of the lumped membrane resistance of the coupled cells; R_s, the tangential resistance, represents the coupling resistance of the network. The voltage drop evoked by a point injection of current i at a distance $r(r\neq 0)$ from the injection site is given by

$$V_i = \frac{iR_s}{2\pi} K_0\left(\frac{r}{\lambda}\right)$$

in which K_0 is a modified Bessel function of zero order. Because K_0 is a monotonically decreasing function, an increase in $V_i(r)$ when λ decreases and when r and i are kept constant cannot result from a decrease in R_m and can only be obtained by an increase of R_s. Thus, the increase of the electrotonic potentials elicited by DA in the current injection experiment together with the DA-induced decrease of λ imply that DA increased the coupling resistance between H1AT, i.e., it decreased the conductance of the gap junctions. A more complete discussion of these results can be found in Piccolino et al. (1984).

Effects of DA on the Diffusion of Lucifer Yellow Dye in the H1AT Network

Figure 3a illustrates the typical pattern of diffusion of Lucifer Yellow obtained in control conditions after intracellular injection of the dye in one H1AT. The most fluorescent structure was the injected axon terminal from where Lucifer Yellow diffused to a rich network of interconnected H1AT. A certain number of

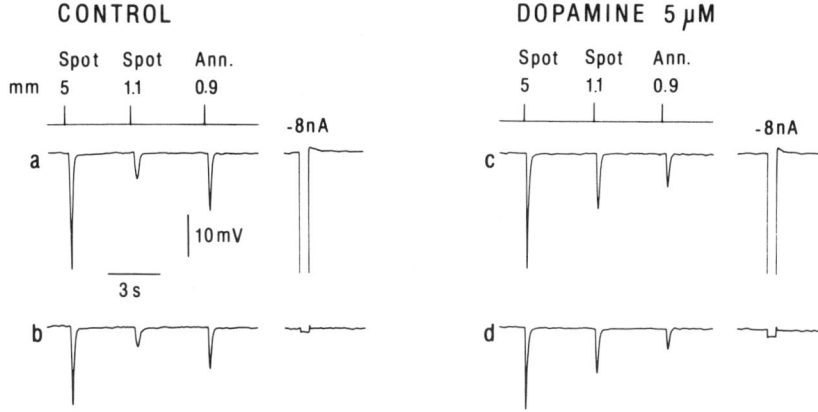

Figure 2
Effects of DA on the electrical coupling between H1AT. (*a,b*) Recordings simultaneously obtained from two H1AT impaled with two different microelectrodes whose tips were separated by less than 100 μm. The H1AT light responses were elicited by the three stimuli indicated above the upper trace. The electrotonic potential recorded in one H1AT (*b*) resulted from the injection of hyperpolarizing current pulse in the other (*a*). (*c,d*) Responses of the same two H1AT to the same light and electrical stimuli recorded after application of 5 μM DA for 10 min. Notice the DA-induced modifications of the responses to the smaller light spot and to the annulus and the DA-induced increase of the amplitude of the electrotonic potential.

cell bodies were also backfilled in these control experiments. Figure 3b shows that when the dye was injected 10 minutes after applying 10 μM DA, the diffusion of Lucifer Yellow became restricted to a few H1AT besides the injected one. The cell body also appeared backfilled by Lucifer Yellow, indicating that the partial staining of the network was not due to a lack of cytoplasmic diffusion of the dye.

The decrease in the passage of dye between coupled cells that are normally stained by Lucifer Yellow is an important indication of the decrease in permeability of the gap junction channels (Spray et al. 1979; Murphy et al. 1983). In our experiments, if the decrease of electrical coupling induced by DA were due to a pure decrease of the extrajunctional membrane resistance of the H1AT, no change in the dye diffusion should have been observed, because the dye does not normally permeate the plasma membrane (Bennett et al. 1978; Stewart 1981). Thus, the decrease of the Lucifer Yellow diffusion strongly indicates, independently from the current-injection experiments, that DA decreases the permeability of the gap junctions between H1AT.

Effects of DA after Suppression of Chemical Synaptic Transmission in the Retina by a Cobalt-containing Medium

Since the H1AT participate in a circuit of neurons establishing synapses at the outer plexiform layer of the retina, it was important to know whether the de-

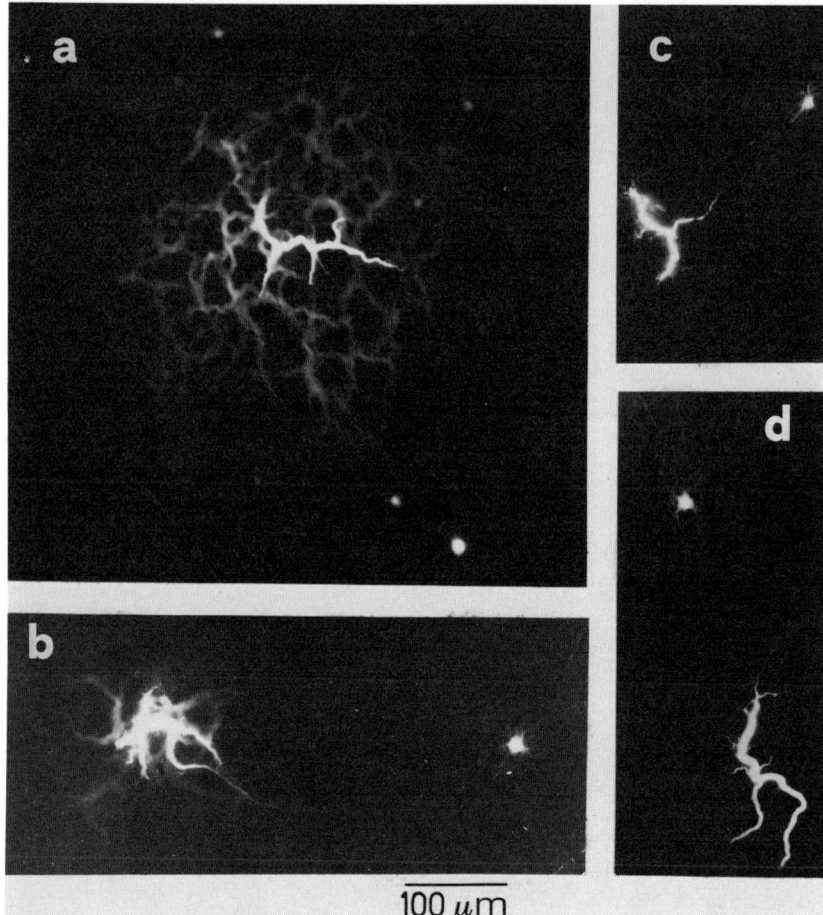

Figure 3
Experiments of Lucifer Yellow injection. (a) Control. The dye injection was performed in a retina bathed in normal Ringer's solution. The most fluorescent structure observed is the injected H1AT from which the Lucifer Yellow diffused into a complex network of H1AT. The dye also backfilled some H1CB through the fine axon fibers. (b) The diffusion of the dye in the H1AT network is restricted in a retina bathed in a saline containing 10 μM DA. (c) Similar effects of 10 μM forskolin on the Lucifer Yellow diffusion in the H1AT network. (d) The same restriction of Lucifer Yellow diffusion is observed when 10 μM DA is added to a saline containing 3 mM cobalt chloride.

scribed effects of DA on the H1AT network were exerted directly on receptors located on the H1AT membrane proper or indirectly, via some other neuron connected to the H1AT. To clarify this problem, the H1AT were disconnected from their chemical synaptic inputs by blocking neurotransmission in the retina through the introduction of Co^{++} ions into the saline (Cervetto and Piccolino 1974). In this condition, the H1AT became hyperpolarized and their light responses were blocked. Therefore we tested the effect of DA on the Lucifer Yellow diffusion. Figure 3d corresponds to an experiment in which DA was applied to the retina in the presence of Co^{++} in the saline evoking a marked reduction of the dye diffusion in the H1AT network. In the absence of DA, Co^{++} only produced a mild decrease of the Lucifer Yellow diffusion in the network (see Piccolino et al. 1984). These results are much in favor of a location of the DA receptors on the membrane of the H1AT proper.

THE DA-INDUCED DECREASE IN THE PERMEABILITY OF THE H1AT GAP JUNCTIONS PROBABLY INVOLVES cAMP AS SECOND MESSENGER

Effect of DA Antagonists on the DA-induced Modification of the H1AT Receptive Field Profile

A series of DA antagonists (see Creese et al. 1982) were assayed to determine their capability either to prevent or reverse the DA effects on the receptive field properties of the H1AT. Table 1 shows that flupenthixol, pifluthixol, fluphenazine, haloperidol, and (+)-butaclamol were able to prevent or block the DA effects whereas a specific antagonist of the D2 receptors, such as (+)-sulpiride, did not affect the DA action. None of the effective antagonists produced a change in the H1AT dark membrane potential. The pharmacological profile shown in Table 1 indicates that the DA receptor present in the H1AT membrane is of the D1 type, which has been shown in both brain and retina (Kebabian et al. 1972; Brown and Makman 1972; Watling and Dowling 1981) to be linked to the activation of adenylate cyclase, the enzyme that synthesizes cAMP. Moreover, DA has been recently reported to stimulate the adenylate cyclase activity of isolated horizontal cells of the fish retina (Van Buskirk and Dowling 1981).

Table 1
Efficacy of DA Antagonists

Compound	Antagonist action	Concentration range (μM)
α-Flupenthixol	+ + +	1–50
α-Pifluthixol	+ + +	1–50
Haloperidol	+ + +	50–100
(+)-Butaclamol	+ + +	50–100
(−)-Butaclamol	−	100
Fluphenazine	+ + +	50–100
Clozapine	+ +	100
Spiperone	+	100
(+)-Domperidone	±	100
(−)-Sulpiride	−	120
(+)-Sulpiride	−	100

This table expresses the qualitative results obtained in at least five cells for each compound. The efficacy was measured as the ability of the DA antagonist to prevent the narrowing of the H1AT receptive field profile induced by 5 μM DA.

Effects of Pharmacological Agents That Produce an Increase of Intracellular cAMP.

The first procedure that we used to increase intracellular cAMP concentration was to stimulate its synthesis by applying forskolin, a diterpene compound known to stimulate directly the adenylate cyclase activity of intact and broken cells.

Figure 4 corresponds to an experiment in which forskolin was applied to the retina. The protocol was the same as in Figure 2. The application of 10 μM forskolin produced, as DA, an increase in the amplitude of the response to the small central spot, a decrease in the peripheral response and an increase in the electrotonic potential resulting from current injection in one of the H1AT. As discussed above, these results indicate that forskolin increases the coupling resistance in the H1AT network. Figure 3c shows that besides its electrophysiological effects, forskolin also induced a restriction of the Lucifer Yellow diffusion in the H1AT network to a few axon terminals around the injected one and its backfilled cell body. These results show that forskolin mimicks the action of DA and very probably decreases the permeability of the gap junctions between the H1AT.

A second procedure used to increase the intracellular cAMP concentration was to inhibit the activity of phosphodiesterase (PDE), the enzyme that inactivates cAMP. For this purpose we used some well-known PDE inhibitors such as isobutylmethylxanthine (IBMX), theophylline, aminophylline, and the nonxanthinic compound RO 20-1724. IBMX was the most potent of them and also mimicked both the DA and forskolin effects on the receptive field profile, the current spread, and the Lucifer Yellow diffusion in the H1AT network.

Our experiments are then much in favor of an involvement of cAMP in the mediation of the DA-induced decrease of the H1AT gap junction conductance.

CONCLUDING REMARKS

The main finding of the work summarized here is that DA induces a narrowing of the receptive field of the turtle H1AT due to a decrease of the permeability of the gap junctions in the network. This effect of DA involves the activation of D1 receptors located on the H1AT surface and is then mediated by a second messenger, probably cAMP.

Two major questions arise immediately. The first concerns the mechanism by which the intracellular increase in cAMP leads to the closing of the gap junction channels. It has been reported in recent years that an increase in intracellular Ca^{++} (Rose and Loewenstein 1976) or an acidification of the internal milieu (Turin and Warner 1977, 1980; Giaume et al. 1980; Spray et al. 1981b) cause cell uncoupling by increasing the junctional resistance, but we do not know whether cAMP modifies either the intracellular Ca^{++} or pH in the H1AT or whether cAMP intervenes by phosphorylating a protein related to the gap junction channel.

A second major question that needs an answer concerns the possible functional signification of the DA-induced uncoupling of the H1AT in the turtle retina. The experimental evidence summarized above would be in favor of a pos-

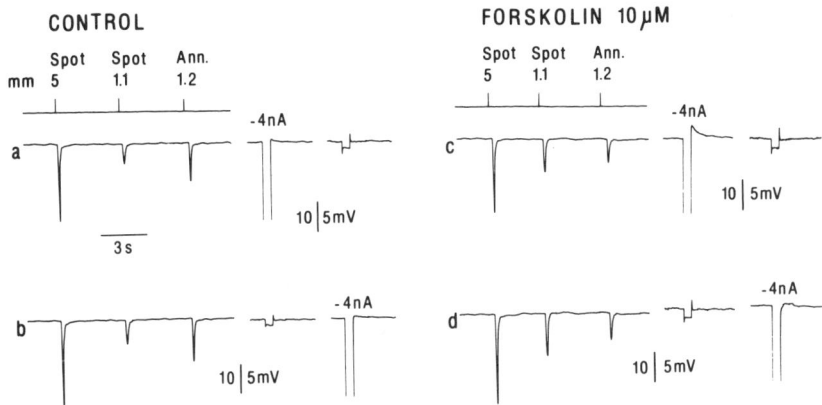

Figure 4
Effects of forskolin on the H1AT light responses and on the current spread in the H1AT network. (a,b) Recordings of the responses of light and to current pulses of two different H1AT impaled with independent microelectrodes whose tips are separated by less than 100 μm. (c,d) The responses of the same H1AT to the same stimuli recorded after bathing the retina for 10 min in a saline containing 10 μM forskolin. Notice the similarity of the effects of forskolin to those of DA recorded in Fig. 2.

sible participation of DA in the modulation of the electrical coupling in the turtle H1AT network in physiological conditions. Pharmacological experiments using drugs that induce the release of DA or increase the amount of DA in the retina induced in the H1AT the same effects as DA, and such effects were blocked by D1 antagonists (Piccolino et al. 1984 and unpubl.). Nevertheless, the only dopaminergic neurons that have been identified in the turtle retina are some amacrine cells (Witkovsky et al. 1984), and no processes of these cells have been visualized in the outer plexiform layer or in the vicinity of the H1 horizontal cells. This means that the turtle retina is at variance with teleost fish, rat, monkey, and human retinas, in that it does not seem to be endowed with a dopaminergic interplexiform cell that would establish synapses at both plexiform layers (see Ehinger 1983). Therefore, it remains to be determined whether the dopaminergic amacrine cells send still undetected processes to the outer plexiform layer or if DA released from the processes of the dopaminergic amacrine cells may diffuse to the distal retina and act on the H1AT.

The main conclusion of this paper is that a neurotransmitter modulates the permeability of gap junctions. This is an important argument in favor of the functional flexibility of these junctions and makes worthwhile the search for other agents and for other physiological or pathological processes that would be able to modify the permeability of gap junctions.

ACKNOWLEDGMENTS

This work was supported by grants from the Centre National de la Recherche Scientifique, France; from the Ministère de la Recherche et de la Technologie,

France; from the Université Pierre et Marie Curie, Paris, France; and from the Consiglio Nazionale delle Ricerche, Italy.

REFERENCES

Bennett, M.V.L. 1977. Electrical transmission: A functional analysis and comparison to chemical transmission. In *Cellular basis of neurons, handbook of physiology:* Section 2, *The nervous system* (ed. E.R. Kandel), vol. 1, p. 357–416. William and Wilkins, Baltimore.

Bennett, M.V.L., M.E. Spira, and D.C. Spray. 1978. Permeability of gap junctions between embryonic cells of *Fundulus:* A reevaluation. *Dev. Biol.* **65:** 114–125.

Brown, J.H. and M.H. Makman. 1972. Stimulation by dopamine of adenylate cyclase in retinal homogenate and of adenosine 3',5'-cyclic monophosphate formation in intact retina. *Proc. Natl. Acad. Sci.* **69:** 539–543.

Byzov, L. 1975. Interaction between horizontal cells in the turtle retina. *Neirofiziologia* **7:** 279–276.

Carew, T.J. and E.R. Kandel. 1976. Two functional effects of decreased conductance EPSP: Synaptic augmentation and increased electrotonic coupling. *Science* **192:** 150–153.

Cervetto, L. and M. Piccolino. 1974. Synaptic transmission between photoreceptors and horizontal cells in the turtle retina. *Science* **183:** 417–419.

Creese, A.L., A.L. Morrow, S.E. Leff, D.R. Sibley, and H.W. Hambley. 1982. Dopamine receptors in the central nervous system. *Int. Rev. Neurobiol.* **23:** 255–301.

Ehinger, B. 1983. Functional role of dopamine. In *Progress in retinal research* (ed. N. Osborne and G. Chader), vol. 2, p. 213–232. Pergamon Press, Oxford, England.

Findlay, I. and O.H. Petersen. 1982. Acetylcholine evoked uncoupling restricts the passage of Lucifer yellow between pancreatic acinar cells. *Cell Tissue Res.* **225:** 633–638.

Gerschenfeld, H.M., J. Neyton, M. Piccolino, and P. Witkovsky. 1982. L-horizontal cells of the turtle: Network organization and coupling modulation. *Biomed. Res.* (suppl.) **3:** 21–32.

Giaume, C., M. Spira, and H. Korn. 1980. Uncoupling of invertebrate electrotonic synapses by carbon dioxide. *Neurosci. Lett.* **17:** 197–202.

Harris, A.L., D.C. Spray, and M.V.L. Bennett. 1983. Control of intercellular communication by voltage dependence of gap junctional conductance. *J. Neurosci.* **3:** 79–100.

Iwatsuki, N. and O.H. Petersen. 1978. Pancreatic acinar cells: Acetylcholine-evoked electrical uncoupling and its ionic dependency. *J. Physiol.* **274:** 81–96.

Kebabian, J.W., G.L. Petzold, and P. Greengard. 1972. Dopamine-sensitive adenylate cyclase in the caudate nucleus of the rat brain and its similarity to the "dopamine receptor." *Proc. Natl. Acad. Sci.* **69:** 2145–2159.

Lamb, T.D. 1976. Spatial properties of horizontal cell responses in the turtle retina. *J. Physiol.* **263:** 239–255.

Leeper, H.F. 1978. Horizontal cells of the turtle retina: I. Light microscopy of Golgi preparations. *J. Comp. Neurol.* **182:** 777–794.

Loewenstein, W.R. 1981. Junctional intercellular communication: The cell to cell membrane channel. *Physiol. Rev.* **61:** 809–913.

Murphy, A.D., R.D. Hadley, and S.B. Kater. 1983. Axotomy-induced increases in electrical and dye coupling between identified neurons of *Helisoma. J. Neurosci.* **3:** 1422–1429.

Obaid, A.M., S.J. Socolar, and B. Rose. 1983. Cell to cell channels with two independently regulated gates in series: Analysis of junctional conductance modulation by membrane potential, calcium and pH. *J. Membr. Biol.* **73:** 69–89.

Piccolino, M., J. Neyton, and H.M. Gerschenfeld. 1984. Decrease of gap junction permeability induced by dopamine and cyclic adenosine 3'-5'-monophosphate. *J. Neurosci.* **4**: 2477-2488.

Piccolino, M., J. Neyton, P. Witkovsky, and H.M. Gerschenfeld. 1982 γ-Aminobutyric acid antagonists decrease junctional communication between horizontal cells of the retina. *Proc. Natl. Acad. Sci.* **79**: 3671-3675.

Rose, B. and W. Loewenstein. 1976. Permeability of a cell junction and the local cytoplasmic free calcium concentration. *J. Membr. Biol.* **28**: 87-119.

Spira, M.E. and M.V.L. Bennett. 1972. Synaptic control of electrotonic uncoupling between neurons. *Brain Res.* **37**: 294-300.

Spira, M.E., D.C. Spray, and M.V.L. Bennett. 1980. Synaptic organization of expansion motoneurons in *Navanax inermis*. *Brain Res.* **195**: 261-269.

Spray, D.C., A.L. Harris, and M.V.L. Bennett. 1979. Voltage dependence of junctional conductance in early amphibian embryos. *Science* **204**: 432-434.

———. 1981a. Equilibrium properties of the voltage dependent junctional conductance. *J. Gen. Physiol.* **77**: 75-94.

———. 1981b. Gap junctional conductance is a simple and sensitive function of pH. *Science* **211**: 712-715.

Spray, D.C., J.H. Stern, A.L. Harris, and M.V.L. Bennett. 1982. Gap junctional conductance: Comparison of senstivities to H and Ca ions. *Proc. Natl. Acad. Sci.* **79**: 441-445.

Spray, D.C., R.L. White, A. Campos de Carvalho, and M.V.L. Bennett. 1984. Gating of gap junction channels. *Biophys. J.* **45**: 219-230.

Stewart, W.W. 1978. Functional connections between cells revealed by dye coupling with a highly fluorescent naphthalimide tracer. *Cell* **14**: 741-759.

———. 1981. Lucifer dyes—Highly fluorescent dyes for biological tracing. *Nature* **292**: 17-21.

Teranishi, T., K. Negishi, and S. Kato. 1983. Dopamine modulates S-potential amplitude and dye coupling between external horizontal cells in carp retina. *Nature* **301**: 243-246.

———. 1984. Regulatory effect of dopamine on spatial properties of horizontal cells of the carp retina. *J. Neurosci.* **4**: 1271-1280.

Turin, L. and A. Warner. 1977. Carbon dioxide reversibly abolishes ionic communication between cells of early amphibian embryo. *Nature* **270**: 56-57.

———. 1980. Intracellular pH in early *Xenopus* embryos: Its effect on current flow between blastomeres. *J. Physiol.* **300**: 489-504.

Van Buskirk, R. and J.E. Dowling. 1981. Isolated horizontal cells from carp retina demonstrate dopamine-dependent accumulation of cAMP. *Proc. Natl. Acad. Sci.* **78**: 7825-7829.

Watling, K.J. and J.E. Dowling. 1981. Dopaminergic mechanisms in the teleost retina. I. Dopamine-sensitive adenylate in homogenates of carp retina; effects of agonists, antagonists and ergots. *J. Neurochem.* **36**: 559-568.

Witkovsky, P., W. Eldred, and H.J. Karten. 1984. Catecholamine- and indolamine-containing neurons in the turtle retina. *J. Comp. Neurol.* **228**: 217-225.

Witkovsky, P., W.G. Owen, and M. Woodworth. 1983. Gap junctions among the perikarya, dendrites, and axon terminals of the luminosity-type horizontal cells of the turtle retina. *J. Comp. Neurol.* **216**: 359-358.

Electrical Coupling between Pairs of Isolated Fish Horizontal Cells Is Modulated by Dopamine and cAMP

Eric M. Lasater and John E. Dowling
*Department of Cellular and Developmental Biology
Harvard University, Cambridge, Massachusetts 02138
and The Marine Biological Laboratory
Woods Hole, Massachusetts 02543*

In the outer plexiform layer of the retina, the spread of lateral inhibitory signals is mediated by horizontal cells (Werblin and Dowling 1969; Baylor et al. 1971; Naka 1972, 1977). In fish, electrical coupling occurs between horizontal cells of the same type (Yamada and Ishikawa 1966; Witkovsky and Dowling 1969; Stell and Lightfoot 1975). This facilitates the lateral spread of potential from one cell to another and thus serves to increase the size of a cell's receptive field. For example, the dendritic diameter of horizontal cells ranges from about 30 to 200 μm whereas their physiologically measured receptive fields vary in size from 200 μm to more than 2 mm in diameter (Naka and Rushton 1967; Norton et al. 1968; Dowling and Ripps 1971; Kaneko 1971; Davis and Naka 1980).

In the teleost retina, horizontal cells receive two synaptic inputs (Stell and Lightfoot 1975). One is from the photoreceptors which are thought to use L-glutamate as their transmitter (Ishida and Fain 1981; Lasater and Dowling 1982). The second input is from the interplexiform cell, a neuron that receives input from amacrine cells and that sends processes to the outer plexiform layer to synapse onto the horizontal cells (Dowling and Ehinger 1978). Dopamine (DA) has been localized to the interplexiform cells and is therefore likely to be the neurotransmitter used by these cells (Dowling and Ehinger 1978).

The exogenous application of DA to the retina has been shown to decrease the size of horizontal cell receptive fields (Negishi and Drujan 1979; Teranishi et al. 1983; Piccolino et al. 1984), thus implying a role for interplexiform cells in the modulation of horizontal cell receptive fields. DA has also been shown to decrease the diffusion of dye between horizontal cells (Teranishi et al. 1983;

Piccolino et al. 1984), suggesting that DA alters the horizontal cell receptive field size by acting on the electrical (gap) junctions. Other experiments have shown that analogs of cAMP and forskolin, a nonspecific cyclase activator, have the same effect as DA on horizontal cells (Teranishi et al. 1983; Piccolino et al. 1984), indicating that DA acts on gap junctions via a cAMP-mediated pathway.

Here we describe experiments using coupled pairs of isolated horizontal cells that show directly that this electrical coupling is modulated by DA and also by cAMP.

ISOLATION AND RECORDING PROCEDURE

Retinas from the white perch (*Roccus americana*) were removed and incubated in tissue culture medium containing the proteolytic enzyme papain. Following washing and trituration, isolated neurons were plated into plastic tissue culture dishes (Lasater and Dowling 1982; Lasater et al. 1984). Isolated perch horizontal cells are of four morphological types termed H1, H2, H3, and H4 cells. H1, H2, and H3 horizontal cells are likely to be cone-related cells whereas H4 may be a rod-related cell. Following isolation, contacting pairs of H2 and H3 cells could be found in most of the culture dishes. Figure 1 is a photomicrograph of such pairs. A pair of H2 cells is shown in Figure 1a and in Figure 1b a pair of H3 horizontal cells is shown. Recordings were made from cell pairs using whole-cell patch electrode recording techniques (Hamill et al. 1981). A patch electrode was placed on each cell of a pair and access was gained to the cell's interior via suction applied to the pipets. Two recording configurations for evaluating the coupling characteristics between cells were used. In the first, both cells were voltage-clamped and the membrane potentials held at -60 mV. Hyperpolarizing command pulses (20 mV) were applied to one cell (driver cell) and the resultant current flowing through the junction into the second cell (follower cell) was measured as the current clamping the follower cell's potential. This approach allowed a measure of junctional resistance and conductance. For the second recording configuration, the driver cell was voltage-clamped at -60 mV while the follower cell was current-clamped. Again, 20-mV hyperpolarizing pulses were passed into the driver cell and the change in membrane potential of the follower cell recorded. This approach enables us to determine a coupling coefficient for a cell pair, i.e., change in potential of the follower cell/change in potential of the driver cell (Harris et al. 1983), and to follow any membrane potential changes that occurred in the follower cell as a result of the application of drugs.

In early experiments we found that the anions in the patch electrodes affected the coupling between pairs of cells. Initially F^- was used as the major anion because consistently good seals can be obtained with high concentrations of F^- in the patch pipets. However, with F^- as the major anion in the pipets, good electrical coupling was observed but DA usually had only weak, if any, effects. Why this should occur is not clear (see Summary). On the other hand, when aspartate, gluconate, or Cl^- were used as the major anion in the pipets, the coupling observed was usually weak. As a result, we tried mixtures of F^- with other anions in the pipet solutions. We found that a mixture of 40%

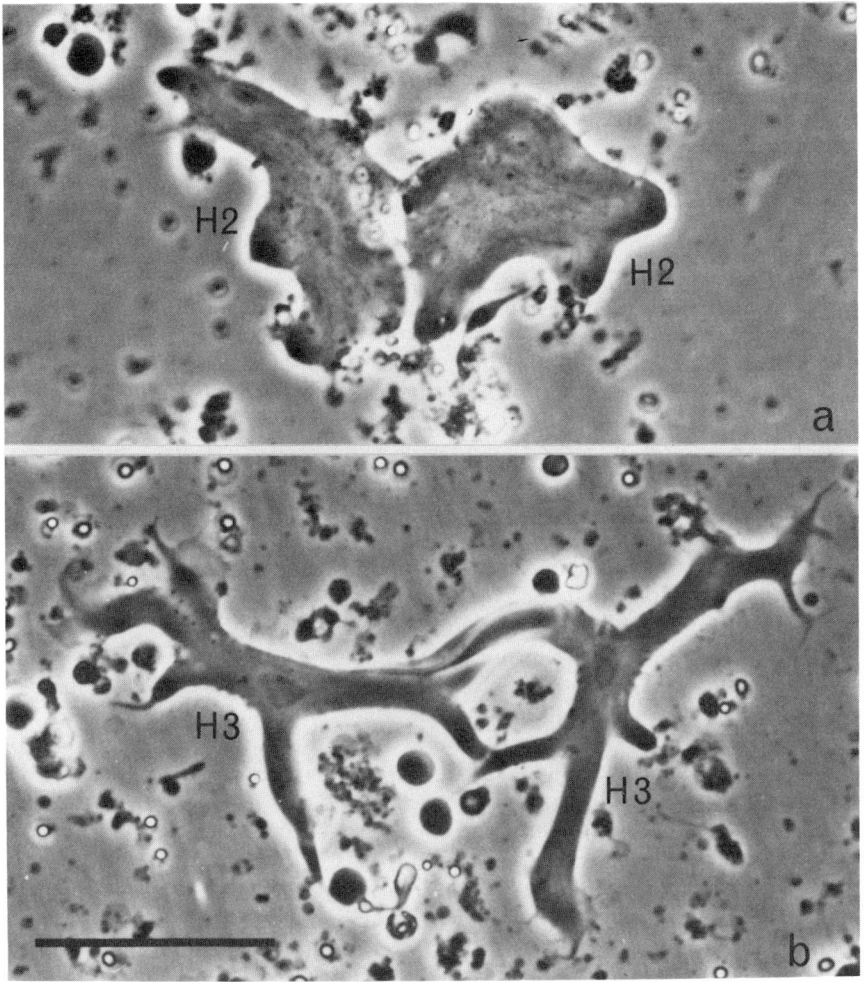

Figure 1
Photomicrographs of coupled horizontal cells in culture. (*a*) A pair of type H2 cells. (*b*) A pair of type H3 horizontal cells. Calibration bar, 50 μm.

KF and 60% K-aspartate in the pipets worked well. With this mixture good coupling was seen and consistent drug effects were obtained. The composition of the pipet solution used was: 72 mM K-aspartate, 48 mM KF, 11 mM EGTA, 1 mM $CaCl_2$, 2 mM $MgCl_2$, 4 mM KCl, 1 mM MgATP, and 10 mM HEPES.

CHARACTERISTICS OF HORIZONTAL CELL COUPLING

We have so far recorded from over 100 pairs of contacting cells of the same morphological type. Of these about 95% showed evidence of coupling and, of the coupled pairs, about 70% had coupling coefficients of 0.6 or better. On the other hand, we found coupling between horizontal cells of different morphological types to be absent or very weak. Thus, H2 cells were coupled only to other H2 cells but not to H1 or H3 cells. Likewise H3 cells were only coupled to other H3 cells and not to H1 cells or H2 cells.

Table 1 is a summary of measurements made on three pairs of tightly coupled cells; i.e., with coupling coefficients of 0.88, 0.95, and 0.99. The resistances of the junctions were determined by shifting the membrane potential of the current-clamped cell 20 mV and measuring the current that flowed in the adjacent voltage-clamped cell. Junctional resistances were calculated to be 30, 15, and 7 MΩ, respectively. Junctional resistance measurements were also made from other pairs of cells when both cells were voltage-clamped. For seven well-coupled cell pairs in which junctional resistance was determined, the average coupling resistance was calculated to be 28.5 MΩ ($\pm$ 13.8 S.D.). In contrast, input resistances for single isolated cells range from 600 MΩ to about 1.2 GΩ. Thus junctional resistance is usually 6% or less of the input resistance of a single cell.

For the three cell pairs included in Table 1, specific resistance was roughly estimated by measuring the length and thickness of the point of contact between the two cells. Interestingly, the specific resistances for the three cell pairs appeared to be similar, i.e., around 3 $\Omega \cdot cm^2$. This value is similar to values reported for coupling between toad rod photoreceptors (Gold 1979) and heart muscle cells (Spira 1971). However it is very unlikely that all of the contact area between horizontal cell pairs is junctional membrane; thus, the specific resistances given are likely to be high estimates. Furthermore, because of the large size of the cells and the fact that some had long thin dendrites, it may well be that the cells were not always adequately clamped. This would also result in an apparently higher junctional resistance.

Figure 2 illustrates an experiment in which the size of the hyperpolarizing voltage shift in the driver cell was altered from 1 to 10 mV. In this experiment, both cells were voltage-clamped. With increasing voltage shifts in the driver cell, the current flowing into the follower cell increased correspondingly. However, with all voltage shifts, current flow into the follower cell remained constant over time and never decreased as a result of transjunctional voltages. In other experiments (not shown) the voltage of the driver cell was changed by 20 mV and 40 mV and a similar result was observed. Also alternating the driver cell and follower cell had no effect on the linearity of coupling. Thus changes in transjunctional voltage did not alter the coupling coefficient; i.e., the junctions do not appear to be voltage sensitive (Spray et al. 1984).

Table 1
Resistance Values of Horizontal Cell Gap Junctions

Coupling coefficient	Resistance	Size of contact	Specific resistance
0.88	30 MΩ	10 μ^2	3.0 Ω cm^2
0.95	15 MΩ	20 μ^2	3.0 Ω cm^2
0.99	7 MΩ	30 μ^2	2.1 Ω cm^2

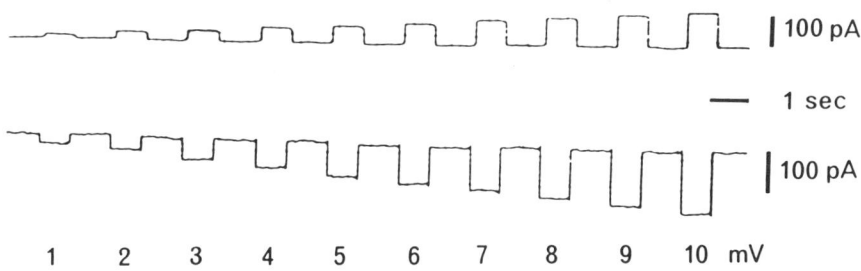

Figure 2
Effect of transjunctional voltage on coupling. Driver cell (*bottom trace*) and follower cell (*top trace*) were voltage-clamped to −60 mV and command potentials of 0–10 mV impressed on the driver cell. The command pulse had a duration of about 850 msec. No change in coupling was observed during the presentation of the pulse here as well as in other experiments in which the voltage was shifted by 20 or 40 mV (not shown). Thus these experiments indicate that the junctional conductance in horizontal cells is not sensitive to the voltage difference across the junction.

EFFECTS OF DOPAMINE ON COUPLING

To assay the effects of test agents on horizontal cell coupling, Ringer's solution containing a test substance was applied to a pair of cells via micropipets brought to within 100 μm of the cells. The Ringer's was expressed by pressure in short pulses of about 1.0-second duration in a volume of 0.5–1.0 μl.

Figure 3 shows the effect of an application of Ringer's containing DA on a pair of H2 cells, one of which, the driver cell, was voltage-clamped whereas the other, the follower cell, was current-clamped. DA was applied at the arrow and within about 10 seconds the cells began to uncouple. Initally the cells had a coupling coefficient of about 0.9, but within 2 minutes the coupling coefficient was reduced to approximately 0.15. The uncoupling also resulted in a hyperpolarization of the follower cell's membrane potential to approximately −80 mV, the usual membrane potential of isolated horizontal cells in culture. The amount of current passing into the voltage-clamped driver cell, in response to the 20-mV hyperpolarizing commands, was reduced during the uncoupling, indicating an increase in the input resistance of the cell (bottom trace). Recovery of coupling required, in this instance, about 8 minutes.

Figure 4 shows the effects of DA on two voltage-clamped H3 cells. Before the application of DA, a junctional resistance of 83 MΩ was measured. After drug application there was a latency of about 30 seconds before the cells began to uncouple. It took about 3 minutes for the uncoupling to reach a peak, at which point the junctional resistance had risen to 660 MΩ. Coupling between the cells remained low for about 1 minute and then slowly increased. After about 8 minutes the junctional resistance had recovered to 109 MΩ. For other cell pairs that were tested in this way, junctional resistance increased from control values of between 20–60 MΩ to between 300–700 MΩ following DA application.

Toward the end of this experiment the follower cell suddenly burst open and

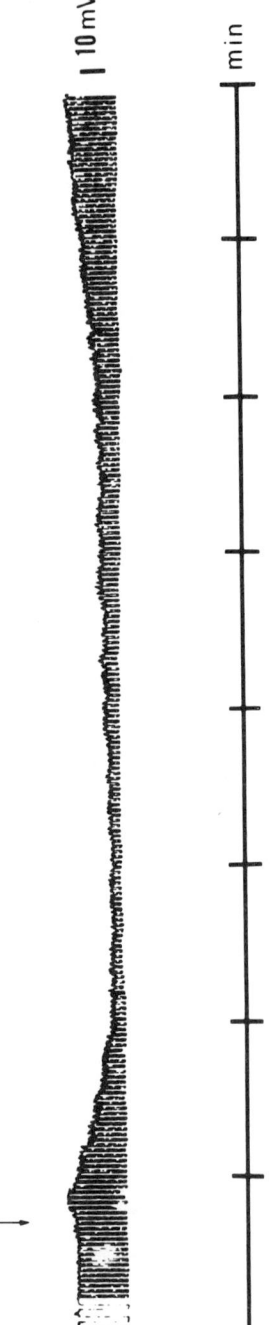

Figure 3
Effects of DA on coupling between two H2 cells. The driver cell (*lower trace*) was voltage-clamped and the resulting voltages were recorded in the follower cell, i.e., the follower cell was current-clamped at zero current (*upper trace*). Membrane potential of the driver cell was shifted by 20-mV hyperpolarizing pulses. A 1-sec pulse of 200 µM DA was applied to the cell pair at the arrow. The cells began to uncouple within about 15–20 sec. As the uncoupling progressed, the membrane potential of the follower cell hyperpolarized from the holding potential of −60 mV to about −78 mV. In addition the current flow into the driver cell decreased, indicating an increase in the input resistance of the cell due to the uncoupling. At this point the cells remained less well coupled for about 1.5 min, then the coupling slowly returned over a period of 5–6 min. A movement artifact can be seen in both traces at the time of dopamine application. The time scale for the first three pulses is 2.5 × the indicated scale.

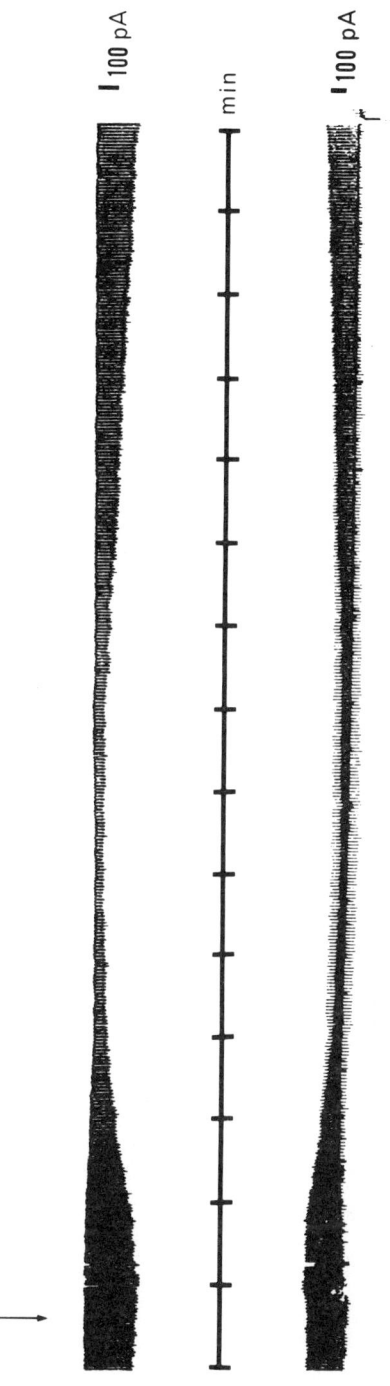

Figure 4
Effects of DA on two voltage-clamped H3 cells. In this experiment the membrane potential of the driver cell (*bottom trace*) was shifted by 20-mV depolarizing pulses and counteracting current pulses were recorded in the follower cell (*top trace*). DA application at the arrow started to uncouple the cells after about 30 sec. After about 4 min the coupling between the cells was relatively weak. The coupling slowly returned over a period of 14 min. At the point on the driver cell trace marked by the arrow, the follower cell was lost. This resulted in a large (300 pA), brief inward current in the driver cell. When the current trace had returned to baseline the measured current pulses were much smaller (too small to see at the gain employed), indicating a large increase in the input resistance of the cell.

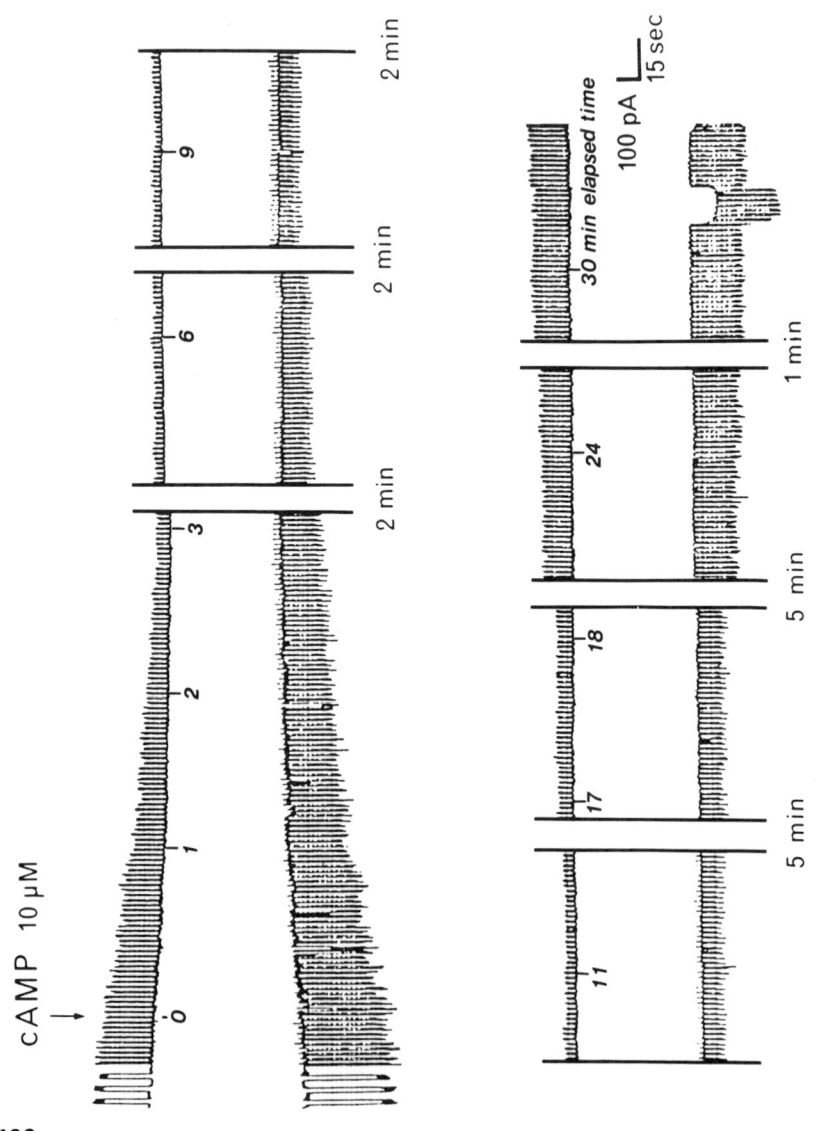

Figure 5 (See facing page for legend.)

died. This resulted in a transient influx of current into the driver cell of about 300 pA. When the current trace returned to baseline, it was now possible to determine the nonjunctional resistance of the cell; it was 1.05 GΩ. Sealing of gap junctional membrane following the loss of one cell of a coupled pair has been observed with many kinds of cells. It has been postulated that Ca^{++} influx through gap junctions after one cell of a pair is ruptured is responsible for the closing down of the junctional channels (Barr et al. 1965; Loewenstein et al. 1967; Rose and Loewenstein 1976). This may very well be what occurred here.

EFFECT OF cAMP

In initial experiments with cAMP, the membrane-permeable analog, 8-bromo-cAMP, was applied to coupled cells extracellularly. We found that 8-bromo-cAMP almost always uncoupled the cells rapidly, but usually the membrane potential fluctuated wildly following the application of this agent. Often this was followed by the death of one or another of the cells. Why this instability occurred is not clear. 8-Bromo-cAMP has little effect on the membrane potential of single isolated carp horizontal cells (Lasater et al. 1983).

To avoid this complication, in a second series of experiments, cAMP was added to the solution that filled the patch pipets and thus cAMP was delivered directly into one of the cells. With concentrations of cAMP of 100 μM or more within a patch electrode, coupling disappeared very rapidly after the recordings began. However, with lower concentrations of cAMP in the pipets, i.e., 10 μM, good coupling was observed initially and it decreased only relatively slowly thereafter. In such situations it was then possible to inject additional cAMP into a cell. Figure 5 shows the results from one such experiment. At the arrow, pressure was applied to the pipet, forcing some of the pipet contents into one H2 cell. Immediately after injection the coupling started to decrease more rapidly and after about 3 minutes it decreased such that the junctional resistance went up to 660 MΩ, from a control value of about 100 MΩ. The coupling remained at this level for about 3 minutes and then slowly started to improve. After a period of about 15 minutes, coupling had recovered almost to its initial values.

Throughout this experiment cAMP probably leaked slowly into the cell. However, the rate of leak at this concentration of cAMP seems not to have been enough to overcome the endogenous phosphodiesterases that inactivate cAMP. Thus coupling remained. However when an additional bolus of cAMP

Figure 5
Effect of cAMP injection into one of a pair of H2 cells. cAMP was added to the intracellular media contained in the patch pipet. At the arrow positive pressure was applied to the lumen of the pipet to inject a bolus of cAMP into the cell. The cAMP began to take effect immediately and within 6 min the coupling had decreased to a coupling coefficient of about 0.1. The coupling stayed low for several minutes then gradually increased over the next 15 min to near initial values. Three initial pulses were recorded at a chart speed 2.5× that indicated.

was forced into the cell, dramatic uncoupling occurred. Coupling then returned with time, suggesting that the phosphodiesterases were once again able to reduce cAMP levels to a point where significant coupling could occur.

SUMMARY

The present experiments describe the properties of electrical coupling between contacting pairs of isolated perch horizontal cells. We observed that cells of the same morphological type are often tightly coupled with coupling ratios of 0.9 or better. In such cells the junctional resistance was typically found to be about 30 MΩ, less than 6% of the input resistance of a single isolated horizontal cell. Furthermore the coupling between horizontal cells maintained in culture does not appear to be voltage sensitive.

An interesting question is whether the coupling we observed is the result of junctions formed in vivo or if the junctions are formed during the culture period. It is likely that most of the coupled cells we recorded from are cells that were originally coupled in the intact retina. This notion is supported by the fact that immediately following the isolation procedure, contacting pairs of cells can be seen settling in the dish. However we cannot rule out the possibility that we recorded from two like cells that settled abutting one another and subsequently coupled.

Coupling between pairs of horizontal cells is decreased by DA. At concentrations of 200 μM, applied DA was found to reduce the coupling coefficient of a pair of cells from 0.9 to 0.1. The reduction in coupling appears to be graded and reversible. Furthermore it seems likely that DA is acting through a cAMP-mediated pathway. The application of the membrane-permeable cAMP analog, 8-bromo-cAMP, to these cells, as well as the addition of cAMP into the recording pipet, had the same effect as DA, a dramatic reduction in electrical coupling.

It is of interest that the uncoupling effects of DA were not seen when F^- was present as the major anion in the patch pipets. This may be because F^- affects the enzymatic steps presumably involved in the modulation of coupling induced by DA. However, this is not likely because in other systems F^- has been shown to stimulate cyclase activity directly as well as to inhibit phosphatase action (see Daley 1977). These two effects would act to uncouple the cells. Also it is doubtful the effect is due to the chelation of Ca^{++} by F^-, since it has been shown that DA stimulation of adenylate cyclase in horizontal cells is not Ca^{++} dependent (Van Buskirk and Dowling 1982). It may be that F^- acts directly on horizontal cell gap junction protein to prevent uncoupling from taking place. However, this action must only occur at high F^- concentrations, since at lower F^- concentrations the coupling is modulated by DA.

Given our present understanding of the interplexiform cell system in the fish retina, it is possible to suggest the chain of events that leads to horizontal cell uncoupling and to speculate on its physiological significance. Activation of the interplexiform cells results in the release of DA onto horizontal cells. The DA interacts with receptors on the horizontal cells that activate adenylate cyclase, resulting in an increase in intracellular levels of cAMP (Van Buskirk and Dowling 1981). The cAMP presumably activates protein kinases (Greengard 1978), which in turn may phosphorylate gap junction proteins (see Johnson, this volume). This results in a reversible decrease in the conductance of the electrical junctions between horizontal cells. The size of the horizontal cell receptive field is reduced as a consequence of the decrease in coupling (Negishi and Drujan 1979), and a reduction in the size of the horizontal cell receptive fields leads to a decrease in the inhibitory surrounds of the receptive fields of bipolar and photoreceptor cells (Heddon and Dowling 1978).

At the present time we do not know what physiological conditions activate the interplexiform cells and cause them to release dopamine. However, it is interesting to note that Barlow et al. (1957) found that in the cat inhibitory surrounds of ganglion cell receptive fields were reduced or altogther eliminated with prolonged dark adaptation. This decrease in the inhibitory surrounds did not appear to relate to a switch from cone to rod vision, and what mechanisms underlie this phenomenon remain unknown. It is possible that such an effect could be mediated by the interplexiform cell system.

ACKNOWLEDGMENTS

We would like to thank M.V.L. Bennett for his particularly helpful comments and criticisms during the course of these experiments and during the drafting of this manuscript. We would also like to thank David Spray and Roy White for their work during the initial phases of this study. This work was supported by NIH grant EY-00824.

REFERENCES

Barlow, H.B., R. Fitzhugh, and S.W. Kuffler. 1957. Change of organization in the receptive fields of the cat's retina during dark adaptation. *J. Physiol.* **137:** 338–354.

Barr, L., M.M. Dewey, and W. Berger. 1965. Propagation of action potentials and the structure of the nexus in cardiac muscle. *J. Gen. Physiol.* **48:** 797–823.

Baylor, D.A., M.G.F. Fourtes, and P.M. O'Bryan. 1971. Receptive fields of cones in the retina of the turtle. *J. Physiol.* **214:** 265–294.

Daly, J., ed. 1977. *Cyclic nucleotides in the nervous system.* Plenum Press, New York.

Davis, G.W. and K.I. Naka. 1980. Spatial organizations of catfish retinal neurons. I. Single and random bar stimulation. *J. Neurophysiol.* **43:** 807–831.

Dowling, J.E. and B. Ehinger. 1978. The interplexiform cell system. I. Synapses of the dopaminergic neurons of the goldfish retina. *Proc. R. Soc. Lond. B* **201:**7–26.

Dowling, J.E. and H. Ripps. 1971. S-potentials in the skate retina. Intracellular recording during light and dark adaptation. *J. Gen. Physiol.* **58:** 163–189.

Gold, G.H. 1979. Photoreceptor coupling in retina of the toad *Bufo marinus*. II. Physiology. *J. Neurophysiol.* **42:** 311–328.

Grengard, P., ed. 1978. *Cyclic nucleotides, phosphorylated proteins, and neuronal function.* Raven Press, New York.

Hamill, O.P., A. Marty, R. Neher, B. Sakmann, and F.J. Sigworth. 1981. Improved patch-clamp techniques for high-resolution current recording from cells and cell-free membrane patches. *Pfluegers Arch.* **391:** 85–100.

Harris, A.L., D.C. Spray, and M.V.L. Bennett. 1983. Control of intercellular communication by voltage dependence of gap junctional conductance. *J. Neurosci.* **3:** 79–100.

Heddon, W.L. and J.E. Dowling. 1978. The interplexiform cell system II. Effects of dopamine on goldfish retinal neurons. *Proc. R. Soc. Lond. B* **201:** 27–55.

Ishida, A.T. and G.L. Fain. 1981. D-aspartate potentiates the effects of L-glutamate on horizontal cells in goldfish retina. *Proc. Natl. Acad. Sci.* **78:** 5890–5894.

Kaneko, A. 1971. Electrical connexions between horizontal cells in the dogfish retina. *J. Physiol.* **213:** 95–105.

Lasater, E.M. and J.E. Dowling. 1982. Carp horizontal cells in culture respond selectively to L-glutamate. *Proc. Natl. Acad. Sci.* **79:** 936–940.

Lasater, E.M., J.E. Dowling, and H. Ripps. 1984. Pharmacological properties of isolated horizontal and bipolar cells from the skate retina. *J. Neurosci.* **4:** 1966–1975.

Lasater, E.M., K.J. Watling, and J.E. Dowling. 1983. Vasoactive intestinal peptide alters membrane potential and cyclic nucleotide levels in retinal horizontal cells. *Science* **221:** 1070–1072.

Loewenstein, W.R., M. Nakas, and S.J. Socolar. 1967. Junctional membrane uncoupling: Permeability transformations at a cell membrane junction. *J. Gen. Physiol.* **50:** 1865–1981.

Naka, K.-I. 1972. The horizontal cells. *Vision Res.* **12:** 573–588.

———. 1977. Functional organization of catfish retina. *J. Neurophysiol.* **40:** 26–43.

Naka, K.-I. and W.A.H. Rushton. 1967. The generation and spread of S-potentials in fish (Cyprinidae). *J. Physiol.* **192:** 437–461.

Negishi, K. and B.D. Drujan. 1979. Reciprocal changes in center and surrounding S-potentials of fish retina in response to dopamine. *Neurochem. Res.* **4:** 313–318.

Norton, A.C., H. Spekreijse, M.L. Wolbarsht, and H.G. Wagner. 1968. Receptive field organization of the S-potential. *Science* **160:** 1021–1022.

Piccolino, M., J. Neyton, and H.M. Gerschenfeld. 1984. Decrease of gap junction permeability induced by dopamine and cyclic adenosine 3':5'-monophosphate in horizontal cells of turtle retina. *J. Neurosci.* **4:** 2477–2488.

Rose, B. and W.R. Loewenstein. 1976. Permeability of a cell junction and the local cytoplasmic free ionized calcium concentration: A study with aequorin. *J. Membr. Biol.* **28:** 87–119.

Spira, A.W. 1971. The nexus in the intercalated disc of the canine heart: Quantitative data for an estimation of its resistance. *J. Ultrastruct. Res.* **34:** 409–425.

Spray, D.C., R.L. White, A. Campos De Carvalho, A.L. Harris, and M.V.L. Bennett. 1984. Gating of gap junction channels. *Biophys. J.* **45:** 219–230.

Stell, W.K. and D.O. Lightfoot. 1975. Color-specific interconnections of cones and horizontal cells in the retina of the goldfish. *J. Comp. Neurol.* **159:** 473–502.

Teranishi, T., K. Negishi, and S. Kato. 1983. Dopamine modulates S-potential amplitude and dye-coupling between external horizontal cells in carp retina. *Nature* **301:** 243–246.

Van Buskirk, R. and J.E. Dowling. 1981. Isolated horizontal cells from carp retina demonstrate dopamine-dependent accumulation of cyclic AMP. *Proc. Natl. Acad. Sci.* **78:** 7825–7829.

———. 1982. Calcium alters the senstivity of intact horizontal cells to dopamine antagonists. *Proc. Natl. Acad. Sci.* **79:** 3350–3354.

Werblin, F.S. and J.E. Dowling. 1969. Organization of the retina of the mudpuppy, *Necturus maculosus*. II. Intracellular recording. *J. Neurophysiol.* **32:** 339–355.

Witkovsky, P. and J.E. Dowling. 1969. Synaptic relationships in the plexiform layers of carp retina. *Z. Zellforsch. Mikrosk. Anat.* **100:** 60–82.

Yamada, E. and T. Ishikawa. 1966. Fine structure of the horizontal cells in some vertebrate retina. *Cold Spring Harbor Symp. Quant. Biol.* **30:** 383–392.

Author Index

Atkinson, M.M., 205

Beers, W.H., 307
Bennett, M.V.L., 1, 139, 231, 355
Biegon, R., 91
Brink, P.R., 123
Buultjens, T.E.J., 77

Caveney, S., 265
Cohan, C.S., 241
Cole, W.C., 215

Dermietzel, R., 67
Dowling, J.E., 393
Dudek, F.E., 325

Finbow, M.E., 77
Frenzel, E., 91
Frixen, U., 67

Garfield, R.E., 215
Gerschenfeld, H.M., 381
Giaume, C., 367
Girsch, S.J., 191

Hall, J.E., 177
Hanna, R.B., 23
Haydon, P.G., 241
Hertzberg, E.L., 57

Janssen-Timmen, U., 67
Johnson, K., 91
Johnson, R., 91

Kam, E., 77
Kater, S.B., 241
Kessler, J.A., 231
Klukas, K., 91
Korn, H., 367

Lampe, P., 91
Larsen, W.J., 289
Lasater, E.M., 393
Leibstein, A., 67
Llinás, R.R., 337
Lo, C.W., 251
Louis, C., 91

Makowski, L., 5
Manjunath, C.K., 49

Neyton, J., 381

Olsiewski, P.J., 307
Ornberg, R.L., 23

Page, E., 49
Paul, D., 67
Paul, D.L., 107
Peracchia, C., 191
Petersen, O.H., 315
Piccolino, M., 381
Pitts, J.D., 77

Rabes, H., 67
Ramón, F., 155
Reese, T.S., 23
Revel, J.P., 33
Rivera, A., 155

Saez, J.C., 231
Safranyos, R., 265
Sas, D., 91
Sheridan, J.D., 205
Shuttleworth, J., 77
Simon, S.A., 13
Snow, R.W., 325

Spira, M.E., 355
Spray, D.C., 1, 57, 139, 231, 355

Traub, O., 67

Verselis, V., 139

Warner, A.E., 275
White, R.L., 139
Willecke, K., 67
Wojtczak, J.A., 167

Yancey, S.B., 33

Zampighi, G.A., 13, 155, 177
Zimering, M.B., 355

Subject Index

Acetylocholine, 317
Adenocarcinoma (SW13), 297
Adrenal gland, 59
Alcohols, 157, 171, 185, 371
Anesthetics, 157, 167, 170, 171, 185, 278
Antibodies
    effects on function, 61, 85, 281
    monoclonal, 70, 93, 112
    polyclonal, 57, 67, 77, 107, 281

Brain, 59, 79

Calmidazolium, 174, 195, 226
Calmodulin, 168, 226, 381
    antagonists, 169, 174, 195, 226, 270
Carbachol, 237
Cardiac junctions, 49, 59, 68, 79, 98, 112, 147, 307
Catecholamines, 307
Cell lines
    BRL (liver), 61
    clone 1D (fibroblast), 61
    hepatoma cells, 73, 97
    HIT (pancreas), 61
    MDCK (kidney), 71
    NRK, 206
    SW13 (adenocarcinoma), 297

Cerebral cortex, 325
Channels. *See* Gating
Chlorpromazine, 226, 270
Choleratoxin, 309
Closure of channels. *See* Gating
Compartmentation, 251, 265, 275
Connexon (definition), 6
Coupling
    coefficient (definition), 139
    dye and tracer, 61, 85, 99, 123, 139, 215, 251, 265, 278, 311, 326, 381
    electrical (electrotonic), 61, 123, 139, 167, 233, 244, 251, 265, 276, 316, 326, 337, 355, 368, 381, 393
    rectifying electrical, 148, 367
    reversed, 359
Crayfish, 79, 155, 367
Cumulus cells, 293, 307

D_2O, 123
Development, 251, 265, 275
Diamide, 149, 269
Diffusion models, 125, 207, 384
Digitalis, 167
Dopamine, 237, 381, 393
    agonists, 383
    antagonists, 387

Drosophila, 255
Dye coupling. *See* Coupling

Earthworm, 123
Electrical coupling. *See* Coupling
Embryonic cells
  amphibia (*Xenopus, Rana*), 139, 281
  fish (*Fundulus*), 150
  mouse, 252
  rat, 307
  squid (*Loligo*), 148
Ephaptic interactions, 325
Epilepsy, 325

Fallopian tubes, 68
Field effects, 325
Filipin, 291
Fixatives, 25, 145
Fluoride, 394
Formation and degradation, 215, 231, 241, 289, 307. *See also* Gating
Forskolin, 226, 234, 309, 388

Gastric mucosa, 59, 115, 291
Gating, 139, 155
  alcohol mediated, 157, 185, 371
  Ca^{++} mediated, 141, 155, 167, 195, 225, 316, 327
  cAMP mediated, 381, 393
  H^+ mediated, 141, 155, 167, 195, 225, 316, 327, 371
  nicotine mediated, 161
  structural change, 8, 19, 23, 193
  voltage mediated, 143, 157, 185, 367, 396
Glutaraldehyde, 145
Granulosa cells, 291, 307
Group-specific protein reagents, 145, 149, 269
Growth cone, 245

Harmaline, 348
Heart, 49, 59, 68, 79, 98, 112, 147, 307
Helisoma, 241
Hepatocyte, 5, 13, 57, 67, 79, 107, 147
Hepatoma cells, 73, 97
Hepatopancreas, 60, 79
Heterologous junctions, 97
Hippocampus, 325, 342
Horizontal cells, 381, 393
Hormones
  ecdysones, 265
  estrogens, 215
  FSH, 307

gonadotropins, 309
insulin, 234
LH, 310
oxytocin, 68
progesterone, 215, 232
prostaglandins, 215
5-Hydroxytryptamine, 244, 350

Imaginal disc, 255
Immunolabeling
  agglutination, 84
  EM, 84, 94, 107
  fluorescence, 60, 72, 94, 107
Inferior olive, 341, 343
Inhibitors of protein and mRNA synthesis
  actinomycin D, 221, 235
  camptothecin, 235
  cycloheximide, 221, 235
Intestine, 68, 112
Isoproterenol, 226

K^+ channels, 318
Kidney, 59, 68, 79, 112

Lens, 33, 59, 68, 91, 112, 148, 177, 193
Limulus, 60
Liver. *See* Hepatocyte
Lobster, 60
Lung, 79
Lysosomes, 295

Mesencephalic nucleus, 339
Molluscs
  Aplysia, 60, 147
  Navanax, 355
  octopus, 84
  squid, 148
Myometrium, 68, 79, 215

N-Ethyl maleimide, 149, 269
Nematode, 60
Nicotine, 161

Octanol, 157, 171, 185, 371
Oncogene, 205
Oncopeltis, 278
Oocyte, 301, 307
Ovary, 293, 307

Pancreas, 59, 112, 315
Pancreatic secretagogues, 316
Pargyline, 237
Pattern formation, 251, 265, 275
Phosphodiesterase inhibitors, 388
  caffeine, 233

IBMX, 233, 388
  theophylline, 226, 388
Phosphorylation, 63, 69, 91, 211
Protease action, 8, 83, 86, 117
Protein composition, 53, 59, 67, 82
Pyramidal cell, 327

Reconstitution, 61, 177, 191
Regulation. See Formation and degradation; Gating
Retina, 59, 148, 381, 393
Retinoic acid, 145
Reversed coupling, 359

Serotonin, 244, 350
Smooth muscle, 215
Stomach, 59, 115, 291
Structure
  annular junctions, 163, 222, 299
  electron microscopy, 13, 23, 49, 94, 162, 194, 222, 299
  effect of Ca, 19, 192
  location of amino and carboxy termini, 39, 93, 200
  models based on protein sequence, 33, 96
Sympathetic neurons, 147, 231
Synapse, chemical, 326, 357
  interactions with electrical, 307, 355, 381
Synapse, electrical (electrotonic), 123, 155, 231, 241, 367
  formation and degradation, 241
  rectifying, 148, 367
Synthesis, in vitro, 118

Temperature effects, 131
Trifluoperazine, 169, 195
Tumor promoters, 81, 299
Tunicate (*Ciona*), 25

Uncoupling. See Gating
Uterus, 68, 79, 215

Vestibular nucleus, 340
Voltage clamp, 131, 139, 316, 369, 393